Differential Inclusions and Optimal Control

Mathematics and its Applications (*East European Series*)

Volume 44

Michał Kisielewicz
Higher School of Engineering, Zielona Góra, Poland

Differential Inclusions and Optimal Control

KLUWER ACADEMIC PUBLISHERS
DORDRECHT / BOSTON / LONDON

PWN—POLISH SCIENTIFIC PUBLISHERS
WARSZAWA

Library of Congress Cataloging-in-Publication Data

Kisielewicz, M. (Michał)
Differential inclusions and optimal control/Michał Kisielewicz.
p. cm.
Includes bibliographical references.
ISBN 0-7923-0675-9
1. Differential inclusions. 2. Control theory. I. Title.

OA371. K54 1990
515'. 35—dc20 90-4116

Published by PWN—Polish Scientific Publishers,
Miodowa 10, 00-251 Warszawa, Poland
in co-edition with Kluwer Academic Publishers,
P.O. Box 17, 3300 AA Dordrecht, The Netherlands

Distributors for the U.S.A. and Canada:
Kluwer Academic Publishers,
101 Philip Drive, Norwell, MA 02061, U.S.A.

Distributors for Albania, Bulgaria, Chinese People's Republic, Cuba, Czechoslovakia, Hungary, Korean People's Republic, Mongolia, Poland, Romania, the U.S.S.R., Vietnam, and Yugoslavia: ARS POLONA, Krakowskie Przedmieście 7, 00-068 Warszawa, Poland

Distributors for all remaining countries:
Kluwer Academic Publishers Group,
P.O. Box 322, 3300 AH Dordrecht, The Netherlands

Printed in Poland by D.N.T.

To my Pupils

Author

Contents

Series Editor's Preface xi
Preface xv
List of symbols xvii

CHAPTER I. Preliminaries 1
1. Set theory and topological preliminaries 1
2. Functional analysis preliminaries 5
3. Vector measures and vector-valued functions 10
4. Special spaces and fixed point theorems 15
5. Notes and remarks 22

CHAPTER II. Set-valued functions 23
1. Spaces of subsets of metric space 23
1.1. Hausdorff distance and Hausdorff topology of Comp(X) 23
1.2. Spaces Comp($\boldsymbol{R}^n$), Conv($\boldsymbol{R}^n$) and support functions 25
2. Continuity concepts 28
2.1. Upper semicontinuous set-valued functions 28
2.2. Lower semicontinuous and continuous set-valued functions 34
3. Measurable set-valued functions and Aumann's integral 37
3.1. Foundations of measure theory 38
3.2. Measurable set-valued functions 41
3.3. Measurable selections 46
3.4. Aumann's integral 49
4. Continuous selections and multivalued fixed point theorems 57
4.1. Michael's selection theorem 57
4.2. Multivalued fixed point theorems 61
5. Notes and remarks 64

CHAPTER III. Subtrajectory and trajectory integrals of set-valued functions 66
1. Fundamental space of Aumann integrable set-valued functions 66
1.1. Metric space ($\mathscr{A}(I, \boldsymbol{R}^n), d$) 66
1.2. Weak convergence in $\mathscr{A}(I, \boldsymbol{R}^n)$ 69
1.3. Fundamental properties of linear mappings $\mathscr{I}$ and $\mathscr{D}$ 75
2. Subtrajectory and trajectory integrals of set-valued functions 77
2.1. Topological properties of subtrajectory integrals 77
2.2. Subtrajectory integrals of set-valued functions depending on parameter 79
2.3. Continuous selection theorem 88
3. Fixed point properties of subtrajectory and trajectory integrals depending on parameters 96

3.1. Existence of fixed points of subtrajectory integrals of set-valued functions with non-convex values 96
3.2. Existence of fixed points of subtrajectory integrals of set-valued functions with convex values 99
3.3. Properties of set of fixed points of subtrajectory integrals 100
4. Subtrajectory and trajectory integrals of σ-selectionable set-valued functions 106
4.1. σ-selectionable approximation theorem 107
4.2. Properties of trajectory integrals of σ-selectionable set-valued functions 112
5. Relaxation theorems 115
5.1. Some properties of Stieltjes integrals 115
5.2. Continuous selection theorem 119
5.3. Relaxation theorems 127
6. Viability theorems 129
6.1. Bouligand's contingent cone 129
6.2. Viability theorem for subtrajectory integrals of set-valued functions with convex values 130
6.3. Viability theorem for subtrajectory integrals of set-valued functions with non-convex values 134
7. Notes and remarks 139

CHAPTER IV. Neutral functional-differential inclusions 141
1. Neutral functional-differential equations and fundamental problems in optimal control theory of systems described by NFDEs 141
1.1. Neutral functional-differential equations 141
1.2. Mayer problem of optimal control theory of systems described by NFDEs 149
1.3. Lagrange problem of optimal control theory of systems described by NFDEs 152
2. Neutral functional-differential inclusions 154
2.1. Neutral functional-differential inclusions—basic concepts 155
2.2. Sufficient conditions for existence of local solutions 159
2.3. Local existence theorems 167
2.4. Continuation of solutions 169
3. Properties of sets of solutions of NFDIs 170
3.1. Compactness and upper semicontinuity 171
3.2. Approximating and acyclic properties 173
3.3. Relaxation theorem 174
4. Controllability theorems for NFDIs with convex-valued functions . . . 175
4.1. Notations and approximation theorem 175
4.2. Viability theorems 179
4.3. η-approximation viability theorems 189
4.4. Controllability theorems 191

4.5. Compactness of set of attainable trajectories 193
5. Relaxation theorem for attainable sets 198
5.1. Controllability theorem for NFDIs with non-convex-valued functions 198
5.2. Relaxation theorem 205
6. Notes and remarks. 208

CHAPTER V. Optimal control problems for systems described by NFDIs 210
1. Existence theorems. 210
1.1. Weak sequential lower semicontinuity of integrals 210
1.2. Existence theorems for Mayer problem of optimal control . . . 216
1.3. Existence theorem for Lagrange problem of optimal control . . 217
1.4. Relaxed optimal controls problem for systems described by NFDIs 218
2. Attainable trajectories with minimal selection property 221
2.1. Dissipative set-valued functions 221
2.2. Integral-dissipative set-valued functions. 225
2.3. Controllability of NFDIs with m-dissipative right-hand sides. . 228
2.4. Existence of attainable trajectories with minimal selection property 231
3. Notes and remarks 233
Bibliography 234
Index 239

Series Editor's Preface

'Et moi, ..., si j'avait su comment en revenir, je n'y serais point allé.'

Jules Verne

The series is divergent; therefore we may be able to do something with it.

O. Heaviside

One service mathematics has rendered the human race. It has put common sense back where it belongs, on the topmost shelf next to the dusty canister labelled 'discarded nonsense'.

Eric T. Bell

Mathematics is a tool for thought. A highly necessary tool in a world where both feedback and nonlinearities abound. Similarly, all kinds of parts of mathematics serve as tools for other parts and for other sciences.

Applying a simple rewriting rule to the quote on the right above one finds such statements as: 'One service topology has rendered mathematical physics ...'; 'One service logic has rendered computer science ...'; 'One service category theory has rendered mathematics ...'. All arguably true. And all statements obtainable this way form part of the raison d'être of this series.

This series, *Mathematics and Its Applications*, started in 1977. Now that over one hundred volumes have appeared it seems opportune to reexamine its scope. At the time I wrote

> "Growing specialization and diversification have brought a host of monographs and textbooks on increasingly specialized topics. However, the 'tree' of knowledge of mathematics and related fields does not grow only by putting forth new branches. It also happens, quite often in fact, that branches which were thought to be completely disparate are suddenly seen to be related. Further, the kind and level of sophistication of mathematics applied in various sciences has changed drastically in recent years: measure theory is used (non-trivially) in regional and theoretical economics; algebraic geometry interacts with physics; the Minkowsky lemma, coding theory and the structure of water meet one another in packing and covering theory; quantum fields, crystal defects and mathematical programming profit from homotopy theory; Lie algebras are relevant to filtering; and prediction and electrical engineering can use Stein spaces. And in addition to this there are such new emerging subdisciplines as 'experimental mathematics', 'CFD', 'completely integrable systems', 'chaos, synergetics and large-scale order', which are almost impossible to fit into the existing classification schemes. They draw upon widely different sections of mathematics."

By and large, all this still applies today. It is still true that at first sight mathematics seems rather fragmented and that to find, see, and exploit the deeper underlying interrelations more effort is needed and so are books that can help mathematicians and scientists do so. Accordingly MIA will continue to try to make such books available.

If anything, the description I gave in 1977 is now an understatement. To the examples of interaction areas one should add string theory where Riemann surfaces, algebraic geometry, modular functions, knots, quantum field theory, Kac-Moody algebras, monstrous moonshine (and more) all come together. And to the examples of things which can be usefully applied let me add the topic 'finite geometry'; a combination of words which sounds like it might not even exist, let alone be applicable. And yet it is being applied: to statistics via designs, to radar/sonar detection arrays (via finite projective planes), and to bus connections of VLSI chips (via difference sets). There seems to be no part of (so-called pure) mathematics that is not in immediate danger of being applied. And, accordingly, the applied mathematician needs to be aware of much more. Besides analysis and numerics, the traditional workhorses, he may need all kinds of combinatorics, algebra, probability, and so on.

In addition, the applied scientist needs to cope increasingly with the nonlinear world and the extra mathematical sophistication that this requires. For that is where the rewards are. Linear models are honest and a bit sad and depressing: proportional efforts and results. It is in the nonlinear world that infinitesimal inputs may result in macroscopic outputs (or vice versa). To appreciate what I am hinting at: if electronics were linear we would have no fun with transistors and computers; we would have no TV; in fact you would not be reading these lines.

There is also no safety in ignoring such outlandish things as nonstandard analysis, superspace and anticommuting integration, p-adic and ultrametric space. All three have applications in both electrical engineering and physics. Once, complex numbers were equally outlandish, but they frequently proved the shortest path between 'real' results. Similarly, the first two topics named have already provided a number of 'wormhole' paths. There is no telling where all this is leading—fortunately.

Thus the original scope of the series, which for various (sound) reasons now comprises five subseries: white (Japan), yellow (China), red (USSR), blue (Eastern Europe), and green (everything else), still applies. It has been enlarged a bit to include books treating of the tools from one subdiscipline which are used in others. Thus the series still aims at books dealing with:

— a central concept which plays an important role in several different mathematical and/or scientific specialization areas;
— new applications of the results and ideas from one area of scientific endeavour into another;
— influences which the results, problems and concepts of one field of enquiry have, and have had, on the development of another.

The traditional, classical, analytical mechanics paradigm tends to yield mathematical models in the form of (coupled) sets of ordinary differential equations $\dot{x} = f(x)$ in which the future is determined uniquely by the present state (positions and velocities) and is independent of the past. However, in many problems in economics, biology, magnetism, communications, the natural formulation does involve influence from the past; for instance in the form of accumulated effects (say, in the form of an integral) over a finite chunk of time stretching back from the present. Then, of course, one can work in terms of an infinite, dimensional state vector and that approach has certainly its merits and has been used frequently. However, this tends to ignore part of the structure of functional-differential equations. Whence, the rise of functional-differential equations as a subject in its own right; a subject, incidentally, which holds a few surprises for those who are primarily used to the classical mechanics paradigm.

Given a model (of anything) with some parameter (or control) variables in it, control and, in particular, optimal control becomes interesting. In the present case this leads to functional-differential inclusions. A set of possible evolution directions is given in terms of the past and present of the system.

There are, so far, but a couple monographs on the topic of differential inclusions and it is therefore definitely a pleasure to welcome the present volume by a top authority in the field in the series MIA.

The shortest path between two truths in the real domain passes through the complex domain.

J. Hadamard

La physique ne nous donne pas seulement l'occasion de résoudre des problèmes ... elle nous fait pressentir la solution.

H. Poincaré

Never lend books, for no one ever returns them; the only books I have in my library are books that other folk have lent me.

Anatole France

The function of an expert is not to be more right than other people, but to be wrong for more sophisticated reasons.

David Butler

Bussum, March 1989

Michiel Hazewinkel

Preface

In the majority of applications, the future behaviour of many phenomena is assumed to be described by the solutions of ordinary differential equations. Implicit in this assumption is that the future behaviour is uniquely determined by the present and independent of the past. Many problems lead in a natural way to functional-differential equations where the past exerts its influence in a significant manner upon the future. In the last twenty years, the subject has been continually investigated at a very rapid pace, and this has mainly been due to development in the theory of control, mathematical biology, mathematical economics and systems theory.

Recently, there has been a great deal of interest in optimal control systems described by functional-differential equations. These optimal control problems lead to functional-differential inclusions and, in particular, to neutral functional-differential inclusions, where the present dynamics of the systems are influenced by its past behaviour. Such systems have been studied in connection with feedback control systems, collisions problems in electrodynamics, computer technology where transmission time between various circuits can be ignored, in the control and in connection with some variational problems.

The present book is devoted to the investigation of the properties, and some of their applications, of neutral functional-differential inclusions of the form $\dot{x}(t) \in F(t, x_t, \dot{x}_t)$. The content is divided into five parts. Chapter I covers basic notations and theorems of topology, functional analysis and vector measure theory which are needed in the text. Chapter II contains the fundamental theory of set-valued functions which are used in the next chapters. The properties of subtrajectory and trajectory integrals of set-valued functions depending on parameters are given in Chapter III. Chapters IV and V deal with neutral functional-differential inclusions and some of their applications in the optimal control theory of systems described by neutral functional-differential equations.

The present book is intended for students, professionals in mathematics, and specialists in optimal control theory and its applications. It complements the monographs of T. Parthasarathy, C. Castaing and M. Valadier, J. P. Aubin and A. Cellina, A. F. Fillipov and A. A. Tolstonogov on the subject of functional-differential inclusions. Selected methods of advanced calculus, functional analysis and the theory of set-valued functions are needed for understanding the text.

Formulae, theorems and lemmas, propositions, remarks, corollaries are numbered separately in each chapter and denoted by a pair of numbers. The first stands for the section number, the second for the formula, theorem, etc., number. If we need to quote some formula or theorem given in the same chapter we always write only this pair. In other cases we will use three numbers instead of two, where the first number will indicate the chapter number. The end of proofs, examples, remarks and corollaries will be denoted by ■.

I would like to express my sincere thanks to Professor Stefan Rolewicz who inspired me to write this book. Part of this book was written during the author's stay at the Institute of Statistical and Mathematical Economics at the University of Karlsruhe. The stay was sponsored by DAAD. The author is sincerely thankful to Professor Diethard Pallaschke for his valuable suggestions and remarks which improved the text. The manuscript of the book was also read by my colleagues Dr. L. Rybiński, Dr. J. Motyl and Dr. K. Przesławski who have given their valuable remarks. It is my pleasure to thank them sincerely for that. Finally, I am much indebted to M.A. R. Kowalczyk for computer-typing the text. His job was not easy.

Michał Kisielewicz

List of symbols

$\boldsymbol{N}$	set of all positive integrals, 1
$\boldsymbol{R}$	set of all real numbers, 1
$\in$	is an element of, 1
$\emptyset$	the void (empty) set, 1
$\subset$	is a subset of (set inclusion relation), 1
$\cup$	union, 1
$\cap$	intersection, 1
$A \setminus B$	complement of B with respect to A (difference of A and B), 1
$A \triangle B$	symmetric difference of A and B, 1
$\notin$	is not an element of, 1
S^0	inferior of set, 3
$\bar{S}$	closure of set, 3
$\mathrm{Seqcl}(S)$	sequential closure, 3
$\mathrm{supp}\,\psi$	support of function ψ, 5
$B^o(x, \eta)$	open ball of a metric space X centred at $x \in X$ with a radius $\eta > 0$, 5
$V^o(A, \eta)$	open neighbourhood of $A \subset X$ of the form $\bigcup_{x \in A} B^o(x, \eta)$, 5
$\mathrm{dist}(x, A)$	distance from $x \in X$ to $A \subset X$, 5
$\boldsymbol{L}(X, Y)$	linear space of all continuous linear operators from X to Y, 6
X^*	dual space of X, 6
$\sigma(X, X^*)$	weak topology in X, generated by X^*, 6
$\boldsymbol{R}^n$	real n-dimensional Euclidean space, 6
$\bar{S}^w$	weak closure of S, 7
$\mathrm{Seqcl}_w(S)$	weak sequential closure of S, 7
$\sigma(X^*, X)$	weak-* topology for X^*, 7
$\mathrm{co}\,S$	convex hull of S, 8
$\overline{\mathrm{co}}\,S$	closed convex hull of S, 8
$(T, \mathscr{F})$	measurable space, 12
$(T, \mathscr{F}, \mu)$	measure space, 10
$L^p(\mu, X)$	space of μ-Bochner integrable vector-valued functions with p-th power, 12
$L^\infty(\mu, X)$	space of essentially bounded μ-Bochner integrable functions, 12
$\mathrm{BV}([\alpha, \beta], \boldsymbol{R}^n)$	space of functions of bounded variation, 14
$C(S, X)$	space of all X-valued bounded continuous functions on a topological space S, 15

$\mathrm{AC}(I, \boldsymbol{R}^n)$	space of all $\boldsymbol{R}^n$-valued absolutely continuous functions on I, 15, 16
$L(I, \boldsymbol{R}^n)$	space of all $\boldsymbol{R}^n$-valued Lebesgue integrable functions on I, 16
$C^0([\alpha, \beta], \boldsymbol{R}^n)$	subspace of $C([\alpha, \beta], \boldsymbol{R}^n)$ of all $x \in C([\alpha, \beta], \boldsymbol{R}^n)$ such that $x(\alpha) = 0$, 18
$\mathrm{AC}^0([\alpha, \beta], \boldsymbol{R}^n)$	subspace of $\mathrm{AC}([\alpha, \beta], \boldsymbol{R}^n)$ of all $x \in \mathrm{AC}([\alpha, \beta], \boldsymbol{R}^n)$ such that $x(\alpha) = 0$, 18
C_{0r}	space $C(I, \boldsymbol{R}^n)$ with $I = [-r, 0]$, $r \geqslant 0$, 18
AC_{0r}	space $\mathrm{AC}(I, \boldsymbol{R}^n)$ with $I = [-r, 0]$, $r \geqslant 0$, 18
L_{0r}	space $L(I, \boldsymbol{R}^n)$ with $I = [-r, 0]$, $r \geqslant 0$, 18
$(\mathscr{X}_r(\Omega), d)$	special metric space of absolutely continuous functions, 18
$(\mathscr{C}_r(\Omega), \varrho)$	special metric space of continuous functions, 18
$(\mathscr{L}_r(\Omega), l)$	special metric space of Lebesgue integrable functions, 19
$\mathrm{Cl}(X)$	space of all nonempty closed bounded subsets of a metric space X, 23
$\mathrm{Comp}(X)$	space of all compact members of $\mathrm{Cl}(X)$, 23
$\mathrm{Conv}(\boldsymbol{R}^n)$	space of all convex members of $\mathrm{Comp}(\boldsymbol{R}^n)$, 25
$V(A, \varepsilon)$	closed ε-neighbourhood of $A \in \mathrm{Cl}(X)$, 23
$s(\cdot, A)$	support function of $A \in \mathrm{Cl}(\boldsymbol{R}^n)$, 25
$\mathscr{P}(X)$	space of all nonempty subsets of X, 28
u.s.c.	upper semicontinuity of, 28
H-u.s.c.	H-upper semicontinuity of, 28
w.-w.u.s.c.	weak-weak upper semicontinuity of, 32
s.-w.u.s.c.	strong-weak upper semicontinuity of, 32
l.s.c.	lower semicontinuity of, 34
H-l.s.c.	H-lower semicontinuity of, 35
w.-w.l.s.c.	weak-weak lower semicontinuity of, 36
w.-s.l.s.c.	weak-strong lower semicontinuity of, 36
$\beta(T)$	Borel σ-algebra on T, 38
$\mathscr{F}_1 \otimes \mathscr{F}_2$	product σ-algebra of $\mathscr{F}_1$ and $\mathscr{F}_2$, 38
$\mathrm{Graph}(F)$	graph of a set-valued function F, 47
$\mathscr{A}(I, \boldsymbol{R}^n)$	some metric space of set-valued functions, 66
$\mathscr{A}^{\mathrm{C}}(I, \boldsymbol{R}^n)$	set of all $G \in \mathscr{A}(I, \boldsymbol{R}^n)$ with convex values, 72
$\mathscr{I}, \mathscr{D}$	some linear mappings, 75
$\overline{[A]}_L$	closure of A in a norm-topology of L, 77
$\overline{[B]}_{\mathrm{AC}}$	closure of B in a norm-topology of AC, 77
$\overline{[B]}_C$	closure of B in a norm-topology of C, 77
$\overline{[A]}_L^w$	closure of A in a weak topology of L, 77
$\overline{[B]}_{\mathrm{AC}}^w$	closure of B in a weak topology of AC, 77
$\mathscr{F}(G)$	subtrajectory integrals of G, 77
$\mathscr{I}\mathscr{F}(G)$	trajectory integrals of G, 77
$\mathscr{K}_\lambda$	closed ball of $C^0(I, \boldsymbol{R}^n)$ centred at the origin with a radius $\lambda > 0$, 79

B_λ	closed ball of $L(I, \boldsymbol{R}^n)$ centred at the origin with a radius $\lambda > 0$, 79
$H_\lambda(I, \boldsymbol{R}^n)$, $\mathscr{H}_\lambda(I, \boldsymbol{R}^n)$	some spaces of set-valued functions, 79
$\boldsymbol{P}_\lambda$, $\boldsymbol{K}_\lambda$	some subsets of $L(I, \boldsymbol{R}^n)$ and $AC(I, \boldsymbol{R}^n)$, 89
$\mathscr{H}_\lambda^C(I, \boldsymbol{R}^n)$	some space of set-valued functions, 100
$T_H(x)$	Bouligand's contingent cone to H at x, 129
$\mathscr{W}^p([\alpha, \beta], \boldsymbol{R}^n)$	Sobolev space, 141
$\mathscr{M}(D, \boldsymbol{R}^n)$	space of all functions $f\colon D \to \boldsymbol{R}^n$ satisfying local measurability condition ($\mathscr{M}$), 141
NFDE(D, f)	neutral functional-differential equation $\dot{x}(t) = f(t, x_t, \dot{x}_t)$ with $f\colon D \to \boldsymbol{R}^n$, 142
NFDE(D, A, f)	some boundary problem for NFDE(D, f), 143
$\mathfrak{M}(D, f, \Omega, C, g)$	Mayer problem of optimal control, 149
$\mathfrak{A}(D, f, \Omega, C)$	set of all admissible pairs of $\mathscr{M}(D, f, \Omega, C, g)$, 149
$\mathfrak{I}(D, f, \Omega, C)$	set of all admissible trajectories of $\mathscr{M}(D, f, \Omega, C, g)$, 149
$\mathfrak{L}(D, f, \Omega, C, f_0)$	Lagrange problem of optimal control, 152
$\mathscr{M}(D, \mathrm{Comp}(\boldsymbol{R}^n))$	set of all set-valued functions $F\colon D \to \mathrm{Comp}(\boldsymbol{R}^n)$ satisfying local measurability condition (M), 155
NFDI(D, F)	neutral functional-differential inclusion, 155
NFDI(D, A, F)	some boundary value problem of NFDI(D, F), 155
$\mathscr{R}$, $\mathscr{S}$	some biaffine mappings, 156, 157
G^F	set-valued function associated to F, 157
$\boldsymbol{K}_\alpha(\Phi, H)$	some special set, 176
$\boldsymbol{B}_\alpha(\varrho, \Phi, H)$	some special set, 177
$\mathrm{K}(\Omega, C)$	some special set, 194

Chapter I

Preliminaries

In this chapter we give a survey of concepts and results from functional analysis that are used in the rest of the book. The greater part of all results is stated without proofs which can be found in the standard monographs.

1. Set theory and topological preliminaries

The sets of positive integers and real numbers will be denoted by $\boldsymbol{N}$ and $\boldsymbol{R}$, respectively. Capital Latin or Greek letters will usually be used to denote sets, collections, families or classes. The symbol $\in$ will indicate membership in a set. The void (empty) set is denoted by $\emptyset$. If every element of a set A is also an element of a set B, then we write $A \subset B$. For given sets A and B, by $A \cup B$, $A \cap B$ and $A \setminus B$ we denote the union, the intersection and the complement of B with respect to A, respectively. The symmetric difference of A and B is denoted by $A \triangle B$ and defined by $A \triangle B := (A \setminus B) \cup (B \setminus A)$. Given sets A and B are written $A = B$ if and only if $A \subset B$ and $B \subset A$. They are called disjoint if $A \cap B = \emptyset$.

If $\mathfrak{U}$ is a set whose elements are sets then $\bigcup \mathfrak{U}$ or $\bigcup \{A: A \in \mathfrak{U}\}$ denotes the union of all elements of $\mathfrak{U}$. Similarly, $\bigcap \mathfrak{U}$ or $\bigcap \{A: A \in \mathfrak{U}\}$ denotes the intersection of all members of $\mathfrak{U}$. Sometimes, the set $\mathfrak{U}$ considered above is said to be a family or class of sets. In particular if $\mathfrak{U} = \{P_t: t \in T\}$, where T and P_t are given sets, we write $\bigcup \{P_t: t \in T\}$ or $\bigcup_{t \in T} P_t$ and $\bigcap \{P_t: t \in T\}$ or $\bigcap_{t \in T} P_t$, instead of $\bigcup \mathfrak{U}$ and $\bigcap \mathfrak{U}$, respectively. If $T = \boldsymbol{N}$ or $T = \{1, \ldots, n\}$ we write $\bigcup_{n=1}^{\infty} P_n$ or $\bigcup_{i=1}^{n} P_i$ and $\bigcap_{n=1}^{\infty} P_n$ or $\bigcap_{i=1}^{n} P_i$ instead of $\bigcup_{t \in T} P_t$ and $\bigcap_{t \in T} P_t$, respectively.

Finally, if $\mathfrak{U} = \{A_n: n \in \boldsymbol{N}\}$ we define $\limsup_{n \to \infty} A_n := \bigcap_{n=1}^{\infty} \bigcup_{k=n}^{\infty} A_k$ and $\liminf_{n \to \infty} A_n := \bigcup_{n=1}^{\infty} \bigcap_{k=n}^{\infty} A_k$.

From an axiomatic viewpoint, a topology $\mathscr{T}$ for a set X is a family of subsets of X satisfying:

(a) the union of any collection of sets in $\mathscr{T}$ is again a set in $\mathscr{T}$,

(b) the intersection of a finite collection of sets of $\mathscr{T}$ is a set in $\mathscr{T}$.

(c) X itself, and the empty set $\emptyset$ belong to $\mathscr{T}$.

The elements of $\mathscr{T}$ are called *open sets* and the set X with topology is called a *topological space*. It is written as a pair $(X, \mathscr{T})$. A *base* for the topology $\mathscr{T}$ is a collection β of open subsets of X such that any element of $\mathscr{T}$ can be written as a union of elements of β. *Closed sets* are defined as the complements with respect to X of open sets. Any open set which contains a point $x \in X$ is called a *neighbourhood of* x. Similarly, any open set which contains a set $A \subset X$ is called a *neighbourhood of* A. If X is a topological space with the property that each pair of distinct points of X has disjoint neighbourhoods, the topology is said to be a *Hausdorff topology* and X a *Hausdorff topological space.*

Let X be a nonempty set and β a nonempty collection of subsets of X which satisfy:

(i) for each $x \in X$ there is a $\beta_x \in \beta$ such that $x \in \beta_x$,

(ii) given two sets $\beta_1, \beta_2 \in \beta$, if $x \in \beta_1 \cap \beta_2$ there is a $\beta_3 \in \beta$ such that $x \in \beta_3 \subset \beta_1 \cap \beta_2$.

We may define the elements of a topology $\mathscr{T}$ for X by taking arbitrary unions of elements of β together with $\emptyset$. It is easily seen that the axioms of a topology are satisfied, and $\mathscr{T}$ will have β as a base.

A nonempty collection S of open subsets of a topological space X is called a *subbase* if the collection of all finite intersections of elements of S is a base for $\mathscr{T}$.

If we have two topologies $\mathscr{T}_1$ and $\mathscr{T}_2$ for a set X, the topology $\mathscr{T}_1$ is said to be *weaker* than the topology $\mathscr{T}_2$ if $\mathscr{T}_1 \subset \mathscr{T}_2$. Equivalently, in this case $\mathscr{T}_2$ is said to be *stronger* than $\mathscr{T}_1$. Two topologies are the same or equivalent if $\mathscr{T}_1 = \mathscr{T}_2$.

Let $(X, \mathscr{T})$ and (Y, T) be topological spaces and $f\colon X \to Y$ be a function. Then f is said to be *continuous at a point* $x_0 \in X$ if to each neighbourhood $V \in T$ of $f(x_0)$ there is a neighbourhood $U \in \mathscr{T}$ of x_0 such that $f(U) \subset V$. A sequence (x_n) of points $x_n \in X$ is said to be *convergent* to $x \in X$ (written $x_n \to x$ or $\lim x_n = x$) if each neighbourhood of x contains all but a finite numbers of elements of (x_n). A function $f\colon X \to Y$ is said to be *sequentially continuous at a point* $x_0 \in X$ if for every sequence (x_n) of X converging to x_0, the sequence $\{f(x_n)\}$ converges to $f(x_0)$. A function $f\colon X \to Y$ is called *continuous* (*seqentially continuous*) if it is continuous (sequentially continuous) at each point $x_0 \in X$.

A function $f\colon X \to \boldsymbol{R}$ is called a *functional.* Given $f\colon X \to \boldsymbol{R}$ and $x_0 \in X$ we define $\limsup_{x \to x_0} f(x) := \inf\{\sup f(B)\colon x_0 \in B \in \mathscr{T}\}$ and $\liminf_{x \to x_0} f(x) := \sup\{\inf f(B)\colon x_0 \in B \in \mathscr{T}\}$. A function $f\colon X \to \boldsymbol{R}$ is said to be *upper semicontinuous* (resp. *lower semicontinuous*) *at* $x_0 \in X$ if $f(x_0) \geqslant \limsup_{x \to x_0} f(x)$ (resp. if $f(x_0) \leqslant \liminf_{x \to x_0} f(x)$). It is easy to see that $f\colon X \to \boldsymbol{R}$ is lower (upper) semicontinuous at $x_0 \in X$ if and only if for every $\alpha < f(x_0)\, (\alpha > f(x_0))$ there exists a neighbourhood $U \in \mathscr{T}$ of x_0 such that for every $x \in U$ we have $\alpha < f(x)\, (\alpha > f(x))$. A function $f\colon X \to \boldsymbol{R}$ is called *lower* (*upper*) *semicontinuous* if it is lower (upper) semicontinuous at each point $x_0 \in X$.

A topological space $(X, \mathscr{T})$ is said to be *normal* if the following two conditions are satisfied:

(i) for every different points $x_1, x_2 \in X$ there is an open set $U \subset X$ such that $x_1 \in U$ and $x_2 \notin U$,

(ii) for every pair $(A\ \ B)$ of disjoint closed subsets of X there are open sets $U, V \subset X$ such that $A \subset U$, $B \subset V$ and $U \cap V = \emptyset$.

We have the following important result dealing with normal topological spaces.

THEOREM (Urysohn). *For every pair* (A, B) *of disjoint closed subset of a normal topological space* $(X, \mathscr{T})$ *there is a continuous function* $f\colon X \to [0, 1]$ *such that* $f(x) = 0$ *for* $x \in A$ *and* $f(x) = 1$ *for* $x \in B$. ■

Let $(X, \mathscr{T})$ be a topological space and S be a subset of X. Then we can define a topology for S to be the collection of sets of the form $U \cap S$, $U \in \mathscr{T}$. This is called the *relative topology on S generated by the topology* $\mathscr{T}$. If X is a topological space and S a subset of X, the union of its all open subsets is said to be the *interior* of S. It is denoted by S^0. The *closure* $\bar{S}$ of S is the intersection of all closed subsets of X containing S. The set S is said to be *sequentially closed* if for every converging sequence of points of S its limit belongs to S. By the *sequential closure* of the set S we mean the set of all limits of converging sequences of points of S. This is denoted by $\mathrm{Seqcl}(S)$. We have of course $\mathrm{Seqcl}(S) \subset \bar{S}$.

A subset S of a topological space X is said to be *connected* if there exists no continuous mapping $f\colon S \to \boldsymbol{R}$ such that $f(S)$ consists of exactly two points. It is not difficult to verify that a set $S \subset X$ is connected if and only if S is not the union of two separated sets $S_1, S_2 \subset X$, i.e., such that $S_1 \neq \emptyset$, $S_2 \neq \emptyset$, $\bar{S}_1 \cap S_2 = \emptyset$ and $S_1 \cap \bar{S}_2 = \emptyset$. It is also easy to see that if $S \subset X$ is such that for every $x, y \in S$ there exists a connected subset S_{xy} of S such that $x, y \in S_{xy}$ then S is connected.

A set S of a topological space X is said to be *dense* in X if $\bar{S} = X$. A topological space X is said to be *separable* if there is a countable dense subset in X. It is called *perfectly separable* if there is a countable basis for its topology. It is clear that if X is perfectly separable it is also separable.

Let X be a topological space and S be a subset of X. A family $\mathscr{F}$ of open sets in X is said to be an *open covering* of S if every point of S belongs to at least one element of $\mathscr{F}$. S is *compact* if every open covering of S contains a finite subfamily which covers S. S is *sequentially compact* if every sequence in S has a subsequence which converges to a point of S. S is *relatively compact* if its closure $\bar{S}$ is compact. S is *relatively sequentially compact* if every sequence in S has a subsequence which converges to a point in X. A topological space X is called *compact* if the set X is compact. It is called *sequentially compact* if X is a sequentially compact subset of X. It is clear that every closed (sequentially closed) subset of a compact (sequentially compact) topological space is compact (sequentially compact). It is also clear that every compact (sequentially compact) subset of a topological space is closed (sequentially closed).

We shall also call a sequence (x_n) of X as *relatively* (sequentially) *compact* if a set $S = \bigcup_{n=1}^{\infty} \{x_n\}$ is relatlvely (sequentially) compact.

If X is a nonempty topological space, it may be possible in various ways to make a new space Y which is compact, which contains X as a dense subset of Y, and which is such that the original topology of X is identical with the relative topology of X generated by the topology of Y. A space Y related to X in this way is called a *compactification of X*. In particular, it can be proved that a perfectly separable noncompact topological space has a perfectly separable compactification.

Let $\{U_i\}_{i \in I}$ and $\{V_j\}_{j \in J}$ be two coverings of a topological space $(X, \mathscr{T})$. $\{U_i\}_{i \in I}$ is said to be a *refinement of* $\{V_j\}_{j \in J}$ if for every $i \in I$ there exists a $j \in J$ such that $U_i \subset V_j$. A covering $\{U_i\}_{i \in I}$ is called *locally finite* if for every $x \in X$, there exists a neighborhood V of x such that $U_i \cap V \neq \emptyset$ only for a finite number of indexes. A topological space $(X, \mathscr{T})$ is called *paracompact* if it is a Hausdorff space and its each open covering has a locally finite open refinement.

Let X be an arbitrary set and S a collection of subsets of X. Then there are certainly topologies for X which contains S, and there is a unique topology for X containing S, which is weaker than any other topology with this property. It is the intersection of all topologies containing S. This topology is called the *topology generated by a family* S.

A more constructive way to characterize the unique weakest topology containing S is by taking all unions of finite intersections of elements of S, together with $\emptyset$ and X, as a topology for X. It has S as a subbase.

As an example, which will be useful later, let X be any set and let Y be a topological space with topology $\mathscr{T}_Y$ and $\{f_\alpha\colon \alpha \in A\}$ a collection of functions, each defined on X with range in Y. The weak topology generated by $\{f_\alpha\colon \alpha \in A\}$ is the weakest topology in X under which each of the function f_α is continuous. This requires that $f_\alpha^{-1}(\theta)$ be an open set in X for each $\alpha \in A$ and $\theta \in \mathscr{T}_Y$, where $f_\alpha^{-1}(\theta) = \{x \in X\colon f_\alpha(x) \in \theta\}$. Let $S = \{f_\alpha^{-1}(\theta)\colon \alpha \in A, \theta \in \mathscr{T}_Y\}$ and use S, as in the previous passage, to generate a topology. This topology with S as a subbase is then the weak topology generated by $\{f_\alpha\colon \alpha \in A\}$.

Let $X_1, \ldots, X_n$ be topological spaces, with β_i as basis for the topology of X_i. Their topological product $X_1 \times X_1 \times \ldots \times X_n$ is defined as the set of all n-tuples $(x_1, \ldots, x_n)$ with $x_i \in X_i$, taking as a base for the topology all products $U_1 \times \ldots \times U_n$ of U_i in β_i. It can be proved that the topological product of compact spaces is compact.

Another very useful way of generating a topology for a set X is via a metric function which is a real-valued function ϱ defined on $X \times X$ satisfying

(i) $\varrho(x, y) = \varrho(y, x)$,

(ii) $\varrho(x, z) \leqslant (x, y) + \varrho(y, z)$,

(iii) $\varrho(x, y) = 0$ if and only if $x = y$ with $x, y, z \in X$.

It follows that the values of ϱ are nonnegative. A set X with a metric ϱ considered as a topological space with a base defined by the "open balls" $\{y\colon \varrho(y, x) < r\}$, $x \in X$, $r > 0$ is called a *metric space.*

An importrant question is: given a set S with a topology $\mathscr{T}$, can one determinate if $\mathscr{T}$ is a metric topology; i.e., can a base for the elements of $\mathscr{T}$ be defined via a metric function ? If this is possible the topology is said to be *metrizable*, if not it is *nonmetrizable.* Necessary and sufficient conditions for a topology to be metrizable can be given. In particular, a compact topological space is metrizable if and only if it is perfectly separable.

In a metric space X, compactness and sequential compactness of a set S are equivalent. A metric space X is paracompact. It is perfectly separable if and only if it is separable. Hence in particular it follows that a separable metric space has a metrizable compactification. It can be proved that to any locally finite open covering $(\Omega_i)_{i \in I}$ of a metric space X we can associate a locally Lipschitzean partition of unity subourdinate to it. Recall, a family $\{\psi_i\}_{i \in I}$ of real-valued functions ψ_i defined on X is said to be a *locally Lipschitz partition* of unity if:

(a) for all $i \in I$ ψ_i is locally Lipschitz and nonnegative,

(b) a family $\{\mathrm{supp}\,\psi_i\}_{i\in I}$ with $\mathrm{supp}\,\psi_i := \overline{\{x\colon \psi_i(x) \neq 0\}}$ is a closed locally finite covering of X and

(c) for each $x \in X$, $\sum_{i\in I} \psi(x) = 1$.

A sequence (x_n) in a metric space (X, ϱ) is called a *Cauchy sequence* if for every $\varepsilon > 0$ there is a positive integer N such that $\varrho(x_n, x_m) < \varepsilon$ whenever $n, m \geqslant N$. A metric space is complete if every Cauchy sequence converges to an element of the space.

Let (X, ϱ) be a metric space and $S \subset X$. A set S is called *bounded* if there are $x_0 \in X$ and a number $M > 0$ such that $\varrho(x_0, x) \leqslant M$ for every $x \in S$. It is clear that every compact subset of a metric space is bounded. It it also clear that every Cauchy sequence (x_n) of a metric space is bounded, i.e., the set $\bigcup_{n=1}^{\infty}\{x_n\}$ is bounded. Given a nonempty set $A \subset X$ we define the *diameter* $\mathrm{diam}(A)$ of A and the distance function d_A on X by setting $\mathrm{diam}(A) := \sup\{\varrho(x, y)\colon x, y \in A\}$ and $d_A(x) := \inf\{\varrho(x, a)\colon a \in A\}$ for $x \in X$, respectively. Often we will write $\mathrm{dist}(x, A)$ instead of $d_A(x)$. For every nonempty bounded set $A \subset X$ we have $|d_A(x_1) - d_A(x_2)| \leqslant \varrho(x_1, x_2)$ for $x_1, x_2 \in X$ and $x \in \bar{A}$ if and only if $\mathrm{dist}(x, A) = 0$. Given a nonempty set $A \subset X$ and a positive number $\eta > 0$ we define a neighbourhood $V^{\circ}(A, \eta)$ of A by $V^{\circ}(A, \eta) := \bigcup_{x\in A} B^{\circ}(x, \eta)$, where $B^{\circ}(x, \eta) := \{z \in X\colon \varrho(x, z) < \eta\}$. Similarly, we define $V(A, \eta) := \bigcup_{x\in A} B(x, \eta)$, where $B(x, \eta) := \{z \in X\colon \varrho((x, z) \leqslant \eta\}$. It is clear that for every $0 < \varepsilon < \eta$ one has $V^{\circ}(A, \varepsilon) \subset V(A, \varepsilon) \subset V^{\circ}(A, \eta)$. Observe that if A is a compact subset of a metric space (X, ϱ) and U an open set in X such that $A \subset U$ then there exists $\eta > 0$ such that $V^{\circ}(A, \eta)$ is contained in U. Finally, let us note the following result.

THEOREM (Cantor). *Let (X, ϱ) be a complete metric space and (S_n) be a sequence of nonempty closed and bounded sets in X such that $S_{n+1} \subset S_n$ for each $n = 1, 2, \ldots$ Suppose also* $\mathrm{diam}(S_n) \to 0$ *as $n \to \infty$. Then there is a unique point in the intersection of all the S_n.* ■

2. FUNCTIONAL ANALYSIS PRELIMINARIES

Let X be a linear space over a field Φ. Assume, X is also a topological space and Φ is either the real or the complex scalar field with their usual topology. Then X is called a *linear topological space* if the mappings $X \times X \in (x_1, x_2) \to x_1 + x_2 \in X$ and $\Phi \times X \ni (\alpha, x) \to \alpha x \in X$ are continuous.

A *normed linear space* is a linear space X together with a function from X into $\boldsymbol{R}$, called the *norm* and denoted $\|\cdot\|$, defined on it which "measures the distance to zero" of elements of X and satisfies:

(i) $\|x\| \geqslant 0$ and $\|x\| = 0$ if and only if $x = 0$,

(ii) $\|x + y\| \leqslant \|x\| + \|y\|$,

(iii) $\|\alpha x\| = \ = |\alpha|\,\|x\|$ for $x, y \in X$ and $\alpha \in \Phi$.

One may note that given a norm, if we define $\varrho(x, y) = \|x - y\|$, then ϱ is a metric, and hence the norm can be used to define a metric topology. It is easily seen that a normed linear space, understood to have the norm-generated topology, is a linear topological space. If the normed linear space is complete it is called a *Banach space.*

If X and Y are normed linear space, a linear operator T: $X \to Y$ is continuous if and only if there is an $M > 0$ such $||Tx|| \leqslant M||x||$ for all $x \in X$. The set of all continuous linear operators from X to Y, with the usual definitions of addition and scalar multiplication, is itself a linear space. It will be denoted by $\boldsymbol{L}(X, Y)$. One may define a norm of an element $T \in \boldsymbol{L}(X, Y)$ to be $||T||_L := \sup\{||Tx||: ||x|| \leqslant 1\}$ and associated norm-generated topology is called the *uniform operator topology* for $\boldsymbol{L}(X, Y)$.

Consider, in particular, the case when X is a normed linear space and Y is the scalar field Φ, which is itself a normed linear space. The set of all continuous linear mappings from X to Y is then the set of all continuous linear functionals on X. With the norm defined as above, it is denoted by X^* and called the *normed conjugate* or *normed dual* of X.

Let X be a normed linear space and X^* its normed dual space. Elements of X^* constitute a set of functions which can be used to generate another topology for X, which is the *weak topology generated by* X^*. This will be the weakest topology for X under which the elements of X^* are still continuous. Taking into consideration that a base for a topology may be formed by taking all finite intersections of elements of a subbase, a base for the weak topology of X consists of neighbourhoods of the form

$$N(x_0, \varepsilon, A) = \{x \in X: |x'(x-x_0)| < \varepsilon, x' \in A\},$$

where $\varepsilon > 0$ and A is a finite subset of X^*. The weak topology of X is sometimes denoted by $\sigma(X, X^*)$, the topology of X generated by the elements of X^*. It can be verified that the weak topology $\sigma(X, X^*)$ is a Hausdorff topology. One may easily show that the weak topology of $\boldsymbol{R}^n := \boldsymbol{R} \times \boldsymbol{R} \times \ldots \times \boldsymbol{R}$ is the same as the usual Euclidean metric topology. However, for an infinite-dimensional Banach space, the weak topology is not metrizable.

A sequence (x_n) in a normed linear space X converges weakly to x (converges in the weak topology) if and only if the scalar sequence $\{x'(x_n - x)\}$ converges to zero for each fixed $x' \in X^*$. Geometrically this means that the distance (in norm) from x_n to each hyperplane through x tends to zero. The point x is called a *weak limit of the sequence* (x_n), and the sequence (x_n) is said to *converge weakly to* x.

Every sequence (x_n) of X such that $\{x'(x_n)\}$ is a Cauchy sequence of scalars for each $x' \in X^*$ is called a *weak Cauchy sequence*. It can be easily verified that every weakly convergent sequence (x_n) of a normed linear space is bounded. It is also clear that a sequence (x_n) of a normed linear space is weakly convergent to x if and only if every subsequence of (x_n) has a subsequence weakly convergent to x.

Given a set S in a normed linear space X it can be considered as a subset of a topological space $(X, \sigma(X, X^*))$. We will say that S is *weakly closed* (*compact*) if it is closed (compact) with respect to the weak topology $\sigma(X, X^*)$ of X. Similarly, the *weak sequential closedness*, the *weak sequential compactness* and the *weak relative sequential compactness* is understood as the sequential closedness, the sequential compactness and the relative sequential compactness in the topological space $(X, \sigma(X, X^*))$. It is clear that every weakly closed subset of a normed linear space is closed in the norm-generated topology. The weak closure of a set S in a normed linear space we will denote $\bar{S}^w$, whereas its weak sequential closure by $\mathrm{Seqcl}_w(S)$. We have of course $S \subset \mathrm{Seqcl}_w(S) \subset \bar{S}^w$. If S is weakly relatively sequentially compact in a Banach space, we have also $\bar{S}^w \subset \mathrm{Seqcl}_w(S)$, i.e., $\bar{S}^w = \mathrm{Seqcl}_w(S)$. It follows from the following Eberlein–Šmulian's and Šmulian's Theorems.

THEOREM (Eberlein–Šmulian). *Let S be a subset of a Banach space. Then the following statments are equivalent:*

(i) *S is weakly relatively sequentially compact*

(ii) *$\bar{S}^w$ is weakly compact, i.e., S is relatively weakly compact.* ■

THEOREM (Šmulian). *Let S be a subset of a Banach space X. If S is relatively weakly compact then for every $x \in \bar{S}^w$ there is a sequence (x_n) of elements of S that converges weakly to x.* ■

If X is a normed linear space, then X^* is also a normed linear space, and hence X^{**} (the space of continuous linear functionals on X^*) is again a normed linear space with $\|x''\| = \sup\{|x''(x')|: \|x'\| = 1\}$. We define a linear mapping $J: X \to X^{**}$, called the *canonical embedding*, by the equation $(Jx)(x') = x'(x)$ for all $x' \in X^*$, $x \in X$. It is clear that J is well defined; i.e., $Jx = x_1''$, $Jx = x_2''$ imply $(x_1'' - x_2'')(x') = 0$ for all $x' \in X^*$. It can be verified that the mapping J is always one-to-one and norm preserving. Therefore, we can consider elements x of X as continuous linear functionals on X^* with $x(x') = (Jx)(x')$. This means that X^* has two quite natural-weak topologies; that generated by X^{**} which is the weak topology for X^* (denoted $\sigma(X^*, X^{**})$), and that generated by the elements of X. The latter is called the *weak-* topology* and denoted $\sigma(X^*, X)$. The normed linear space X is said to be *norm reflexive* if J is onto. If X is reflexive the weak and weak-* topologies for X^* are the same. If X is not reflexive, the weak-* topology of X^* is weaker then its weak topology. One may easily show that a normed linear space, with either the weak or the weak-* topology is a linear topological space.

THEOREM (Alaoglu). *Let X be a real normed linear space and $Q \subset X^*$ be bounded in the norm topology and closed in the weak-* topology of X^*. Then Q is compact in the weak-* topology $\sigma(X^*, X)$.* ■

Let X be a linear space and K a set in X. K is said to be *convex* if, whenever $x, y \in K$, the "line segment" $\alpha x + (1-\alpha)y$, $0 \leqslant \alpha \leqslant 1$, joining x and y also belongs to K. It is clear that every convex set K in a normed linear space is connected because, for every $x, y \in K$ it contains a connected subset $L_{xy} = \{\alpha x + (1-\alpha)y: 0 \leqslant \alpha \leqslant 1\}$ such that $x, y \in L_{xy}$. It is also clear that the intersection of an arbitrary family of convex subsets of a linear space is convex.

A linear topological space is said to be *locally convex* if it possesses a base for its topology consisting of convex sets.

Given a subset S of a linear space X, the *convex hull* of S is the intersection of all convex sets of X containing S. It is denoted by $\operatorname{co} S$. If X is a linear topological space, the set $\overline{\operatorname{co}}\, S$, called the *closed convex hull* of S is the intersection of all closed convex subsets of X containing S. It can be proved that if $K_1, K_2 \subset X$ are convex then αK_1 and $K_1 + K_2$ are convex, where $\alpha K_1 = \{\alpha x\colon x \in K_1\}$ for $\alpha \in \boldsymbol{R}$ and $K_1 + K_2 = \{x_1 + x_2\colon x_1 \in K_1;\, x_2 \in K_2\}$.

We have the following results.

LEMMA 2.1. *For arbitrary sets A, B in a linear space X*

(i) $\operatorname{co}(\alpha A) = \alpha \operatorname{co} A$, $\operatorname{co}(A+B) = \operatorname{co} A + \operatorname{co} B$.

If X is a linear topological space, then

(ii) $\overline{\operatorname{co}}\, A = \overline{\operatorname{co} A}$,

(iii) $\overline{\operatorname{co}}(\alpha A) = \alpha\, \overline{\operatorname{co}}\, A$,

(iv) *If $\overline{\operatorname{co}}\, A$ is compact, then $\overline{\operatorname{co}}(A+B) = \overline{\operatorname{co}}\, A + \overline{\operatorname{co}}\, B$.*

(v) *If $X = \boldsymbol{R}^n$ and $A \subset \boldsymbol{R}^n$ is compact then also $\operatorname{co} A$ is compact.* ■

THEOREM (Carathéodory). *Let A be a subset of $\boldsymbol{R}^n$. Then every point $x \in \operatorname{co} A$ is the convex combination of at most $n+1$ points in A.* ■

THEOREM (Minkowski). *For every nonempty compact convex disjoint sets $B, C \subset \boldsymbol{R}^n$ there exist $u \in \boldsymbol{R}^n$ and numbers $\gamma_1, \gamma_2 \in \boldsymbol{R}$ such that $b \cdot u < \gamma_1 < \gamma_2 < c \cdot u$ for every $b \in B$ and $c \in C$, where "$\cdot$" denotes the inner product of $\boldsymbol{R}^n$.* ■

It can be proved that if S is a convex set in a real normed linear space X and x_0 is any point in X, then $x_0 \in \bar{S}$ if and only if for each $x' \in X^*$ with $\|x'\| = 1$ we have $x'(x_0) \leqslant \sup\{x'(x)\colon x \in S\}$. Hence in particular it follows that a convex set S in a real normed linear space is closed in the norm topology if and only if it is weakly closed.

We have the following theorems.

THEOREM (Mazur). *The closed convex hull of a norm compact subset of a Banach space is norm compact.* ■

THEOREM (Krein–Šmulian). *The closed convex hull of a weakly compact subset of a Banach space is weakly compact.* ■

THEOREM (Banach–Mazur). *If X is a Banach space and (x_n) is a sequence of elements of X converging weakly to x, then some sequence of convex combinations of the elements x_n converges to x in the norm topology of X.* ■

Let X be a linear space and K a set in X. A point $x \in K$ is an *extreme point* of K if whenever $x = \alpha x_1 + (1-\alpha) x_2$ for $x_1, x_2 \in X$, $0 < \alpha < 1$, then $x_1 = x_2$. If K is convex this is equivalent to stating that x is an extreme point of K if x is not the midpoint of any nondegenerate line segment in K. Although the concept of extreme point does not depend upon the topology, we often have to use topological properties of a set to establish their existence. We have the following Krein–Milman theorem.

THEOREM (Krein–Milman). *Let X be a real linear topological space with the property that for any two distinct points x_1 and x_2 of X there is a continuous linear functional x' with $x'(x_1) \neq x'(x_2)$. Then each nonempty compact set K of X has at least one extreme point.* ■

There is a concept, acyclicity, weaker then the convexity but stronger then the connectedness. The first very important example of acyclicity is given by convex sets which are acyclic. As a topological property of *acyclic sets* we know that they are connected. Furthermore it can be proved that if (K_n) is a sequence of nonempty compact acyclic subsets of a topological space X such that $K_{n+1} \subset K_n$ for all $n \in N$, then also $\bigcap_{n=1}^{\infty} K_n$ is acyclic. We have also the following Vietoris–Begle theorem.

THEOREM (Vietoris–Begle). *Let Y and Z be two metric spaces and $f\colon Y \to Z$ a given continuous surjective mapping which is supposed to be closed. i.e., for every closed set A in Y, $f(A)$ is closed in Z. Let us suppose that for any $z \in Z$, $f^{-1}(z)$ is acyclic. Then the acyclicity of Y or Z implies the acyclicity of the other one.* ■

Now, consider again a linear mapping T of a Banach space X into a Banach space Y. We have the following theorem.

THEOREM (Dunford–Schwartz). *Let T be a linear mapping of a Banach space X into a Banach space Y. Then T is continuous with respect to the metric topologies in X and Y if and only if it is continuous with respect to the weak topologies.* ■

Let $T\colon X \to Y$ be an affine mapping, i.e., such that $T[\alpha x + (1-\alpha)y] = \alpha T(x) + (1-\alpha)T(y)$ for every $x, y \in X$ and $\alpha \in [0, 1]$. We shall prove that Dunford–Schwartz theorem is also true for affine mappings. For this it is enough only to prove the following proposition.

PROPOSITION 2.1. *A mapping T from a Banach space X into a Banach space Y is affine if and only if there are a linear mapping $\hat{T} \in L(X, Y)$ and a point $y_0 \in Y$ such that $T(x) = \hat{T}(x) + y_0$ for every $x \in X$.*

Proof. Suppose T is affine and let $\hat{T}\colon X \to Y$ be defined by $\hat{T}(x) = T(x) - T(0)$ and put $y_0 = T(0)$. We have $\hat{T}(0) = 0$ and $\hat{T}$ is affine. Hence, in particular it follows $\hat{T}(2z) = 2\hat{T}(z)$ for every $z \in X$, because $\hat{T}(z) = \hat{T}(\frac{1}{2} \cdot 2z + \frac{1}{2} \cdot 0) = \frac{1}{2}\hat{T}(2z) + \frac{1}{2}\hat{T}(0) = \frac{1}{2}\hat{T}(2z)$. Therefore, for every $n \in N$ we obtain $\hat{T}(2^n z) = 2^n \hat{T}(z)$ for every $z \in X$. Now, for every $x, y \in X$ we have $\hat{T}(x+y) = \hat{T}(\frac{1}{2} \cdot 2x + \frac{1}{2} \cdot 2y) = \frac{1}{2}\hat{T}(2x) + \frac{1}{2}\hat{T}(2y) = \hat{T}(x) + \hat{T}(y)$. Since $\hat{T}(0) = 0$, then $\hat{T}(x + (-x)) = \hat{T}(x) + \hat{T}(-x) = 0$ for every $x \in X$ and therefore, for every $x \in X$ we have $\hat{T}(-x) = -\hat{T}(x)$. Let $\alpha \in [0, 1]$. For every $x \in X$ we obtain $\hat{T}(\alpha x) = \hat{T}(\alpha x + (1-\alpha)0) = \alpha \hat{T}(x)$. For $\alpha > 1$ we can find $n \in N$ such that $\alpha/2^n \in [0, 1\}$. Therefore, for every $x \in X$ we obtain $\hat{T}(\alpha x) = \hat{T}(2^n \cdot \alpha/2^n \cdot x) = 2^n \hat{T}(\alpha/2^n \cdot x) = 2^n \cdot \alpha/2^n \cdot \hat{T}(x) = \alpha \hat{T}(x)$. Then $\hat{T} \in L(X, Y)$. Finally, it is clear that for every linear mapping $\hat{T} \in L(X, Y)$ and every $y_0 \in Y$, a mapping T: $X \to Y$ defined by $T(x) = \hat{T}(x) + y_0$ is affine. ■

Now as a corollary of the Dunford–Schwartz theorem and Proposition 2.1 the following result follows.

THEOREM 2.2. *An affine mapping T of a Banach space X into a Banach space Y is continuous with respect to the metric topologies in X and Y if and only if it is continuous with respect to the weak topologies.* ■

3. Vector measures and vector-valued functions

We present here some properties of vector measures and vector-valued functions. Some properties of vector-valued functions are present only for the finite-dimensional case. The basis for this material is a finite measure space $(T, \mathscr{F}, \mu)$ with a positive measure μ and a Banach space $(X, \|\cdot\|)$. We deal here with functions taking values in a Banach space X. We call them *vector-valued functions*. Some of them will be called *vector measures*.

There are two forms of measurability of vector-valued functions, i.e., strong or μ-measurability and weak or weak μ-measurability. We shall consider here only strong measurable vector-valued functions.

A function f: $T \to X$ is called *simple* if there exists $x_1, \ldots, x_n \in X$ and $E_1, \ldots, E_n \in \mathscr{F}$ such that $f = \sum_{i=1}^{n} x_i \chi_{E_i}$, where $\chi_{E_i}(t) = 1$ if $t \in E_i$ and $\chi_{E_i}(t) = 0$ if $t \notin E_i$. A function f: $T \to X$ is called *μ-measurable* if there exists a sequence of simple functions (f_n) with $\lim_{n\to\infty} \|f_n(t) - f(t)\| = 0$ for μ-almost all $t \in T$, i.e., for all $t \in T \setminus E$, with $\mu(E) = 0$. The usual facts regarding the stability of measurable functions under sums, scalar multiples and pointwise (almost everywhere) limits hold. Also, for every μ-measurable function f: $T \to X$, a real-valued function $T \ni t \to \|f(t)\| \in R^*$ is measurable. Moreover, if (f_n) is a sequence of μ-measurable functions f_n: $T \to X$ μ-almost everywhere weakly convering to f: $T \to X$, then f is also μ-measurable. Furthermore, we also have the following theorem.

THEOREM (Egoroff). *Let (f_n) be a sequence of μ-measurable vector-valued functions f_n: $T \to X$, μ-almost everywhere converging to f: $T \to X$. Then, for every $\varepsilon > 0$ there is a measurable set $E \in \mathscr{F}$ with $\mu(E) < \varepsilon$ such that* $\lim_{n\to\infty} \sup_{t\in T\setminus E} \|f_n(t) - f(t)\| = 0$. ■

We can also define an integral, called the *Bochner integral*, for some μ-measurable vector-valued functions. It is a straight-forward abstraction of the Lebesgue integral.

A μ-measurable vector-valued function f: $T \to X$ is called *Bochner integrable* if there exists a sequence (f_n) of simple functions f_n: $T \to X$ such that $\lim_{n\to\infty} \int_T \|f - f\| d\mu = 0$. In this case, $\int_E f d\mu$ is defined for each $E \in \mathscr{F}$ by $\int_E f d\mu :=$

$\lim_{n\to\infty}\int_E f_n d\mu$, where $\int_E f_n d\mu$ is defined in the obvious way. It is not difficult to verify that a μ-measurable function $f\colon T\to X$ is Bochner integrable if and only if $\int_T \|f\| d\mu < \infty$. Further elementary facts about the Bochner integrals are collected in the following theorems.

THEOREM 3.1. *If $f\colon T\to X$ is Bochner integrable, then*

(i) $\lim_{\mu(E)\to 0}\int_E f d\mu = 0$,

(ii) $\|\int_E f d\mu\| \leqslant \int_E \|f\| d\mu$ *for all* $E\in\mathscr{F}$,

(iii) *if (E_n) is a sequence of pairwise disjoint members of $\mathscr{F}$ and $E = \bigcup_{n=1}^{\infty} E_n$, then $\int_E f d\mu = \sum_{n=1}\int_{E_n} f d\mu$, where the sum on the right is absolutely convergent.* ■

THEOREM 3.2. *Let f be Bochner integrable on $[\alpha,\beta]$ with respect to Lebesgue measure. Then for almost all $s\in[\alpha,\beta]$ one has*

$$\lim_{h\to 0}\frac{1}{h}\int_s^{s+h}\|f(t)-f(s)\| dt = 0.$$

Consequently, for almost all $s\in[\alpha,\beta]$ one has

$$\lim_{h\to 0}\frac{1}{h}\int_s^{s+h} f(t) dt = f(s). \quad ■$$

We have also for the Bochner integral the following mean-valued and convergence theorems.

THEOREM (mean-value theorem). *Let $f\colon T\to X$ be Bochner integrable. Then for each $E\in\mathscr{F}$ with $\mu(E)>0$ one has*

$$\frac{1}{\mu(E)}\int_E f d\mu \in \overline{\mathrm{co}}\big(f(E)\big). \quad ■$$

THEOREM (Vitali convergence theorem). *Let $(T,\mathscr{F},\mu)$ be a finite measure space and (f_n) be a sequence of Bochner integrable X-valued functions on T converging μ-almost everywhere to a function $f\colon T\to X$. If furthermore, $\lim_{\mu(E)\to 0}\int_E\|f_n\| d\mu = 0$ uniformy with respect to $n\in N$ then, f is Bochner integrable and $\lim_{n\to\infty}\int_E f_n d\mu = \int_E f d\mu$ for each $E\in\mathscr{F}$. In fact, $\lim_{n\to\infty}\int_E\|f_n - f\| d\mu = 0$.* ■

As a corollary of the above Vitali's theorem we obtain the following known result.

THEOREM (Lebesgue dominated convergence theorem). *Let $(T, \mathscr{F}, \mu)$ be a finite measure space and (f_n) be a sequence of Bochner integrable X-valued functions on T converging μ-almost everywhere to a function $f\colon T \to X$. Suppose there exists a real-valued Lebesgue integrable function m on T with $\|f_n\| \leqslant m$ μ-almost everywhere, then f is Bochner integrable and* $\lim_{n\to\infty} \int_E f_n d\mu = \int_E f d\mu$ *for each* $E \in \mathscr{F}$. *In fact,* $\lim_{n\to\infty} \int_T \|f_n - f\| d\mu = 0$. ■

If $1 \leqslant p < \infty$, the symbol $L^p(T, \mathscr{F}, \mu, X)$ or shortly $L^p(\mu, X)$ will stand for all (equivalence classes of) Bochner integrable functions $f\colon T \to X$ such that $\|f\|_p := (\int_T \|f\|^p d\mu)^{1/p} < \infty$. Normed by the functional $\|\cdot\|_p$ defined above, $L^p(\mu, X)$ becomes a Banach space, a fact whose proof is the same as in the real-valued case. The symbol $L^\infty(\mu, X)$ will stand for all (equivalence classes of) essentially bounded Bochner integrable functions $f\colon T \to X$. Normed by the functional $\|\cdot\|_\infty$ defined for $f \in L^\infty(\mu, X)$ by $\|f\|_\infty := \text{ess sup}\|f\|$, $L^p(\mu, X)$ becomes a Banach space. The symbols $L^p(\mu)$ with $1 \leqslant p < \infty$ will always mean $L^p(\mu, X)$ for X = scalars. If $(T, \mathscr{F}, \mu)$ is a finite measure space, then two basic theorems of measure theory, the Riesz representation theorem and the Radon–Nikodym theorem, guarantee that $[L^1(\mu)]^* = L^\infty(\mu)$ and that if λ is a finite μ-continuous scalar-valued measure on $\mathscr{F}$, then there exists $f \in L^1(\mu)$ such that $\lambda(E) = \int_E f d\mu$ for all $E \in \mathscr{F}$. Each of these theorems can be derived from the other and it is not surprising that their vector-valued extensions are related in the most intimate of ways.

It follows immediately from the properties of the space $L^1(\mu)$ that if X is finite-dimensional then $L^1(\mu, X)$ is weakly complete and a sequence (f_n) of $L^1(\mu, X)$ converges weakly to an element f in $L^1(\mu, X)$ if and only if it is bounded and $\int_E f d\mu = \lim_{n\to\infty} \int_E f\, d\mu$ for every $E \in \mathscr{F}$. It is also true in this case that every bounded and uniformly integrable (sequence) subset of $L(\mu, X)$ is relatively weakly compact. Recall, that a subset K of $L^1(\mu, X)$ is uniformly integrable if $\lim_{\mu(E)\to 0} \int_E \|f\| d\mu = 0$ uniformly in $f \in K$. The last result follows from the following Dunford's theorem

THEOREM (Dunford). *Let $(T, \mathscr{F}, \mu)$ be a finite measure space. A subset K of $L^1(\mu)$ is relatively sequentially weakly compact if and only if it is bounded and uniformly integrable.* ■

Given a measurable space $(T, \mathscr{F})$ and a Banach space $(X, \|\cdot\|)$ a function $\nu\colon \mathscr{F} \to X$ is said to be *finitely additive vector measure*, or simply a *vector measure*, if whenever E_1 and E_2 are disjoint members of $\mathscr{F}$ then $\nu(E_1 \cup E_2) = \nu(E_1) + \nu(E_2)$. If, in addition $\nu(\bigcup_{n=1}^\infty E_n) = \sum_{n=1}^\infty \nu(E_n)$ in the norm topology of X for all sequences (E_n) pairwise disjoint members of $\mathscr{F}$, then it is termed a *countable additive vector measure*.

Let $\nu\colon \mathscr{F} \to X$ be a vector measure. The variation of ν is the extended nonnegative function $|\nu|$ whose value on a set $E \in \mathscr{F}$ is given by $|\nu|(E) = \sup_{\pi} \sum \|\nu(A)\|$, where the supremum is taken over all partitions π of E into a finite number of pairwise disjoint members of $\mathscr{F}$. If $|\nu|(T) < +\infty$, then ν will be called a *measure of bounded variation.*

Let $(T, \mathscr{F}, \mu)$ be a measure space with a nonnegative measure μ and suppose $\nu: \mathscr{F} \to X$ is a vector measure. Then ν is said to be *μ-continuous* if $\lim_{\mu(E)\to 0} \nu(E) = 0$. It can be proved that if $f: T \to X$ is Bochner integrable then $\nu: \mathscr{F} \to X$ defined by $\nu(E) = \int_E f d\mu$ for $E \in \mathscr{F}$ is μ-continuous countable additive vector measure of bounded variation and $|\nu|(E) = \int_E \|f\| d\mu$ for $E \in \mathscr{F}$.

We have the following results.

THEOREM (Radon–Nikodym). *Let $(T, \mathscr{F}, \mu)$ be a positive measure space and X a finite-dimensional Banach space. If $\nu: \mathscr{F} \to X$ is μ-continuous vector measure of bounded variation, then there is a unique $g \in L^1(\mu, X)$ such that $\nu(E) = \int_E g d\mu$ for all $E \in \mathscr{F}$.* ■

THEOREM (Riesz representation theorem). *Let $(T, \mathscr{F}, \mu)$ be a positive measure space and X a Banach space. If $L: L^1(\mu) \to X$ is a continuous linear operator then there exists $g \in L^\infty(\mu, X)$ such that $Lf = \int_T fg d\mu$ for all $f \in L^1(\mu)$.* ■

The unique function $g \in L^1(\mu, X)$ defined in Radom–Nikodym theorem is called the *Radom–Nikodym derivative of a vector measure ν with respect to μ* and is denoted by $d\nu/d\mu$.

We shall deal now with nonatomic vector measures. Similarly as in the scalar case an atom of a vector measure $\mu: \mathscr{F} \to X$ is an element $E \in \mathscr{F}$ for which $\mu(E) \neq 0$ and such that if $E_1 \subset E$ and $E_1 \in \mathscr{F}$ then $\mu(E_1) = 0$ or $\mu(E \setminus E_1) = 0$. A vector measure $\mu: \mathscr{F} \to X$ with no atoms is called *nonatomic*.

We shall prove the following theorem.

THEOREM (Lyapunov). *Let $(T, \mathscr{F})$ be a measure space and $\nu: \mathscr{F} \to \boldsymbol{R}^n$ be a countable additive vector measure. If ν is nonatomic then the range of ν is a compact convex subset of $\boldsymbol{R}^n$.*

Proof. Let $\nu = (\mu_1, \ldots, \mu_n)$. Obviously μ_i is for each $i = 1, 2, \ldots, n$ a finite (not necessarily positive) nonatomic scalar measure on $(T, \mathscr{F})$. By decomposing each μ_i into its positive and negative parts we reduce the case of general measures μ_i to that of positive measure. The proof will be done by induction on n. The proof for $n = 1$ is the same as that of the induction step so we present only induction step.

Let $\mu = \sum_{i=1}^{n} \mu_i$, W be the subset $\{g: 0 \leqslant g \leqslant 1\}$ of $L^\infty(\mu)$ and let $\mathscr{I}: L^\infty(\mu) \to \boldsymbol{R}^n$

be defined by

$$\mathscr{I}g = \left(\int_T g d\mu_1, \int_T g d\mu_2, \ldots, \int_T g d\mu_n\right)$$

for each $g \in L^\infty(\mu)$. Since $\mathscr{I}$ is linear and W is convex and weakly-* compact then by Dunford–Schwartz theorem $\mathscr{I}(W)$ is a convex and compact subset of $\boldsymbol{R}^n$.

It is clear that $\nu(T) \subset \mathscr{I}(W)$. In order to complete the proof, it suffices to show that these two sets coincide, i.e. that, for every $(a_1, a_2, \ldots, a_n) \in \mathscr{I}(W)$, $W_0 = \mathscr{I}^{-1}(a_1, a_2, \ldots, a_n) \cap W$ contains a characteristic function. The set W_0 is weakly-* compact and convex; thus it has extreme points, by Krein–Milman theorem, and it is enough to prove that any extreme point g of W_0 must be a characteristic function.

Assume that g is an extreme point of W_0 and is not a characteristic function. Then there is an $\varepsilon > 0$ and a subset $E_0 \in \mathscr{F}$ so that $\mu(E_0) > 0$ and $\varepsilon \leqslant g(t) \leqslant 1-\varepsilon$ for $t \in E_0$. Since μ is non-atomic there exists $E_1 \subset E_0$ so that $\mu(E_1) > 0$ and $\mu(E_2) > 0$, where $E_2 = E_0 \setminus E_1$. By the induction hypothesis, there are $F_1 \subset E_1$ and $F_2 \subset E_2$ so that $\mu_i(F_1) = \mu_i(E_1)/2$ and $\mu_i(F_2) = \mu_i(E_2)/2$ for $i = 2, 3, \ldots, n$. Pick real s, t so that $|s|$, $|t| < \varepsilon$, $s^2+t^2 > 0$ and

$$s[\mu_1(E_1)-2\mu_1(F_1)]+t[\mu_1(E_2)-2\mu_1(F_2)] = 0.$$

Let $h = s(\chi_{E_1}-2\chi_{F_1})+t(\chi_{E_2}-2\chi_{F_2})$. Then $\int_T h d\mu_i = 0$, for $1 \leqslant i \leqslant n$ and $|h| \leqslant g \leqslant 1-|h|$ on T. Hence $g \pm h \in W_0$ and, since $h \neq 0$, this contradicts the assumption that g is an extreme point of W_0. ■

COROLLARY 3.1. *Let $(T, \mathscr{F})$ be a measurable space and let ν: $\mathscr{F} \to \boldsymbol{R}^n$ be such as in Lyapunov's theorem. Then for an arbitrary $B \in \mathscr{F}$ there is $A \in \mathscr{F}$ such that $A \subset B$ and $\nu(A) = \frac{1}{2}\nu(B)$.* ■

We can also define a Stieltjes integral for vector-valued functions with respect to a scalar function of bounded variation. Given a partition $\Delta = \{\alpha = t_0 < t_1 < \ldots < t_n = \beta\}$ of a compact interval $[\alpha, \beta]$ and a function g: $[\alpha, \beta] \to \boldsymbol{R}^n$ by $T(\Delta, g)$ we denote the sum $\sum_{i=1}^{n} |g(t_i)-g(t_{i-1})|$. The supremum $\sup_\Delta T(\Delta, g)$ taken over all partitions Δ of $[\alpha, \beta]$ we denote by $V(g)$ and call the *total variation of g on* $[\alpha, \beta]$. If $V(g) < \infty$ we say that g is of bounded variation on $[\alpha, \beta]$. We shall denote by $\mathrm{BV}([\alpha, \beta], \boldsymbol{R}^n)$ the class of all functions g: $[\alpha, \beta] \to \boldsymbol{R}^n$ of bounded variation on $[\alpha, \beta]$. It is known that every $g \in \mathrm{BV}([\alpha, \beta], \boldsymbol{R}^n)$ is differentiable, a.e., on $[\alpha, \beta]$.

Let (Δ_n) be a sequence of partitions $\Delta_n = \{\alpha = t_0^n < t_n^1 < \ldots < t_{N_n}^n = \beta\}$ of $[\alpha, \beta]$ such that $\|\Delta_n\| \to 0$ as $n \to 0$, where $\|\Delta_n\| = \max_{1 \leqslant i \leqslant N_n} |t_i^n - t_{i-1}^n|$. Denote by $\pi(\alpha, \beta)$ the collection of all such sequences (Δ_n). Let f: $[\alpha, \beta] \to X$ and g: $[\alpha, \beta] \to \boldsymbol{R}$ be given functions. We say that f is *Stieltjes integrable with respect to g* if for every sequence $(\Delta_n) \in \pi(\alpha, \beta)$ the limit $\lim_{n\to\infty} \sum_{i=1}^{N_n} f(\tau_i)[g(t_i^n)-g(t_{i-1}^n)]$ exists in the norm topology in X independent of (Δ_n) and $\tau_i \in [t_{i-1}^n, t_i^n]$. In this case this limit is denoted by $\int_\alpha^\beta f(t)dg(t)$ and called the *Stieltjes integral of f with respect to g*. Similarly as in the scalar case we have the following theorem.

THEOREM 3.3. *Suppose f: $[\alpha, \beta] \to X$ is continuous and $g \in \mathrm{BV}([\alpha, \beta], \boldsymbol{R})$. Then the Stieltjes integral $\int_\alpha^\beta f(t)dg(t)$ exists and $\|\int_\alpha^\beta ft)dg(t)\| \leqslant V(g) \max_{\alpha \leqslant t \leqslant \beta} \|f(t)\|$.* ■

Given a compact interval $I := [\alpha, \beta]$ and a Banach space $(X, \|\cdot\|)$ by $C(I, X)$ we denote the space of all X-valued continuous functions defined on I. Similarly as in the case of $L^p(\mu, X)$, $C(I, X)$ normed by the functional $C(I, X) \ni x \to |x|_C := \sup_{t \in I} \|x(t)\| \in \boldsymbol{R}^1$ becomes a Banach space. Generally, for a given topological space S by $C(S, X)$ we denote a Banach space of all X-valued bounded continuous functions on S with the usual supremum norm $|\cdot|_C$.

Given a set $K \subset C(I, X)$, a function ω_K: $(0, \infty] \to (0, \infty]$ such that $\|x(t_1) - x(t_2)\| \leqslant \omega_K(|t_1 - t_2|)$ for every $t_1, t_2 \in I$, $x \in K$ and $\lim_{\delta \to 0^+} \omega_K(\delta) = 0$ is said to be a *modulus of continuity of a set K*. A set $K \subset C(I, X)$ is said to be *equicontinuous* if it has a modulus of continuity. We have the following Ascoli theorem.

THEOREM (Ascoli). *A set $K \subset C(I, X)$ is relatively compact if and only if it is bounded, equicontinuous and such that the set $K(t) := \{x(t)\colon x \in K\}$ is relatively compact in X.* ■

Finally, let us observe that the Banach spaces $L^1(\mu, X)$ and $C(I, X)$ are separable whenever X is separable.

4. SPECIAL SPACES AND FIXED POINT THEOREMS

Let $(T, \mathscr{F}, \mu)$ be a positive σ-finite measure space. We shall now consider a partially ordered space of real measurable functions on T. We have the following theorem.

THEOREM 4.1. *Let $(T, \mathscr{F}, \mu)$ be a positive σ-finite measure space. Then the partially ordered space of real measurable functions on T, where $f \geqslant g$ means that $f(t) \geqslant g(t)$ for μ-almost all t in T, is a complete lattice. Furthermore, the least upper bound of every bounded set B in this lattice is the least upper bound of a suitably chosen countable subset of B.* ■

We shall now consider the space $\mathrm{AC}(I, \boldsymbol{R}^n)$ of all absolutely continuous functions f: $I \to \boldsymbol{R}^n$, with $I := [\alpha, \beta]$. Recall, that f: $[\alpha, \beta] \to \boldsymbol{R}^n$ is called *absolutely continuous* if to each $\varepsilon > 0$ corresponds a $\delta > 0$ such that for any countable collection of disjoint subintervals $[\alpha_k, \beta_k]$ of $[\alpha, \beta]$ such that $\sum_{k=1}^\infty (\beta_k - \alpha_k) < \delta$ we have $\sum_{k=1}^\infty |f(\beta_k) - f(\alpha_k)| < \varepsilon$. It is easy to see that every absolutely continuous function is of bounded variation and therefore it is differentiable almost everywhere in $[\alpha, \beta]$. Let us also note that any absolutely continuous function f: $I \to \boldsymbol{R}^n$ determines a unique of bounded variation vector measure ν_f on the σ-algebra $\mathscr{L}(I)$ of all Lebesgue measurable sets in I such that $\nu_f([a, b)) = f(b) - f(a)$ for each $[a, b] \subset I$.

Immediately from Theorem 3.1 and Radon–Nikodym theorem the following theorem follows.

THEOREM 4.2. *A function f: $[\alpha, \beta] \to \boldsymbol{R}^n$ is absolutely continuous on $[\alpha, \beta]$ if and only if there exists a Lebesgue integrable function g: $[\alpha, \beta] \to \boldsymbol{R}^n$ such that $f(t) = f(\alpha) + \int_\alpha^t g(\tau)d\tau$ for $t \in [\alpha, \beta]$.* ■

Given a compact interval $I := [\alpha, \beta]$ the symbol $\mathrm{AC}(I, \boldsymbol{R}^n)$ will stand for all absolutely continuous functions f: $I \to \boldsymbol{R}^n$. Normed by functional $\|\cdot\|$ defined for $f \in \mathrm{AC}(I, \boldsymbol{R}^n)$ by $\|f\| := |f(\alpha)| + \int_\alpha^\beta |f'(t)|dt$, $\mathrm{AC}(I, \boldsymbol{R}^n)$ becomes a Banach space. It is clear that $\mathrm{AC}(I, \boldsymbol{R}^n) \subset C(I, \boldsymbol{R}^n)$. Furthermore, it can be proved that $\mathrm{AC}(I, \boldsymbol{R}^n)$ is isometrically isomorphic to the direct sum of $L(I, \boldsymbol{R}^n)$ and an n-dimensional space. Consequently $\mathrm{AC}(I, \boldsymbol{R}^n)$ is weakly complete. Here and later $L(I, \boldsymbol{R}^n)$ denote the space $L(I, \mathscr{L}(I), \mu, \boldsymbol{R}^n)$, where μ is a Lebesgue measure on the σ-algebra $\mathscr{L}(I)$ of Lebesgue measurable subsets of I.

A given subset K of $\mathrm{AC}(I, \boldsymbol{R}^n)$ is said to be *equiabsolutely continuous* if for every $\varepsilon > 0$ there is a $\delta > 0$ such that $\sum_{i=1}^N |f(\beta_i) - f(\alpha_i)| \leqslant \varepsilon$ for all $f \in K$ and all finite systems of nonoverlapping intervals $[\alpha_i, \beta_i]$, $i = 1, \ldots, N$, in $[\alpha, \beta]$ with $\sum_{i=1}^N (\beta_i - \alpha_i) < \delta$. We shall prove now the following lemma.

LEMMA 4.3. *Let $I := [\alpha, \beta]$ and let $K \subset \mathrm{AC}(I, \boldsymbol{R}^n)$ be given and put $\dot{K} = \{f' : f \in K\}$, where for each $f \in K$, f' denotes a Lebesgue integrable function g: $[\alpha, \beta] \to \boldsymbol{R}^n$ such that $f(t) = f(\alpha) + \int_\alpha^t g(\tau)d\tau$ for $t \in [\alpha, \beta]$. Then K is equiabsolutely continuous if and only if $\dot{K} \subset L(I, \boldsymbol{R}^n)$ is uniformly integrable.*

Proof. Suppose $\dot{K}$ is uniformly integrable. Given $\varepsilon > 0$, let $\delta > 0$ be such that $\int_E |f'(t)|dt \leqslant \varepsilon$ for every $f \in K$ and each $E \in \mathscr{L}(I)$ with $\mu(E) \leqslant \delta$. Let $[\alpha_i, \beta_i]$, $i = 1, \ldots, N$, be any system of nonoverlapping intervals in $I := [\alpha, \beta]$ with $\sum_i (\beta_i - \alpha_i) \leqslant \delta$. If $E = \bigcup_{i=1}^N [\alpha_i, \beta_i]$, then $\mu(E) \leqslant \delta$ and $\sum_{i=1}^N |f(\beta_i) - f(\alpha_i)| = \sum | \int_{\alpha_i}^{\beta_i} f'(t)dt| \leqslant \int_E |f'(t)|dt \leqslant \varepsilon$ and this holds for every $f \in K$. Conversely, assume that K is equiabsolutely continuous. Given $\varepsilon > 0$, let $\delta = \delta(\varepsilon/6)$ be the number given in the definition of equiabsolute continuity of K. Let $f \in K$ and let E be any measurable subset of $I := [\alpha, \beta]$ with $\mu(E) \leqslant \delta/2$. Let E_j^+, E_j^- be the subset of all $t \in E$ where $f_j'(t)$ is defined and $f_j'(t) \geqslant 0$ or $f_j'(t) \leqslant 0$, respectively with $f = (f_1, \ldots f_n)$ and $j = 1, \ldots n$. Then $\mu(E_j^+) \leqslant \delta/2$

and $\mu(E_j^-) \leqslant \delta/2$. Since $f' \in L(I, \boldsymbol{R}^n)$, then there is $\sigma := \sigma(\varepsilon, f) > 0$ such that $\int_F |f'(t)| dt \leqslant \varepsilon/6$ for every measurable subset F of I with $\mu(F) \leqslant \sigma$. It is not restrictive to take $\sigma \leqslant \delta/2$. Now E_j^+ with $\mu(E_j^+) \leqslant \delta/2$ is certainly covered by some open set G with $\mu(G) \leqslant \mu(E_j^+)+\sigma$. Let (α_i, β_i), $i = 1, 2, \ldots$ denote the disjoint open intervals which are the components of G, and note that $\sum_i^{\infty} (\beta_i - \alpha_i) = \mu(G) \leqslant \mu(E_j^+)+\sigma \leqslant \delta/2+ +\delta/2 = \delta$. Then, the same holds for finite system (α_i, β_i), $i = 1, \ldots, N$, whose union we denote by G_N, N arbitrary. Thus $\mu(G_N) \leqslant \delta$, $\mu(G_N \setminus E_j^+) \leqslant \sigma$, and for N large enough also $\mu(G \setminus G_N) \leqslant \sigma$. Then

$$\begin{aligned}
\int_{E_j^+} |f_j'(t)| dt &= \Big(\int_{E_j^+ \cap G_N} + \int_{E \cap (G \setminus G_N)} \Big) f_j'(t) dt \\
&= \Big(\int_{E \cap G_N} + \int_{G_N \setminus E_j^+} \Big) f_j'(t) dt - \int_{G_N \setminus E_j^+} f_j'(t) dt + \int_{E \cap (G \setminus G_N)} f_j'(t) dt \\
&\leqslant \int_{G_N} f_j'(t) dt + \int_{G_N \setminus E_j^+} |f_j'(t)| dt + \int_{G \setminus G_N} |f_j'(t)| dt \\
&\leqslant \sum_{i=1}^{N} [f_j(\beta) - f_j(\alpha)] + \varepsilon/6 + \varepsilon/6 \leqslant \varepsilon/6 + \varepsilon/6 + \varepsilon/6 = \varepsilon/2.
\end{aligned}$$

Analogously, we can prove that $\int_{E_j^-} |f_j'(t)| dt \leqslant \varepsilon/2$ for each $j = 1, \ldots, n$ and thus $\int_E |f'(t)| dt \leqslant \varepsilon$, and this holds for every measurable subset E of I with $\mu(E) < \delta/2$. ∎

Let us observe that every uniformly integrable subset of $L(I, \boldsymbol{R}^n)$ is bounded in $L(I, \boldsymbol{R}^n)$. Therefore, also every equiabsolutely continuous subset of $\mathrm{AC}(I, \boldsymbol{R}^n)$ is also bounded in $\mathrm{AC}(I, \boldsymbol{R}^n)$. In particular, by Lemma 4.3 it follows that a sequence (f_n) of $\mathrm{AC}(I, \boldsymbol{R}^n)$ weakly converges to any $f \in \mathrm{AC}(I, \boldsymbol{R}^n)$ if and only if $\lim_{\mu(E) \to 0} \int_E |f_n'(t)| dt = 0$ uniformly in $n = 1, 2, \ldots$, with $E \in \mathscr{L}(I)$ and $\lim_{n \leftarrow \infty} |f_n(t) - f(t)| = 0$ for all $t \in I$. We have the following gauge for the relative weak sequential compactness in $\mathrm{AC}(I, \boldsymbol{R}^n)$.

THEOREM 4.4. *A bounded subset K of* $\mathrm{AC}(I, \boldsymbol{R}^n)$ *is relatively weakly compact if and only if it is equiabsolutely continuous.* ∎

We have also the following lemma.

LEMMA 4.5. *Let (x_m) be a sequence of* $\mathrm{AC}(I, \boldsymbol{R}^n)$ *with $I := [\alpha, \beta]$ and suppose $x \in C(I, \boldsymbol{R}^n)$ is such that $\sup_{\alpha \leqslant t \leqslant \beta} |x_m(t) - x(t)| \to 0$ as $m \to \infty$. If furthermore, $(\dot{x}_m)$ is uniformly integrable on I then $x \in \mathrm{AC}(I, \boldsymbol{R}\ )$ and $\dot{x}_m \rightharpoonup \dot{x}$ in $L(I, \boldsymbol{R}^n)$ as $m \to \infty$, where "$\rightharpoonup$" denotes the weak convergence in $L(I, \boldsymbol{R}^n)$.*

Proof. Let $(\dot{x}_n)$ be an arbitrary subsequence of $(\dot{x}_m)$. Since $(\dot{x}_n)$ is uniformly integrable then it is relatively weakly compact in $L(I, \boldsymbol{R}^n)$. Then there are $u \in L(I, \boldsymbol{R}^n)$ and a subsequence, say (x_k), of (x_n) such that $\dot{x}_k \rightharpoonup u$ in $L(I, \boldsymbol{R}^n)$ as $k \to \infty$. Since for every $t \in I$ and $k = 1, 2, \ldots$ we have $x_k(t) = x_k(\alpha) + \int_\alpha^t \dot{x}\,(\tau)d\tau$ and $x_k(\alpha) \to x(\alpha)$ and $\int_\alpha^t x\,(\tau)d\tau \to \int_\alpha^t u(\tau)d\tau$ as $k \to \infty$ then $x_k(t) \to x(\alpha) + \int_\alpha^t u(\tau)d\tau$ for each $t \in I$ as $k \to \infty$. On the other hand we also have $x_k(t) \to x(t)$ for each $t \in I$ as $k \to \infty$. Therefore $x(t) = x(\alpha) + \int_\alpha^t u(\tau)d\tau$ for $t \in I$. Thus $x \in \mathrm{AC}(I, \boldsymbol{R}^n)$ and $\dot{x} = u$. Since every subsequence $(\dot{x}_n)$ of $(\dot{x}_m)$ has a subsequence weakly converging to $\dot{x}$ then also $\dot{x}_m \rightharpoonup \dot{x}$ as $m \to \infty$. ■

In what follows we shall also consider Banach subspaces $C^0(I, \boldsymbol{R}^n)$ and $\mathrm{AC}^0(I, \boldsymbol{R}^n)$ of $C(I, \boldsymbol{R}^n)$ and $\mathrm{AC}(I, \boldsymbol{R}^n)$, respectively of all x: $[\alpha, \beta] \to \boldsymbol{R}^n$ satisfying $x(\alpha) = 0$. In the sequel the Banach spaces $L(I, \boldsymbol{R}^n)$, $C(I, \boldsymbol{R}^n)$ and $\mathrm{AC}(I, \boldsymbol{R}^n)$ with $I := [-r, 0]$ for any $r \geqslant 0$ will be denoted by L_{0r}, C_{0r} and AC_{0r}, respectively.

Let Ω be a nonempty closed and bounded subset of $\boldsymbol{R} \times \mathrm{AC}_{0r}$ and let I_Ω denote the image of Ω under the projection Π_R of Ω into the real line $\boldsymbol{R}^1$. Denote by $S(\Omega)$ the family of all closed intervals $[t_1, t_2]$ contained in I_Ω. Suppose Ω is such that $S(\Omega) \neq \emptyset$ and define for fixed $r \geqslant 0$, $\mathscr{X}_r(\Omega)$ by

$$\mathscr{X}_r(\Omega) = \bigcup_{[t_1, t_2] \in S(\Omega)} \mathrm{AC}([t_1 - r, t_2], \boldsymbol{R}^n).$$

Thus, for every $x \in \mathscr{X}_r(\Omega)$ there is a closed interval $[t_1^x, t_2^x] \subset I_\Omega$ such that $x \in \mathrm{AC}([t_1^x - r, t_2^x], \boldsymbol{R}^n)$. Therefore, we will denote elements $x \in \mathscr{X}_r(\Omega)$ also by $(x;\ [t_1^x - r, t_2^x])$. Since Ω is assumed to be bounded, then I_Ω is contained in some compact interval $[t_0, T]$.

We now introduce the structure of a metric space on $\mathscr{X}_r(\Omega)$ by introducing, as usual, a metric function d: $\mathscr{X}_r(\Omega) \times \mathscr{X}_r(\Omega) \to \boldsymbol{R}^+$. For this purpose, let $(x;\ [t_1^x - r, t_2^x])$, $(y;\ [t_1^y - r, t_2^y]) \in \mathscr{X}_r(\Omega)$. Denote by x_Ω and y_Ω extensions of x and y, recpectively on $[t_0 - r, T]$ by taking $x_\Omega(t) := (t_1^x - r)$ for $t \in [t_0 - r, t_1^x - r)$, $y(t) := y(t_1^y - r)$ for $t \in [t_0 - r, t_1^y - r)$, $x_\Omega(t) := x(t_2^x)$ for $t \in (t_2^x, T]$ and $y_\Omega(t) := y(t_2^y)$ for $t \in (t_2^y, T]$. We then define the distance function d by

$$d(x, y) = \max\{|t_1^x - t_1^y|,\ |t_2^x - t_2^y|,\ ||x_\Omega - y_\Omega||_r\},$$

where $||\cdot||_r$ denotes the norm of the Banach space $\mathrm{AC}([t_0 - r, T], \boldsymbol{R}^n)$, i.e. $||z||_r = |z(t_0 - r)| + \int_{t_0 - r}^T |\dot{z}(t)|dt$ for $z \in \mathrm{AC}([t_0 - r, T], \boldsymbol{R}^n)$.

We shall consider $\mathscr{X}_r(\Omega)$ as a subset of the space $\mathscr{C}_r(\Omega)$ defined by

$$\mathscr{C}_r(\Omega) = \bigcup_{[t_1, t_2] \in S(\Omega)} C([t_1 - r, t_2], \boldsymbol{R}^n)$$

considered with the metric ϱ of the form

$$\varrho(x, y) = \max\{|t_1^x - t_1^y|,\ |t_x^2 - t_2^y|,\ |x_\Omega - y_\Omega|_r\},$$

where $|\cdot|_r$ denotes the supremum norm of the Banach space $C([t_0 - r, T], \boldsymbol{R})$ and $x, y \in \mathscr{C}_r(\Omega)$.

In a similar way we define the metric space $(\mathscr{L}_r(\Omega), l)$, with

$$\mathscr{L}_r(\Omega) = \bigcup_{(t_1, t_2] \in S(\Omega)} L([t_1 - r, t_2], \boldsymbol{R}^n)$$

and

$$l(u, v) = \max\{|t_1^u - t_1^v|, |t_2^u - t_2^v|, |u_\Omega - v_\Omega|_r\}$$

for $u, v \in \mathscr{L}_r(\Omega)$, where $|\cdot|_r$ denotes the usual norm of $L([t_0 - r, T], \boldsymbol{R}^n)$, $u_\Omega(t) = 0$ for $t \in [t_0 - r, t_1^u) \cup (t_2^u, T]$ and $= v_\Omega(t) = 0$ for $t \in [t_0 - r, t_1^v - r) \cup (t_2^v, T]$.

We have the following result.

THEOREM 4.6. *$(\mathscr{X}_r(\Omega), d)$, $(\mathscr{C}_r(\Omega), \varrho)$, $\mathscr{L}_r(\Omega), l)$ are complete metric spaces.*

Proof. Suppose $\{(x^n, [t_1^n - r, t_2^n])\}$ is a Cauchy sequence of $\mathscr{X}_r(\Omega)$, i.e. suppose that $\lim_{n, m \to \infty} d(x^n, x^m) = 0$. Then $\{t_1^n\}$, $\{t_n^2\}$ and $\{x_\Omega^n\}$ are Cauchy sequences of $\boldsymbol{R}$ and $\mathrm{AC}([t_0 - r, T], \boldsymbol{R}^n)$, respectively. By the completeness of $(\boldsymbol{R}, |\cdot|)$ and $(\mathrm{AC}([t_0 - - r, T], \boldsymbol{R}^n), \|\cdot\|_r)$ there exist $t_1, t_2 \in [t_0, T]$ and $x_\Omega \in \mathrm{AC}([t_0 - r, T], \boldsymbol{R}^n)$ such that $\max\{|t_1^n - t_1|, |t_2^n - t_2|, \|x_\Omega^n - x_\Omega\|_r\} \to 0$ as $n \to 0$. Now, let x be the restriction of x_Ω to $[t_1 - r, t_2]$. We have $x \in \mathscr{X}_r(\Omega)$ and $\lim_{n \to \infty} d(x^n, x) = 0$. Thus $(\mathscr{X}_r(\Omega), d)$ is complete. The completeness of $(\mathscr{C}_r(\Omega), \varrho)$ and $(\mathscr{L}_r(\Omega), l)$ can be obtained similarly. ■

A sequence (x^n) of $\mathscr{X}_r(\Omega)$ converging in the d-metric topology to $x \in \mathscr{X}_r(A)$ is said to be *d-convergent to x*. Similarly we define *ϱ-convergence* and *l-convergence* of sequences of $\mathscr{C}_r(\Omega)$ and $\mathscr{L}_r(\Omega)$, respectively. Often, we will be interested in any weaker mode of convergence of sequences of $\mathscr{L}_r(\Omega)$ and $\mathscr{X}_r(\Omega)$ than the convergences mentioned above. We call them *l-weak* and *d-weak convergences*, respectively. A more distinctly, we call a sequence $\{(u^n; [t_1^n - r, t_2^n])\}$ of $\mathscr{L}_r(\Omega)$ *l-weak convergent* to $(u; [t_1 - r, t_2])$ if $t_1^n \to t_1$, $t_2^n \to t_2$ and $u_\Omega^n \rightharpoonup u_\Omega$ as $n \to \infty$, where "$\rightharpoonup$" denotes as usual the convergence of the sequence (u_Ω^n) to u_Ω in the weak topology of $L([t_0 - - r, T], \boldsymbol{R}^n)$. Similarly, a bounded sequence $\{(x^n; [t_1^n - r, t_2^n])\}$ of $\mathscr{X}_r(\Omega)$ is said to be *d-weak convergent* to $(x; [t_1 - r, t_2])$ if the sequence (x_Ω^n) is equiabsolutely continuous on $[t_0 - r, T]$ and $\lim_{n \to \infty} \max\{|t_1^n - t_1|, |t_2^n - t_2|, |x_\Omega^n(t) - x_\Omega(t)|\} = 0$ for every $t \in [t_0 - r, T]$.

We will say that a set $B \subset \mathscr{C}_r(\Omega)$ is *ϱ-closed* (*ϱ-compact*) if it is closed (compact) in ϱ-metric topology of $\mathscr{C}_r(\Omega)$. Similarly *d-closedness*, *l-closedness*, *d-compactness* and *l-compactness* of subset of $\mathscr{X}_r(\Omega)$ and $\mathscr{L}_r(\Omega)$, respectively can be defined. A set $B \subset \mathscr{L}_r(\Omega)$ is said to be *relatively l-weakly sequentially compact* if for every sequence (u^n) of B we can select its subsequence l-weakly convergent to any $u \in \mathscr{L}_r(\Omega)$. It is said to be *l-weakly sequentially closed* if every l-weakly convergent sequence of B has its l-weakly limit in B. Similarly we can define *relative d-weak sequential compactness* and *closedness* of subsets of $\mathscr{X}_r(\Omega)$. We have the following gauges for the relative ϱ-compactness (l-weak and d-weak sequential compactness) of subsets of $\mathscr{C}_r(\Omega)$ ($\mathscr{L}_r(\Omega)$ and $\mathscr{X}_r(\Omega)$, respectively).

THEOREM 4.7. *A bounded set B of $\mathscr{C}_r(\Omega)$ is relatively ϱ-compact if and only if for every $\varepsilon > 0$ there is a $\delta > 0$ such that for every $(x; [t_1^x - r, t_2^x]) \in B$ and every $s, t \in [t_1^x - r, t_2^x]$ satisfying $|s-t| < \delta$ we have $|x(t) - x(s)| \leqslant \varepsilon$.*

Proof. Let $B_\Omega = \{x_\Omega : x \in B\}$. Since B is bounded in $\mathscr{C}_r(\Omega)$ then in particular B_Ω is a bounded subset of $C([t_0 - r, T], \boldsymbol{R}^n)$. Let $\varepsilon > 0$ be given and suppose $\delta > 0$ is such that conditions given above are satisfied. For every $t, s \in [t_0 - r, T]$ satisfying $|t-s| < \delta$ we have $|x_\Omega(t) - x_\Omega(s)| \leqslant \varepsilon$ for every $x_\Omega \in B_\Omega$. Indeed, if $t, s \in [t_0 - -r, t_1^x - r)$ or $t, s \in (t_2^x, T]$ we have $0 = |x_\Omega(t) - x_\Omega(s)| \leqslant \varepsilon$. For $t, s \in [t_1^x - r, t_2^x]$ we have $|x_\Omega(t) - x_\Omega(s)| = |x(t) - x(s)| \leqslant \varepsilon$. Finally, if $t \in [t_0 - r, t_1^x - r)$ and $s \in [t_1 - -r, t_2^x]$, or $t \in [t_1^x - r, t_2^x]$ and $s \in (t_2^x, T]$ or $t \in [t_0 - r, t_1^x - r)$ and $s \in (t_2, T]$ we have respectively $|x_\Omega(t) - x_\Omega(s)| = |x(t_1^x - r) - x(s)| \leqslant \varepsilon$, $|x_\Omega(t) - x_\Omega(s)| = |x(t) - x(t_2^x)| \leqslant \varepsilon$ and $|x_\Omega(t) - x_\Omega(s)| = |x(t_1^x - r) - x(t_2^x)| \leqslant \varepsilon$ because in each above case we have $|t_1^x - r - s| \leqslant \delta$, $|t - t_2^x| \leqslant \delta$ and $|t_1^x - r - t_2^x| \leqslant \delta$, respectively. Now, by virtue of Ascoli's theorem, B_Ω is relatively compact subset of $C([t_0 - r, T], \boldsymbol{R}^n)$.

Let $\{(x^n; [t_1^n - r, t_2^n])\}$ be any sequence of B. Since (x_Ω^n) is a sequence of B_Ω then there is a subsequence, say (x_Ω^k) of (x_Ω^n) and a function $x_\Omega \in C([t_0 - r, T], \boldsymbol{R}^n)$ such that $|x_\Omega^k - x_\Omega|_r \to 0$ as $k \to \infty$. Furthermore, (t_1^n) and (t_2^n) have converging subsequences, say again (t_1^k) and (t_2^k), because $t_1^n, t_2^n \in [t_0 - r, T]$ for each $n = 1, 2, \ldots$ Suppose $t_1, t_2 \in [t_0 - r, T]$ are such that $t_i^k \to t_i$ for $i = 1,2$ as $k \to \infty$ and let x be the restriction of x_Ω to $[t_1 - r, t_2]$. Then, there exists $(x; [t_1 - r, t_2]) \in \mathscr{C}_r(\Omega)$ and subsequence $\{(x^k; [t_1^k - r, t_2^k])\}$ of $\{(x^n; [t_1^n - r, t_2^n])\}$ such that $\varrho(x^k, x) \to 0$ as $k \to +\infty$. Thus B is relatively ϱ-compact.

Suppose, B is relatively ϱ-compact. In particular it implies that B_Ω is relatively compact in $C([t_0 - r, T], \boldsymbol{R}^n)$. Therefore, by Ascoli's theorem for every $x_\Omega \in B_\Omega$ and $s, t \in [t_0 - r, T]$ satisfying $|t-s| < \delta$ we have $|x_\Omega(t) - x_\Omega(s)| \leqslant \varepsilon$. Then for every $t, s \in [t_1^x - r, t_2^x]$ satisfying $|t-s| < \delta$ we have $|x(t) - x(s)| = |x_\Omega(t) - x_\Omega(s)| \leqslant \varepsilon$. ■

THEOREM 4.8. *A bounded subset B of $\mathscr{L}_r(\Omega)$ is relatively l-weakly sequentially compact if and only if for every $\varepsilon > 0$ there is a $\delta > 0$ such that for every $(u; [t_1^u - r, t_2^u]) \in B$ and every measurable set $E \subset [t_1^u - r, t_2^u]$ satisfying $\mu(E) < \delta$ one has $\int_E |u(t)| dt \leqslant \varepsilon$.*

Proof. Let $B_\Omega = \{u_\Omega : u \in B\}$ and suppose $\varepsilon > 0$. Let $\delta > 0$ be such that conditions given above are satisfied. For every $u_\Omega \in B_\Omega$ and a measurable set $E \subset [t_0 - -r, T]$ satisfying $\mu(E) < \delta$ we have

$$\int_E |u_\Omega(t)| dt = \int_{E \cap [t_1^u - r, t_2^u]} |u(t)| dt \leqslant \varepsilon$$

because $E \cap [t_1^u - r, t_2^u] \subset [t_1^u - r, t_2^u]$ satisfies $\mu(E \cap [t_1^u - r, t_u^2]) < \delta$. Then B_Ω is an uniformly integrable subset of $L([t_0 - r, T], \boldsymbol{R}^n)$. Hence, by Dunford's theorem it follows that B_Ω is relatively sequentially weakly compact in $L([t_0 - r, T], \boldsymbol{R}^n)$ because B_Ω is also bounded. Then, for every sequence (u_Ω^n) of B_Ω there are a subsequence, say (u_Ω^k), of (u_Ω^n) and a function $u \in L([t_0 - r, T], \boldsymbol{R}^n)$ such that $u_\Omega^k \rightharpoonup u_\Omega$ as $k \to \infty$. Similarly as in the proof of Theorem 4.7, hence it follows that the sequence $\{(u^n; [t_1^n -$

$-r, t_2^n])\}$ of B has a subsequence $\{(u^k; [t_1^k-r, t_2^k])\}$ l-weakly convergent to the restriction $(u; [t_1-r, t_2]) \in \mathscr{L}_r(\Omega)$ of u to any interval $[t_1-r, t_2] \subset [t_0-r, T]$. Therefore, B is relatively l-weakly sequentially compact in $\mathscr{L}_r(\Omega)$.

If B is relatively l-weakly sequentially compact in $\mathscr{L}_r(\Omega)$, in particular, it implies that B_Ω is relatively sequentially weakly compact in $L([t_0-r, T], \boldsymbol{R}^n)$. Therefore, by Dunford's theorem, B_Ω is uniformly integrable. Hence, in particular it follows that for every $\varepsilon > 0$ there is a $\delta > 0$ such that for every $(u; [t_1^u-r, t_2^u]) \in B$ and every measurable set $E \subset [t_1^u-r, t_2^u]$ satisfying $\mu(E) < \delta$ we have $\int_E |u(t)|dt < \varepsilon$. ■

In a similar way we can also prove the following theorem.

THEOREM 4.9. *A bounded subset B of $\mathscr{X}_r(\Omega)$ is relatively d-weakly sequentially compact if and only if for every $\varepsilon > 0$ there is a $\delta > 0$ such that for every $(x; [t_1^x-r, t_2^x]) \in B$ and every measurable set $E \subset [t_1^x-r, t_2^x]$ with $\mu(E) \leqslant \delta$ one has $\int_E |\dot{x}(t)|dt \leqslant \varepsilon$.* ■

Given $S \subset \mathscr{X}_r(\Omega)$ a functional I: $S \to \boldsymbol{R}$ is said to be *d-sequentially weakly lower semicontinuous at* $x \in S$ if for every sequence (x_n) of S *d-weakly* converging to x one has $I(x) \leqslant \liminf_{n\to\infty} I(x_n)$.

We have the following modification of the Weierstrass theorem.

THEOREM (Weierstrass). *Let S be a nonempty subset of $\mathscr{X}_r(\Omega)$ ϱ-closed in $\mathscr{C}_r(\Omega)$ and relatively d-weakly sequentially compact in $\mathscr{X}_r(\Omega)$. If I: $S \to \boldsymbol{R}$ is d-sequentially weakly lower or ϱ-lower semicontinuous on S then I is bounded below in S and has an absolutely minimum in S.*

Proof. Let $m = \inf_{x\in S} I(x)$. We have $-\infty \leqslant m < +\infty$. Take any sequence (x_n) of S such that $I(x_n) \to m$ as $n \to \infty$. We may well assume that $I(x_n) \leqslant m+1/n$ if m is finite, and $I(x_n) \leqslant -n$ if $m = -\infty$. Now, by the properties of S we can select a subsequence, say (x_k) of (x_n) such that (x_k) is ϱ-converging and d-weakly converging to any $\bar{x} \in S$. Since I is defined on S, then $I(\bar{x}) \in \boldsymbol{R}$ and $I(\bar{x}) \leqslant \liminf_{k\to\infty} I(x_k) = m < +\infty$. Thus $I(x)$ is finite, and so is m. Since $\bar{x} \in S$ also $m \leqslant I(\bar{x})$. Therefore, $I(\bar{x}) = m$. ■

Let (X, ϱ) be a metric space and F a mapping from X into X. A point $x \in X$ such that $x = F(x)$ is called a *fixed point of* F. We say that F is a contraction mapping if there exists a number $k < 1$ such that $\varrho(F(x), F(y)) \leqslant k\varrho(x, y)$ for every $x, y \in X$. We have the following basic fixed point theorems.

THEOREM (Banach). *Any contraction mapping of a complete nonempty metric space X into X has a unique fixed point in X.* ■

THEOREM (Schauder). *Let S be a nonempty closed convex subset of a Banach space X and let F: $S \to S$ be completely continuous, i.e., it is continuous and maps bounded subsets of S into compact sets. Then F has a fixed point in S.* ■

THEOREM (Schauder–Tikhonov). *Let X ba a linear, locally convex topological space. Let S be a compact, convex subset of X and F a continuous mapping of S into itself. Then F has a fixed point in S.* ■

5. NOTES AND REMARKS

The definitions and most results of this chapter are classical. They are selected from Alexiewicz [1], Hermes and LaSalle [1], Dunford and Schwartz [1], Edwards [1], Diestel and Uhl [1] and Cesari [1]. In particular, topological preliminary results are taken from Hermes and LaSalle [1], whereas functional analysis preliminaries are based on Hermes and LaSalle [1], Dunford and Schwartz [1] and Edwards [1]. The definitions and results dealing with vector measures and vector-valued functions are selected from Diestal and Uhl [1], Dunford and Schwartz [1] and Alexiewicz [1]. The proof of Lyapunov's theorem is taken from Lindenstrauss [1]. Finally, some results concerning special spaces can be found in Dunford and Schwartz [1] and Cesari [1]. The properties of the metric spaces $\mathscr{L}_r(\Omega)$, $\mathscr{C}_r(\Omega)$ and $\mathscr{X}_r(\Omega)$ are modifications of some results contained in Cesari [1]. Fixed point theorems are selected from Edwards [1]. For a precise definition of the acyclicity see Bourgin [1]. Proposition 2.1 is in general true for linear vector spaces.

In the text of the book we also apply the classical Fatou's and Gronwall's lemmas (see Taylor [1] and Cesari [1], respectively).

LEMMA (Fatou). *Suppose $\eta \in L(I, \boldsymbol{R})$ and a sequence (η_k) of $L(I, \boldsymbol{R})$ are such that*

(i) $\eta(t) \leqslant \eta_k(t)$ *for* $k = 1, 2, \ldots$ *and a.e.* $t \in I$,

(ii) $\liminf_{k\to\infty} \int_I \eta_k(t)\,dt < +\infty$

and let $f(t) = \liminf_{k\to\infty} \eta_k(t)$ *for a.e.* $t \in I$. *Then* $f \in L(I, \boldsymbol{R})$ *and*

$$\int_I f(t)\,dt \leqslant \liminf_{k\to\infty} \int_E \eta_k(t)\,dt. \quad ■$$

LEMMA (Gronwall). *Let $I := [\alpha, \beta]$ and suppose $u \in L(I, \boldsymbol{R}^+)$, $v \in C(I, \boldsymbol{R}^+)$ and $\eta > 0$ are such that $v(t) \leqslant \eta + \int_\alpha^t u(s)\,v(s)\,ds$ for $t \in I$. Then $v(t) \leqslant \eta \exp\left(\int_\alpha^t u(s)\,ds\right)$ for each $t \in I$.* ■

Chapter II

Set-valued functions

In this chapter we gather the basic properties of set-valued functions needed for the study of functional-differential inclusions. We begin with various continuity properties and then the measurability of set-valued functions is investigated. The basic properties of the Aumann's integrals and some fixed point theorems are contained in the last part of this chapter.

1. Spaces of subsets of metric space

We shall consider here some spaces of subsets of a metric space. We begin with the space $\mathrm{Comp}(X)$ of all nonempty compact subsets of a metric space (X, ϱ) and then we consider some properties of the space $\mathrm{Comp}(\boldsymbol{R}^n)$.

1.1. Hausdorff distance and Hausdorff topology of Comp(X)

Let $\mathrm{Cl}(X)$ and $\mathrm{Comp}(X)$ denote the family of all nonempty closed bounded and nonempty compact, respectively subsets of a metric space (X, ϱ). Given $A, B \in \mathrm{Cl}(X)$ let $h(A, B) := \max\{\bar{h}(A, B), \bar{h}(B, A)\}$, where $\bar{h}(A, B) := \sup_{a\in A} \mathrm{dist}(a, B)$, and $\bar{h}(B, A) := \sup_{b\in B} \mathrm{dist}(b, A)$.

PROPOSITION 1.1. *The function* $h\colon \mathrm{Cl}(X)\times \mathrm{Cl}(X) \to \boldsymbol{R}^+$ *is a metric on* $\mathrm{Cl}(X)$.

Proof. It is clear that $h(A, B) \geqslant 0$ and $h(A, B) = 0$ if and only if $A = B$ for every $A, B \in \mathrm{Cl}(X)$. Furthermore, for every $A, B \in \mathrm{Cl}(X)$ we have $h(A, B) = h(B, A)$. Let $A, B, C \in \mathrm{Cl}(X)$ be fixed. For every $x \in A$ and $y \in B$ one has

$$\mathrm{dist}(x, C) \leqslant \varrho(x, y)+\mathrm{dist}(y, C) \leqslant \varrho(x, y)+h(B, C).$$

Hence, if follows $\mathrm{dist}(x, C) \leqslant h(A, B)+h(B, C)$ for every $x \in A$. Therefore, $\bar{h}(A, C) \leqslant h(A, B)+h(B, C)$. In a similar way we obtain $\bar{h}(C, A) \leqslant h(A, B)+h(B, C)$. ∎

The metric h defined on $\mathrm{Cl}(X)$ in Proposition 1.1 is called the *Hausdorff distance* or *metric* in $\mathrm{Cl}(X)$.

PROPOSITION 1.2. *For every* $A, B \in \mathrm{Cl}(X)$, $h(A, B) = \inf\{\varepsilon > 0,\ A \subset V(B, \varepsilon)$ *and* $B \subset V(A, \varepsilon)\}$, *where given* $\varepsilon > 0$ *and* $S \in \mathrm{Cl}(X)$, $V(S, \varepsilon) = \{x \in X\colon \mathrm{dist}(x, S) \leqslant \varepsilon\}$.

Proof. Let us observe that for given $\varepsilon > 0$ and $A, B \in \mathrm{Cl}(X)$ we have $A \subset V(B, \varepsilon)$ and $B \subset V(A, \varepsilon)$ if and only if $\bar{h}(A, B) \leqslant \varepsilon$ and $\bar{h}(B, A) \leqslant \varepsilon$, respectively. Furthermore $\inf\{\varepsilon > 0: \bar{h}(A, B) \leqslant \varepsilon$ and $\bar{h}(B, A) \leqslant \varepsilon\} = \inf(\{\varepsilon > 0: \varepsilon \geqslant h(A, B)\}$ $= h(A, B)$. ■

Note that the topology on $\mathrm{Cl}(X)$ derived from the Hausdorff distance h is not determined by the metric topology of (X, ϱ). Two topologically equivalent metrics ϱ and ϱ' on X may lead to very different topologies on $\mathrm{Cl}(X)$ by the Hausdorff distance procedure. It follows from the example given below.

Example 1.1. Let $X = \boldsymbol{R}^+ := [0, \infty]$, $\varrho(x, y) = |x/(1+x) - y/(1+y)|$ and $\varrho'(x, y) = \min(1, |x-y|)$ for $x, y \in X$. The metrics ϱ and ϱ' define the same topology on $\boldsymbol{R}^+$ but the topologies of the Hausdorff distance on $\mathrm{Cl}(\boldsymbol{R}^+)$ are different, i.e., the set $\boldsymbol{N}$ of positive integers belongs to the closure of the set of all finite subsets of $\boldsymbol{N}$ in the first space but not in the second. ■

However, two metrics ϱ and ϱ' which define the same uniformity (i.e., if for every function which is uniformly continuous with respect to ϱ is also uniformly continuous with respect to ϱ' and vice versa) lead to the same Hausdorff distance topology on $\mathrm{Cl}(X)$.

THEOREM 1.1. $(\mathrm{Cl}(X), h)$ *is a complete metric space whenever* (X, ϱ) *is complete.*

Proof. Let (A_n) be a Cauchy sequence of $\mathrm{Cl}(X)$. We shall prove first that $A := \bigcap_{n=1}^{\infty} \overline{\bigcup_{m=n}^{\infty} A_m} \neq \emptyset$ and then that $A_n \to A$.

Let $\varepsilon > 0$. For each $k \in \boldsymbol{N}$ there exists N_k such that $n, m \geqslant N_k$ implies $h(A_n, A_m) < 2^{-k}\varepsilon$. Let (n_k) be a strictly increasing sequence of $\boldsymbol{N}$ such that $n_k \geqslant N_k$. Let $x_0 \in A_{n_0}$. Suppose we have chosen $x_0, \ldots, x_k$ with the properties $x_i \in A_{n_i}$, $\varrho(x_i, x_{i+1}) < 2^{-i}\varepsilon$ for $i = 1, \ldots, k-1$. Then x_{k+1} is chosen in $A_{n_{k+1}}$ in order to satisfy $\varrho(x_k, x_{k+1}) < 2^{-(k+1)}\varepsilon$. Observe that such x_{k+1} exists because $\mathrm{dist}(x_k, A_{n_{k+1}}) \leqslant h(A_{n_k}, A_{n_{k+1}}) < 2^{-(k+1)}\varepsilon$. It is easy to see that (x_k) is a Cauchy sequence of X. Then there is $x \in X$ such that $\varrho(x_k, x) \to 0$ as $k \to \infty$. We have, of course, $x \in A$. Furthermore, $\varrho(x_0, x) \leqslant 2\varepsilon$. Therefore, for every $n_0 \geqslant N_0$ and $x_0 \in A_{n_0}$ there exists a point $x \in A$ such that $\varrho(x_0, x) \leqslant 2\varepsilon$. Hence $\bar{h}(A_{n_0}, A) \leqslant 2\varepsilon$ for $n_0 \geqslant N_0$.

Now we will show that $\bar{h}(A, A_n) \to 0$ as $n \to \infty$ which together with the above will prove that $h(A_n, A) \to 0$ as $n \to \infty$. Let N be such that $m, n \geqslant N$ implies $h(A_n, A_m) \leqslant \varepsilon$. Let $x \in A$. Then $x \in \overline{\bigcup_{m=n}^{\infty} A_m}$. Therefore there exists $n_0 \geqslant N$ and $y \in A_{n_0}$ such that $\varrho(x, y) \leqslant \varepsilon$. For each $m \geqslant N$ we have $\mathrm{dist}(x, A_m) \leqslant \mathrm{dist}(x, A_{n_0}) + \bar{h}(A_{n_0}, A) \leqslant 2\varepsilon$. Hence $\bar{h}(A, A_m) \leqslant 2\varepsilon$. ■

THEOREM 1.2. *The Hausdorff topology on the space* $(\mathrm{Comp}(X), h)$ *is generated by sets* $\{K \in \mathrm{Comp}(X)\colon K \subset U\}$ *and* $\{K \in \mathrm{Comp}(X)\colon K \cap V \neq \emptyset\}$ *for every open sets* $U, V \subset X$.

Proof. Let us observe first that for an arbitrary $\varepsilon > 0$ and $K_0 \in \mathrm{Comp}(X)$ we have $\{K \in \mathrm{Comp}(X)\colon h(K, K_0) < \varepsilon\} = \{K \in \mathrm{Comp}(X)\colon K \subset V^{\circ}(K_0, \varepsilon)\}$, where $V^{\circ}(K_0, \varepsilon)$

is a neighbourhood of K_0 defined in I.1. For the proof it is enough only to check that for every open set $U, V \subset X$ the sets $\theta := \{K \in \mathrm{Comp}(X)\colon K \subset U\}$ and $\Sigma := \{K \in \mathrm{Comp}(X)\colon K \cap V \neq \emptyset\}$ are open.

Let $K_0 \in \theta$. Since $K_0 \subset U$, U is open and K_0 compact then there is $\varepsilon > 0$ such that $V^{\circ}(K_0, \varepsilon) \subset U$. Then $h(K, K_0) < \varepsilon$ implies $K \subset U$. Thus $K \in \theta$. Hence, it follows that for every $K_0 \in \theta$ there is $\varepsilon > 0$ such that an open ball $S(K_0, \varepsilon) := \{K \in \mathrm{Comp}(X)\colon h(K, K_0) < \varepsilon\}$ of $(\mathrm{Comp}(X), h)$ with $\varepsilon > 0$ given above is contained in θ.

Suppose, $K_0 \in \Sigma$. There exists an open ball of X with the centre $x_0 \in K_0 \cap V$ and a radius $\varepsilon > 0$ which is contained in V. Then, if $h(K, K_0) < \varepsilon$, K meets the ball, hence $K \cap V \neq \emptyset$ and $K \in \Sigma$. ■

REMARK 1.1 *If* (X, ϱ) *is a separable metric space, then also* $(\mathrm{Comp}(X), h)$ *is separable.*

Indeed, let (x_n) be a dense sequence in X. Let K be the set of all finite sets $\{x_{i_1}, \ldots, x_{i_n}\}$. Then K is countable part of $\mathrm{Comp}(X)$, and it is dense in $\mathrm{Comp}(X)$. ■

REMARK 1.2. *If* (X, ϱ) *is a compact metric space then a sequence* (A_n) *of* $\mathrm{Comp}(X)$ *converges to* A *in* $(\mathrm{Comp}(X), h)$ *if and only if* $\liminf_{n\to\infty} A_n = A = \limsup_{n\to\infty} A$. ■

1.2. Spaces $\mathrm{Comp}(\boldsymbol{R}^n)$, $\mathrm{Conv}(\boldsymbol{R}^n)$ *and support functions*

Given a field, Φ, of scalars and a set, K, of vectors, together with functions $+\colon K \times K \to K$ and $\otimes\colon \Phi \times K \to K$, K is called a *quasilinear space over* Φ if and only if all axioms for linear space are satisfied except

(i) the distributivity of $\otimes$ over scalar addition and

(ii) the existence of an inverse under $+$.

For $\alpha \in \boldsymbol{R}$, $A, B \in \mathrm{Comp}(\boldsymbol{R}^n)$, $A+B$ and αA we define by $A+B := \{a+b\colon a \in A; b \in B\}$ and $\alpha A := \{\alpha a, a \in A\}$, respectively. Apart from $\mathrm{Comp}(\boldsymbol{R}^n)$ we shall also consider its subspace $\mathrm{Conv}(\boldsymbol{R}^n)$ containing all convex members of $\mathrm{Comp}(\boldsymbol{R}^n)$. It can be easily verified that with the foregoing definition of addition and scalar multiplication, $\mathrm{Comp}(\boldsymbol{R}^n)$ and $\mathrm{Conv}(\boldsymbol{R}^n)$ are quasilinear spaces over the real field. Furthermore, it also can be verified that $\mathrm{Conv}(\boldsymbol{R}^n)$ is a closed subset of $(\mathrm{Comp}(\boldsymbol{R}^n), h)$. Therefore, $(\mathrm{Conv}(\boldsymbol{R}^n), h)$ is a complete metric space. It is clear that for a given $A \in \mathrm{Comp}(\boldsymbol{R}^n)$ and $\varepsilon > 0$ we have $V(A, \varepsilon) = A + \varepsilon B$ and $V^{\circ}(A, \varepsilon) = A + \varepsilon B^{\circ}$, where $B = \{x \in \boldsymbol{R}^n\colon |x| \leqslant 1\}$ and $B^{\circ} = \{x \in \boldsymbol{R}^n\colon |x| < 1\}$.

Given a set $A \in \mathrm{Conv}(\boldsymbol{R}^n)$, a real-valued function $s(\cdot, A)$ defined on $\boldsymbol{R}^n$ by setting $s(x, A) := \sup\{x \cdot a;\ a \in A\}$ for $x \in \boldsymbol{R}^n$, where "$\cdot$" denotes the inner product of $\boldsymbol{R}^n$ is called the *support function of A*.

Immediately from the above definition it follows that for every $x \in \boldsymbol{R}^n$, $\alpha > 0$ and $A, B \in \mathrm{Conv}(\boldsymbol{R}^n)$ one has $s(x, A+B) = s(x, A) + s(x, B)$, $s(x, \alpha A) = \alpha s(x, A)$ and $s(x, A) \leqslant s(x, B)$, whenever $A \subset B$. Furthermore, by the above definition it follows that for given $x \in \boldsymbol{R}^n$ with $|x| \neq 0$ and $A \in \mathrm{Conv}(\boldsymbol{R}^n)$, $l(x) := \{u \in \boldsymbol{R}^n\colon u \cdot x = s(x, A)\}$ is a hyperplane of $\boldsymbol{R}^n$ supporting the set A. The number $s(x/|x|, A)$ defines the distance of $l(x)$ to the origin of $\boldsymbol{R}^n$. Hence, in particular it follows that $s(x, \varepsilon B) = \varepsilon |x|$ for every $x \in \boldsymbol{R}^n$ with $|x| \neq 0$. It is also possible to define $s(p, A)$

for given $p \in \boldsymbol{R}^n$ and a nonempty set $A \subset \boldsymbol{R}^n$. It is clear that $\operatorname{co} A = \{a \in \boldsymbol{R}^n: p \cdot a \leqslant s(p, A)$ for every $p \in \boldsymbol{R}^n\}$. We have the following result.

PROPOSITION 1.3. *For every $A, B \in \operatorname{Conv}(\boldsymbol{R}^n)$, $A \subset B$ is equivalent to $s(x, A) \leqslant s(x, B)$ for every $x \in \boldsymbol{R}^n$.*

Proof. Let $A, B \in \operatorname{Conv}(\boldsymbol{R}^n)$ be given. Immediately from the definition of the support function it follows that $A \subset B$ implies $s(x, A) \leqslant s(x, B)$ for every $x \in \boldsymbol{R}^n$.

Suppose $s(x, A) \leqslant s(x, B)$ is satisfied for every $x \in \boldsymbol{R}^n$ and there exists $a \in A$ such that $a \notin B$. By virtue of Minkowski theorem there are $u \in \boldsymbol{R}^n$ and numbers $\gamma_1, \gamma_2 \in \boldsymbol{R}$ such that $b \cdot u < \gamma_1 < \gamma_2 < a \cdot u$ for every $b \in B$. Therefore, we also have $s(u, B) < a \cdot u \leqslant s(u, A)$. Contradiction. ■

Now, we prove the following lemmas.

LEMMA 1.3. *For every sequence (A_n) of $\operatorname{Conv}(\boldsymbol{R}^n)$ and $A \in \operatorname{Conv}(\boldsymbol{R}^n)$ one has*

(i) $\lim_{n\to\infty} \bar{h}(A_n, A) = 0$ *if and only if* $\limsup_{n\to\infty} s(x, A_n) \leqslant s(x, A)$ *for every* $x \in \boldsymbol{R}^n$,

(ii) $\lim_{n\to\infty} \bar{h}(A, A_n) = 0$ *if and only if* $s(x, A) \leqslant \liminf_{n\to\infty} s(x, A_n)$ *for every* $x \in \boldsymbol{R}^n$.

Proof. Let us observe that for every A, $C \in \operatorname{Comp}(\boldsymbol{R}^n)$ we have $\bar{h}(C, A) = \inf\{\varepsilon > 0: C \subset A + \varepsilon B\}$, where B is a closed unit ball of $\boldsymbol{R}^n$.

Suppose $\lim_{n\to\infty} \bar{h}(A_n, A) = 0$. For every $\varepsilon > 0$ there is $N \geqslant 1$ such that for $n \geqslant N$ one has $\bar{h}(A_n, A) < \varepsilon$, i.e., $A_n \subset A + \varepsilon B$. Hence it follows that $s(x, A\ ) \leqslant s(x, A) + \varepsilon|x|$ for every $x \in \boldsymbol{R}^n$ and $n \geqslant N$. Therefore, also for every $x \in \boldsymbol{R}^n$ we have $\limsup_{n\to\infty} s(x, A_n) \leqslant s(x, A)$.

Conversely, suppose $\limsup_{n\to\infty} s(x, A_n) \leqslant s(x, A)$ for every $x \in \boldsymbol{R}^n$. For every $x \in \boldsymbol{R}^n$ and $\varepsilon > 0$ there is $N_\varepsilon(x) \geqslant 1$ such that $s(x, A_n) \leqslant s(x, A) + \varepsilon|x|$ for $n \geqslant N_\varepsilon(x)$, i.e., $s(x, A\ ) \leqslant s(x, A) + s(x, \varepsilon B)$ for $n \geqslant N\ (x)$ and $x \in \boldsymbol{R}^n$. Hence, by Proposition 1.3, it follows $A_n \subset A + \varepsilon B$ for $n \geqslant N_\varepsilon(x)$, i.e. $\bar{h}(A_n, A) \leqslant \varepsilon$ for $n \geqslant N_\varepsilon(x)$.

Assume $\lim_{n\to\infty} \bar{h}(A, A_n) = 0$ and let $\varepsilon > 0$ be fixed. Select $N \geqslant 1$ such that $A \subset A_n + \varepsilon B$ for $n \geqslant N$. Hence, for every $x \in \boldsymbol{R}^n$ one obtains $s(x, A) \leqslant \liminf_{n\to\infty} s(x, A_n)$. Conversely, if the last inequality is satisfied then for fixed $x \in \boldsymbol{R}^n$ and $\varepsilon > 0$ there is $N_\varepsilon(x) \geqslant 1$ such that $s(x, A) - \varepsilon|x| \leqslant s(x, A_n)$. Therefore, $s(x, A) \leqslant (x, A_n) + s(x, \varepsilon B)$ for every $x \in \boldsymbol{R}^n$ and $n \geqslant N_\varepsilon(x)$. Hence it follows $A \subset A_n + \varepsilon B$, i.e., $\bar{h}(A, A_n) \leqslant \varepsilon$ for $n \geqslant N_\varepsilon(x)$. ■

LEMMA 1.4. *For every $A, C \in \operatorname{Conv}(\boldsymbol{R}^n)$ one has*

(i) $\bar{h}(A, C) = \max\{s(x, A) - s(x, C): |x| = 1\}$,

(ii) $h(A, C) = \max\{|s(x, A) - s(x, C): |x| = 1\}$.

Proof. By Proposition 1.3 for every $\varepsilon > 0$ $A \subset C + \varepsilon B$ is equivalent to $s(x, A) \leqslant s(x, C) + \varepsilon|x|$ for every $x \in \boldsymbol{R}^n$, i.e., to $\sup_{x \in \boldsymbol{R}^n, |x|=1} [s(x, A) - s(x, C)] \leqslant \varepsilon$. Thus,

$\bar{h}(A, C) = \inf\{\varepsilon: A \subset C+\varepsilon B\} = \inf\{\varepsilon: \sup_{x\in R^n, |x|=1} [s(x, A)-s(x,C)] \leqslant \varepsilon\}$ $= \max\{s(x, A)-s(x, C): |x| = 1\}$. In a similar way (ii) can be obtained. ■

COROLLARY 1.1. *For every $A, B, C \in \mathrm{Conv}(R^n)$ one has $\bar{h}(A+B, A+C) = \bar{h}(B, C)$.*

Indeed, by (i) of Lemma 1.4, we obtain

$$\begin{aligned}\bar{h}(A+B, A+C) &= \max\{s(x, A+B)-s(x, A+C): \|x\| = 1\}\\ &= \max\{s(x, A)+s(x, B)-s(x, A)-s(x, C): \|x\| = 1\}\\ &= \max(s(x, B)-s(x, C): \|x\| = 1\} = \bar{h}(B, C). \ ■\end{aligned}$$

LEMMA 1.5. *For every $A, B, C, D \in \mathrm{Comp}(R^n)$ and $\eta \in R^1$ one has*

(i) $\bar{h}(\eta A, \eta B) = |\eta|\bar{h}(A, B)$,

(ii) $\bar{h}(\mathrm{co}\,B, \mathrm{co}\,C) \leqslant \bar{h}(A+B, A+C) \leqslant \bar{h}(B, C)$,

(iii) $\bar{h}(A+B, C+D) \leqslant \bar{h}(A, C)+\bar{h}(B, D)$.

Proof. The proof of (i) is trivial. Part (iii) is an easy consequence of (ii). The second inequality of (ii) follows readily from the definitions and only the first inequality remains to be proved. By Lemma I.2.1. we obtain $\bar{h}(\mathrm{co}\,A+\mathrm{co}\,B, \mathrm{co}\,A+\mathrm{co}\,C)$ $= \bar{h}(\mathrm{co}(A+B), \mathrm{co}(A+C))$. Hence, by Corollary 1.1, it follows $\bar{h}(\mathrm{co}\,B, \mathrm{co}\,C)$ $= \bar{h}(\mathrm{co}(A+B), \mathrm{co}(A+C))$. Now for $D, E \in \mathrm{Comp}(R^n)$ we have $D \subset E+\eta B$, if $\eta := \bar{h}(D, E)$ and B denotes a closed unit ball of R^n. Hence $\mathrm{co}\,D \subset \mathrm{co}\,E+\eta B$ which means that $\bar{h}(\mathrm{co}\,D, \mathrm{co}\,E) \leqslant \eta := \bar{h}(D, E)$. Setting $D := A+B$, $E := A+C$, the first inequality of (ii) follows from this result and the last formula line. ■

Given $A \in \mathrm{Comp}(R^n)$ we put $\|A\| := h(A, \{0\})$. It is clear that $\|A\| = \bar{h}(A, \{0\})$ because $\bar{h}(A, \{0\}) = \sup\{|x|: x \in A\}$ and $\bar{h}(\{0\}, A) = \inf\{|x|: x \in A\}$.

COROLLARY 1.2. *If $\eta, \gamma \in R^1$ and $A, B \in \mathrm{Comp}(R^n)$ then*

(i) $\|\eta A\| = |\eta|\|A\|$,

(ii) $\|A\| \geqslant 0$ *and* $\|A\| = 0$ *if and only if* $A = \{0\}$,

(iii) $\|A+B\| \leqslant \|A\|+\|B\|$,

(iv) $|\|A\|-\|B\|| \leqslant h(A, B) \leqslant \|A\|+\|B\|$,

(v) $\bar{h}(\eta A, \gamma A) \leqslant |\eta-\gamma|\|A\|$.

Indeed, (i) through (iv) follow easily from the definitions and Lemma 1.5. For (v) we have from Lemma 1.5 (i), (ii) $\bar{h}(\eta A, \gamma A) = |\eta-\gamma|\bar{h}[(1+\gamma/(\eta-\gamma))A, \gamma A/(\eta-\gamma)]$ $\leqslant |\eta-\gamma|F(A, \{0\}) = |\eta-\gamma|\,\|A\|$. ■

COROLLARY 1.3. $\mathrm{Conv}(R^n)$ *is a closed subset of the metric space* $(\mathrm{Comp}(R^n), h)$.

Indeed, suppose (A_n) is a sequence of $\mathrm{Conv}(R^n)$ converging in $(\mathrm{Comp}(R^n), h)$ to $A \in \mathrm{Comp}(R^n)$, Since $h(A, \mathrm{co}\,A) \leqslant h(A, A_n)+h(A_n, \mathrm{co}\,A)$ for each $n = 1, 2, \ldots$ then by (ii) of Lemma 1.5 for $n = 1, 2, \ldots$ we have $h(A, \mathrm{co}\,A) \leqslant 2h(A_n, A)$. Therefore $A = \mathrm{co}\,A$. ■

2. Continuity concepts

We shall study various continuity properties of set-valued functions defined and having their values, usually in linear normed spaces. These spaces will be considered as linear topological spaces with their norm topologies as well as with their weak topologies. Then one obtains various continuity concepts for given set-valued functions.

2.1. Upper semicontinuous set-valued functions

Let X and Y be topological Hausdorff spaces and F a mapping defined on X with values in the space $\mathscr{P}(Y)$ of all nonempty subsets of Y. Such mappings we will call *set-valued functions* or *multifunctions*.

We will say that $F\colon X \to \mathscr{P}(Y)$ is *upper semicontinuous* (*u.s.c.*) *at* $x \in X$ if for every neighbourhood U of $F(x)$ there exists a neighbourhood V of $\bar{x}$ such that $F(x) \subset U$ for every $x \in V$. A set-valued function $F\colon X \to \mathscr{P}(Y)$ is called *upper semi-continuous* (*u.s.c.*) *on* X if it is u.s.c. at every $x \in X$.

Example 2.1. Let $f\colon X \to Y$ and $g\colon Y \to X$ be given functions and $F, G\colon X \to \mathscr{P}(Y)$ be defined by $F(x) = \{f(x)\}$ and $G(x) = g^{-1}(x)$ for $x \in X$. F (resp. G) is u.s.c. if and only if f is continuous (resp. g is closed). ■

Let (X, ϱ) and (Y, d) be metric spaces and $F\colon X \to \mathscr{P}(Y)$. We say that F is *H-upper semicontinuous* (*H-u.s.c.*) *at* $\bar{x} \in X$ if for every $\varepsilon > 0$ there exists a $\delta > 0$ such that $F(B^{o}(\bar{x}, \delta)) \subset V(F(\bar{x}), \varepsilon)$, where $B^{o}(\bar{x}, \delta) = \{x \in X\colon \varrho(x, x) < \delta\}$ and for a given $A \subset X$, $F(A) := \bigcup_{x \in A} F(x)$. We say that F is *H-upper semicontinuous* (*H-u.s.c.*) *on* X if it is H-u.s.c. at every $\bar{x} \in X$.

Clearly, if $F\colon X \to \mathscr{P}(Y)$ is u.s.c. at $\bar{x} \in X$ it is also H-u.s.c. at this point. Indeed, taking for fixed $\varepsilon > 0$, $U = [V(F(\bar{x}), \varepsilon)]^{0}$ we can find a neighbourhood O_{ε} of $\bar{x}$ such that for every $x \in O_{\varepsilon}$ one has $F(x) \subset U$. Let $\delta > 0$ be such that $B^{o}(\bar{x}, \delta) \subset O_{\varepsilon}$. Obviously we have $F(B^{o}(\bar{x}, \delta)) \subset [V(F(\bar{x}), \varepsilon)]^{0} \subset V(F(\bar{x}), \varepsilon))$.

The converse is not true, i.e., a set-valued function that is H-u.s.c. at any $\bar{x} \in X$ is not necessarily be u.s.c. at this point.

Example 2.2. Let $X = \boldsymbol{R}$, $Y = \boldsymbol{R}^2$ and $F(t) := \{(x, y)\colon x = t\}$ for $t \in \boldsymbol{R}$. F is not u.s.c. at $t = 0$, because for $U = \{(x, y)\colon |y| < 1/|x| \text{ or } x = 0\}$ we have $F(0) \subset U$ but for every $t \neq 0$, $F(t) \not\subset U$. On the other hand for every $\varepsilon > 0$ there is $\delta = \varepsilon$ such that for $|t| \leqslant \delta$ we have $F(t) \subset V^{o}(F(0), \varepsilon)$, i.e. $F(B^{o}(0, \varepsilon)) \subset V^{o}(F(0), \varepsilon) \subset V(F(0), \varepsilon)$. ■

In the case when the image $F(\bar{x})$ of F at $\bar{x} \in X$ is compact, the two definitions coincide. In fact, it is enough to observe that, given any open set U containing a compact set $F(\bar{x})$ there is $\eta > 0$ such that $V^{o}(F(\bar{x}), \eta) \subset U$. Furthermore, for every $\varepsilon > 0$ we have $[V(F(\bar{x}), \varepsilon)]^{0} = V^{o}(F(\bar{x}), \varepsilon)$. Since $V^{o}(F(\bar{x}), \eta) = \{z \in Y\colon \operatorname{dist}(z, F(\bar{x})) < \eta\}$ then $V(F(\bar{x}), \varepsilon) \subset V^{o}(F(\bar{x}), \eta) \subset U$ for every $\varepsilon \in (0, \eta)$. If F is H-u.s.c. at

$\bar{x} \in X$ then, for a given above U and to every $\varepsilon \in (0, \eta)$ there is $\delta > 0$ such that $F(x) \subset [V(F(\bar{x}), \varepsilon)]^0 \subset U$ for every $x \in V := B^0(\bar{x}, \delta)$.

Immediately from the above definition it follows that $F\colon X \to \mathrm{Cl}(Y)$ is H-u.s.c. at $\bar{x} \in X$ if and only if for every sequence (x_n) of X converging to $\bar{x}$ one has $\lim_{n\to\infty} \bar{h}(F(x_n), F(\bar{x})) = 0$.

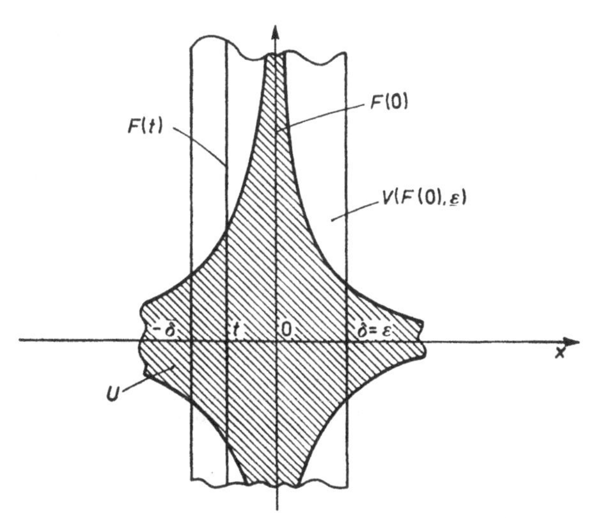

Fig. 1. The mapping H-u.s.c. but not u.s.c. at $t = 0$.

PROPOSITION 2.1. *Given metric spaces (X, ϱ) and (Y, d), a continuous function $f\colon X \to Y$ and a H-u.s.c. set-valued function $F\colon X \to \mathrm{Cl}(Y)$ a real-valued function $\delta\colon X \ni x \to \delta(x) := \mathrm{dist}(f(x), F(x)) \in \boldsymbol{R}^+$ is lower semicontinuous on X.*

Proof. Let $x \in X$ be fixed and suppose (x_n) is an arbitrary sequence of X converging to x. For every $\varepsilon > 0$ there exists a number $N \in \boldsymbol{N}$ such that $\varrho(f(x), f(x_n)) < \varepsilon/2$ and $\bar{h}(F(x_n), F(x)) < \varepsilon/2$ for $n \geqslant N$. Therefore, for $n \geqslant N$ one obtains $\delta(x) = \mathrm{dist}(f(x), F(x)) \leqslant \varrho(f(x), f(x_n)) + \mathrm{dist}(f(x_n), F(x_n)) + \bar{h}(F(x_n), F(x)) < \varepsilon + \delta(x_n)$.

Thus, $\delta(x) \leqslant \liminf_{n\to\infty} \delta(x_n) + \varepsilon$ for every $\varepsilon > 0$. Therefore, $\delta(x) \leqslant \liminf_{n\to\infty} \delta(x_n)$. ∎

Let X, Y and Z be topological Hausdorff spaces. We obtain the following properties of upper semicontinuous set-valued functions.

THEOREM 2.1. *Let $F\colon X \to \mathscr{P}(Y)$ be given. The following assertions are equivalent*:

(i) *F is u.s.c. on X,*

(ii) *the set $F_-(G) := \{x \in X\colon F(x) \subset G\}$ is open for every open set $G \subset Y$,*

(iii) *the set $F^-(M) := \{x \in X\colon F(x) \cap M \neq \emptyset\}$ is closed for every closed set $M \subset Y$.*

Proof. (i) → (ii): Let G be open in Y and $x \in F_-(G)$; we show that x is an inferior point of $F_-(x)$. Since G is a neighbourhood of $F(x)$ it follows by (i) that there exists a neighbourhood V of x such that $F(V) \subset G$, i.e. $V \subset F_-(G)$.

(ii) → (i): Let U be a neighbourhood of $F(x)$ for any $x \in X$ and put $V = F_-(U)$. By (ii), $F_-(U)$ is open. Furthermore $x \in V$ and $F(V) \subset U$. Therefore F is u.s.c. on X.

(ii) ↔ (iii): Indeed, for every $A \subset Y$ we have $X \setminus F^-(A) = F_-(Y \setminus A)$. ■

COROLLARY 2.1. *Let F: $X \to \mathscr{P}(Y)$ and G: $Y \to \mathscr{P}(\mathscr{Z})$ be u.s.c. on X and Y, respectively. Then the composition $G \circ F$: $X \to \mathscr{P}(\mathscr{Z})$ is u.s.c. on X.*

Indeed, since $(G \circ F)^-(M) = F^-(G^-(M))$ for $M \subset \mathscr{Z}$ then the result follows immediately from (iii) of Theorem 2.1. ■

Given metric spaces X and Y the set-valued function F: $X \to \mathscr{P}(Y)$ is said to be *closed at* $\bar{x} \in X$ if for every sequence $\{(x_n, y_n)\}$ in $X \times Y$ such that $(x_n, y_n) \to (\bar{x}, \bar{y})$ and $y_n \in F(x_n)$ for $n = 1, 2, \ldots$ it follows $\bar{y} \in F(\bar{x})$. It is said to be *closed* if F is closed at every $x \in X$.

REMARK 2.1. *If a set-valued function F is closed at $\bar{x}$, then it is not necessarily u.s.c. at $\bar{x}$ (even if $F(\bar{x})$ is compact).*

Indeed, a set-valued function F from $[0, \infty)$ into $\mathscr{P}(\boldsymbol{R})$ defined by

$$F(x) = \begin{cases} \{0, 1/x\} & \text{for} \quad x > 0 \\ \{0\} & \text{for} \quad x = 0, \end{cases}$$

is closed at $\bar{x} = 0$, but it is not u.s.c. there. ■

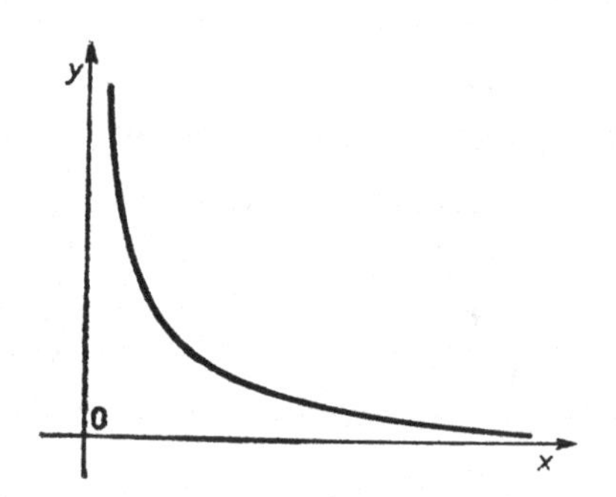

Fig. 2. The mapping F.

However, if Y is compact, then a set-valued function F: $X \to \mathscr{P}(Y)$ which is closed at x is also u.s.c. at x. We have the following result.

PROPOSITION 2.2. *Let (X, ϱ) and (Y, d) be metric spaces and suppose F: $X \to \mathrm{Cl}(Y)$ it H-u.s.c. on X. Then F is closed on X.*

Proof. Let $x \in X$ be fixed and suppose $\{(x_n, y_n)\}$ is a sequence of $X \times Y$ such that $(x_n, y_n) \to (x, y)$ and $y_n \in F(x_n)$ for $n = 1, 2, \ldots$ Since F is H-u.s.c. at x, then $\lim_{n\to\infty} \bar{h}(Fx_n), F(x)) = 0$. Since $\mathrm{dist}(y, F(x)) \leqslant d(y, y_n) + \mathrm{dist}(y_n, F(x_n)) + \bar{h}(F(x_n), F(x))$ for $n = 1, 2, \ldots$, then $y \in F(x)$. ■

PROPOSITION 2.3. *Let $F: X \to \text{Comp}(Y)$ be u.s.c. on X. Then for every compact set $K \subset X$ the image $F(K) := \bigcup_{x \in K} F(x)$ is compact in Y.*

Proof. Let $\{G_i\}_{i \in I}$ be an open covering of $F(K)$. Since for every $x \in K$ the set $F(x)$ is compact there exists a finite set $I_x \subset I$ such that $F(x) \subset \bigcup_{i \in I_x} G_i := G_x$. Since G_x is open, then by Theorem 2.1 $F^-(G_x)$ is open and we have trivially $x \in F^-(G_x)$. Therefore, $\{F^-(G_x)\}_{x \in X}$ is an open covering of K. Since by assumption K is compact there exist $x_1, \ldots, x_m \in K$ such that $K \subset \bigcup_{k=1}^{m} F^-(G_{x_k})$. Clearly, $\{G_i\}_{i \in I_{x_k}, k=1,\ldots,m}$ is a finite subcovering of $\{G_i\}_{i \in I}$. ■

Now we can prove the following characterization of the upper semicontinuity of compact-valued set-valued functions.

THEOREM 2.2. *Let X and Y be metric spaces. A set-valued function $F: X \to \text{Comp}(Y)$ is u.s.c. on X if and only if for every $x \in X$ and every sequence (x_n) of X converging to x and every sequence (y_n) of Y with $y_n \in F(x_n)$, there is a converging subsequence of (y_n) whose limit belongs to $F(x)$.*

Proof. Let F be u.s.c. on X, $x \in X$ and (x_n) be any sequence of X converging to x. The set $K = \{x, x_1, x_2, \ldots\}$ is compact and the restriction of F on K is u.s.c. Thus, by Proposition 2.3, the set $F(K)$ is compact and hence the sequence (y_n) has a conveging subsequence, say $(y_{n_k}) \to y$ as $k \to \infty$. Assume $y \notin F(x)$. Then there is a closed neighbourhood $\bar{U}$ of $F(x)$ not containing y. But for n large enough we have $F(x_n) \subset \bar{U}$ since F is u.s.c. at x. Thus $y_n \in \bar{U}$ for n large enough and hence $y \in \bar{U}$; a contradiction.

To prove the converse, we assume that F is not u.s.c. at x, i.e. there exists an open neighbourhood U of $F(x)$ such that every neighbourhood V of x contains a point z with $F(z) \not\subset U$. Therefore there is a sequence (x_n) converging to x and $y_n \in F(x_n)$ and $y_n \notin U$. By assumption there exists a converging subsequence (y_{n_k}) and we have $\lim_{k \to \infty} y_{n_k} \in F(x)$. But $y_n \notin U$ for all n implies $\lim_{k \to \infty} y_{n_k} \notin U$ and therefore $\lim_{k \to \infty} y_{n_k} \notin F(x)$. Contradiction. ■

We have also the following generalization of Dugundji's extension theorem.

THEOREM 2.3. *Let A be a closed subset of a metric space X, Y a separable normed linear space, and* $\text{Conv}(Y)$ *the family of all nonempty compact convex subsets of Y. If $F: A \to \text{Conv}(Y)$ is u.s.c. on A then F has an upper semicontinuous extension $G: X \to \text{Conv}(Y)$ such that $G(X) \subset \text{co}\, F(A)$.*

Proof. Let d be a metric of X. For each $y \in X \setminus A$, let $V(y)$ be a ball with the centre y and a radius $0 < r < \frac{1}{2} \text{dist}(y, A)$. Let $\mathscr{V} = \{V_j: j \in J\}$ be a locally finite open refinement of $\{V(y): y \in X \setminus A\}$ such that $\mathscr{V}$ covers $X \setminus A$. Let $\{\mathscr{P}_j: j \in J\}$ be a partition of unity on $X \setminus A$ subourdinated to $\mathscr{V}$. Let us pick up $x_j \in V_j$ for every $j \in J$. There exists an $a_j \in A$ with $d(a_j, x_j) < 2\text{dist}(x_j, A)$. Define $G: X \to \text{Conv}(Y)$

as follows:

$$G(x) = \begin{cases} F(x) & \text{for } x \in A, \\ \sum_{j \in J} \mathscr{P}_j(x) F(a_j) & \text{for } x \in X \setminus A. \end{cases}$$

Then G is an extension of F and $G(X) \subset \operatorname{co} F(A)$. We want to show that G is u.s.c. at each point $a \in X$. If $a \in X \setminus A$, since $\mathscr{V}$ is locally finite, there exists an open neighbourhood N of $a \in X$ such that $N \subset X \setminus A$ and N meets only a finite numbers of sets in $\mathscr{V}$.

It is clear that $G|_N$ is u.s.c. on N and hence G is u.s.c. at $a \in X \setminus A$. On the other hand, if $a \in A$, let B be an open set in Y with $G(a) \subset B$. Since $G(a) = F(a)$ is compact there exists a convex neighbourhood U of $O \in Y$ with $G(a)+U \subset B$. Since F is u.s.c., there exists an open ball V in X with centre $a \in A$ and a radius $9\varepsilon > 0$ such that $F(V \cap A) \subset G(a)+U$. It suffices to show that $G(W) \subset G(a)+U$, where W is an open ball in X with a centre $a \in A$ and a radius $\varepsilon > 0$. If $x \in W \cap A$, then $x \in V \cap A$ and hence $G(x) = F(x) \subset G(a)+U$. If $x \in W \setminus A$, there exists only a finite number of sets in $\mathscr{V}$, say $\{V_{j_1}, \ldots, V_{j_m}\}$, containing x. If λ is one of the indices $j_1, \ldots, j_m$, then $x \in W \cap V_\lambda$. Since $\mathscr{V}$ is an refinement of $\{V(y)\colon y \in X \setminus A\}$, we can write $V_\lambda \subset V(y)$ for some $y \in X \setminus A$. Now $\varepsilon > d(a, x) \geqslant d(a, y) - d(y, x) \geqslant \operatorname{dist}(y, A) - r \geqslant r$. Thus $d(x, x_\lambda) \leqslant 2r \leqslant 2\varepsilon$. Also $d(a_\lambda, x_\lambda) < \operatorname{dist}(x_\lambda, A) \leqslant 2d(x_\lambda, a)$. Hence

$$\begin{aligned} d(a, a_\lambda) &\leqslant d(a_\lambda, x_\lambda) + d(x_\lambda, a_\lambda) \leqslant 3d(a, x_\lambda) \leqslant 3[d(a, x) + d(x, x_\lambda)] < \\ &< 3(\varepsilon + 2\varepsilon) = 9\varepsilon. \end{aligned}$$

Therefore, $a_\lambda \in A \cap V$ and $F(a_\lambda) \subset G(a)+U$. Since $G(a)+U$ is convex, we have

$$G(x) = \sum_{j \in J} \mathscr{P}_j(x) F(a_j) = \sum_{\nu=1}^{m} \mathscr{P}_{j_\nu}(x) F(a_{j_\nu}) \subset G(a)+U.$$

Thus, $G(W) \subset G(a)+U \subset B$. Consequently, G is u.s.c. at $a \in X$. ∎

Assume now X and Y are normed linear spaces. They will be considered as locally convex topological Hausdorff space with their weak topologies $\sigma(X, X^*)$ and $\sigma(Y, Y^*)$, respectively.

We will say that $F\colon X \to \mathscr{P}(Y)$ is *weakly-weakly upper semicontinuous* (*w.-w.u.s.c.*) *on* X if for every weakly closed set $M \subset Y$ the set $F^-(M) := \{x \in X\colon F(x) \cap M \neq \emptyset\}$ is sequentially weakly closed in X. We say that $F\colon X \to \mathscr{P}(Y)$ is *weakly-strongly upper semicontinuous* (*w.-s.u.s.c.*) *on* X if for every weakly closed set $M \subset Y$ the set $F^-(M)$ is closed (in the norm topology) in X.

Similarly we can define mappings s.-w.u.s.c. and s.-s.u.s.c. on X. Of course the case of mappings s.-s.u.s.c. covers with u.s.c. mappings defined above.

Let us observe that we have in fact defined above weak-weak, weak-strong etc. sequential upper semicontinuities. But in the case considered in this book they will be reduced to upper semicontinuities involving strong and weak topologies some normed spaces. In a similar way as above we obtain the following results.

THEOREM 2.4. *Let D be a nonempty weakly compact subset of a separable Banach space and let $F: D \to \mathscr{P}(Y)$. The following assertions are equivalent:*

(i) *F is w.-w.u.s.c. on D,*

(ii) *for every weakly open set $G \subset Y$ the set $F_{-}(G) := \{x \in D: F(x) \subset G\}$ is weakly open in D (in the relative weak topology of D),*

(iii) *for every sequence (x_n) of D weakly converging to $x \in D$ and each weakly open set $G \subset Y$ containing $F(x)$ there is a positive integer N such that $F(x_n) \subset G$ for each $n \geqslant N$.* ■

REMARK 2.2. The separability of X and weak completeness of D are needed only for the proof of (iii) → (i). In this case the weak topology of D is a metric topology. ■

THEOREM 2.5. *A set-valued function $F: X \to \mathscr{P}(Y)$ is w.-w.u.s.c. (w.-s.u.s.c.) on X if for every sequence (x_n) of X weakly converging (converging in the norm topology of X, resp.) to x and every sequence (y_n) of Y with $y_n \in F(x_n)$ for $n = 1, 2, \ldots$ there exists a subsequence of (y_n) weakly convergent to any $y \in F(x)$. If furthermore, there is a weakly compact set $C \subset Y$ such that $F(x) \subset C$ for $x \in X$, then the above conditions are also necessary for F to be w.-w.u.s.c. on X.* ■

THEOREM 2.6. *Let X and Y be Banach spaces and let C be a weakly compact subset of Y and suppose $F: X \to \mathscr{P}(Y)$ is w.-w.u.s.c. on X and such that $F(x) \subset C$ for every $x \in X$. Then for every weakly compact set $B \subset X$ the set $F(B)$ is weakly compact in Y.*

Proof. Let B be a weakly compact subset of X. Since $F(B) \subset C$ and C is weakly compact then $F(B)$ is relatively weakly sequentially compact in Y. By Eberlein–Šmulian's theorem, $F(B)$ is relatively weakly compact. Then, by Šmulian's theorem, for every $y \in \overline{[F(B)]}^w$ there exists a sequence (y_n) of $F(B)$ weakly converging to $y \in Y$ Since $y_n \in F(B)$, then for every $n = 1, 2, \ldots$ there is $x \in B_n$ such that $y_n \in F(x_n)$ for each $n = 1, 2, \ldots$ By the weak compactness of B there exists a subsequence, say (x_k) of (x_n) weakly converging to any $x \in B$. Since F is w.-w.u.s.c. on X and $y_k \in F(x_k)$ for $k = 1, 2, \ldots$ then by Theorem 2.5 there exists a subsequence, say again (y_k) of (y_n) weakly converging to $y \in F(x)$. Then $\overline{[F(B)]}^w \subset F(B)$ and therefore, $F(B)$ is weakly closed. ■

Finally, as a corollary of Lemmas 1.3 and 1.5 and properties of H-upper semicontinuous mappings we obtain the following result.

THEOREM 2.7. *Let (X, ϱ) be a metric space and $F: X \to \mathrm{Comp}(\boldsymbol{R}^n)$. Then*

(i) *if F is u.s.c. on X then also a set-valued function $\mathrm{co}\,F: X \in x \to \mathrm{co}\,F(x) \subset \boldsymbol{R}^n$ is u.s.c. on X,*

(ii) *if F has furthermore convex values then F is u.s.c. on X if and only if for every $p \in \boldsymbol{R}^n$ a real-valued function $s(p, F(\cdot))$ is u.s.c. on X.*

Proof. By virtue of (v) of Lemma I.2.1. it suffices only to show that $\mathrm{co}\,F$ is H-u.s.c. on X. Let $x \in X$ be fixed and (x_n) be an arbitrary sequence of X converging to x. Since, $\bar{h}(\overline{\mathrm{co}}\,F(x_n), \overline{\mathrm{co}}\,F(\bar{x})) \leqslant \bar{h}(F(x_n), F(x))$ for $n = 1, 2, \ldots$ then (i) follows.

If F has convex values then, by Lemma 1.3, $\lim_{n\to\infty} h(F(x_n), F(x)) = 0$ if and only if $\limsup_{n\to\infty} s(p, F(x_n)) \leqslant s(p, F(x))$ for every $p \in \mathbf{R}^n$. Therefore, (ii) is satisfied. ■

2.2. *Lower semicontinuous and continuous set-valued functions*

Let X and Y be topological Hausdorff spaces and $F\colon X \to \mathscr{P}(Y)$ be given. A set-valued function F is said to be *lower semicontinuous* (*l.s.c.*) *at* $\bar{x} \in X$ if for every open set U in Y with $F(x) \cap U \neq \emptyset$ there exists a neighbourhood V of x such that $F(x) \cap U \neq \emptyset$ for every $x \in V$. F is called *lower semicontinuous* (*l.s.c.*) *on* X if it is *l.s.c.* at every $x \in X$.

THEOREM 2.3. Let $f\colon X \to Y$ and $g\colon Y \to X$ be given functions and $F, G\colon X \to \mathscr{P}(Y)$ be defined by $F(x) = \{f(x)\}$ and $G(x) = g^{-1}(x)$ for $x \in X$. F (resp. G) is l.s.c. on X if and only if f is continuous (resp. g is open). ■

EXAMPLE 2.4. Let $F^+, F^-\colon \mathbf{R} \to \mathscr{P}(\mathbf{R})$ be defined by

$$F^+(x) = \begin{cases} \{0\} & \text{for } x \neq 0, \\ [-1, +1] & \text{for } x = 0 \end{cases}$$

and

$$F^-(x) = \begin{cases} [-1, +1] & \text{for } x \neq 0, \\ \{0\} & \text{for } x = 0. \end{cases}$$

It is easy to see that F^+ is u.s.c. at $x = 0$ and is not l.s.c., while F^- is l.s.c. at $x = 0$ and is not u.s.c. ■

Immediately from the above definition it follows that $F\colon X \to \mathscr{P}(Y)$ is l.s.c. on X if and only if a set-valued function $F\colon X \ni x \to \overline{F(x)} \in \mathscr{P}(Y)$ is l.s.c. on X.

Indeed, this follows from the above definition and the obvious fact that, if U is an open subset of Y and $B \subset Y$, then $B \cap U \neq \emptyset$ if and only if $\bar{B} \cap U \neq \emptyset$.

Similarly as in the case of upper semicontinuous mappings the following result is obtained.

THEOREM 2.8. *Let* $F\colon X \to \mathscr{P}(Y)$ *be given. The following assertions are equivalent*:

(i) *F is l.s.c. on X,*

(ii) *the set* $F_-(G) := \{x \in X\colon F(x) \subset G\}$ *is closed for every closed set* $G \subset Y$,

(iii) *the set* $F^-(M) := \{x \in X\colon F(x) \cap M \neq \emptyset\}$ *is open for every open set* $M \subset Y$.

Furthermore, if X and Y are metric spaces then $F\colon X \to \mathscr{P}(Y)$ is l.s.c. on X if and only if for every $z \in Y$ a functional $X \ni x \to \operatorname{dist}(z, F(x)) \in \mathbf{R}$ *is upper semicontinuous on X.* ■

Now, we can give the following characterization of l.s.c. mappings.

THEOREM 2.9. *Let X and Y be metric spaces. A set-valued function $F\colon X \to \mathscr{P}(Y)$ is l.s.c. at $\bar{x} \in X$ if and only if for every sequence (x_n) of X converging to x and any $\bar{y} \in F(\bar{x})$ there exists a sequence (y_n) of Y converging to $\bar{y}$ and such that $y_n \in F(x_n)$ for $n = 1, 2, \ldots$*

Proof. Let F be l.s.c. at $\bar{x}$ and let (x_n) converge to $\bar{x}$ and let $\bar{y} \in F(\bar{x})$. For every integer r, let $B_r^\circ(y)$ denote the open ball with radius $1/r$ and centre y. Since F is l.s.c. at $\bar{x}$, there exists for every r a neighbourhood V_r of $\bar{x}$ such that for every $z \in V_r$ we have $F(z) \cap B_r^\circ(\bar{y}) \neq \emptyset$. Let the subsequence of integers (n_r) be such that $n_r < n_{r+1}$ and $x_n \in V_r$ if $n \geqslant n_r$. For n with $n_r \leqslant n < n_{r+1}$ let us choose y_n in the set $F(x_n) \cap B_r^\circ(\bar{y})$. The sequence (y_n), so constructed, converges to $\bar{y}$. Conversely, assume that for every sequence (x_n) converging to $\bar{x}$ and any $\bar{y} \in F(\bar{x})$ there exists a sequence (y_n) converging to $\bar{y}$ with $y_n \in F(x_n)$ for $n = 1, 2, \ldots$ and suppose F is not l.s.c. at $\bar{x}$. Then there exists an open set G with $G \cap F(\bar{x}) \neq \emptyset$ such that every neighbourhood V of $\bar{x}$ contains a point z such that $F(z) \cap G = \emptyset$. Therefore, there exists a sequence (x_n) converging to $\bar{x}$ with $F(x_n) \cap G = \emptyset$ for $n = 1, 2, \ldots$ Let $\bar{y} \in G \cap F(\bar{x})$. By the above assumption there exists a sequence (y_n) converging to $\bar{y}$ with $y_n \in F(x_n)$. For n large enough we have $y_n \in G$. Thus $F(x_n) \cap G \neq \emptyset$; a contradiction. ■

Given a metric spaces X and Y a set-valued function F: $X \to \mathscr{P}(Y)$ is said to be *H-lower semicontinuous (H-l.s.c.) at* $\bar{x} \in X$ if for every $\varepsilon > 0$ there exists a $\delta > 0$ such that $F(\bar{x}) \subset V(F(x), \varepsilon)$ for every $x \in B^\circ(\bar{x}, \delta)$. F is called *H-lower semicontinuous (H-l.s.c.) on* X if it is H-l.s.c. at each $x \in X$.

Clearly, if F: $X \to \mathscr{P}(Y)$ is H-l.s.c. at $\bar{x} \in X$ it is also l.s.c. at $\bar{x}$. Indeed, suppose F is H-l.s.c. and is not l.s.c. at $\bar{x} \in X$. There exists an open set $U \subset Y$ with $F(\bar{x}) \cap U \neq \emptyset$ such that in every neighbourhood V of $\bar{x}$ there is $\tilde{x} \in V$ such that $F(\tilde{x}) \cap U = \emptyset$. Then, we can find a sequence (x_n) of X converging to $\bar{x}$ and such that $F(\bar{x}_n) \cap U = \emptyset$ for every $n = 1, 2, \ldots$ On the other hand, for every $\varepsilon > 0$ there is $N_\varepsilon \geqslant 1$ such that for every $n \geqslant N_\varepsilon$ we have $F(\bar{x}) \subset V^\circ(F(\bar{x}_n), \varepsilon)$. Hence, in particular, it follows that $F(\bar{x}) \cap U \subset V^\circ(F(x_n), \varepsilon)$ for $n \geqslant N_\varepsilon$. Let $y \in F(\bar{x}) \cap U$ and select, for every $k = 1, 2, \ldots$ $\ldots$, $y_k \in F(x_{n_k})$ such that $|y_k - y| < 1/k$, where $n_k := N_{1/k}$. For k sufficiently large we have $y_k \in U$ and therefore $F(x_{n_k}) \cap U \neq \emptyset$; a contradiction.

In the case when the image $F(\bar{x})$ of F at $\bar{x} \in X$ is compact, F is l.s.c. at x if and only if it is H-l.s.c. at $\bar{x}$.

Indeed, let y_i, $i = 1, \ldots, m$ be such that $\{B^\circ(y_i, \frac{1}{2}\varepsilon);\ i = 1, \ldots, n\}$ covers $F(\bar{x})$ and let δ_i, $i = 1, \ldots, m$ be such that $\varrho(x, \bar{x}) < \delta_i$ implies $F(x) \cap B^\circ(y_i, \frac{1}{2}\varepsilon) \neq \emptyset$. Set $\delta = \inf_{1 \leqslant i \leqslant m} \delta_i$. Then $\varrho(x, \bar{x}) < \delta$ implies $y_i \in V^\circ(F(x), \frac{1}{2}\varepsilon)$, for all $i = 1, \ldots, m$, i.e., $B^\circ(y_i, \frac{1}{2}\varepsilon) \subset V^\circ(F(x), \varepsilon)$ for all $i = 1, \ldots, m$. Therefore $F(\bar{x}) \subset \bigcup_{i=1}^{m} B(y_i, \frac{1}{2}\varepsilon)$ $\subset V^\circ(F(x), \varepsilon)$ for $x \in B^\circ(\bar{x}, \delta)$.

Immediately from the above definition it follows that for given metric spaces X and Y, F: $X \to \mathrm{Cl}(Y)$ is H-l.s.c. at $\bar{x} \in X$ if and only if for every sequence (x_n) of X converging to $\bar{x}$ one has $\lim_{n \to \infty} h(F(\bar{x}), F(x_n)) = 0$.

In what follows we shall need the following result.

PROPOSITION 2.4. *Let* (X, d) *and* (Y, ϱ) *be metric spaces,* G: $X \to \mathscr{P}(Y)$ *be l.s.c. on* X *and* g: $X \to Y$ *be a continuous function on* X. *Let the real-valued function* $X \ni x \to \varepsilon(x) \in \boldsymbol{R}^+$ *be lower semicontinuous on* X. *Then the mapping* Φ: $X \ni x \to \Phi(x) \in \mathscr{P}(Y)$ *defined by* $\Phi(x) := B^\circ[g(x), \varepsilon(x)] \cap G(x)$ *is l.s.c. at every* $x \in X$ *such that* $\Phi(x) \neq \emptyset$.

Proof. Let $\bar{x} \in X$ be such that $\Phi(\bar{x}) \neq \emptyset$ and let $\bar{y} \in \Phi(\bar{x})$ and $\eta > 0$. For some $\sigma > 0$, $\varrho(\bar{y}, g(\bar{x})) = \varepsilon(\bar{x}) - \sigma$. There exist $\delta_1 > 0$ such that to any $x \in X$ with $d(x, \bar{x}) < \delta_1$ we can associate $y_x \in G(x)$ so that $\varrho(y_x, \bar{y}) < \min(\eta, \frac{1}{3}\sigma)$, $\sigma_2 > 0$ such that $d(x, \bar{x}) < \sigma_2$ implies $\varepsilon(x) > \varepsilon(\bar{x}) - \frac{1}{3}\sigma$ and $\sigma_3 > 0$ such that $d(x, \bar{x}) < \sigma_3$ implies $\varrho(g(\bar{x}), g(x)) < \frac{1}{3}\sigma$. Then, when $d(x, \bar{x}) < \min(\sigma_1, \sigma_2, \sigma_3)$ at once $\varrho(y_x, g(x)) \leqslant \varrho(y_x, \bar{y}) + \varrho(\bar{y}, g(\bar{x})) + \varrho(g(\bar{x}), g(x)) < \frac{1}{3}\sigma + \varepsilon(\bar{x}) - \sigma + \frac{1}{3}\sigma = \varepsilon(\bar{x}) - \frac{1}{3}\sigma < \varepsilon(x)$, i.e., $y_x \in \Phi(x)$ and $\varrho(y_x, \bar{y}) < \eta$. ■

Similarly as in the case of upper semicontinuous mappings we can also define some weak forms of lower semicontinuities. To do this let X and Y be normed linear spaces and $F\colon X \to \mathscr{P}(Y)$ be given.

We will say that F is *weakly-weakly lower semicontinuous* (*w.-w.l.s.c.*) *on* X if for every weakly closed subset $M \subset Y$ the set $F_-(M) := \{x \in X\colon F(x) \subset M\}$ is sequentially weakly closed in X. F is called *weakly-strongly lower semicontinuous* (*w.-s.l.s.c.*) *on* X if for every closed (in the norm topology of Y) set $M \subset Y$ the set $F_-(M)$ is sequentially weakly closed in X. Similarly we can define mappings s.-w.l.s.c. and s.-s.l.s.c. on X. Analogously as in the case of upper semicontinuous mappings we also obtain.

THEOREM 2.10. *Let* (X, ϱ) *be a metric space and* $F\colon X \to \mathrm{Comp}(\boldsymbol{R}^n)$. *Then*

(i) *if* F *is l.s.c. on* X *then also a set-valued function* $\mathrm{co}\,F\colon X \ni x \to \mathrm{co}\,F(x) \subset \boldsymbol{R}^n$ *is l.s.c. on* X,

(ii) *if* F *has furthermore convex values, then* F *is l.s.c. on* X *if and only if for every* $p \in \boldsymbol{R}^n$ *a real-valued function* $s(p, F(\cdot))$ *is lower semicontinuous on* X. ■

Let X and Y be topological Hausdorff spaces and let $F\colon X \to \mathscr{P}(Y)$ be given. A set-valued function F is said to be *continuous on* X if it is u.s.c. and l.s.c. on X. If X and Y are metric spaces it is called H-continuous (H-c.) on X if F is H-u.s.c. and H-l.s.c. on X.

COROLLARY 2.2. *A set valued-function* $F\colon X \to \mathrm{Cl}(Y)$ *is H-c. on* X *if and only if it is a continuous mapping of a metric space* (X, ϱ) *into a metric space* $(\mathrm{Cl}(Y), h)$. *In particular* $F\colon X \to \mathrm{Comp}(Y)$ *is continuous on* X *if and only if it is H-c. on* X. ■

COROLLARY 2.3. *Let* X *be a metric space. If* $r\colon X \to \boldsymbol{R}^+$ *is continuous on* X *and* $S \in \mathrm{Comp}(\boldsymbol{R}^n)$, *then a set-valued function* $F\colon X \to \mathrm{Comp}(\boldsymbol{R}^n)$ *defined by* $F(x) = r(x)S$ *is continuous on* X.

Indeed, by Corollary 1.2, for $\bar{x}, x \in X$ we have $h(F(\bar{x}), F(x)) = h(r(\bar{x})B, r(x)B) \leqslant |r(\bar{x}) - r(x)|\,\|S\|$. Hence, by Corollary 2.2, it follows that F is continuous on X. ■

In optimal control theory we have to deal with "parametrized" set-valued functions of the form $F(x) = \{f(x, u)\colon u \in U\}$, where $f\colon X \times U \to Y$ is given.

PROPOSITION 2.5. *Assume that* X *and* Y *are topological Hausdorff spaces and let* $f\colon X \times \times U \to Y$ *be given. Then*

(i) *if* $f(\cdot, u)$ *is continuous on* X *for every* $u \in U$ *then* $F\colon X \to \mathscr{P}(Y)$ *defined by* $F(x) = f(x, U)$ *is l.s.c. on* X,

(ii) *if U is a compact topological space and f: $X \times U \to Y$ is continuous on $X \times U$ then F defined by $F(x) = f(x, U)$ is continuous on X.*

Proof. (i) Let $\bar{x} \in X$ be fixed and N be open set of Y. Suppose $\bar{u} \in U$ is such that $f(\bar{x}, \bar{u}) \in N$. By continuity of $f(\cdot, u)$ at $\bar{x}$ there is a neighbourhood V of $\bar{x}$ such that $f(x, \bar{u}) \in N$ for every $x \in V$. Therefore, for every $x \in V$ one has $F(x) \cap N \neq \emptyset$.

(ii) Let us observe that $F(x) \in \mathrm{Comp}(Y)$ for every $x \in X$. Let $x \in X$ be arbitrarily fixed and suppose V is a neighbourhood of $F(x)$. By the continuity of f for every $u \in U$ there exist neighbourhoods W^u and O^u of x and u, respectively, such that $f(W^u \times O^u) \subset V$. The family $\{W^u;\ u \in U\}$ is an open covering of U. Then there exists its finite subfamily $\{W_i;\ i = 1, \ldots, n\}$ which also covers U. One can easily observe that $F(\bigcap_{i=1}^{n} W_i) \subset V$. ■

We have also the following generalization of Dugundji's extension theorem.

THEOREM 2.12 (Antosiewicz–Cellina). *Let A be a nonempty closed subset of a metric space (X, d) and Y a normed linear space. If F: $A \to \mathrm{Cl}(Y)$ is continuous then F has a continuous extension G: $X \to \mathrm{Cl}(Y)$ such that $G(X) \subset \mathrm{co}\, F(A)$.*

Proof (sketch of the proof). Let us observe that for every $F_1, F_2 \in \mathrm{Cl}(Y)$ with $F_1, F_2 \subset H := \mathrm{co}\, F(A)$ and given $r_1, r_2 \geqslant 0$ we have $h(V(F_1, r_1) \cap H,\ V(F_2, r_2) \cap \cap H) \leqslant \varepsilon_1 + \varepsilon_2$ whenever $h(F_1, F_2) \leqslant \varepsilon_1$ and $|r_1 - r_2| \leqslant \varepsilon_2$.

Denote by W_x, for $x \in X$ an open ball of X with centre x and radius $\frac{1}{2}\mathrm{dist}(x, A)$. Let $\{U_x\colon x \in X\}$ be a locally finite open refinement of covering $\{W_x\colon x \in X\}$ of X and let $(p_x)_{x \in X}$ be a continuous partition of unity subordinated to $\{V_x\colon x \in X\}$. Next, for every $x \in X$, select a point $g(x) \in A$ with $d(g(x), x) < 2\mathrm{dist}(x, A)$ and put $\Delta(x) := \sup\{h(F(g(w_1)), F(g(w_2)))\colon w_1, w_2 \in L(x)\}$, where $L(x) \subset X$ is a finite set such that for a given $x \in X$ there is its open neighbourhood $U_x \subset X$ such that $U_x \cap \cap \mathrm{supp}(p_w) \neq \emptyset$ if and only if $w \in L(x)$.

Now, relative to the covering $(U_x)_{x \in X}$ we can define a function Λ: $X \to \boldsymbol{R}^+$ and a mapping z: $X \to X$ such that $x \in U_{z(x)}$ and $\Lambda(x) \geqslant \Delta(z(x))$ for $x \in X$. Put $p(x) = \sup\{p_w(x)\colon w \in X\}$ for every $x \in X$, which implies that p is continuous in X and $p(x) > 0$ for every $x \in X$.

Introduce $r_w(x) = \Lambda(x) p_w(x)/p(x)$ for every $w \in X$ and every $x \in X$, which implies that the family $(r_w)_{w \in X}$ is equicontinuous in X.

We verify that the set-valued function G: $X \to \mathrm{Cl}(X)$ defined by setting $G(a) = F(a)$ for every $a \in A$ and $G(x) = \bigcup_{w \in L(x)} V(F(g(w)), r_w(x)) \cap H$ for every $x \in X$ is the desired extension of F to X. ■

3. Measurable set-valued functions and Aumann's integral

We present here the basic properties of measurable set-valued functions. They will be given in the general situation where T is an abstract measurable space and X is a separable metric space. We begin with the basic notations of the general measure theory. The basic properties of the Aumann's integral of finite-dimensional set-valued functions are given in the last part of this section.

3.1. Foundations of measure theory

Let $(T, \mathscr{F})$ be a measurable space and S a class of subsets of T. The intersection of all σ-algebras containing S is called the *σ-algebra generated by S*. In particular, if T is a topological space, the σ-algebra $\beta(T)$ generated by the open sets of T is called the *Borel σ-algebra.*

Given measurable spaces $(T_1, \mathscr{F}_1)$ and $(T_2, \mathscr{F}_2)$ a set $A_1 \times A_2$ with $A_1 \in \mathscr{F}_1$ and $A_2 \in \mathscr{F}_2$ is called a *measurable rectangle* in $T_1 \times T_2$. The σ-algebra of subsets of $T_1 \times T_2$ generated by all measurable rectangles is denoted by $\mathscr{F}_1 \otimes \mathscr{F}_2$ and is called the *product σ-algebra* of $\mathscr{F}_1$ and $\mathscr{F}_2$; the measurable space $(T_1 \times T_2, \mathscr{F}_1 \otimes \otimes \mathscr{F}_2)$ then is called the *product of the spaces* $(T_1, \mathscr{F}_1)$ and $(T_2, \mathscr{F}_2)$. In particular, if T_1 an T_2 are topological spaces then $\beta(T_1) \otimes \beta(T_2) = \beta(T_1 \times T_2)$. The product of a finite family $\{(T_i, \mathscr{F}_i)_{i=1,\ldots,n}\}$ of measurable spaces is defined analogously.

Let $(T. \mathscr{F})$ be a measurable space and let $E \in \mathscr{F}$. A dyadic structure of E is a collection of measurable sets $E_{\varepsilon_1\varepsilon_2\ldots\varepsilon_k}$, where $\varepsilon_i = 0, 1$ and $k = 1, 2, \ldots$ such that

(i) $E_{\varepsilon_1\varepsilon_2\ldots\varepsilon_k 0} \cup E_{\varepsilon_1\varepsilon_2\ldots\varepsilon_k 1} = E_{\varepsilon_1\varepsilon_2\ldots\varepsilon_k}$ and $E_0 \cup E_1 = E$ and

(ii) $E_{\varepsilon_1\varepsilon_2\ldots\varepsilon_k 0} \cap E_{\varepsilon_1\varepsilon_2\ldots\varepsilon_k 1} = \emptyset$ and $E_0 \cap E_1 = \emptyset$.

The standard dyadic structure of the unit interval $I = [0, 1]$ is the collection $\{I_{\varepsilon_1\varepsilon_2\ldots\varepsilon_k}: k = 1, 2, \ldots\}$ described by $I_0 = [0, \frac{1}{2})$, $I_{00} = [0, \frac{1}{4})$, $I_{01} = [\frac{1}{4}, \frac{1}{2})$ and generally $I_{\varepsilon_1\varepsilon_2\ldots\varepsilon_k} = [\sum_{i=1}^{k} \varepsilon_i 2^{-i}, 2^{-k} + \sum_{i=1}^{k} \varepsilon_i 2^{-i})$.

Deonte by $\mathscr{M}$ the collection of all Lebesgue measurable sets in $\boldsymbol{R}^n$ which have finite Lebesgue measure. It can be proved that for every $E \in \mathscr{M}$ and every $\varepsilon > 0$ there is a set of the form $G = \bigcup_{n=1}^{N} I_n$, where $I_1, \ldots, I_N$ are open intervals of $\boldsymbol{R}^n$, such that $\mu(E \triangle G) < \varepsilon$. This property of elements of $\mathscr{M}$ has an interpretation in some metric space $\mathscr{M}_\mu$ associated with $\mathscr{M}$. To define $\mathscr{M}_\mu$ let us define in $\mathscr{M}$ an equivalence relation "$\sim$" by setting for $A, B \in \mathscr{M}$, $A \sim B$ if and only if $\mu(A \triangle B) = 0$. Denote by $[A]$ the equivalence class in $\mathscr{M}$ which contains A. If $A \sim E$ and $B \sim F$ then $\mu(A \triangle B) = \mu(E \triangle F)$. We can see this as follows. In the first place, for any two sets A_1, A_2 in $\mathscr{M}$, $|\mu(A_1) - \mu(A_2)| \leqslant \mu(A_1 \triangle A_2)$. Furthermore, the operation $\triangle$ is commutative and associative, and therefore $(A \triangle B) \triangle (E \triangle F) = (A \triangle E) \triangle (B \triangle F)$. Hence and the inequality $\mu(A \triangle C) \leqslant \mu(A \triangle C) + \mu(B \triangle C)$ we obtain $|\mu(A \triangle B) - -\mu(E \triangle F)| \leqslant \mu(A \triangle E) + \mu(B \triangle F)$. Thus $\mu(A \triangle B) = \mu(E \triangle F)$ if $A \sim E$ and $B \sim F$.

Let us now denote by $\mathscr{M}_\mu$ the set of all equivalence classes formed in $\mathscr{M}$ by the relation "$\sim$". We define a function d: $\mathscr{M}_\mu \times \mathscr{M}_\mu \to \boldsymbol{R}$ as follows: $d([A], [B]) = \mu(A \triangle B)$. It is quickly evident that d is a metric on $\mathscr{M}_\mu$.

The property of a set $E \in \mathscr{M}$ mentioned above asserts that, if N is the set of elements in $\mathscr{M}_\mu$ such that each member of N has a representative which is a finite union of open intervals, then N is dense in $\mathscr{M}_\mu$.

It can be proved that every element of $\mathscr{M}_\mu$ has a representative which is a Bore set. It follows from the fact that every set E in $\boldsymbol{R}^n$ has measurable cover and kernel which are Borel sets, i.e. there are Borel sets F and G in $\boldsymbol{R}^n$ such that $F \subset E \subset G$ and $\mu(F) = \mu(E) = \mu(G)$.

Finally, let us observe that for a given closed interval of I there is a sequence of Borel sets in I that are dense in the Borel field of I.

Given two measurable spaces $(T_1, \mathscr{F}_1)$ and $(T_2, \mathscr{F}_2)$, a mapping f: $T_1 \to T_2$ is said to be $(\mathscr{F}_1, \mathscr{F}_2)$-*measurable* or *simple measurable* if $f^{-1}(E) \in \mathscr{F}_1$ for every $E \in \mathscr{F}_2$. In particular if $T = \boldsymbol{R}$ or $T \subset \boldsymbol{R}$, we usually take for the σ-algebra $\mathscr{F}$ for T the family $\mathscr{L}(T)$ of all Lebesgue measurable subsets of T. In such case real-valued functions defined on T are called *L-measurable.*

REMARK 3.1. *If f: $T_1 \to T_2$ and S is a class of subsets of T_2 which generates T_2 then f is measurable if and only if $f^{-1}(C) \in \mathscr{F}_1$ for every $C \in S$.*

REMARK 3.2. *If $(T, \mathscr{F})$ is a measurable space, X is a metric space and if (f_n) is a sequence of $(\mathscr{F}, \beta(X))$-measurable functions from T to X such that $\lim_{n\to\infty} f_n(t)$ exists at each $t \in T$ then a function f: $T \ni t \to \lim_{n\to\infty} f_n(t) \in X$ is $(\mathscr{F}, \beta(X))$-measurable.*

REMARK 3.3. *If $(T_1, \mathscr{F}_1)$, $(T_2, \mathscr{F}_2)$ and $(X, \mathscr{F})$ are measurable space, g: $T_1 \to X$ is $(\mathscr{F}_1, \mathscr{F})$-measurable, f: $X \to T_2$ is $(\mathscr{F}, \mathscr{F}_2)$-measurable then $f \circ g$ is $(\mathscr{F}_1, \mathscr{F}_2)$-measurable.*

Indeed, for every $E \in \mathscr{F}_2$ one has $f^{-1}(E) \in \mathscr{F}$ and $g^{-1}(B) \in \mathscr{F}_1$ for every $B \in \mathscr{F}$. Then, in particular, for every $E \in \mathscr{F}_2$ we have $(f \circ g)^{-1}(E) = g^{-1}[f^{-1}(E) \in \mathscr{F}_1$. ∎

Let T be a topological Hausdorff space. A positive Radon measure on T is a positive measure μ: $\beta(T) \to \boldsymbol{R}$ such that:

(i) for every $t \in T$ there exists an open neighbourhood of t of finite measure and

(ii) for every $A \in \beta(T)$ one has $\mu(A) = \sup\{\mu(K)\colon K \in \mathrm{Comp}(A)\}$.

A function f: $T \to X$, where X is a metric space is called μ-*measurable* if it is $(\Sigma^*, \beta(X))$-measurable, where (T, Σ^*, μ^*) is a Lebesgue extension of $(T, \beta(T), \mu)$.

Let T be a compact topological space with a positive Radon measure μ and let X be complete separable metric space. We have the following result.

THEOREM (Lusin). *A function f: $T \to X$ is μ-measurable if and only if for every $\varepsilon > 0$ there exists a closed subset E_ε of T with $\mu(T \setminus E_\varepsilon) \leqslant \varepsilon$ and such that the restriction of f to E_ε is continuous.* ∎

PROPOSITION 3.1. *Let (X, ϱ) be a separable metric pace and let* Comp(X) *be a topological space of all compact subset of X with the Hausdorff topology. Then the Borel σ-algebra of* Comp(X) *is generated by sets $\{K \in \mathrm{Comp}(X)\colon K \subset U\}$ for every open sets $U \subset X$. It is also generated by sets $\{K \in \mathrm{Comp}(X)\colon K \cap V \neq \emptyset\}$ for every open sets $V \subset X$.*

Proof. By Theorem 1.2, the sets $\{K \in \mathrm{Comp}(X)\colon K \subset U\}$ and $\{K \in \mathrm{Comp}(X)\colon K \cap V \neq \emptyset\}$ generate the Hausdorff topology of the space $(\mathrm{Comp}(X), h)$. Let us observe that for a given $K_0 \in \mathrm{Comp}(X)$ a basis of neighbourhoods of K_0 consists of sets $\{K \in \mathrm{Comp}(X)\colon K \cap V_1 \neq \emptyset, \ldots, K \cap V_n \neq \emptyset, K \subset U\}$ for open sets $U, V_1, \ldots$ $\ldots, V_n$ of X which contain K_0. Indeed, if $K_0 \in \mathrm{Comp}(X)$ and $\varepsilon > 0$ are given, the ball of centre K_0 and a radius ε contains a set $\{K \in \mathrm{Comp}(X)\colon K \subset U\} \cap \{K \in \mathrm{Comp}(X)\colon K \cap V_1 \neq \emptyset\} \cap \ldots \cap \{K \in \mathrm{Comp}(X)\colon K \cap V_n \neq \emptyset\}$ which contains K_0, because taking $U := \{x\colon \mathrm{dist}(x, K_0) < \varepsilon\}$ and open balls $V_1, \ldots, V_n$ of radius $\frac{1}{2}\varepsilon$ which cover K_0 we obtain $\bar{h}(K, K_0) < \varepsilon$ if $K \subset U$ and $\bar{h}(K_0, K) \leqslant \varepsilon$ if K meets $V_1, \ldots, V_n$.

Consider the set $\{K \in \mathrm{Comp}(X)\colon K \cap V \neq \emptyset\}$. Observe that $V = \bigcup_{n=1}^{\infty} F_n$ with $F_n = \{x\colon \mathrm{dist}(x, X \setminus V) \geqslant 1/n\}$. Then $\{K \in \mathrm{Comp}(X)\colon K \cap V \neq \emptyset\} = \bigcup_{n=1}^{\infty} [\mathrm{Comp}(X) \setminus \{K \in \mathrm{Comp}(X)\colon K \subset X \setminus F_n\}]$. Hence the σ-algebra generated by the sets $\{K \in \mathrm{Comp}(X)\colon K \cap V \neq \emptyset\}$ is included in the σ-algebra generated by the sets $\{K \in \mathrm{Comp}(X)\colon K \subset U\}$.

Consider now the set $\{K \in \mathrm{Comp}(X)\colon K \subset U\}$. Put $V_n = \{x\colon \mathrm{dist}(x, X \setminus U) < 1/n\}$. Then $X \setminus U = \bigcap_{n=1}^{\infty} V_n$. It is easy to see that $[K \cap (X \setminus U) \neq \emptyset] \to [K \cap V_n \neq \emptyset$ for $n = 1, 2, \ldots]$. That proves that σ-algebra generated by the sets $\{K \in \mathrm{Comp}(X)\colon K \subset U\}$ is included in σ-algebra generated by the sets $\{K \in \mathrm{Comp}(X)\colon K \cap V \neq \emptyset\}$.

We shall now prove that any open subset Σ of $\mathrm{Comp}(X)$ belongs to the σ-algebra generated by the sets presented above. Indeed Σ is a union of a family $\mathscr{A}$ of finite intersections of sets $\{K \in \mathrm{Comp}(X)\colon K \subset U\}$ and $\{K \in \mathrm{Comp}(X)\colon K \cap V \neq \emptyset\}$ by virtue of the remark in the beginning of the proof. But, by the Remark 1.1, $\mathrm{Comp}(X)$ is separable; then Σ is also the union of a countable subfamily of $\mathscr{A}$. ■

PROPOSITION 3.2. *Let (X, ϱ) be a separable metric space, (Y, d) a metric space, $(T, \mathscr{F})$ measurable space and $f\colon T \times X \to Y$ such that $f(\cdot, x)$ is measurable for fixed $x \in X$ and $f(t, \cdot)$ is continuous for fixed $t \in T$. Then f is $(T \otimes \beta(X), \beta(Y))$-measurable. In fact, for every closed subset B of Y, $f^{-1}(B)$ is the countable intersection of countable unions of measurable rectangels of $T \times X$.*

Proof. Let B be a closed subset of Y, and A be a countable dense subset of X. Let $B_n = \{y\colon \mathrm{dist}(y, B) < 1/n\}$. Then $f(t, x) \in B$ if and only if for every n there exists $a \in A$ such that $\varrho(x, a) < 1/n$ and $f(t, a) \in B_n$. Hence $f^{-1}(B) = \bigcap_{n=1}^{\infty} \bigcup_{a \in A} \{t\colon f(t, a) \in B\} \times \{x \in X\colon \varrho(x, a) < 1/n\}$, and so $f^{-1}(B)$ is the countable intersection of countable unions of measurable rectangles of $T \times X$. ■

Let $\mathscr{H}$ be a family of all finite sequences of positive integers and put for given sequence $\mu := (\mu_k)$ of N and for $n \in N$, $\mu|_n := (\mu_1, \ldots, \mu_n)$.

Given a metric space X and a mapping $A\colon \mathscr{H} \to \mathscr{P}(X) \cup \{0\}$, the *Souslin operator*

$\mathscr{A}$ is defined by

$$\mathscr{A}(A) := \bigcup_{\mu \in \mathscr{N}} \bigcap_{n=1}^{\infty} A_{\mu|_n},$$

where $A_{\mu|_n} := A(\mu_1, \ldots, \mu_n)$ and $\mathscr{N}$ denotes the family of all N-valued sequences of N. It is clear that the operator of countable union and countable intersection are both special cases of the Souslin operation.

Let $(T, \mathscr{F})$ be a measurable space, and S be finite-dimensional real vector space. A set-valued measure Φ on $\mathscr{F}$ is a function from $\mathscr{F}$ to the nonempty subsets of S, which is countable additive, i.e., $\Phi(\bigcup_{j=1}^{\infty} E_j) = \sum_{j=1}^{\infty} \Phi(E_j)$ for every sequence $E_1, E_2, \ldots$ of mutually disjoint elements in $\mathscr{F}$, where the series is absolutely convergent.

A set-valued measure Φ is bounded if $\Phi(T)$ is a bounded set. The additivity implies that $\Phi(T)$ includes a translation of every $\Phi(E)$, thus Φ is bounded if and only if $\Phi(E)$ is bounded for every $E \in \mathscr{F}$. The additivity of Φ also implies that either $\Phi(\emptyset) = \{0\}$ or $\Phi(\emptyset)$ is an unbounded set. Thus if Φ is bounded then $\Phi(\emptyset) = \{0\}$. It can be proved that if Φ is a bounded set-valued measure, then the range of Φ, i.e., $\bigcup_{E \in \mathscr{F}} \Phi(E)$ is a bounded set.

An atom of the set-valued measure Φ is an element $E \in \mathscr{F}$, for which $\Phi(E) \neq \{0\}$, and such that if $E_1 \subset E$ and $E_1 \in \mathscr{F}$ then either $\Phi(E_1) = \{0\}$ or $\Phi(E \setminus E_1) = \{0\}$. A set-valued measure with no atoms is called *nonatomic*.

Let $(T, \mathscr{F}, \lambda)$ be a measure space. A set-valued measure Φ on $\mathscr{F}$ is absolutely continuous with respect to λ (or λ-continuous) if $\lambda(E) = 0$ implies $\Phi(E) = \{0\}$. We denote this by $\Phi \ll \lambda$.

3.2. Measurable set-valued functions

Let $(T, \mathscr{F})$ be a measurable space, X be a separable metric space and $F\colon T \to \mathscr{P}(X)$. A set-valued function F is said to be *measurable* (*weakly measurable*, *β-measurable*) if $F^{-}(E) := \{t \in T\colon F(t) \cap E \neq \emptyset\} \in \mathscr{F}$ for every closed (open, Borel) set $E \subset X$. If $F\colon Y \to \mathscr{P}(X)$, where Y is a topological space, then the assertion that F is measurable (weakly-, β-measurable) means that F is measurable (weakly-, β-measurable) when Y is assigned the σ-algebra $\beta(Y)$ of Borel subsets of Y. Likewise, if $F\colon T \times Y \to \mathscr{P}(X)$, then the various kinds of measurability of F are always defined in terms of the product σ-algebra $\mathscr{F} \otimes \beta(Y)$ on $T \times Y$. In particular if $T = \boldsymbol{R}$ or $T \subset \boldsymbol{R}$, usually we will consider T together with the σ-algebra of all its Lebesgue measurable subsets. If $X = \boldsymbol{R}^n$, then often a measurable set-valued function $F\colon T \to \mathscr{P}(\boldsymbol{R}^n)$ we will also call *L-measurable.*

If T is a topological space with a positive Radon measure μ we shall consider T with the σ-algebra Σ^*, where (T, Σ^*, μ^*) is a Lebesgue extension of $(T, \beta(T), \mu)$. In this case a measurable set-valued function $F\colon T \to \mathscr{P}(X)$ will be called *μ-measurable.*

Immediately from the above definition it follows that β-measurability implies measurability and measurability implies the weak measurability. Moreover, if

$F\colon T \to \mathscr{P}(X)$ is measurable (weakly measurable, β-measurable) and Z is closed (resp. open, Borel) subset of X such that $F(t) \cap Z \neq \emptyset$ for $t \in T$, then the set-valued function $F_z\colon T \to \mathscr{P}(X)$ defined by $F_z(t) := F(t) \cap Z$ for $t \in T$ is measurable (resp. weakly measurable, β-measurable). Let us observe that if X is σ-compact, i.e. $X = \bigcup_{n=1}^{\infty} X_n$, where each X_n is compact, then for $F\colon T \to \mathscr{P}(X)$ to be measurable, it sufficies that $F^-(C) \in \mathscr{F}$ for every compact set $C \subset X$. Indeed, every closed set in X is a union of countable family of compact sets, and the operation F^- has the property $F^-(\bigcup_{j=1}^{\infty} C_j) = \bigcup_{j=1}^{\infty} F^-(C_j)$. If Y is a subspace of separable metric space X, then $F\colon T \to \mathscr{P}(Y)$ is (weakly) measurable as a mapping into $\mathscr{P}(Y)$ if and only if F is (weakly) measurable as a mapping into $\mathscr{P}(X)$. Notice also that a point-valued function $f\colon T \to X$ is measurable if and only if the set-valued function $F(t) := \{f(t)\}$ is measurable (equivalently weakly measurable, β-measurable).

Let us observe, that for the measurability of set-valued functions it is not needed to assume that they take nonempty values. In such case, if $F\colon T \to \mathscr{P}(X)$ is measurable, then the set $\{t\colon F(t) \neq \emptyset\} = F^-(X)$ is measurable.

Finally, let us observe that $F\colon T \to \mathscr{P}(X)$ is weakly measurable if and only if the set-valued function $\bar{F}$, defined by $\bar{F}(t) = \overline{F(t)}$ for $t \in T$ is weakly measurable because for every open set $U \subset X$ we have $F^-(U) = (\bar{F})^-(U)$.

Theorem 3.1. *Let $F\colon T \to \operatorname{Comp}(X)$. Then F is measurable if and only if F is weakly measurable.*

Proof. It was pointed out that measurability implies weak measurability. Suppose F is weakly measurable and let E be a closed subset of X. We have $X \setminus E = \bigcup_{n=1}^{\infty} E_n$, where $E_n = \{x\colon \operatorname{dist}(x, E) \geqslant 1/n\}$. By hypothesis, $F^-(X \setminus E_n)$ is measurable. So also $T \setminus F^-(X \setminus E_n) = \{t\colon F(t) \subset E_n\} := F_-(E_n)$ is measurable. Applying the compactness of each $F(t)$ we obtain $F^-(E) = T \setminus F_-(X \setminus E) = T \setminus F_-(\bigcup_{n=1}^{\infty} E_n) = T \setminus \bigcup_{n=1}^{\infty} F_-(E_n) = T \setminus \bigcup_{n=1}^{\infty} F^-(X \setminus E_n)$. Hence $F^-(E) \in \mathscr{F}$. ∎

Theorem 3.2. *Let $F\colon T \to \operatorname{Comp}(X)$. Then F is measurable if and only if it is measurable as a function from a measurable space $(T, \mathscr{F})$ to a metric space* $(\operatorname{Comp}(X), h)$.

Proof. Assume a function $F\colon T \to \operatorname{Comp}(X)$ is measurable. Then, by virtue of Theorem 1.2, for every open set $U \subset X$, in particular we have $F^{-1}(\{K \in \operatorname{Comp}(X)\colon K \cap U \neq \emptyset\}) \in \mathscr{F}$. Since $F^-(U) = F^{-1}(\{K \in \operatorname{Comp}(X)\colon K \cap U \neq \emptyset\})$ for every open set $U \subset X$, then F is weakly measurable and hence, by Theorem 3.1, measurable. Suppose F is measurable and therefore also weakly measurable. By virtue of Proposition 3.1, sets $\{K \in \operatorname{Comp}(X)\colon K \cap U \neq \emptyset\}$ generate the Borel σ-algebra of $\operatorname{Comp}(X)$. By the weak measurability of F for every open $U \subset X$ we have $F^{-1}(\{K \in \operatorname{Comp}(X)\colon$

$K \cap U \neq \emptyset\}) = F^{-}(U) \in \mathscr{F}$. Then F is a measurable mapping of $(T, \mathscr{F})$ into a topological space Comp(X). ■

THEOREM 3.3. *Let F: $T \to \mathscr{P}(X)$. Then F is weakly measurable if and only if the function $T \ni t \to \operatorname{dist}(x, F(t)) \in R^+$ is measurable for each fixed $x \in X$.*

Proof. Let us observe that F is weakly measurable if and only if $F^{-}(B^{\circ}(x, \varepsilon)) \in \mathscr{F}$ for each open ball $B^{\circ}(x, \varepsilon)$ in X. On the other hand the function $T \in t \to \operatorname{dist}(x, F(t)) \in R^+$ is measurable for fixed $x \in X$ if and only if $\{t\colon \operatorname{dist}(x, F(t)) < \varepsilon\} \in \mathscr{F}$ for each $\varepsilon > 0$. Since $F^{-1}(B^{\circ}(x, \varepsilon)) = \{t\colon F(t) \cap B^{\circ}(x, \varepsilon) \neq \emptyset\} = \{t\colon \operatorname{dist}(x, F(t)) < \varepsilon\}$ then the proof is complete. ■

THEOREM 3.4. *Let $F : T \to$ Comp(X) be measurable for each $n \in J \subset N$. Then $F = \bigcap_{n \in J} F_n$ is measurable. If $\overline{\bigcup_{n \in J} F_n(t)}$ is compact then also $T \in t \to \overline{\bigcup_{n \in J} F_n(t)} \in$ Comp(X) is measurable.*

Proof. Assume $J = \{1, 2\}$ and let us prove that the mapping $(K_1, K_2) \to K_1 \cap K_2$ from $\overset{\circ}{\text{Comp}}(X) \times \overset{\circ}{\text{Comp}}(X)$ to $\overset{\circ}{\text{Comp}}(X)$, where $\overset{\circ}{\text{Comp}}(X) :=$ Comp$(X) \cup \emptyset$ is measurable. Let U be an open set in X. Let us observe that $\{(K_1, K_2)\colon K_1 \cap K_2 \subset U\}$ is open. Indeed, if $S_1 \cap S_2 \subset U$, $S_1 \setminus U$ and $S_2 \setminus U$ are disjoint then there exist two open sets U_1 and U_2 such that $S_i \setminus U \subset U_i$ $(i = 1, 2)$ and $U_1 \cap U_2 = \emptyset$. By Theorem 1.2, $\{(K_1, K_2)\colon K_1 \subset U_1 \cup U\ K_2 \subset U_2 \cup U\}$ is a neighbourhood of (S_1, S_2). For such a (S_1, S_2) one has $S_1 \cap S_2 \subset (U_1 \cup U) \cap (U_2 \cup U) = U$. Therefore, by Proposition 3.1, the mapping $(K_1, K_2) \to K_1 \cap K_2$ is measurable. Hence and Remark 3.3, he mapping $T \ni t \to (F_1 \cap F_2)(t) \in$ Comp(X) is measurable. Then for every $m \in N$ a set-valued function G_m defined by $G_m = \bigcap_{n=1}^{m} F_n$ is measurable. Since $h(G_m(t), \bigcap_{n=1}^{\infty} F_n(t)) \to 0$ as $m \to \infty$, then by Remark 3.2 and Theorem 3.2, $F = \bigcap_{n=1}^{\infty} F_n$ is also measurable.

Let U be an open set of X. Then

$$\{t\colon [\overline{\bigcup_{n \in J} F_n(t)}] \cap U \neq \emptyset\} = \{t\colon [\bigcup_{n \in J} F_n(t)] \cap U \neq \emptyset\}$$
$$= \bigcup_{n \in J} \{t\colon F_n(t) \cap U \neq \emptyset\} \in \mathscr{F}$$

because F_n is for $n = 1, 2, \ldots$ weakly measurable. Thus the mapping $T \ni t \to \overline{\bigcup_{n \in J} F_n(t)}$ $\in$ Comp(X) is weakly measurable and therefore by Theorem 3.2 also measurable. ■

REMARK 3.4. The set-valued function $\bigcap_{n=1}^{\infty} F_n$ is also measurable if F_n: $T \to \mathscr{P}(X)$ has closed values and there is $N \in N$ such that F_N: $T \to$ Comp(X). Indeed, let Y be a metrizable compactification of X, and define $\bar{F}_n$: $T \to Y$ by $\bar{F}_n(t) = [F_n(t)]_Y$. Then each F_n is weakly measurable and therefore $\bigcap_{n=1}^{\infty} \bar{F}_n$ is measurable. But, as it is easily seen, $\bigcap_{n=1}^{\infty} \overline{F}_n(t) = \bigcap_{n=1}^{\infty} F_n(t)$, for each $t \in T$, since $\overline{[F_N(t)]_Y} = F_N(t)$. Thus F is measurable. ■

Now we shall consider measurable set-valued functions defined on a compact interval of the real line with values in Comp($\boldsymbol{R}^n$). Immediately from Remark 3.2 it follows that a pointwise limit of a sequence of such set-valued functions is still measurable. Now we shall prove that it is also true if a sequence converges almost everywhere.

LEMMA 3.5. *Let I be a compact interval of the real line $\boldsymbol{R}^1$ and F_k: $I \to$ Comp($\boldsymbol{R}^n$), $k = 1, 2, \ldots$ be measurable. Suppose F: $I \to$ Comp($\boldsymbol{R}^n$) is such that $\lim_{k\to\infty} h(F_k(t), F(t)) = 0$ for a.e. $t \in I$. Then F is measurable.*

Proof. Let $\mathcal{P}$ be the class of all open balls of $\boldsymbol{R}^n$ having positive rational radii and centres with rational coordinates. It is clear by Remark 3.1 and Theorem 3.1 that a set-valued function G: $I \to$ Comp($\boldsymbol{R}^n$) is measurable if and only if the set $G^-(E)$ is L-measurable for every $E \in \mathcal{P}$. Let a, r be fixed such that $B^o(a, r) \in \mathcal{P}$. For positive integers m satisfying $mr > 1$ define $T_m^k := F_k^-(B^o(a, r-m^{-1}))$, $k = 1, 2, \ldots$ and $Z_m^n := \bigcap_{k \geq n} T_m^k$ for $n = 1, 2, \ldots$ We shall prove that

$$F^-(B^o(a, r)) = \bigcup_{n,m} Z_m^n. \tag{3.1}$$

Certainly T_m^k is L-measurable, by hypothesis; thus Z_m^n and the right members of (3.1) are L-measurable. Then, (3.1) implies the measurability of F.

Let $t_0 \in F^-(B^o(a, r))$; then $F(t_0) \cap B^o(a, r) \neq \emptyset$ and there exists an integer m_0, $m_0 r > 2$, such that $F(t_0) \cap B^o(a, r-2m^{-1}) \neq \emptyset$. Since $\bar{h}(F(t_0), F_k(t_0)) \to 0$ as $k \to \infty$, it follows that

$$\bar{h}(F(t_0) \cap B^o(a, r-2m_0^{-1}), F_k(t_0)) \to 0.$$

Consequently, there exists $n_0 = n_0(m_0)$ such that if $k \geq n_0$ then $F_k(t_0) \cap B^o(a, r-$ $-m^{-1}) \neq \emptyset$. Hence $t_0 \in T_{m_0}^k$ for $k \geq n_0$ which implies $t_0 \in Z_{m_0}^{n_0}$ and then of course $t_0 \in \bigcup_{n,m} Z_m^n$.

Now let $t_0 \in \bigcup_{n,m} Z_m^n$; then there exists n_0, m_0 such that $t_0 \in Z_{m_0}^{n_0}$. Hence $t_0 \in T_{m_0}^k$ for $k \geq n_0$(i.e., $F_k(t_0) \cap B^o(a, r-m_0^{-1}) \neq \emptyset$ for $k \geq n_0$. Now since $\bar{h}(F_k(t_0) \cap F(t_0))$ $\to 0$ it follows that $\bar{h}(F_k(t_0) \cap B^o(a, r-m_0^{-1}), F(t_0)) \to 0$ as $k \to \infty$. This in turn implies that $B^o(a, r-m)_0^{-1} \cap F(t_0) \neq \emptyset$ so that certainly $F(t_0) \cap B^o(a, r) \neq \emptyset$. Thus $t_0 \in F^-(B^o(a, r))$ and (3.1) follows. ■

Immediately from Corollary 2.2, Theorem 3.2 and Lusin's theorem we obtain

THEOREM 3.6 (Lusin–Pliś). *Let I be a compact interval of $\boldsymbol{R}^1$ and μ be Lebesgue measure on $\boldsymbol{R}^1$. A set-valued function F: $I \to$ Comp($\boldsymbol{R}^n$) is L-measurable if and only if for every $\varepsilon > 0$ there exists a closed set $E_\varepsilon \subset I$ with $\mu(I \setminus E_\varepsilon) \leq \varepsilon$ and such that the restriction of F to E_ε is continuous.* ■

The above necessary and sufficient condition for the L-measurability of $F\colon I \to \text{Comp}(\boldsymbol{R}^n)$ is called the *Lusin's property* for F. Observe that immediately from Antosiewicz–Cellina continuous extension theorem follows that the Lusin's property for $F\colon I \to \text{Comp}(\boldsymbol{R}^n)$ is a sufficient condition for its L-measurability. Indeed, suppose F has the Lusin's property. Let $\varepsilon = 1/2^n$ and $E_n := E_{1/2^n}$ be a closed subset of I such that the restriction of F to each E_n is continuous. By Antosiewicz–Cellina continuous extension theorem there is a continuous extension of F from E_n on the whole interval I. Denote it by F_n. Put $E = \bigcap_{n=1}^{\infty} \bigcup_{k=n}^{\infty} (I \setminus E_k)$. We have $\mu(E) \leqslant \sum_{k=n}^{\infty} \mu(I \setminus E_k) \leqslant 1/2^{n-1}$ for $n = 1, 2, \ldots$ Then $\mu(E) = 0$. Furthermore, for every $t \in I \setminus E$ there exists N_t such that $t \in \bigcap_{k=N_t}^{\infty} E_k$. Therefore, $F_k(t) = F(t)$ for $k \geqslant N_t$ and $t \in I \setminus E$. Then we have $\lim_{k \to \infty} h(F_k(t), F(t)) = 0$ for a.e. $t \in I$. Hence, by Corollary 2.2, Theorem 2.1 and Lemma 3.5 the L-measurability of F follows.

We have also the following important theorem.

THEOREM 3.7 (Scorza–Dragoni–Castaing). *Let T be a compact metric space with a positive Radon measure μ, X a separable complete metric space and (T, Σ^*, μ^*) a Lebesgue extension of $(T, \beta(T), \mu)$. Suppose $F\colon T \times X \to \text{Comp}(\boldsymbol{R}^n)$ is such that F is $(\Sigma^* \otimes \beta(T), \beta(\text{Comp}(\boldsymbol{R}^n))$-measurable and $F(t, \cdot)$ is l.s.c. for fixed $t \in T$. Then for every $\varepsilon > 0$ there exists a closed set $E_\varepsilon \subset T$ with $\mu(T \setminus E_\varepsilon) \leqslant \varepsilon$ and such that F is l.s.c. on $E_\varepsilon \times X$.*

If F is such that $F(\cdot, x)$ is μ-measurable for fixed $x \in X$ and $F(t, \cdot)$ is continuous for fixed $t \in T$ then the above set $E_\varepsilon \subset T$ can be such taken that F is continuous on $E_\varepsilon \times X$. ■

Finally, as a corollary from Lusin–Pliś' theorem and Theorems 2.7 and 2.10 we obtain the following theorems.

THEOREM 3.8. *Let I be a compact interval of the real line $\boldsymbol{R}$ and let $F\colon I \to \text{Conv}(\boldsymbol{R}^n)$. Then F is measurable if and only if for every $p \in \boldsymbol{R}^n$ the function $I \ni t \to s(p, F(t)) \in \boldsymbol{R}$ is measurable.*

Proof. By Lusin–Pliś' theorem F is measurable if and only if for every $\varepsilon > 0$ there is a closed set $E_\varepsilon \subset I$ with $\mu(I \setminus E_\varepsilon) \leqslant \varepsilon$ such that the restriction F_ε of F to E_ε is continuous on E_ε. By Theorems 2.7 and 2.10, the continuity of F_ε is equivalent to the continuity on E_ε for every $p \in \boldsymbol{R}^n$ of the real-valued function $s(p, F_\varepsilon(\cdot))$. But, by Lusin's theorem for real-valued functions the last is equivalent to measurability of $s(p, F(\cdot))$ on I for every $p \in \boldsymbol{R}^n$. ■

THEOREM 3.9. *Let I be a compact interval of the real line $\boldsymbol{R}$ and let $F\colon I \to \text{Comp}(\boldsymbol{R}^n)$ be measurable. Then $\text{co}\, F\colon I \to \text{Conv}(\boldsymbol{R}^n)$ defined by $(\text{co}\, F)(t) := \text{co}\, F(t)$ is also measurable.*

Proof. By Lusin–Pliś' theorem for every $\varepsilon > 0$ there is a closed set $E_\varepsilon \subset I$ with

$\mu(I \setminus E_\varepsilon) \leqslant \varepsilon$ and such that F_ε is continuous. It implies that also $\operatorname{co} F_\varepsilon$ is continuous on E_ε and therefore, by Lusin–Pliś' theorem, $\operatorname{co} F$ is measurable. ■

REMARK 3.5. *If $F: I \to \operatorname{Comp}(\boldsymbol{R}^n)$ is measurable on I then also the real-valued function $I \ni t \to s(p, F(t)) \in \boldsymbol{R}$ is measurable on I for every $p \in \boldsymbol{R}^n$.*

Indeed, let $r \in \boldsymbol{R}$ and $C_k = \{x: p \cdot x \geqslant r + 1/k\}$ for $k \in \boldsymbol{N}$. We have $s(p, F(t)) > r$ if and only if $F(t) \cap C_k \neq \emptyset$ for some k. Thus, the set $\{t: s(p, F(t)) > r\}$ $= \bigcup_{k=1}^{\infty} \{t: F(t) \cap C_k \neq \emptyset\}$ is measurable. ■

COROLLARY 3.1. *If I is a compact interval, $r: I \to \boldsymbol{R}^+$ is L-measurable function and $S \in \operatorname{Comp}(\boldsymbol{R}^n)$, then a set-valued function $F: I \to \operatorname{Comp}(\boldsymbol{R}^n)$ defined by $F(t) := r(t)^+$ for $t \in I$ is measurable.* ■

3.3. Measurable selections

Given a set-valued function $F: T \to \mathscr{P}(X)$ a function $f: T \to X$ is said to be a *selector* for F if $f(t) \in F(t)$ for all $t \in T$.

The existence of selectors follows immediately from the axiom of choice. It is one from several equivalent principles which have been found to be useful and important in mathematical analysis. Another is called the principle (or theorem) of well-ordering. Still another is a maximality principle, which may assume various forms. We recall here the axiom of the maximality principle of Kuratowski and Zorn and then we will show how the principle of choice of Zermelo can be deducible.

KURATOWSKI–ZORN'S LEMMA. *Let P be a nonempty partially ordered set with the property that every completely ordered subset of P has an upper bound in P. Then F contains at least one maximal element.* ■

ZERMELO'S PRINCIPLE OF CHOICE. *Let $\mathscr{E}$ be a nonempty family of nonempty subsets of a set X. Then there exists a function $f: \mathscr{E} \to X$ such that $f(E) \in E$ for each E in $\mathscr{E}$.*

Proof. Consider the class P of all functions $p: \mathscr{D}(p) \to X$ such that the domain $\mathscr{D}(p)$ of p is a subset of $\mathscr{E}$ and $p(E) \in E$ for each E in $\mathscr{D}(p)$. This is a nonempty class, because $\mathscr{E}$ contains a nonempty set E and if $x \in E$ the function with the domain $\{E\}$ and range $\{x\}$ is a member of P. We order P by the inclusion relation in $\mathscr{E} \times X$. It can be verified that P satisfies the conditions of Kuratowski–Zorn's lemma. Therefore we infer that there exists a function $f: \mathscr{E} \to X$ such that $f(E) \in E$ for each $E \in \mathscr{E}$ and Zermelo's principle of choice has been obtained. ■

To see that the last principle of choice implies the existence of selectors of set-valued functions let us observe that for a given set-valued function $F: T \to \mathscr{P}(X)$ we can define $\mathscr{E} = \{F(t)\}_{t \in T}$ and by Zermelo's principle of choice there exists $g: \mathscr{E} \to X$ such that $g(F(t)) \in F(t)$ for $t \in T$. Thus, $f(t) := g(F(t))$ is a selector for F.

We shall present here the basic theorems giving the existence of measurable selectors.

THEOREM 3.10 (Kuratowski–Ryll-Nardzewski). *If (X, ϱ) is a separable complete metric space and $F\colon T \to \mathscr{P}(X)$ is measurable and has closed values then F has a measurable selector.*

Proof. Let $\{x_1, x_2, \ldots\}$ be a countable dense subset in X and let $B_n(i) := \{x \in X\colon \varrho(x, x_i) \leqslant 1/n\}$ for $i, n \in N$. We shall now define inductively a sequence (F_n) of measurable set-valued functions such that $\bigcap_{n=1}^{\infty} F_n$ will be the desired selector.

Let $F_0 = F$ and define inductively $F_{n+1}(t) = F_n(t) \cap B_{n+1}(I_n(t))$, where $I_n(t) := \min\{i \in N\colon F(t) \cap B_{n+1}(i) \neq \emptyset\}$. For every $t \in T$, the sequence $(F_n(t))$ of $\mathrm{Cl}(X)$ is decreasing and the diameter of $F_n(t)$ tends to zero. Therefore, by Cantor's theorem, $\bigcap_{n=1}^{\infty} F_n(t)$ consists of exactly one point. Then we define $f(t) = \bigcap_{n=1}^{\infty} F_n(t)$ for $t \in T$. Clearly f is a selector of F. It remains to show that f is measurable. First we show that F_n is measurable, i.e., that $\{t \in T\colon F_n(t) \cap E \neq \emptyset\} \in \mathscr{F}$ for every closed set $E \subset X$. By assumption, F_0 is measurable. Assume F_n is measurable. Then

$$\{t \in T\colon F_{n+1}(t) \cap E \neq \emptyset\} = \{t \in T\colon F_n(t) \cap B_{n+1}(I_n(t)) \cap E \neq \emptyset\}$$

$$= \bigcup_{i=1}^{\infty} [\{t \in T\colon F_n(t) \cap B_{n+1}(i) \cap E \neq \emptyset\} \cap \{t \in T\colon I_n(t) = i\}].$$

But the last set belongs to $\mathscr{F}$ by induction hypothesis and since

$$\{t \in T\colon I_n(t) = i\}$$

$$= \bigcap_{j=1}^{i-1} [\{t \in T\colon F_n(t) \cap B_{n+1}(i) = \emptyset\} \cap \{t \in T\colon F_n(t) \cap B_{n+1}(i) \neq \emptyset\}]$$

belong to $\mathscr{F}$. Since the metric space X is complete one easily verifies that for every closed set $E \subset X$ we have $f^{-1}(E) = \bigcap_{n=0}^{\infty} \{t \in T\colon F_n(t) \cap E \neq \emptyset\}$ and hence $f^{-1}(E) \in \mathscr{F}$, which proves that the selection f is measurable. ■

We have also the following more general measurable selection theorem.

THEOREM 3.11. *Let $(T, \mathscr{F}, \mu)$ be a measure space and X a separable complete metric space. If $F\colon T \to \mathscr{P}(X)$ is such that* $\mathrm{Graph}\,F := \{(t, x)\colon t \in T, x \in F(t))$ *belongs to $\mathscr{F} \otimes \beta(X)$ then there exists a measurable function $f\colon T \to X$ such that $f(t) \in F(t)$ for a.e. $t \in T$.*

Proof. (Sketch of the proof in the case $T := [0, 1]$, $\mathscr{F} := \beta([0, 1])$, μ—Lebesgue measure). Let us observe that it sufficies only assume that the set $\mathrm{Graph}(F)$ is a analytic set, that is to say, a subset in $T \times X$ which is obtained by applying on the class $\mathscr{F} \to \beta(X)$ of measurable rectangles the Souslin's operation. Since the operation of countable union and countable intersection are both special cases of the Souslin's operation it follows that the product σ-algebra $\mathscr{F} \otimes \beta(X)$ is contained in the class of all $(\mathscr{F} \times \beta(X))$-analytic sets in $T \times X$.

Let $\Phi: P \to \text{Graph}(F)$ be a continuous mapping of a separable complete metric space P onto Graph(F), and define a multifunction $G: T \to \mathscr{P}(P)$ by $G(t) := \Phi^{-1}(\{t\} \times F(t))$ if $t \in T$. Clearly, G has closed values. Moreover, G is measurable if $\mathscr{F}$ is replaced by the completion $\mathscr{F}^{\sim}$ of $\mathscr{F}$. To see this let B be a closed subset of P. Then $G^{-}(B) = \{t\colon (\{t\} \times F(t)) \cap \Phi(B) \neq \emptyset\} = \Pi_T(\Phi(B))$, where Π_T is the projection of $T \times X$ onto T. Since $\Phi(B)$ is a Suslin set then $\Pi_T(\Phi(B)) \in \mathscr{F}^{\sim}$. Thus $G^{-}(B)$ is measurable with respect to $\mathscr{F}^{\sim}$. By Theorem 3.10 there is a selector $g: T \to P$ for G which is measurable with respect to $\mathscr{F}^{\sim}$. Then, if Π_X denotes projection of $T \times X$ onto X, $f = \Pi_X \circ \Phi \circ g\colon T \to X$ is a selector for F, measurable with respect to $\mathscr{F}^{\sim}$. Finally, by a routine use of Lusin's theorem, change the values of f on a set A of measure 0 to obtain a function f measurable with respect to $\mathscr{F}$ and satisfying $f(t) \in F(t)$ for all $t \in T \setminus A$. ■

We use now Theorem 3.10 to prove the general implicit function theorems. We begin with the following result.

PROPOSITION 3.3. *Let X be a separable metric space, Y a metric space, $f\colon T \times X \to Y$ measurable in t and continuous in x, and U an open subset of Y. Then a set-valued function $F\colon T \to \mathscr{P}(X)$ defined by $F(t) := \{x \in X\colon f(t, x) \in U\}$ is measurable. In particular if f is real-valued, then $t \to \{x\colon f(t, x) > \lambda\}$ and $t \to \{x\colon f(t, x) < \lambda\}$ are measurable.*

Proof. Let B be a closed subset of X and let A be a countable dense subset of B. The set $F^{-}(B)$ is measurable, since

$$\begin{aligned} F^{-}(B) &= \{t \in T\colon F(t) \cap B \neq \emptyset\} = \{t \in T\colon f(t, x) \in U \text{ for some } x \in B\} \\ &= \{t \in T\colon f(t, a) \in U \text{ for some } a \in A\} = \bigcup_{a \in A} \{t \in T\colon f(t. a) \in U\}. \end{aligned}$$ ■

Now, we obtain the following implicit function theorem.

THEOREM 3.12. *Let X be a separable complete metric space, (Y, d) a metric space, $f\colon T \times X \to Y$ a function measurable in t and continuous in x, $\Gamma\colon T \to \text{Comp}(X)$ a measurable set-valued function, and $g\colon T \to Y$ a measurable function such that $g(t) \in f(t, \Gamma(t))$ for $t \in T$. Then there exists a measurable selector $\gamma\colon T \to X$ for Γ such that $g(t) = f(t, \gamma(t))$ for all $t \in T$.*

Proof. Define $H\colon T \to \text{Comp}(X)$ *by setting* $H(t) := \Gamma(t) \cap \{x \in X\colon d(f(t, x), g(t)) = 0\}$. By virtue of Theorem 3.10, it suffices only to verify that H is measurable. To do this define $F_n\colon T \to \mathscr{P}(X)$ by $F_n(t) := \{x \in X\colon d(f(t, x), g(t)) < 1/n\}$ for $t \in T$ and $n \in N$. F_n is measurable for each n, by Proposition 3.3 and so $\overline{F_n}$ is weakly measurable. Clearly $\{x \in X\colon d(f(t, x), g(t)) = 0\} = \bigcap_{n=1}^{\infty} \overline{F_n(t)}$, because $\overline{F_n(t)} \subset \{x \in X\colon d(f(t, x), g(t)) \leqslant 1/n\}$ for each $n \in N$. Thus, by Remark 3.4, H is measurable. ■

In a similar way one obtains

THEOREM 3.13. *Let (X, ϱ) be a separable complete metric space, Γ: $T \to \mathrm{Comp}(X)$ and g: $T \to X$ measurable. There exists a measurable selector φ: $T \to X$ of Γ such that $\varrho(g(t), \varphi(t)) = \mathrm{dist}(g(t), \Gamma(t))$ for all $t \in T$.*

Proof. Define $f(t, x) := |\varrho(g(t), x) - \mathrm{dist}(g(t), \Gamma(t))|$ for $t \in T$ and $x \in X$. By virtue of Theorem 3.3, a real-valued function $T \ni t \to \mathrm{dist}(g(t), \Gamma(t)) \in R^+$ is measurable. Therefore, f is measurable in t and continuous in x. Let $H(t) := \Gamma(t) \cap \{x \in X: f(t, x) = 0\}$ for $t \in T$. Similarly as in the proof of Theorem 3.12 it is sufficient only to show that H is measurable. To end this define F_n: $T \to \mathscr{P}(X)$ by $F_n(t) := \{x \in X: f(t, x) < 1/n\}$ for $t \in T$ and $n \in N$. By virtue of Proposition 3.3, for each $n \in N$, F_n is measurable. Now, similarly as in the proof of Theorem 3.12 we infer that H is measurable. Therefore, by Theorem 3.10 there exists a measurable selector φ: $T \to X$ for H. Then $\varphi(t) \in \Gamma(t)$ and $f(t, \varphi(t)) = 0$, for all $t \in T$. Thus φ is a measurable selector for Γ such that $\varrho(g(t), \varphi(t)) = \mathrm{dist}(g(t), \Gamma(t))$ for all $t \in T$. ■

3.4. Aumann integral

Let I be a compact interval of the real line R and F: $I \to \mathrm{Comp}(R^n)$ be measurable. It will be also assumed that F is integrably bounded, i.e. that there exists a Lebesgue integrable function m: $I \to R$ such that $\|F(t)\| \leqslant m(t)$ for a.e. $t \in I$, where $\|F(t)\| := h(F(t), \{0\})$. It is easy to see that a measurable set-valued function F: $I \to \mathrm{Comp}(R^n)$ is integrably bounded if and only if a real-valued function $\|F(\cdot)\|$: $I \ni t \to \|F(t)\| \in R$ is L-integrable.

Given a measurable set-valued function F: $I \to \mathrm{Comp}(R^n)$ the set $\mathscr{F}(F)$ of all L-integrable selectors of F is said to be *subtrajectory integrals of F*. It follows immediately from Theorem 3.10, that for every measurable and integrably bounded set-valued function F: $I \to \mathrm{Comp}(R^n)$ we have $\mathscr{F}(F) \neq \emptyset$.

A measurable set-valued function F: $I \to \mathrm{Comp}(R^n)$ is said to be *Aumann integrable on I* if $\mathscr{F}(F) \neq \emptyset$. In this case we define the *Aumann integral* $\int_I F(t)dt$ of F over I by $\int_I F(t)dt := \{\int_I f(t)dt: f \in \mathscr{F}(F)\}$.

We will investigate here fundamental properties of the Aumann integral. We begin with the following lemmas.

LEMMA 3.14. *For every closed and convex set $A \subset R^n$ and $t_1, t_2 \in I$ with $t_1 < t_2$ we have $\int_{t_1}^{t_2} A\,dt = (t_2 - t_1)A$.*

Proof. It is obvious, in fact, that $(t_2 - t_1)A \subset \int_{t_1}^{t_2} A\,dt$. Let z be in $\int_{t_1}^{t_2} A\,dt$, i.e., $z = \int_{t_1}^{t_2} f(t)dt$ with a measurable function f: $[t_1, t_2] \to R^n$ such that $f(t) \in A$ for a.e. $t \in [t_1, t_2]$. By mean-value theorem for Lebesgue integrals it follows that there exists $\xi \in \overline{\text{co}}\{f(t): t_1 \leqslant t \leqslant t_2\} \subset A$ such that $z = (t_2 - t_1)\xi$. Therefore $z \in (t_2 - - t_1)A$. Thus $\int_{t_1}^{t_2} A\,dt \subset (t_2 - t_1)A$. ■

Let σ^r be the simplex in the real $(r+1)$-dimensional space $\boldsymbol{R}^{r+1}$, $\sigma^r = \{(\xi_0, \ldots, \xi^r) \in \boldsymbol{R}^{r+1}: 0 \leqslant \xi_i \leqslant 1, \sum_{i=0}^{r} \xi_i = 1\}$ and let $V(\sigma^r)$ be the set of $r+1$ vertices of σ^r. We will use the notation $L^\infty(I, \boldsymbol{R}^{r+1})$ to mean the topological product of $L^\infty(I, \boldsymbol{R}^1)$ taken with itself $r+1$ times. Then $u \in L^\infty(I, \boldsymbol{R}^{r+1})$ implies that each component $u_i \in L^\infty(I, \boldsymbol{R}^1)$.

LEMMA 3.15 *Let $Y(t)$ be an $n \times (r+1)$-matrix-valued function with components in $L(I, \boldsymbol{R}^1)$ and let $\Psi = \{u \in L^\infty(I, \boldsymbol{R}^{r+1}): u(t) \in \sigma^r \text{ for } t \in I\}$ and $\Psi_0 = \{u \in L^\infty(I, \boldsymbol{R}^{r+1}): u(t) \in V(\sigma^r) \text{ for } t \in I\}$. Then $\{\int_I Y(t)u(t)dt: u \in \Psi\} = \{\int_I Y(t)u(t)dt: u \in \Psi_0\}$ and both of these sets are compact and convex.*

Proof. Define $T: L^\infty(I, \boldsymbol{R}^{r+1}) \to \boldsymbol{R}^n$ by $T(u) = \int_I Y(t) \cdot u(t)dt$. Clearly Ψ is convex and bounded in the norm topology; hence if we can show Ψ is weak-* closed, it will be weak-* compact. Suppose u^0 is a weak-* limit of Ψ which does not belong to Ψ. Then there is a set $E \subset I$ of positive measure such that $u^0(t) \notin \sigma^r$ for $t \in E$. One may readily establish the existence of an $\varepsilon > 0$ and $\eta \in \boldsymbol{R}^{r+1}$ such that the inner product $\eta \cdot \xi \geqslant C$ if $\xi \in \sigma^r$ and $\eta \cdot u^0(t) < C - \varepsilon$ for t in a subset E_1 of E having positive measure $\mu(E_1)$. Define a function $w(t) = (w_0(t), \ldots, w_r(t))$ in $L^\infty(I, \boldsymbol{R}^{r+1})$ by

$$w_i(t) = \begin{cases} \eta_i/\mu(E_1) & \text{for } t \in E_1, \\ 0 & \text{for } t \notin E_1. \end{cases}$$

Then w separates u^0 and Ψ, contradicting u^0 being a weak-* limit of Ψ. Thus Ψ is closed, convex, and weak-* compact. It is easily seen that T is weak-* continuous, because the weak topology was defined so that the linear functionals which were continuous on given normed space X with its norm topology are still continuous when X has its weak topology. In particular if $T = (T_1, \ldots, T_n)$ is continuous linear mapping from X^*, with its topology, to $\boldsymbol{R}^n$ such that the components T_i of T are representable as elements of X, then T is continuous as a mapping of X^*, with the weak-* topology, to $\boldsymbol{R}^n$. Therefore, $T\Psi = \{Tu: u \in \Psi\}$ is a compact, convex subset of $\boldsymbol{R}^n$. Clearly $T\Psi_0 \subset T\Psi$. Similarly as in the proof of Lyapunov's theorem we can show that $T\Psi \subset T\Psi_0$. ■

A given set $K \subset L(I, \boldsymbol{R}^n)$ is said to be *decomposable* if for all $u, v \in K$ and each measurable set $E \subset I$ one has $\chi_E \cdot u + \chi_{I \setminus E} \cdot v \in K$, where χ_E and $\chi_{I \setminus E}$ denote the characteristic functions of E and $I \setminus E$, respectively.

It is clear that for every measurable and integrably bounded set-valued function $F: I \to \text{Comp}(\boldsymbol{R}^n)$, $\mathscr{F}(F)$ is decomposable.

We have the following lemma.

LEMMA 3.16. *If $K \subset L(I, R^n)$ is decomposable then the set $\mathcal{I}(K) = \{\int_I f(t)dt: f \in K\}$ is convex in R^n.*

Proof. Suppose $K \neq \emptyset$ and let $z_1, z_2 \in \mathcal{I}(K)$ and let $0 < \lambda < 1$. There are $f_1, f_2 \in K$ such that $z_1 = \int_I f_1(t)dt$ and $z_2 = \int_I f_2(t)dt$. Let $\mathcal{L}(I)$ be a family of all L-measurable subsets of I and put $\nu(E) := (\int_I f_1(t)dt, \int_I f_2(t)dt)$ for each $E \in \mathcal{L}(I)$. By Lyapunov's theorem, $\nu(\mathcal{L}(I))$ is a convex compact subset of R^{2n}. Since $(0, 0)$ and (z_1, z_2) belong to $\nu(\mathcal{L}(I))$ then also $(\lambda z_1, \lambda z_2) \in \nu(\mathcal{L}(I))$. Therefore there is $H \in \mathcal{L}(I)$ such that $(\lambda z_1, \lambda z_2) = \nu(H)$.

Let $f = \chi_H \cdot f_1 + \chi_{I \setminus H} \cdot f_2$. Since K is decomposable then $f \in K$. Therefore $\int_I f(t)dt \in \mathcal{I}(K)$. But $\int_I f(t)dt = \lambda z_1 + (1-\lambda)z_2$. Then $\lambda z_1 + (1-\lambda)z_2 \in \mathcal{I}(K)$. ■

COROLLARY 3.2. *If F: $I \to \mathrm{Comp}(R^n)$ is measurable and integrably bounded then $\int_I F(t)dt$ is a nonempty convex subset of R^n.* ■

COROLLARY 3.3. *If $K \subset L(I, R^n)$ is decomposable then a set-valued mapping $\mathcal{L}(I) \in E \to \mathcal{I}(K)(E) \subset R^n$, where $\mathcal{I}(K)(E) := \{\int_E f(t)dt: f \in K\}$ is convex-valued and additive, i.e., for every disjoint sets $A, B \in \mathcal{L}(I)$ one has $\mathcal{I}(K)(A \cup B) = \mathcal{I}(K)(A) + \mathcal{I}(K)(B)$.* ■

LEMMA 3.17. *Let F: $I \to \mathrm{Comp}(R^n)$ be measurable and integrably bounded. Then $\int_I F(t)dt = \int_I \mathrm{co}\, F(t)dt$ and both are nonempty, convex subset of R^n.*

Proof. Nonemptness and convexity follow from Corollary 3.2. We next show the equality. Certainly $\int_I F(t)dt \subset \int_I \mathrm{co}\, F(t)dt$. Suppose $y \in \int_I \mathrm{co}\, F(t)dt$. Then $y = \int_I f(t)dt$ for some $f \in \mathcal{F}(\mathrm{co}\, F)$. By Carathéodory's theorem for each $t \in I$ the point $f(t) \in \mathrm{co}\, F(t)$ may be written as a convex combination of $n+1$ points of $F(t)$; i.e., $f(t) = \sum_{i=0}^{n} \xi_i(t) f^i(t)$, $f^i(t) \in F(t)$, $0 \leqslant \xi\ (t) \leqslant 1$, $\sum_{i=0}^{n} \xi_i(t) = 1$. We let $\xi(t)$ denote the vector function $(\xi_0(t), \ldots, \xi_n(t)) \in \sigma^n$.

Let us observe that functions ξ_i and f^i can be chosen as measurable. Indeed, let $g(t, \xi, \beta^0, \ldots, \beta^n) := \sum_{i=0}^{n} \xi_i \cdot \beta^i$ and $\Gamma(t) := \sigma^{n+1} \times F(t) \times \ldots \times F(t)$ with $F(t)$ appearing $n+1$ times in the product. Since f is measurable and $f(t) \in g(t, \Gamma(t))$ for a.e. $t \in I$, then by Theorem 3.12 there is a measurable function $I \ni t \to (\xi_0(t), \ldots$ $\ldots, \xi_n(t), f^0(t), \ldots, f^n(t)) \in \Gamma(t)$ such that $f(t) = g(t, \xi_0, \ldots, \xi_n(t), f^0(t), \ldots, f^n(t))$ for a.e. $t \in I$.

Let the vectors $f^i(t)$ be the columns of an $n \times (n+1)$-matrix Y. By Lemma 3.15

there exists a measurable vector function $\xi^* = (\xi_0^*, \ldots, \xi_n^*)$ on I taking values in the vertices of the simplex σ^n such that $\int_I f(t)dt = \int_I Y(t)\xi(t)dt = \int_I Y(t)\xi^*(t)dt$. Now $\xi_i^*(I) \subset \{0, 1\}$ for all $i = 0, \ldots, n$ and $\sum_{i=0}^n \xi_i^*(t) = 1$. Let $I_i = \{t \in I\colon \xi^*(t) = 1\}$. Then I_i is measurable, $\bigcup_{i=0}^n I_i = I$, $I_i \cap I_j = \emptyset$ for $i \neq j$. Define $f^*(t) = f^i(t)$ for $t \in I_i$. Then f^* is measurable, $f^*(t) \in F(t)$ and $\int_I f^*(t)dt = \int_I f(t)dt$, showing $\int_I F(t)dt = \int_I \operatorname{co} F(t)dt$. ■

LEMMA 3.18. *If (G_n) is a sequence of set-valued functions $G_n\colon I \to \mathscr{P}(R^n)$ that are all bounded by the same integrable point-valued function $\lambda\colon I \to R$, then $\int_U \limsup_{n\to\infty} G_n(t)dt \supset \limsup_{n\to\infty} \int_U G_n(t)dt$ for every measurable set $U \subset I$.*

Proof. Suppose $x \in \limsup_{n\to\infty} \int_U G_n(t)dt$. Then x is a limit point of a sequence $\int_U g_n(t)dt$, where $g_n(t) \in G_n(t)$ for each n and a.e. $t \in U$; that is, there is a subsequence of $\int_U g_n(t)dt$ converging to x. We wish to show that $x \in \int_U \limsup_{n\to\infty} G_n(t)dt$; for this purpose we may assume without loss of generality that x is actually the limit of the $\{\int_U g_n(t)dt\}$, i.e. that the subsequence converging to x is the whole original sequence. Because the g_n are all bounded by the integrable function λ, then the sequence (g_n) is uniformly integrable. Therefore, by Dunford's theorem there is its any subsequence with a weak limit, which we call g.

Let $G(t) = \{g_1(t), g_2(t), \ldots\}$ for $t \in U$. Without loss of generality one may assume that $\lambda(t) < +\infty$ for each $t \in U$. Therefore, $\overline{G(t)} \in \operatorname{Comp}(R^n)$ for each $t \in U$. Hence, by Theorem 3.4, it follows that a set-valued function $U \ni t \to \overline{G(t)} \in \operatorname{Comp}(R^n)$ is measurable, because $\overline{G(t)} = \overline{\bigcup_{n=1}^\infty F_n(t)}$ for $t \in U$, where $F_n(t) = \{g_n(t)\}$ for $t \in U$ and $n = 1, 2, \ldots$ It is clear that $\overline{G}$ is also integrably bounded. Therefore, by virtue of Lemma 3.17 one has $\int_U \operatorname{co}[\overline{G(t)}]dt = \int_U \overline{G(t)}dt$. Hence, it follows that $\int_U g(t)dt \in \int_U \overline{G(t)}dt$, because by virtue of Banach–Mazur theorem we have $g(t) \in \operatorname{co}[\overline{G(t)}]$ for $t \in U$. Therefore, there is a Lebesgue integrable function $f\colon U \to R^n$ such that $f(t) \in \overline{G(t)}$ for a.e. $t \in U$ and so that $\int_U g(t)dt = \int_U f(t)dt$. But since $g_n(t) \in G_n(t)$ for a.e. $t \in U$, it follows that every limit point of $G(t)$ will be a member of $\limsup_{n\to\infty} G_n(t)$. Then $\int_U g(t)dt = \int_U f(t)dt \in \int_U \limsup_{n\to\infty} G_n(t)dt$. But from the weak convergence of (g_n) to g it follows that $\int_U g(t)dt = \lim_{n\to\infty} \int_U g_n(t)dt = x$, so that $x \in \int_U \limsup_{n\to\infty} G_n(t)dt$. ■

LEMMA 3.19. *If* $G\colon I \to \mathrm{Comp}(\boldsymbol{R}^n)$ *is measurable and integrably bounded, then* $\int_U G(t)\,dt$ *is a nonempty compact subset of* $\boldsymbol{R}$ *for every measurable set* $U \subset I$.

Proof. We have $\int_U G(t)\,dt \neq \emptyset$ for every measurable set $U \subset I$. Set $G_1 = G_2 = \ldots = G$. Then $\limsup_{n\to\infty} G_n(t) = \mathrm{cl}\, G(t) = G(t)$ and $\limsup_{n\to\infty} \int_U G_n(t)\,dt = \mathrm{cl} \int_U G(t)\,dt$ for a measurable set $U \subset I$. So by Lemma 3.18 we have

$$\int_U G(t)\,dt = \int_U \limsup_{n\to\infty} G_n(t)\,dt \supset \limsup_{n\to\infty} \int_U G_n(t)\,dt = \mathrm{cl} \int_U G(t)\,dt.$$

Therefore, $\int_U G(t)\,dt = \mathrm{cl} \int_U G(t)\,dt$. Since $\int_U G(t)\,dt$ is bounded by the integral of the function λ that bounds G, it follows that it is compact. ■

Now we can state the following Aumann's theorem.

THEOREM 3.20 (Aumann). *If* $F\colon I \to \mathrm{Comp}(\boldsymbol{R}^n)$ *is measurable and* integrably bounded then $\int_I F(t)\,dt = \int_I \mathrm{co}\, F(t)\,dt$ and both integrals are nonempty, convex, compact subsets of $\boldsymbol{R}^n$. ■

We shall prove the following theorem.

THEOREM 3.21. *Suppose that* $G\colon I \to \mathrm{Comp}(\boldsymbol{R}^n)$ *is measurable and that* $\int_U G(t)\,dt$ *is not empty. Then* $\int_U s(p, G(t))\,dt = s\big(p, \int_U G(t)\big)dt$ *for every* $p \in \boldsymbol{R}^n$ *and measurable set* $U \subset I$.

Proof. Let us observe that the integral $\int_U s(p, G(t))\,dt$ is well defined and may take $+\infty$ as its value. Indeed, by Theorem 3.8, the function $I \ni t \to s(p, G(t))$ is measurable on I for every $p \in \boldsymbol{R}^n$. Now, by the definition of $s(p, A)$ for a given nonempty set $A \subset \boldsymbol{R}^n$ we get

$$s\Big(p, \int_U G(t)\,dt\Big) \leqslant \int_U s(p, G(t))\,dt.$$

Consider a real number $\alpha < \int_U \mathscr{S}(t)\,dt$, where $\mathscr{S}(t) := s(p, G(t))$. We will show that there is an integrable function $g \in \mathscr{F}(G)$ such that $\alpha < p \cdot \int_U g(t)\,dt$. For this we choose an integrable selector $h \in \mathscr{F}(G)$ and consider for every integer n the truncated set-valued function G_n, $G_n(t) = \{x \in G(t)\colon |x - h(t)| < n\}$. By Proposition 3.3, G_n and therefore also $\overline{G}_n$, is measurable. Hence and Theorem 3.8 the function $\mathscr{S}_n\colon I \ni t \to s(p, G_n(t)) \in \boldsymbol{R}$ is measurable because $s(p, G_n(t)) = s(p, \overline{G_n(t)})$. It is also integrable, because h is integrable. Since $\mathscr{S}_n(t) \to \mathscr{S}(t)$ as $n \to \infty$ then we obtain $\lim_{n\to\infty} \int_U \mathscr{S}_n(t)\,dt = \int_U \mathscr{S}(t)\,dt$. Consequently, for n large enough we have

$\alpha < \int_U \mathscr{S}_n(t)dt$. Then there is an integrable function $f\colon I \to \boldsymbol{R}$ such that $\alpha < \int_U f(t)dt$ and $f(t) < \mathscr{S}_n(t)$ for $t \in I$. Let $F(t) = \{x \in \overline{G(t)}\colon p \cdot x > f(t)\}$. Clearly, $F(t) \neq \emptyset$ and Graph F is measurable. Consequently, by Theorem 3.11, there exists a measurable selection g of F, and hence of $\bar{G}$, which is even integrable. Since $f(t) < p \cdot g(t)$ we obtain $\int_U f(t)dt < p \cdot \int_U g(t)dt$ and consequently $\alpha < p \cdot \int_U g(t)dt$. Therefore, for every $\alpha < \int_U s(p, G(t))dt$ we obtain $\alpha < s\big(p, \int_U G(t)dt\big)$. In particular, taking for $n = 1, 2, \ldots$ $\alpha_n = \int_U s(p, G(t))dt - 1/n$ we get $\alpha_n < s\big(p, \int_U G(t)dt\big)$ for $n = 1, 2, \ldots$ Thus, we also have $\int_U s(p, G(t))dt \leqslant s\big(p, \int_U G(t)dt\big)$. ■

Let $F\colon I \to \mathrm{Comp}(\boldsymbol{R}^n)$ be such that for each $t \in I$ the function $h(F(\cdot), F(t))$ is L-integrable on I. A point $t \in I$ for which

$$\lim_{\eta \to 0} \eta^{-1} \int_t^{t+\eta} h\big(F(\tau), F(t)\big)d\tau = 0$$

is called a *Lebesgue point of F*.

A set-valued function $F\colon I \to \mathrm{Comp}(\boldsymbol{R}^n)$ is said to be *approximately continuous at* $t \in I$ if there exists a measurable set $B \subset I$ for which t is a point of density, i.e. such that $\lim_{h \to 0} \left(\frac{1}{2h}\right) \mu([t-h, t+h] \cap B) = 1$ and such that the restriction of F to B is continuous at t.

Immediately from Lusin–Pliś' theorem it follows that if $F\colon I \to \mathrm{Comp}(\boldsymbol{R}^n)$ is measurable, then F is approximately continuous a.e. on I. Indeed, let $\varepsilon > 0$ be given. By Lusin–Pliś' theorem there exists a closed set $B \subset I$ with $\mu(I \setminus B) < \varepsilon$ and such that the restriction of F to B is continuous. Let $D \subset B$ be the set of points of density of B. Then $\mu(D) = \mu(B) > \mu(I) - \varepsilon$. If $t \in D$, then F is approximately continuous at t, hence the set $H \subset I$ of all points of approximate continuity of F has inner measure greater then $\mu(I) - \varepsilon$. Since $\varepsilon > 0$ is arbitrary, the inner measure of H is greater or equal to $\mu(I)$. But $H \subset I$, hence its outer measure is less then or equal to $\mu(I)$ showing H is measurable with measure $\mu(I)$. We shall prove the following theorems.

THEOREM 3.22. *If $F\colon I \to \mathrm{Comp}(\boldsymbol{R}^n)$ is measurable and integrably bounded then almost all $t \in I$ are Lebesgue points of F.*

Proof. By Lusin–Pliś' theorem it easily follows that a function $h(F(\cdot), F(t))$ is measurable for each $t \in I$. Let $m\colon I \to \boldsymbol{R}$ be L-integrable function such that $\|F(t)\| \leqslant m(t)$ for a.e. $t \in I$. Without loss of generality one may suppose that $m(t) > 0$ on I. We have $h(F(\tau), F(t)) \leqslant m(\tau) + m(t)$ for all $t, \tau \in I$. Hence $h(F(\cdot), F(t))$ is L-integrable on I. Clearly, we can assume that almost all points of I are, at once, points of approximate continuity of F and Lebesgue points of m. Let t be such point and let $B \subset I$ be a measurable set for which t is a point of density and such that the restriction of F to B is continuous at t. For $\eta > 0$ set $B_1(\eta) = [t, t+\eta] \cap B$, $B_2(\eta) = [t, t+\eta] \cap (I \setminus B)$. Thus, given $\varepsilon > 0$, one may choose $\eta = \eta(\varepsilon, t) > 0$ sufficiently small that following three conditions are satisfied:

(i) for $\tau \in B_1(\eta)$, $h(F(\tau), F(t)) < \frac{1}{6}\varepsilon$,

(ii) $\mu(B_2(\eta)) < \frac{1}{6}\varepsilon\eta m(t)$,

(iii) $\int_t^{t+\eta} |m(\tau) - m(t)| d\tau < \frac{1}{3}\eta\varepsilon$.

Now, we have

$$\begin{aligned}
&\eta^{-1} \int_t^{t+\eta} h(F(\tau), F(t)) d\tau \\
&= \eta^{-1} \int_{B_1(\eta)} h(F(\tau), F(t)) d\tau + \eta^{-1} \int_{B_2(\eta)} h(F(\tau), F(t)) d\tau \\
&< \tfrac{1}{3}\varepsilon + \eta^{-1} \int_{B_2(\eta)} [\|F(\tau)\| + \|F(t)\|] d\tau \\
&< \tfrac{1}{3}\varepsilon + \eta^{-1} \int_t^{t+\eta} |m(\tau) - m(t)| d\tau + 2m(t)\eta^{-1}\mu(B_2(\eta)) < \tfrac{1}{3}\varepsilon + \tfrac{1}{3}\varepsilon + \tfrac{1}{3}\varepsilon = \varepsilon.
\end{aligned}$$

Thus, $\lim_{\eta \to 0+} \eta^{-1} \int_t^{t+\eta} h(F(\tau), F(t)) d\tau = 0$, and a similar argument shows that the left hand limit is also zero. ■

THEOREM 3.23. *If $F\colon I \to \mathrm{Comp}(R^n)$ is measurable and integrably bounded then for a.e. $t \in I$ we have* $\lim_{h \to 0} h\left(\eta^{-1} \int_t^{t+\eta} F(\tau) d\tau, \operatorname{co} F(t)\right) = 0$.

Proof. By virtue of Aumann's theorem it is sufficient only to show that $\lim_{\eta \to 0} h\left(\eta^{-1} \int_t^{t+\eta} \operatorname{co} F(\tau) d\tau, \operatorname{co} F(t)\right) = 0$ for a.e. $t \in I$.

By virtue of Lemma 3.14, for $\eta > 0$ we have $\eta \operatorname{co} F(t) = \int_t^{t+\eta} \operatorname{co} F(t) d\tau$. Therefore, for fixed $t \in I$ and $\eta > 0$ we obtain

$$h\left(\eta^{-1} \int_t^{t+\eta} \operatorname{co} F(\tau) d\tau, \operatorname{co} F(t)\right) = \eta^{-1} h\left(\int_t^{t+\eta} \operatorname{co} F(\tau) d\tau, \int_t^{t+\eta} \operatorname{co} F(t) d\tau\right).$$

Hence, by Lemmas 1.4, 1.5 and Theorem 3.21 one has

$$\begin{aligned}
&h\left(\eta^{-1} \int_t^{t+1} \operatorname{co} F(\tau) d\tau, \operatorname{co} F(t)\right) \\
&= \eta^{-1} \max\left\{\left| s\left(p, \int_t^{t+\eta} \operatorname{co} F(\tau) d\tau\right) - s\left(p, \int_t^{t+\eta} \operatorname{co} F(t) d\tau\right| : |p| = 1\right\}
\end{aligned}$$

$$= \eta^{-1}\max\left\{\left|\int_t^{t+\eta} s(p, \mathrm{co}\, F(\tau))d\tau - \int_t^{t+\eta} s(p, \mathrm{co}\, F(t))d\tau\right| : |p| = 1\right\}$$

$$\leqslant \eta^{-1}\int_t^{t+\eta} \max\{|s(p, \mathrm{co}\, F(\tau)) - s(p, \mathrm{co}\, F(t))| : |p| = 1\}d\tau$$

$$\leqslant \eta^{-1}\int_t^{t+\eta} h(\mathrm{co}\, F(\tau), \mathrm{co}\, F(t))d\tau \leqslant \eta^{-1}\int_t^{t+\eta} h(F(\tau), F(t))d\tau$$

because, by (ii) of Lemma 1.5, for every $A, B \in \mathrm{Comp}(\boldsymbol{R}^n)$ we have $h(\mathrm{co}\, A, \mathrm{co}\, B) \leqslant h(A, B)$. Hence, by Theorem 3.22, for a.e. $t \in I$ we have $\lim_{\eta\to 0} \eta^{-1}\int_t^{t+\eta} h(F(\tau), F(t))d\tau = 0$ and therefore also $\lim_{\eta\to 0} h\left(\eta^{-1}\int_t^{t+\eta} \mathrm{co}\, F(\tau)d\tau, \mathrm{co}\, F(t)\right) = 0$. ■

COROLLARY 3.4. *If $F: I \to \mathrm{Comp}(\boldsymbol{R}^n)$ is measurable and integrably bounded and $f: I \to \boldsymbol{R}^n$ is L-integrable function then $f(t) \in \mathrm{co}\, F(t)$ for a.e. $t \in I$ if and only if $\int_E f(\tau)d\tau \in \int_E F(\tau)d\tau$ for every measurable set $E \subset I$.*

Indeed, for each fixed $t \in I$ and $\eta > 0$ such that $t+\eta \in I$ we have

$$\mathrm{dist}\left(f(t), \mathrm{co}\, F(t)\right) \leqslant \left|f(t) - \eta^{-1}\int_t^{t+\eta} f(\tau)d\tau\right| +$$

$$+ \mathrm{dist}\left(\eta^{-1}\int_t^{t+\eta} f(\tau)d\tau, \eta^{-1}\int_t^{t+\eta} F(\tau)d\tau\right) + h\left(\eta^{-1}\int_t^{t+\eta} F(\tau)d\tau, \mathrm{co}\, F(t)\right).$$

Hence, by Theorem 3.23, it follows that $\mathrm{dist}(f(t), \mathrm{co}\, F(t)) = 0$ for a.e. $t \in I$. ■

Finally, we present here Radon–Nikodym theorem for set-valued measures. Let $(T, \mathscr{F}, \lambda)$ be a finite nonnegative measure space and let Φ be a set-valued measure on $(T, \mathscr{F})$ such that $\Phi \ll \lambda$. The set-valued function F from T to the subset of $\boldsymbol{R}^n$, is a *Radom–Nikodym derivative of Φ with respect to λ* if $\int_E F(t)d\lambda(t) = \Phi(E)$ for every $E \in \mathscr{F}$. For two set-valued measures we write $\Psi_1 \subset \Psi_2$ if $\Psi_1(E) \subset \Psi'_2(E)$ for every $E \in \mathscr{F}$. Let Φ be a fixed set-valued measure, with convex values. Denote by $\boldsymbol{M}$ the collection of all set-valued measures Ψ with convex values such that $\mathrm{cl}\,\Psi(E) = \mathrm{cl}\,\Phi(E)$ for every E. Denote by $\hat{\Phi}$ and $\check{\Phi}$, respectively, the greatest and smallest elements in $\boldsymbol{M}$ with respect to the order $\subset$, if such elements exist. We have the following theorem.

THEOREM 3.24. *Let $(T, \mathscr{F}, \lambda)$ be a finite, nonnegative measure space. Let Φ be a set-valued measure with convex values such that $\Phi \ll \lambda$. Then $\boldsymbol{M}$ has a greatest element Φ such that Φ has a measurable Radon–Nikodym derivative with closed and convex values.* ■

COROLLARY 3.5. *Let $(T, \mathscr{F}, \lambda)$ be a finite, nonnegative measure space and Φ be a set-valued measure with closed convex values such that $\Phi \ll \lambda$. Then Φ has a measurable Radon–Nikodym derivative with closed and convex values.* ■

4. CONTINUOUS SELECTIONS AND MULTIVALUED FIXED POINT THEOREMS

The fundamental problem deals with the existence of continuous selections. The following example shows that continuous set-valued functions need not have, in general, continuous selections.

EXAMPLE 4.1. Let F be a set-valued function defined on the interval $(-1, 1)$ by setting

$$F(t) = \begin{cases} \left\{(v_1, v_2)\colon v_1 = \cos\theta, v_2 = t\sin\theta \text{ and } \dfrac{1}{t} \leqslant \theta \leqslant \dfrac{1}{t} + 2\pi - |t|\right\} \\ \qquad\qquad\qquad\qquad\qquad\qquad\qquad \text{for } t \in (-1, 1)\setminus\{0\}, \\ \{(v_1, v_2)\colon -1 \leqslant v_1 \leqslant 1,\ v_2 = 0\} \qquad \text{for } t = 0. \end{cases}$$

For $t \neq 0$ and $t \in (-1, 1)$, $F(t)$ is a subset of an ellipse in $\boldsymbol{R}^2$, whose small axis shrinks to zero as $t \to 0$, so that the ellipse collapses to a segment, $F(0)$. The subset of the ellipse given by $F(t)$ is obtained by removing from it a section, from the angle $1/t - |t|$ to the angle $1/t$. As t gets smaller, the arc length of this hole decreases while the initial angel increases as $1/t$, i.e., it spins around the origin with increasing angular speed. However F is continuous at the origin while any continuous selection u: $(-1, 0) \to \boldsymbol{R}^2$ or u: $(0, +1) \to \boldsymbol{R}^2$ (for instance, $v_1 = \cos(1/t)$, $v_2 = t\sin(1/t)$ could not be continuously extended to the whole interval $(-1, 1)$. In fact the hole in the ellipse would force this selection to rotate around the origin with an angle $\varrho(t)$ between $1/t$ and $1/t + 2\pi - |t|$ and $\lim_{t \to 0} u(t)$ cannot exist. ■

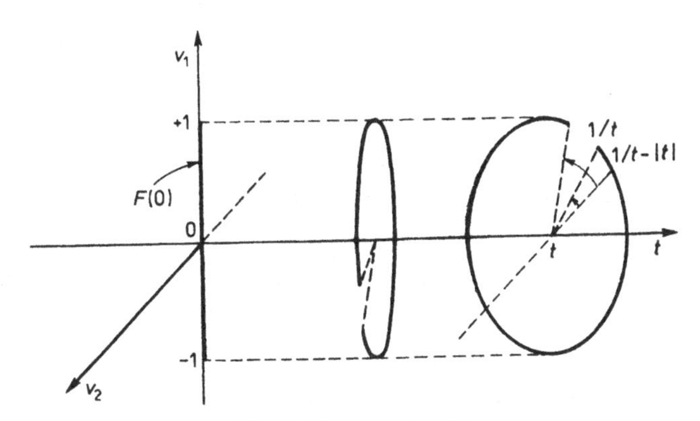

Fig. 3. The mapping F.

4.1. Michael's selection theorem

We have the following famous continuous selection theorem.

THEOREM 4.1 (Michael). *Let X be a metric space, Y a Banach space and F from X*

into the closed convex subsets of Y be l.s.c. on X. Then there exists f: $X \to Y$, a continuous selector of F.

Proof. We shall prove first that for any convex-valued (not necessarily closed valued) lower semicontinuous set-valued function Φ: $X \to \mathscr{P}(Y)$ and every $\varepsilon > 0$, there exists a continuous function $\mathscr{S}$: $X \to Y$ such that for ξ in X, $\operatorname{dist}(\mathscr{S}(\xi), \Phi(\xi)) \leqslant \varepsilon$.

Indeed, for every $x \in X$, let $y_x \in \Phi(x)$ and let $\delta_x > 0$ be such that $(y_x+\varepsilon B^o)\cap \cap\Phi(x') \neq \emptyset$ for $x' \in B^o(x, \delta_x)$, where $B^o(x, \delta_x)$ denotes an open ball of X centred at x with a radius $\delta_x > 0$ and B^o is an open unit ball of Y centred at 0. Since X is paracompact then there exists a locally finite refinement $\{\mathscr{U}_x\}_{x\in\Lambda}$ of $\{B^o(x, \delta_x)\}_{x\in\Lambda}$. Let $\{\Pi_x(\cdot)\}_{x\in\Lambda}$ be a partition of unity subordinate to it. The mapping $\mathscr{S}$: $X \to Y$ given by $\mathscr{S}(\xi) = \sum_{x\in\Lambda} \Pi_x(\xi) y_x$ is continuous since it is locally a finite sum of continuous functions. Fix ξ. Whenever $\Pi_x(\xi) > 0$, $\xi \in \mathscr{U}_x \subset B^o(x, \delta_x)$, hence $y_x \in \Phi(\xi)+\varepsilon B^o$. Since this latter set is convex, any convex combination of such y's (in particular, $\mathscr{S}(\xi)$) belongs to it.

Take now $\Phi = F$ and $\varepsilon = \frac{1}{2}$ and let f_1: $X \to Y$ be continuous function such that $\operatorname{dist}(f_1(x), F(x)) \leqslant \frac{1}{2}$. Put $\Phi_1(x) := [f_1(x)+\frac{1}{2} B^o]\cap F(x)$ for $x \in X$. Obviously $\Phi_1(x) \neq \emptyset$ and Φ_1 is a convex subset of Y. By virtue of Proposition 2.4, the mapping $\xi \to \Phi_1(\xi)$ is l.s.c. on X. Therefore, for $\varepsilon = (\frac{1}{2})^2$ there is a continuous function f_2: $X \to Y$ such that $\operatorname{dist}(f_2(x), \Phi_1(x)) \leqslant (\frac{1}{2})^2$ for $x \in X$. Thus $\operatorname{dist}(f_2(x), F(x)) \leqslant (\frac{1}{2})^2$ and $\operatorname{dist}(f_2(x), [f_1(x)+\frac{1}{2} B^o]) \leqslant (\frac{1}{2})^2$, i.e. $f_2(x)-f_1(x) \in (\frac{1}{2}+(\frac{1}{2})^2)B^o$ for $x \in X$.

Let $\Phi_2(x) = [f_2(x)+(\frac{1}{2})^2 B^o]\cap F(x)$ for $x \in X$. Similarly as above we deduce that there exists a continuous function f_3: $X \to Y$ such that $\operatorname{dist}(f_3(x), F(x)) \leqslant (\frac{1}{2})^3$ and $f_3(x)-f_2(x) \in (\frac{1}{2})^2+(\frac{1}{2})^3)B$ for $x \in X$.

Continuing this procedure we can find a sequence (f_n) of continuous functions f_n: $X \to Y$ such that $\operatorname{dist}(f_n(x), F(x)) \leqslant (\frac{1}{2})^{n+1}$ and $f_{n+1}(x)-f_n(x) \in ((\frac{1}{2})^n+(\frac{1}{2})^{n+1})B^o$, i.e., $\|f_{n+1}(x)-f_n(x)\| \leqslant (\frac{1}{2})^{n-1}$ for $x \in X$, where $\|\cdot\|$ is the norm of Y.

Since the series $\sum_0^\infty (\frac{1}{2})^n$ converges, (f_n) is a Cauchy sequence of the Banach space $C(X, Y)$. Thus there is a continuous function f: $X \to Y$ such that $f_n(x) \to f(x)$ uniformly on X as $n \to \infty$. Since $F(x)$ is closed and $\operatorname{dist}(f(x), F(x)) = 0$ for $x \in X$ then $f(x) \in F(x)$ for $x \in X$. ■

COROLLARY 4.1. *If X, Y and F are such as in Theorem 4.1 and furthermore $0 \in F(\bar{x})$ for $\bar{x} \in X$ then there exists a continuous function f: $X \to Y$ such that $f(x) \in F(x)$ for $x \in X$ and $f(\bar{x}) = 0$.*

Indeed, let G: $X \to \mathscr{P}(Y)$ be a set-valued function defined by

$$G(x) = \begin{cases} F(x) & \text{for } x \in X\setminus\{\bar{x}\}, \\ \{0\} & \text{for } x = \bar{x}. \end{cases}$$

Since G satisfies the assumptions of Theorem 4.1 then there exists a continuous function f: $X \to Y$ such that $f(x) \in G(x)$, i.e. such that $f(x) \in F(x)$ for $x \in X$ and $f(\bar{x}) = 0$. ■

We shall need the following results.

PROPOSITION 4.1. *Let $(X, |\cdot|)$ and $(Y, \|\cdot\|)$ be Banach spaces, F from X into the closed convex subsets of Y be l.s.c. and H-u.s.c. on X. Then for every continuous function $g\colon X \to Y$ and a $\lambda > 1$, there exists $f\colon X \to Y$, a continuous selection of F such that $\|f(x)-g(x)\| \leqslant \lambda \operatorname{dist}(g(x), F(x))$ for $x \in X$.*

Proof. Let $\varepsilon(x) := \lambda \operatorname{dist}(g(x), F(x))$ for $x \in X$. By Proposition 2.1, it follows that a real-valued function $X \ni x \to \varepsilon(x) \in \mathbf{R}^+$ is lower semicontinuous. By Proposition 2.4, a set-valued function Φ defined on X by $\Phi(x) := B^\circ[g(x), \varepsilon(x)] \cap F(x)$ is l.s.c. on X. Then $\bar{\Phi}(x) := \overline{B^\circ[g(x), \varepsilon(x)] \cap F(x)}$ is also l.s.c. on X. Therefore, by Theorem 4.1, there exists a continuous function $f\colon X \to Y$ such that $f(x) \in \bar{\Phi}(x) = \overline{B^\circ[g(x), \varepsilon(x)] \cap F(x)} \subset \overline{B^\circ[g(x), \varepsilon(x)]} \cap F(x)$ for $x \in X$. Then $f\colon X \to Y$ is a continuous selection of F such that $\|f(x)-g(x)\| \leqslant \varepsilon(x) := \lambda \operatorname{dist}(g(x), F(x))$. ∎

LEMMA 4.2. *Let $(X, |\cdot|)$ and $(Y, \|\cdot\|)$ be Banach spaces, S a nonempty bounded closed and convex subset of X and $\mathrm{CCl}(S)$ a family of all nonempty closed convex subset of S. Suppose $F\colon Y \times S \to \mathrm{CCl}(S)$ is l.s.c. and H-u.s.c. on $Y \times S$. If furthermore, $F(y, \cdot)$ is a contraction uniformly with respect to $y \in Y$, i.e., if there is a number $K < 1$ such that $h(F(y,u), F(y,v) \leqslant K|u-v|$ for each fixed $y \in Y$ and $u, v \in S$, then there exists a continuous function $f\colon Y \to S$ such that $f(y) \in F(y, f(y))$ for each $y \in Y$.*

Proof. Suppose $K \in (0, 1)$ is such that $h(F(y, x_1), F(y, x_2)) \leqslant K|x_1 - x_2|$ for $y \in Y$ and every $x_1, x_2 \in S$. Let $K_0 \in (K, 1)$ and put $\lambda = K^{-1}K_0$ and let $f_0\colon Y \to S$ be a continuous mapping. By Proposition 4.1, there exists a continuous selection $f_1\colon Y \to X$ of a set-valued mapping $Y \ni y \to F(y, f_0(y))$ such that

$$|f_1(y) - f_0(y)| \leqslant \lambda \operatorname{dist}\big(f_0(y), F\big(y, f_0(y)\big)\big)$$

for $y \in Y$. Since $f_1(y) \in F(y, f_0(y) \subset S$ then we can again apply Proposition 4.1 to mappings $f_1\colon Y \to S$ and $Y \ni y \to F(y, f_1(y)) \subset S$. Continuing this procedure we define a sequence (f_n) of continuous functions $f_n\colon Y \to S$ such that $f_n(y) \in F(y, f_{n-1}(y))$ and $|f_n(y) - f_{n-1}(y)| \leqslant \lambda \operatorname{dist}(f_{n-1}(y), F(y, f_{n-1}(y))$ for $y \in Y$ and $n = 1, 2, \ldots$ Hence $|f_{n+1}(y) - f_n(y)| \leqslant \lambda h(F(y, f_n(y)), F(y, f_{n-1}(y))) \leqslant (\lambda K)^n |f_1(y) - f_0(y)|$ for $n = 1, 2, \ldots$ and $y \in Y$. Since $f_1(y), f_0(y) \in S$ for $y \in Y$ and S is bounded, there exists $M > 0$ such that $|f_1(y) - f_0(y)| \leqslant M$ for $y \in Y$. Thus $|f_{n+1}(y) - f_n(y)| \leqslant K_0^n M$, where $K_0 \in (K, 1)$. Then there exists a continuous function $f\colon Y \to S$ such that $\sup\{|f_n(y) - f(y)|\colon y \in Y\} \to 0$ as $n \to \infty$. For every $y \in Y$ we have

$$\begin{aligned}
&\operatorname{dist}\big(f(y), F\big(y, f(y)\big)\big) \leqslant |f(y) - f_n(y)| + \operatorname{dist}\big(f_n(y), F\big(y, f(y)\big)\big) \\
&\leqslant |f(y) - f_n(y)| + h\big(F\big(y, f_{n-1}(y)\big), F\big(y, f(y)\big)\big) \leqslant |f(y) - f_n(y)| \\
&\quad + K|f_{n-1}(y) - f(y)|.
\end{aligned}$$

Hence it follows $\operatorname{dist}(f(y), F(y, f(y))) = 0$ for $y \in Y$. Since $F(y, f(y))$ is closed then $f(y) \in F(y, f(y))$ for every $y \in Y$. ∎

There are closed convex-valued upper semicontinuous mappings that do not possess continuous selections.

EXAMPLE 4.2. Let F be a set-valued function defined on $\boldsymbol{R}$ by setting

$$F(x) = \begin{cases} \{-1\} & \text{for } x < 0, \\ \langle -1, 1\rangle & \text{for } x = 0, \\ \{+1\} & \text{for } x > 0. \end{cases}$$

It is clear that there is no continuous real-valued function defined on the real line $\boldsymbol{R}$ that is a selector of F. ■

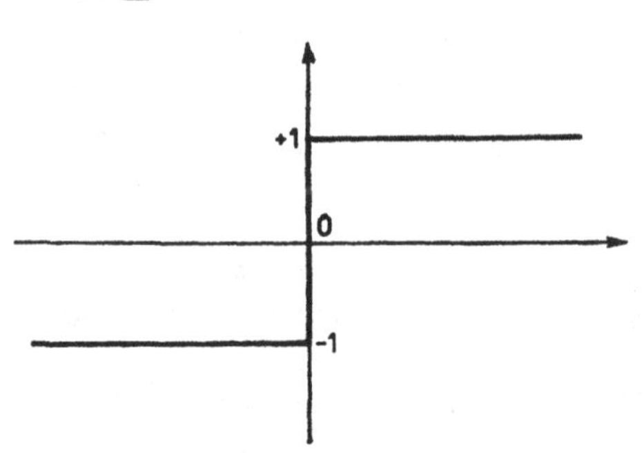

Fig. 4. The mapping F.

We have the following continuous approximate selection theorem.

THEOREM 4.3. *Let $(X, |\cdot|)$ be a normed space, $(Y, \|\cdot\|)$ a Banach space, F a set-valued function from X into the convex subsets of Y, upper semicontinuous on X. Then for every $\varepsilon > 0$ there exists a locally Lipschitzean function f_ε: $X \to Y$ such that $f_\varepsilon(X) \subset \operatorname{co} F(X)$ and* $\operatorname{Graph}(f_\varepsilon) \subset \operatorname{Graph}(F)+\varepsilon\beta^o$, *where β^o is an open unit ball of $X \times Y$.*

Proof. Fix $\varepsilon > 0$. For every $x \in X$ there exists $\delta(x) > 0$ such that for $x^* \in B^o(x, \delta(x))$ we have $F(x^*) \subset F(x)+\frac{1}{2}\varepsilon B^o$, where $B^o(x, \delta(x))$ denotes an open ball of X with a radius $\delta(x)$ and centred at x and B^o is an open unit ball of Y. We can take $\delta(x) < \frac{1}{2}\varepsilon$.

The family $\{B^o(x, \delta(x)/4\}_{x\in X}$ covers the paracompact space X. Let $\{U_i\}_{i\in I}$ be a locally finite refinement and $\{p_i(\cdot)\}$ a locally Lipschitzean partition of unity subordinate to it. Choose for each $i \in I$ an $\bar{x}_i \in U_i$ and define f_ε by setting

$$f_\varepsilon(x) := \sum_{i\in I} p_i(x) m_i$$

for $x \in X$, where $m_i \in F(\bar{x}_i)$. It is obvious that f_ε is well defined and locally Lipschitzean and such that $f_\varepsilon(X) \subset \operatorname{co} F(X)$.

Fix $x \in X$. Then $p_i(x)$ is strictly positive only for a finite subset $I(x) \subset I$. For $i \in I(x)$, set x_i to be such that $U_i \subset B^o(x_i, \frac{1}{4}\delta(x_i)$. Set $\delta_i := \delta(x_i)$ and let $j \in I(x_i)$ be such that $\delta_j := \max_{i\in I(x)} \delta_i$. Then $x_i \in B^o(x_j, \frac{1}{2}\delta_j)$ and thus $U_j \subset B^o(x_j, \delta_j)$. Therefore, for any $i \in I(x)$ one has $m_i \in F(U_i) \subset F(B^o(x_j, \delta_j)) \subset F(x_j)+\frac{1}{2}\varepsilon B^o$.

Since the latter set is convex, we deduce that $f_\varepsilon(x) \in F(x_j)+\frac{1}{2}\varepsilon B^o$. So there exists $y_j \in F(x_j)$ such that $\|f_\varepsilon(x)-y_j\| \frac{1}{2}\varepsilon$. Then

$$\mathrm{dist}[(x, f_\varepsilon(x)),\ (x_j, y_j)] \leqslant d(x, x_j)+\|f_\varepsilon(x)-y_j\| \leqslant \varepsilon,$$

i.e., $(x, f_\varepsilon(x)) \in \mathrm{Graph}(F)+\varepsilon\beta^\circ$ for every $x \in X$. Thus $\mathrm{Graph}(f_\varepsilon) \subset \mathrm{Graph}(F)+$ $+\varepsilon\beta^\circ$. ■

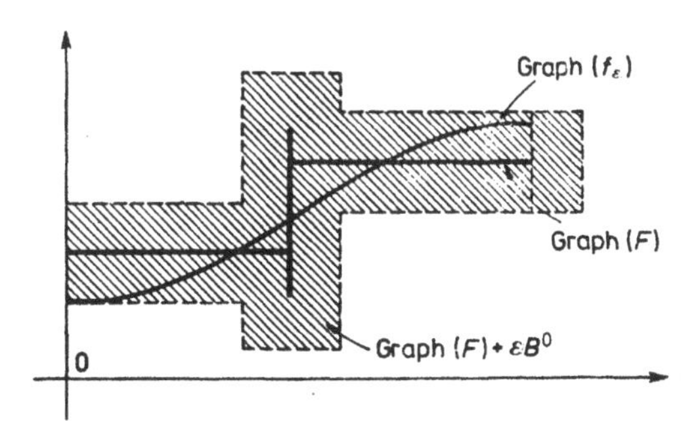

Fig. 5. Approximate selection of a set-valued mapping.

4.2. *Multivalued fixed point theorems*

We give here some fixed point theorems for set-valued mappings. We begin with the following Covitz–Nadler fixed point theorem.

THEOREM 4.4 (Covitz–Nadler). *Let (X, ϱ) be a complete metric space and $F\colon X \to \mathrm{Cl}(X)$ a set-valued contraction mapping, i.e. such that $h(F(x), F(y)) \leqslant K\varrho(x, y)$ for very $x, y \in X$ with $K \in [0, 1)$. Then there exists $x \in X$ such that $x \in F(x)$.*

Proof. Let $L \in (K, 1)$ and $\lambda = K^{-1}L$. Now for some $x \in X$ we have $\overline{B^\circ(x, \lambda \mathrm{dist}(x, F(x)))}\cap F(x) \neq \emptyset$. Then we can select $x_1 \in F(x)$ such that $\varrho(x, x_1) \leqslant \lambda \mathrm{dist}(x, F(x))$. Then for such $x_1 \in X$ select $x_2 \in F(x_1)$ such that $\varrho(x_1, x_2) \leqslant \lambda \mathrm{dist}(x_1, F(x_1))$. Continuing this procedure we can find a sequence (x_n) of X such that $\varrho(x_n, x_{n+1}) \leqslant \lambda \mathrm{dist}(x_n, F(x_n))$ for $n = 1, 2, \ldots$ Hence it follows

$$\varrho(x_n, x_{n+1}) \leqslant \lambda \mathrm{dist}(x_n, F(x_n)) \leqslant \lambda h(F(x_{n-1}), F(x_n))$$
$$\leqslant L\varrho(x_{n-1}, x_n) \leqslant L^n \mathrm{dist}(x, F(x)).$$

Now, similarly as in the proof of Banach fixed point theorem we can easily verify that the sequence (x_n) defined above has a limit, say z, belonging to X. Since F is H-continuous and

$$\mathrm{dist}(z, F(z)) \leqslant \varrho(z, x_n)+\mathrm{dist}(x_n, F(x_{n+1}))+h(F(x_{n+1}), F(z))$$

for $n = 1, 2, \ldots$, then $z \in F(z)$. ■

Immediately from Schauder's fixed point theorem and Theorem 4.3 we obtain the following Kakutani's fixed point theorem.

THEOREM 4.5 (Kakutani). *Let X be a Banach space, S a nonempty compact convex subset of X and $\mathrm{CCl}(S)$ a family of all nonempty closed convex subsets of S. If $F\colon S \to \mathrm{CCl}(S)$ is u.s.c. on S, then there is $x \in S$ such that $x \in F(x)$.*

Proof. Let β^o be an open unit ball of $X \times X$ and $\{f_n\}$ be a sequence of continuous functions from S to S such that $\text{Graph}(f_n) \subset \text{Graph}(F_n) + \varepsilon B^o$, where (ε_n) is a real-valued sequence converging to zero as $n \to \infty$. By Schauder's fixed point theorem there are $x_n \in S$ such that $x_n = f_n(x_n)$ for each $n \in N$. By the compactness of S there is a subsequence, say (x_k) of (x_n) converging to any $x \in S$. Thus $\text{dist}[(x, x), \text{Graph}(F)] = \lim_{k \to \infty} \text{dist}[(x_k, f_k(x_k)), \text{Graph}(F)] = 0$. By Proposition 2.2, hence it follows $(x, x) \in \text{Graph}(F)$, i.e., $x \in F(x)$. ■

A much more sophisticated fixed point theorem due to Kakutani and Ky Fan.

THEOREM 4.6 (Kakutani–Ky Fan). *Let X be a locally convex topological Hausdorff space, S a nonempty compact convex subset of X and* CCl(S) *a family of all nonempty closed convex subsets of S. If F: $S \to \text{CCl}(S)$ is u.s.c. on S then there is $x \in S$ such that $x \in F(x)$.* ■

Now as a corollary of Kakutani–Ky Fan's fixed point theorem we obtain the following its weak form.

THEOREM 4.7. *Let X be a Banach space and let S be weakly compact convex subset of X. Suppose F is a set-valued mapping of S into the space of all nonempty weakly closed convex subsets of S. If F is w-w.u.s.c. on S then there exists $x \in S$ such that $x \in F(x)$.*

Proof. Let us observe that the existence of a fixed point of F will follow immediately from Kakutani–Ky Fan's theorem if we are able to show, that F is u.s.c. on S as a mapping defined on a subset of a locally convex topological Hausdorff space $(X, \sigma(X, X^*))$, where as usual $\sigma(X, X^*)$ denotes the weak topology of X. By virtue of Theorem 2.1 it suffices only to show that for every weakly closed subset B of S, the set $F^-(B)$ is weakly closed.

Let B be a weakly closed subset of S. Since F is w-w.u.s.c. on S, $F^-(B)$ is sequentially weakly closed. But $F^-(B) \subset S$ and S is weakly compact. Then $F^-(B)$ is relatively sequentially compact and hence by, Eberlein–Šmulian's theorem, $\overline{[F^-(B)]}^w$ is weakly compact.

Let $x \in \overline{[F^-(B)]}^w$. By Šmulian's theorem there exists a sequence, say (x_n), of $F^{-1}(B)$ weakly converging to x. But $F^-(B)$ is sequentially weakly closed. Then $x \in F^-(B)$, i.e. $\overline{[F^-(B)]}^w \subset F^-(B)$. Thus for every weakly closed set $B \subset S$, $F^-(B)$ is weakly closed. Then F is u.s.c. from compact convex set S of locally convex topological Hausdorff space $(X, \sigma(X, X^*))$ into a space its nonempty compact convex subsets. Thus, by Kakutani–Ky Fan's fixed point theorem there is at least one fixed point of F. ■

COROLLARY 4.2. *If Λ is a nonempty convex weakly compact subset of the Banach space X and G: $\Lambda \to \Lambda$ is w-w.s.c. on Λ, i.e. such that for every $x \in \Lambda$ and every sequence (x_n) of Λ weakly converging to x a sequence $\{G(x_n)\}$ is also weakly converging to $G(x)$, then there is $x \in \Lambda$ such that $x = G(x)$.*

Indeed, by Theorem 2.5, the set-valued function F defined on Λ by setting $F(x) = \{G(x)\}$ is w.-w.u.s.c. on Λ. It takes also weakly compact and convex values. Then, by Theorem 4.7, there is $x \in \Lambda$ such that $x = G(x)$. ■

Now, we shall prove some extension of Krasnosielski's fixed point theorem.

THEOREM 4.8. *Let X and Y be Banach spaces, S a nonempty bounded closed and convex subset of X and* $\mathrm{CCl}(S)$ *a family of all nonempty closed convex subsets of S. Assume that F: $Y \times S \to \mathrm{CCl}(S)$ is l.s.c. and H-u.s.c. on $Y \times S$. If, furthermore, F is such that $F(y, \cdot)$ is a contraction uniformly with respect to $y \in Y$ and Γ: $S \to Y$ is completely continuous then there exists $x \in S$ such that $x \in F(\Gamma(x), x)$.*

Proof. By Lemma 4.2 there exists a continuous function f: $Y \to S$ such that $f(y) \in F(y, f(y))$ for $y \in Y$. Let $g(x) = f(\Gamma(x))$ for $x \in S$. It is easy to see that the mapping $S \ni x \to g(x) \in S$ is completely continuous, since Γ is completely continuous and f is continuous. Now, by Schauder's fixed point theorem there exists $x \in S$ such that $x = g(x)$. Then we have $x = g(x) := f(\Gamma(x)) \in F(\Gamma(x), f(\Gamma(x)) = F(\Gamma(x), g(x)) = F(\Gamma(x), x)$. ■

Finally, we shall prove a fixed point theorem for σ-selectionable set-valued functions.

Given normed linear spaces X and Y we say that F: $X \to \mathscr{P}(Y)$ is *σ-selectionable* if there exists a decreasing sequence (F_n) of u.s.c. mappings F_n: $X \to \mathrm{Comp}(Y)$ such that F_n has for each $n = 1, 2, \dots$ a continuous selection and for each $x \in X$, $F(x) = \bigcap_{n=1}^{\infty} F_n(x)$.

LEMMA 4.9. *Let $(X, \|\cdot\|)$ be a Banach space, K a compact convex subset of X, f a continuous mapping from K into X. Then there is $x \in K$ such that $\|f(x) - x\| = \mathrm{dist}(f(x), K)$.*

Proof. Let $H := f(K)$. Given $z \in H$ let $B_K(z) = \{y \in X\colon \|y - z\| \leqslant \mathrm{dist}(z, K)\}$. Observe that for every $z \in H$ there is $x_z \in K$ such that $\|x_z - z\| = \mathrm{dist}(z, K)$, i.e. such that $x_z \in B_K(z)$. Then for every $z \in H$ we have $K \cap B_K(z) \neq \emptyset$. Define now a set-valued mapping Π_K: $K \to \mathrm{Conv}(K)$ by taking $\Pi_K(x) := K \cap B_K(f(x))$ for each $x \in K$. It is easy to see that Π_K sends each $x \in K$ into the space $\mathrm{Conv}(K)$ of all nonempty compact and convex subsets of K. We shall show that Π_K is u.s.c. on K. Indeed, let x be any point of K and (x_n) any sequence of K converging to x. Suppose furthermore (z_n) is a sequence of K such that $z_n \in \Pi_K(x_n)$ for $n = 1, 2, \dots$ By the compactness of K there is a subsequence, say (z_k) of (z_n) and $z \in K$ such that $z_k \in \Pi_K(x_k)$ and $z_k \to z$ as $k \to \infty$. In particular, for every $k = 1, 2, \dots$ we have $\|z_k - f(x_k)\| \leqslant \mathrm{dist}(f(x_k), K)$. Hence, by the continuity of f it follows $\|z - f(x)\| \leqslant \mathrm{dist}(f(x), K)$ i.e. $z \in B_K(f(x))$. Then, $z \in \Pi_K(x)$ and therefore Π_K is u.s.c. at $x \in K$.

Now by, Kakutani's fixed point theorem there is $x \in K$ such that $x \in \Pi_K(x)$, i.e. such that $\|x - f(x)\| \leqslant \mathrm{dist}(f(x), K)$. Hence it follows $\|x - f(x)\| = \mathrm{dist}(f(x), K)$ since $x \in K$. ■

Now, we can prove the following fixed point theorem.

THEOREM 4.10. *Let K be a convex compact subset of a Banach space $(X, \|\cdot\|)$ and let F be a σ-selectinable map from K to K. Then F has a fixed point.*

Proof. Let (F_n) be a sequence of set-valued functions corresponding to F by virtue of its σ-selectionability and let f_n be a continuous selection of F_n. By Lemma 4.9, there exists x_n such that $\|x_n - f_n(x_n)\| = \operatorname{dist}(f(x_n), K)$. For any $\nu \leqslant n$ hence we have

$$\begin{aligned}\operatorname{dist}\big(x_n, F_\nu(x_n)\big) &\leqslant \operatorname{dist}\big(x_n, F_n(x_n)\big) \leqslant \|x_n - f_n(x_n)\| \\ &= \operatorname{dist}\big(f_n(x_n), K\big) \leqslant \bar{h}\big(F_n(x_n), K\big) \leqslant \bar{h}\big(F(x_n), K\big).\end{aligned}$$

By the compactness of K we can assume that (x_n) converges to $x \in K$, and taking into account Proposition 2.1 by the upper semicontinuity of F_ν we can write

$$\begin{aligned}\operatorname{dist}\big(x, F_\nu(x)\big) &\leqslant \liminf_{n\to\infty} \operatorname{dist}\big(x_n, F_\nu(x_n)\big) \\ &\leqslant \liminf_{n\to\infty} \bar{h}\big(F_\nu(x_n), K\big) \leqslant \bar{h}\big(F_\nu(x), K\big).\end{aligned}$$

Since $F_\nu(x)$ decreases to $F(x)$ contained in K, then by Remark 1.2 the last term converges to zero, and $\operatorname{dist}(x, F(x)) = \lim_{\nu\to\infty} \operatorname{dist}(x, F_\nu(x)) = 0$. ■

5. NOTES AND REMARKS

The definitions and results of this chapter are mainly based on Aubin and Cellina [1], Castaing and Valadier [1], Himmelberg [1] and Hildenbrand [1]. In particular, Example 1.1 is taken from Hildenbrand [1], whereas Theorems 1.1 and 1.2 from Castaing and Valadier [1]. The proof of Lemma 1.5 and Corollary 1.2 are taken from Bridgland [1]. The proof of Remark 1.2 can be found in Hildenbrand [1].

The results concerning upper and lower semicontinuous mappings are mainly taken from Aubin and Cellina [1] and Hildenbrand [1]. In particular Examples 1.1 and 2.3 can be find in Hildenbrand [1], whereas Examples 2.2 and 2.4 in Aubin and Cellina [1]. The proofs of Theorems 2.1, 2.2 and 2.9 are taken from Hildenbrand [1]. Some remarks concerning weak sequential upper semicontinuous set-valued mappings are given in Arino, Gautier and Penat [1]. Extensions of results of Theorems 2.1 and 2.2 on the case of sequentially weakly-weakly upper semicontinuous mappings are containd in Kisielewicz [4]. The proof of Proposition 2.2 is taken from Aubin and Cellina [1], whereas Proposition 4.1 is taken from Michael [1]. The generalized Dungundji's Theorem 2.4 is taken from Tsoy-Wo Ma [1] and Theorem 2.12 from Antosiewicz and Cellina [1]. The definitions and basic properties of mesurable set-valued functions are taken from Himmelberg [1]. Theorems 3.2 and 3.6 are contained in Castaing and Valadier [1]. The concept of the weak measurability and its basic properties has been in fact first done in Pliś [1]. The proofs of Theorems 3.3 and 3.4 can be found in Bridgland [1] and Himmelberg and Van Vleck [1], respectively, whereas the proofs of Lusin's and Scorza–Dragoni–Castaing's theorems given in Sections 3.1 and 3.2 can be found in Schwartz [1], Castaing [1] and Artstein and Prikry [1], respectively. More general forms of Lusin–Pliś' theorem can be finded in Pliś [1] and Jacobs [1]. Furthermore in Jacobs [1] some Lusin's properties of set-valued mappings are investigated.

The proof of Kuratowski–Ryll-Nardzewski measurable selection theorem presented in Section 3.3 arises from Hilderbrand [1]. It is a special case of the original Kuratowski–Ryll-Nardzewski selection theorem contained in Kuratowski and Ryll-Nardzewski [1]. The proof of Theorem 3.11 can be found in Hildenbrand [1]. Theorem 3.12 is in fact the generalized version of a implicit function lemma of Fillipov [2]. Much more and general measurable selection theorems may be found in Wagner [1], where also substantial historical comments and an extensive bibliography are included. The main property of Aumann's integral, given by Theorem 3.20 is in the original Aumann's paper obtained on the base of the results of Richter's and von Neumann. The proof of Aumann's theorem contained in Section 3.4 is taken from Hermes and LaSalle [1] and Aumann [1], whereas the proof of Lemma 3.21 from Hilderbrand [1]. More details concerning decomposability can find in Olech [3]. In particular Lemma 3.16 is a part of Theorem 3 given in Olech [3]. The first proof of this theorem has been done in Olech [1] (see also Olech [2]). A general properties of the Aumann's type integrals of Banach space valued multifunctions are given in Hiai and Unegaki [1] and Papageorgiou [1]. Theorems 3.22 and 3.23 are taken from Bridgland [1]. Similar results are also contained in Hermes [1]. The proofs of Michael's continuous selection theorem, and continuous approximate selection theorem presented in Section 4.1 and Examples 4.1 and 4.2 arise from Aubin and Cellina [1]. Also the proof of Lemma 4.3 is a modification of a proof given in Aubin and Cellina [1]. Fixed point Theorem 4.4 is a special case of a more general fixed point theorem given in Covitz and Nadler [1]. The proof of Kakutani fixed point theorem is taken from Aubin and Cellina [1] and Kakutani–Ky Fan fixed point theorem is taken from Edwards [1]. Theorem 4.8 is in fact generalization of Melvin's results given in Melvin [2]. This last fixed point theorem has been first extended in Kisielewicz [3] on the case of set-valued mappings with values in the Hilbert space. Later on, in Kisielewicz and Rybiński [1] it was extended to uniformly convex Banach spaces. Finally, it was proved in Rybiński [1] and Rzepecki [1] for the general case. Theorem 4.8 is also true if $F(y, \cdot)$ is a contraction uniformly with respect to $y \in Y$ and for every $s \in S$ and every $z \in S$ a real-valued function $Y \ni y \to \mathrm{dist}(z, F(y, s)) \in \boldsymbol{R}^+$ is lower semicontinuous on Y. Further generalizations of this theorem are given in Rybiński [2] and Papageorgiou [2]. The proof of fixed point Theorem 4.10 arises from Aubin and Cellina [1].

Chapter III

Subtrajectory and trajectory integrals of set-valued functions

Let I be a compact interval of the real line and let $G\colon I \to \mathscr{P}(\boldsymbol{R}^n)$ be a set-valued function, where as usual $\mathscr{P}(\boldsymbol{R}^n)$ denotes the space of all nonempty subsets of $\boldsymbol{R}^n$. Denote by $\mathscr{F}(G)$ the collection of all L-integrable functions $u\colon I \to \boldsymbol{R}^n$ having the property that $u(t) \in G(t)$ almost everywhere in I. The set $\mathscr{F}(G)$ is said to be the *subtrajectory integrals of the set-valued function* G. Especially interesting, from the point of view of functional differential inclusions, are subtrajectory integrals of set-valued functions $G(z)\colon I \to \mathscr{P}(\boldsymbol{R}^n)$ depending on a parametr z. The subtrajectory integrals of such multifunctions will be denoted by $\mathscr{F}(G)(z)$. In this chapter we are going to investigate topological properties of subtrajectory integrals of set-valued mappings and properties of some of their images by the linear isometry $\mathscr{I}$ defined in Section 2. Furthermore, we shall here consider the properties of set-valued functions $\mathscr{F}(G)\colon z \to \mathscr{F}(G)(z)$.

1. Fundamental space of Aumann integrable set-valued functions

Let I be an interval in the real line and let $G\colon I \to \mathscr{P}(\boldsymbol{R}^n)$ be such that $\mathscr{F}(G) \neq \emptyset$. As usual by $\int_U G(t)\,dt$ we denote the Aumann integral of the set-valued function G over the measurable subset U of I, i.e. $\int_U G(t)\,dt = \left\{\int_U u(t)\,dt\colon u \in \mathscr{F}(G)\right\}$. We shall consider here the space $\mathscr{A}(I, \boldsymbol{R}^n)$ of all Aumann integrable set-valued functions $G\colon I \to \mathscr{P}(\boldsymbol{R}^n)$ with compact values. Furthermore, several equivalent definitions for a weak convergence of sequences of $\mathscr{A}(I, \boldsymbol{R}^n)$ are given.

1.1. The metric space $(\mathscr{A}(I, \boldsymbol{R}^n), d)$

Consider now a family of all set-valued functions $G\colon I \to \mathrm{Comp}(\boldsymbol{R}^n)$ that are L-measurable and integrably bounded. It follows immediately from Kuratowski–Ryll-Nardzewski measurable selection theorem that for every such set-valued function G we have $\mathscr{F}(G) = \mathscr{M}(G) \neq \emptyset$, where $\mathscr{M}(G)$ denotes the family of all measurable selectors of G. We will deal here with equivalence classes of members of the above family of set-valued functions with the equivalence defined by the equality almost everywhere, and as usual we will not distinguish between a function and its equivalence class. The space of all such equivalence classes is denoted by $\mathscr{A}(I, \boldsymbol{R}^n)$. We shall define a metric on $\mathscr{A}(I, \boldsymbol{R}^n)$ which will be analogous to the L-metric on the space of L-integrable functions, and get similar results. A complete analogy can not be achieved since the space of compact subsets in $\boldsymbol{R}^n$ is not a vector space. We begin with the following lemma.

LEMMA 1.1. *A mapping* $d\colon \mathscr{A}(I, \boldsymbol{R}^n)\times\mathscr{A}(I, \boldsymbol{R}^n) \to \boldsymbol{R}^+$ *defined by*

$$d(F, G) = \int_I h\big(F(t), G(t)\big)dt \tag{1.1}$$

for every $F, G \in \mathscr{A}(I, \boldsymbol{R}^n)$ *is a metric on* $\mathscr{A}(I, \boldsymbol{R}^n)$.

Proof. First we have to show that d is well defined. Let $F, G \in \mathscr{A}(I, \boldsymbol{R}^n)$ be given. By Lusin–Pliś' theorem, for every $\varepsilon > 0$ there is a closed set $E \in I$ with $\mu(I\setminus E < \varepsilon$ and such that the restrictions $F|_E$ and $G|_E$ of F and G to E are continuous. Then a real-valued function $\psi\colon I \ni t \to h(F(t), G(t)) \in \boldsymbol{R}^+$ is such that its restriction to E is continuous for every $\varepsilon > 0$. Therefore, by Lusin's theorem it is L-measurable. Furthermore, for a.e. $t \in I$ we have $|\psi(t)| \leqslant 2\max(\|F(t)\|, \|G(t)\|)$, where $\|F(\cdot)\|$ and $\|G(\cdot)\|$ are L-integrable, because F and G are integrably bounded. Then ψ is L-integrable on I, and therefore d is well defined. Obviously, d is a metric on $\mathscr{A}(I, \boldsymbol{R}^n)$. ■

THEOREM 1.2. *$(\mathscr{A}(I, \boldsymbol{R}^n), d)$ is a complete metric space. Furthermore, if (G_k) is a sequence of $\mathscr{A}(I, \boldsymbol{R}^n)$ converging in the metric d to G, then there is a subsequence, say (G_{k_n}) of (G_k) such that $h\big(G_{k_n}(t), G(t)\big) \to 0$ for a.e. $t \in I$ as $n \to \infty$.*

Proof. Suppose (G_n) is a Cauchy sequence of $(\mathscr{A}(I, \boldsymbol{R}^n), d)$. Then for every $\varepsilon > 0$ there exists a positive integer N such that for every $n, m > N$ we have $d(G_n, G_m) < \varepsilon$. Taking $\varepsilon = 3^{-k}$ we can define an increasing sequence (n_k) of positive integer such that $d(G_m, G_{n_k}) < 3^{-k}$ for $m \geqslant n_k$ and $k = 1, 2, \ldots$ In particular, we have $d(G_{n_{k+1}}, G_{n_k}) < 3^{-k}$ for $k = 1, 2, \ldots$

We shall show that there is a measurable set-valued function $G\colon I \to \mathrm{Comp}(\boldsymbol{R}^n)$ so that $h(G_{n_k}(t), G(t)) \to 0$ for a.e. $t \in I$ as $k \to \infty$. Indeed, let $A_k = \{t \in I\colon h(G_{n_{k+1}}(t), G_{n_k}(t)) \geqslant 2^{-k}\}$. Then

$$2^{-k}\mu(A_k) \leqslant \int_{A_k} h\big(G_{n_{k+1}}(t), G_{n_k}(t)\big)dt \leqslant d(G_{n_{k+1}}, G_{n_k}) < 3^{-k}.$$

Thus $\mu(A_k) < (\frac{2}{3})^k$. Putting $A = \bigcap_{m=1}^{\infty}\bigcup_{k=m}^{\infty} A_k$ we obtain

$$\mu(A) \leqslant \mu\Big(\bigcup_{k=m}^{\infty} A_k\Big) \leqslant \sum_{k=m}^{\infty} \mu(A_k) < \sum_{k=m}^{\infty} (\tfrac{2}{3})^k = 3(\tfrac{2}{3})^m$$

for every $m = 1, 2, \ldots$ Then $\mu(A) = 0$. Moreover $I\setminus A = \bigcup_{m=1}^{\infty}\bigcap_{k=m}^{\infty}(I\setminus A_k)$ and therefore for every $t \in I\setminus A$ there is a positive integer m such that $t \notin A_k$ for each $k \geqslant m$. Thus for a.e. $t \in I$ there is $m \in \boldsymbol{N}$ such that $h(G_{n_{k+1}}(t), G_{n_k}(t) < 2^{-k}$ for $k \geqslant m$. Hence it follows that for every $t \in I\setminus A$, $\{G_{n_k}(t)\}$ is a Cauchy sequence of a metric space $(\mathrm{Comp}(\boldsymbol{R}^n), h)$. By the completeness of the latter space for every $t \in I\setminus A$ there exists $\tilde{G}(t) \in \mathrm{Comp}(\boldsymbol{R}^n)$ such that $h\big(G_{n_k}(t), \tilde{G}(t)\big) \to 0$ as $k \to \infty$. We can define a mapping $G: I \to \text{Comp}(\boldsymbol{R}^n)$ taking $G(t) = \tilde{G}(t)$ for $t \in I \setminus A$ and $G(t) = \{0\}$ for $t \in A$. We have $h(G_{n_k}(t), G(t)) \to 0$ for a.e. $t \in I$ as $k \to \infty$. By Lemma II.3.5 $G: I \to \text{Comp}(\boldsymbol{R}^n)$ is measurable.

We shall now show that $d(G_{n_k}, G) \to 0$ as $k \to \infty$. Let $k \in N$ be fixed and let $g_i(t) = h(G_{n_k}(t), G_{n_i}(t))$ for $t \in I$ and $i \geqslant k$. By Fatou's lemma we have

$$d(G_{n_k}, G) = \int_I \lim_{i \to \infty} g_i(t)dt \leqslant \liminf_{i \to \infty} \int_I g_i(t)dt = \liminf_{i \to \infty} d(G_{n_k}, G_{n_i}) \leqslant 3^{-k}$$

because $n_i \geqslant n_k$. Then $\lim_{k \to \infty} d(G_{n_k}, G) = 0$.

Since (G_n) is a Cauchy sequence of $(\mathscr{A}(I, \boldsymbol{R}^n), d)$ then $\lim_{n, k \to \infty} d(G_n, G_{n_k}) = 0$. Hence and an inequality $d(G_n, G) \leqslant d(G_n, G_{n_k}) + d(G_{n_k}, G)$ it follows that $d(G_n, G) \to 0$ as $n \to \infty$.

If (G_k) is a sequence of $\mathscr{A}(I, \boldsymbol{R}^n)$ converging in the metric d to G then the sequence (ψ_k) of real-valued functions $\psi_k: I \to \boldsymbol{R}^1$ defined by $\psi_k(t) = h(G_k(t), G(t))$ for $k = 1, 2, \ldots$ and $t \in I$ is converging in $L(I, \boldsymbol{R}')$ to the origin. Thus there is a subsequence, say (ψ_{k_n}) of (ψ_k) such that $\psi_{k_n}(t) \to 0$ for a.e. $t \in I$ as $n \to \infty$. ■

We can also introduce in the case of the space $\mathscr{A}(I, \boldsymbol{R}^n)$ the concept of the uniform integrability of the family of $\mathscr{A}(I, \boldsymbol{R}^n)$.

A family $\{G_\lambda\}_{\lambda \in \Lambda} \subset \mathscr{A}(I, \boldsymbol{R}^n)$ is called *uniformly integrable* if for every $\varepsilon > 0$ there is a $\delta = \delta(\varepsilon) > 0$ such that $\left\| \int_E G_\lambda(t)dt \right\| \leqslant \varepsilon$ holds for every $\lambda \in \Lambda$ and all measurable sets $E \in I$ with $\mu(E) \leqslant \delta$.

We have the following theorem.

THEOREM 1.3. *A family $\{G_\lambda\}_{\lambda \in \Lambda} \subset \mathscr{A}(I, \boldsymbol{R}^n)$ is uniformly integrable if and only if the family $\{m_\lambda\}_{\lambda \in \Lambda}$ of real-valued functions $m_\lambda: I \to \boldsymbol{R}$ defined by $m_\lambda(t) = \|G_\lambda(t)\|$ for $t \in I$ is uniformly integrable.*

Proof. Suppose the family $\{m_\lambda\}_{\lambda \in \Lambda}$ is uniformly integrable. Then, given $\varepsilon > 0$, there is $\delta(\varepsilon) > 0$ such that for every $\lambda \in \Lambda$ and a measurable set $E \in I$ satisfying $\mu(E) \leqslant \delta(\varepsilon)$ we have $\int_E m_\lambda(t)dt \leqslant \varepsilon$. Let $g \in \mathscr{F}(G_\lambda)$ for fixed $\lambda \in \Lambda$. We have $\left| \int_E g(t)dt \right| \leqslant \int_E |g(t)|\, dt \leqslant \int_E m_\lambda(t)dt \leqslant \varepsilon$. Thus $\left\| \int_E G_\lambda(t)dt \right\| = \sup_{g \in \mathscr{F}(G_\lambda)} \left| \int_E g(t)dt \right| \leqslant \varepsilon$ for every $\lambda \in \Lambda$ and every measurable set $E \subset I$ satisfying $\mu(E) \leqslant \delta(\varepsilon)$.

Suppose now $\{G_\lambda\}_{\lambda \in \Lambda}$ is uniformly integrable and there exists $\varepsilon_0 > 0$ such that for every $\delta > 0$ there is a measurable set $E \subset I$ with $\mu(E) \leqslant \delta$ and there is a $\lambda \in \Lambda$ such that $\int_E m_\lambda(t)dt \geqslant \varepsilon_0$. Let $g_\lambda \in \mathscr{F}(G_\lambda)$ be such that $|g_\lambda(t)| = m_\lambda(t)$ for a.e. $t \in I$. Such selector exists by virtue of Theorem II.3.12. Indeed, let $f(t, x) := |x|$ for each $(t, x) \in I \times \boldsymbol{R}^n$ and put for fixed $\lambda \in \Lambda$ and $t \in I$, $\Gamma(t) := G_\lambda(t)$ and $m_\lambda(t) := \|G_\lambda(t)\|$. We have $m_\lambda(t) \in f(t, \Gamma(t))$ for $t \in I$. Therefore, by Theorem II.3.12 there exists a measurable selector $\gamma: I \to \boldsymbol{R}^n$ for Γ such that $m_\lambda(t) = f(t, \gamma(t) := |\gamma(t)|$. Taking $g_\lambda := \gamma$ we obtain $|g_\lambda(t)| = m_\lambda(t)$ for $t \in I$. Let $E_1, \ldots, E_{2n}$ be the partition of E

such that the values of g_λ restricted to each E_i are in the same orthant of $\boldsymbol{R}^n$. Then the measure of each E_i still satisfies $\mu(E_i) \leqslant \delta$, but at least for one i it is true that $\left|\int_{E_i} g_\lambda(t)dt\right| \geqslant \varepsilon/2^n$. This contradicts the existence of $\delta = \delta(\varepsilon 2^{-n})$ for establishing the uniform integrability of G_λ. ■

COROLLARY 1.1. *The family* $\{G_\lambda\}_{\lambda\in\Lambda} \subset \mathscr{A}(I, \boldsymbol{R}^n)$ *is uniformly integrable if and only if the collection of all selectors of all its members is uniformly integrable.* ■

1.2. Weak convergence in $\mathscr{A}(I, \boldsymbol{R}^n)$

The following theorem gives the equivalence of three conditions which will serve as the definition of the weak convergence in $\mathscr{A}(I, \boldsymbol{R}^n)$. Recall, that by $s(\cdot, A)$ we denote the support function defined for a given nonempty set $A \subset \boldsymbol{R}^n$ on $\boldsymbol{R}^n$ by $s(p, A) = \sup\{p \cdot a\colon a \in A\}$, where $p \cdot a$ denotes the inner product of p and $a \in A$.

THEOREM 1.4. *Let* (G_k) *be a sequence of* $\mathscr{A}(I, \boldsymbol{R}^n)$ *and let* $G \in \mathscr{A}(I, \boldsymbol{R}^n)$. *The following three statements are equivalent*:

(i) *For every bounded and measurable function* $p\colon I \to \boldsymbol{R}^n$ *the sequence* $\left\{\int_I p(t)\cdot G_k(t)dt\right\}$ *of sets in* $\boldsymbol{R}$ *converges to* $\int_I p(t)\cdot G(t)dt$.

(ii) *For every bounded and measurable function* $p\colon I \to \boldsymbol{R}^n$ *the sequence* $\left\{\int_I s(p(t), G_k(t))\,dt\right\}$ *of real numbers converges to* $\int_I s(p(t), G(t))\,dt$.

(iii) *For every* $p \in \boldsymbol{R}^n$ *and every measurable set* $U \subset I$ *the sequence* $\left\{\int_U s(p, G_k(t))\,dt\right\}$ *converges to* $\int_U s(p, G(t))\,dt$.

Proof. (i) → (ii). If $\int_I p(t)\cdot G_k(t)dt$ *converges to* $\int_I p(t)\cdot G(t)dt$, then, in particular, the sequence of numbers $b_k = \sup\left(\int_I p(t)\cdot G_k(t)dt\right)$ converges to to the number $b = \sup\left(\int_I p(t)\cdot G(t)dt\right)$. Indeed, let $A_k = \int_I p(t)\cdot G_k(t)dt$ and $A = \int_I p(t)\cdot G(t)dt$. By virtue of Theorem II.3.20, A_k $(k = 1, 2, \ldots)$ and A are compact convex subsets of $\boldsymbol{R}$. Then there are numbers $a_k \leqslant b_k$ $(k = 1, 2, \ldots)$ and $a \leqslant b$ such that $A_k = [a_k, b_k]$ for $k = 1, 2, \ldots$ and $A = [a, b]$. Since $h(A_k, A) = \max(|a_k - a|, |b_k - b|)$ for $k = 1, 2, \ldots$ and $h(A_k, A) \to 0$ as $k \to \infty$ then, in particular, $|b_k - b| \to 0$ as $k \to \infty$.

We shall show that $b_k = \int_I s(p(t), G_k(t))dt$ and $b = \int_I s(p(t), G(t))dt$ and this will complete the proof. By Remark II.3.5, it is clear that $s(p(\cdot), G(\cdot))$ is measurable. By Theorem II.3.12 there exists a measurable selector $\gamma\colon I \to \boldsymbol{R}^n$ for G such that $\int_I s(p(t), G(t))dt = \int_I p(t)\cdot\gamma(t)dt$. Indeed, taking for each $(t, x) \in I \times \boldsymbol{R}^n$, $f(t, x) := p(t)\cdot x$ and $g(t) := s(p(t), G(t))$ we have $f(t, G(t)) = p(t)\cdot G(t)$ and $g(t) \in f(t, G(t))$ for $t \in I$. Therefore, by virtue of Theorem II.3.12 there exists a measurable selector $\gamma\colon I \to \boldsymbol{R}^n$ for G such that $g(t) = f(t, \gamma(t))$. On the other hand, by the definitions of $s(p(t), G(t))$ and b for every $r \in \int_I p(t) \cdot G(t) dt$ we have $r \leqslant \int_I p(t) \cdot \gamma(t) dt \leqslant b$. Then $b = \int_I p(t) \cdot \gamma(t) dt$. Therefore, $b = \int_I s(p(t), G(t)) dt$. The claim for b_k is proved by adding the subscript k whenever G and b appear in the proof.

(i) $\rightarrow$ (iii). Obvious.

(iii) $\rightarrow$ (i). It is clear that (iii) implies that the sequence of real-valued functions $s(p, G_k(\cdot))$ is weakly convergent to $s(p, G(\cdot))$. Then, by Dunford's theorem, $\{s(p, G_k(\cdot))\}_{k \in N}$ is uniformly integrable. Let $p_1, \ldots, p_{2n}$ be the $2n$ vectors $(0, \ldots, 0, \pm 1, 0, \ldots, 0) \in \boldsymbol{R}^n$. Then the collection of functions $s(p_i, G_k(\cdot))$, where $i = 1, 2, \ldots, 2n$, $k = 1, 2, \ldots$ is uniformly integrable. Since the support function $s(\cdot, A)$ is convex in $p \in \boldsymbol{R}^n$ and since $||G_k(t)|| = \sup\{(x/|x|) \cdot x \colon x \in G_k(t)\}$ it follows that $||G_k(t)|| \leqslant c \sup\{p_i \cdot x \colon x \in G_k(t),\ i = 1, 2, \ldots, 2n\}$ for a certain fixed c ($c = n^{1/2}$ will fit). Thus the collection of real-valued functions $m_k(\cdot) = ||G_k(\cdot)||$, $k = 1, 2, \ldots$ is uniformly integrable. In particular, a sequence $\left(\int_I m_k(t) dt\right)$ is bounded, say by M.

Let $p \colon I \rightarrow \boldsymbol{R}^n$ be a measurable and bounded function. For each $\varepsilon > 0$ we can find a function $q \colon I \rightarrow \boldsymbol{R}^n$, such that q has only a finite number of values, q is measurable and $|p(t) - q(t)| < \varepsilon$ for a.e. $t \in I$. Then

$$h\left(\int_I p(t) \cdot G_k(t) dt, \int_I q(t) \cdot G_k(t) dt\right)$$

$$\leqslant \left|\int_I (p(t) - q(t)) \cdot G_k(t) dt\right| \leqslant \int_I |p(t) - q(t)| \cdot ||G_k(t)|| dt \leqslant \varepsilon M.$$

Condition (iii) implies that for each measurable set $U \subset I$ and a fixed vector $p \in \boldsymbol{R}^n$ the sets $\int_U p \cdot G_k(t) dt$ converge to $\int_U p \cdot G(t) dt$. Indeed, $\int_U p \cdot G_k(t) dt$ is the interval whose extremes are given by $\int_U s(-p, G_k(t)) dt$ and $\int_U s(p, G_k(t)) dt$ while $\int_U p \cdot G(t) dt$ is the interval with extremes $\int_U s(-p, G(t)) dt$ and $\int_U s(p, G(t)) dt$. Since q has only a finite numbers of values also $\int_U q(t) \cdot G_k(t) dt$ converges to $\int_U q(t) \cdot G(t) dt$. Since ε was arbitrary small the inequalities above show the desired convergence also for $p(t)$. Finally, it is clear that (ii) $\rightarrow$ (iii). ■

If for the sequence (G_k) of $\mathscr{A}(I, \boldsymbol{R}^n)$ and $G \in \mathscr{A}(I, \boldsymbol{R}^n)$ condition (i) [and hence also (ii) and (iii)] of Theorem 1.4 holds, we say that the sequence (G) *converges weakly to* G.

It is easily seen that for singleton-set-valued functions, the definition coincides with the usual definition of weak convergence of point-valued functions.

Each of (i), (ii) and (iii) of Theorem 1.4 can be stated in terms of Cauchy sequences (of $\{\int_I p(t) \cdot G_k(t) dt\}$, $\{\int_I s(p(t), G_k(t)) dt\}$ and $\{\int_I s(p, G_k(t)) dt\}$, respectively) rather then in terms of convergence. It is also clear that the proof shows the equivalence also in this presentation. Therefore, we have also the concept of a weak Cauchy sequence, namely a sequence of elements such that one of (i), (ii) or (iii) (and hence the other two) is satisfied in the Cauchy formulation. We shall see below that every weak Cauchy sequence has a weak limit.

Notice that the limit of a weakly convergent sequence is not unique. Indeed, the support functions of two sets with the same convex hulls are the same and the convergence criterion (iii) is given in terms of the support functions. Thus if (G_k) converges weakly to G, then it converges weakly to every $F \in \mathscr{A}(I, \boldsymbol{R}^n)$ provided the convex hull of $F(t)$ equals a.e. the convex hull of $G(t)$. However, it is clear that the convex-valued limit is unique. Indeed, for every $p \in \boldsymbol{R}^n$ the function $s(p, G(\cdot))$ is the weak limit in $L(I, \boldsymbol{R})$ of the functions $s(p, G_k(\cdot))$ and in $L(I, \boldsymbol{R})$ the weak limit is unique.

We have the following consequences of the weak convergence of sequences in $\mathscr{A}(I, \boldsymbol{R}^n)$.

THEOREM 1.5. *If a sequence (G_k) of $\mathscr{A}(I, \boldsymbol{R}^n)$ converges weakly to $G \in \mathscr{A}(I, \boldsymbol{R}^n)$ or if (G_k) is a weak Cauchy sequence of $\mathscr{A}(I, \boldsymbol{R}^n)$, then the family $\{m_k\}_{k \in N}$ of functions $m_k\colon I \to \boldsymbol{R}$ defined by $m_k(t) = \|G_k(t)\|$ for $t \in I$ is uniformly integrable.*

Proof. The proof follows immediately from the definition of the weak convergence in $\mathscr{A}(I, \boldsymbol{R}^n)$ and Dunford's theorem. Indeed, if (G_k) is weakly converging to $G \in \mathscr{A}(I, \boldsymbol{R}^n)$ or a weak Cauchy sequence, then for every $p \in \boldsymbol{R}^n$ $\{s(p, G_k(\cdot))\}$ converges weakly to $s(p, G(\cdot))$ in $L(I, \boldsymbol{R}^n)$. Then by Dunford's theorem, for every fixed $p \in \boldsymbol{R}^n$ the family $\{s(p, G_k(\cdot))\}_{k \in N}$ is uniformly integrable. As it was proved in the last part of the proof of Theorem 1.4, hence it follows that the family $\{m_k\}_{k \in N}$ is uniformly integrable. ■

COROLLARY 1.2. *A weakly convergent sequence of $\mathscr{A}(I, \boldsymbol{R}^n)$ is bounded in $\mathscr{A}(I, \boldsymbol{R}^n)$ and uniformly integrable. Furthermore, the collection of all selectors of the members of a weakly converging sequence of $\mathscr{A}(I, \boldsymbol{R}^n)$ is uniformly integrable in $L(I, \boldsymbol{R}^n)$.*

Indeed, let (G_k) be a weakly convergent sequence of $\mathscr{A}(I, \boldsymbol{R}^n)$. The uniform integrability of the family $\{\|G_k(\cdot)\|\}_{k \in N}$ implies the boundedness of a sequence $\{\|G_k(\cdot)\|\}$ in $L(I, \boldsymbol{R})$. The rest of the claims follows from Theorem 1.5 and Corollary 1.1. ■

THEOREM 1.6. *Let (G_k) be a sequence of $\mathscr{A}(I, \boldsymbol{R}^n)$ and let $G \in \mathscr{A}(I, \boldsymbol{R}^n)$. The sequence (G_k) converges weakly to G if and only if for every measurable set $U \subset I$ a sequence $\{\int_U G_k(t)\,dt\}$ of* Comp$(\boldsymbol{R}^n)$ *converges in the Hausdorff metric h to $\int_U G(t)\,dt$.*

Proof. By Theorem II.3.20 the sets $\int_U G_k(t)\,dt$ and $\int_U G(t)\,dt$ are convex. In this case the convergence of a sequence $\{\int_U G_k(t)\,dt\}$ to $\int_U G(t)\,dt$ is equivalent to the convergence of a sequence $\{s(p, \int_U G_k(t)\,dt)\}$ to $s(p, \int_U G(t)\,dt)$ for every $p \in \boldsymbol{R}^n$

(see Lemma II.1.3). By Theorem II.3.21 the equalities

$$s\left(p, \int_U G_k(t)\,dt\right) = \int_U s(p, G_k(t))\,dt \quad \text{and} \quad s\left(p, \int_U G(t)\,dt\right) = \int_U s(p, G(t))\,dt$$

hold, but the convergence of the right-hand side of these equalities is the definition via condition (iii) of the weak convergence. ■

It is clear that in terms of Cauchy sequences the last proposition should be read as: (G_k) *is a weak Cauchy sequence if and only if for every measurable set* $U \subset I$ *the sequence of sets* $\left\{\int_U G_k(t)\,dt\right\}$ *is a Cauchy sequence of* $(\mathrm{Comp}(\mathbf{R}^n), h)$.

The following results are analogous to the properties of weak convergence in $L(I, \mathbf{R})$.

THEOREM 1.7. *The space* $\mathscr{A}(I, \mathbf{R}^n)$ *is weakly sequentially complete, i.e., every weak Cauchy sequence of* $\mathscr{A}(I, \mathbf{R}^n)$ *has a weak limit in* $\mathscr{A}(I, \mathbf{R}^n)$.

Proof. Suppose that (G_k) is a weak Cauchy sequence of $\mathscr{A}(I, \mathbf{R}\ )$. In view of the above remark for every measurable set $U \subset I$ the sets $\int_U G_k(t)\,dt$ form a Cauchy sequence of a complete metric space $(\mathrm{Comp}(\mathbf{R}^n), h)$. Therefore a sequence $\left\{\int_U G_k(t)\,dt\right\}$ has a limit, denoted by $\Phi(U)$. By virtue of Corollary II.3.3 $\int_U G_k(t)\,dt$ is additive in U and convex-valued. Therefore $\Phi(U)$ is additive and convex-valued too. Since the family $\{G_k\}_{k \in N}$ is uniformly integrable it follows that $\Phi(U)$ is countably additive and bounded, i.e., it is a set-valued measure (see II.3.1). Since Φ has closed and convex values it follows from Corollary II.3.5 that Φ has a measurable Radon–Nikodyn derivative with closed convex values, i.e., there is a measurable set-valued function G such that for every measurable $U \subset I$, $\int_U G(t)\,dt = \Phi(U)$. Hence, in particular it also follows that for a.e. $t \in I$, $\|G(t)\| < \infty$. Then, $G(t)$ is for a.e. $t \in I$ a bounded, and therefore also compact subset of $\mathbf{R}^n$. ■

COROLLARY 1.3. *A set* $\mathscr{A}^{\mathrm{C}}(I, \mathbf{R}^n)$ *of* $\mathscr{A}(I, \mathbf{R}^n)$ *containing all* $G \in \mathscr{A}(I, \mathbf{R}^n)$ *such that* $G(t) \in \mathrm{Conv}(\mathbf{R}^n)$ *for a.e.* $t \in I$ *is a closed subset of* $\mathscr{A}(I, \mathbf{R}^n)$.

Indeed, suppose a sequence (G_n) of $\mathscr{A}^{\mathrm{C}}(I, \mathbf{R}^n)$ and $G \in \mathscr{A}(I, \mathbf{R}^n)$ are such that $d(G_n, G) \to 0$ as $n \to \infty$. Therefore, there exists a subsequence, say (G_k) of (G_n) such that $h(G_k(t), G(t)) \to 0$ for a.e. $t \in I$ as $k \to \infty$. Hence it follows that $G(t) \in \mathrm{Conv}(\mathbf{R}^n)$ for a.e. $t \in I$, because by Corollary II.1.3 $\mathrm{Conv}(\mathbf{R}^n)$ is a closed subset of the metric space $(\mathrm{Comp}(\mathbf{R}^n), h)$. ■

COROLLARY 1.4. *A set* $\mathscr{A}^{\mathrm{C}}(I, \mathbf{R}^n)$ *is a sequentially weakly complete subset of* $\mathscr{A}(I, \mathbf{R}^n)$.

Indeed, if (G_n) is a weak Cauchy sequence of $\mathscr{A}^{\mathrm{C}}(I, \mathbf{R}^n)$ then, by the proof of Theorem 1.7, there exists $G \in \mathscr{A}(I, \mathbf{R}^n)$ with convex values such that (G_n) is weakly convergent to G. By Corollary II.1.3 $G \in \mathscr{A}^{\mathrm{C}}(I, \mathbf{R}^n)$ and (G_n) converges weakly to G. ■

THEOREM 1.8. *A set in* $\mathscr{A}(I, \boldsymbol{R}^n)$ *is relatively weakly sequentially compact, i.e., each of its sequence has a subsequence weakly convergent to any* $G \in \mathscr{A}(I, \boldsymbol{R}^n)$, *if and only if its members form an uniformly integrable family of* $\mathscr{A}(I, \boldsymbol{R}^n)$.

Proof. If the set is not uniformly integrable, then a sequence can be found in it such that every subsequence of it is not uniformly integrable and hence (see Corollary 1.2) no subsequence converges weakly.

Suppose the set is uniformly integrable and let (G_k) be a sequence in it. Let $U_1, U_2, \dots$ be a sequence of Borel sets in I that are dense in the Borel field of I. The sequence $\{\int_{U_1} G_k(t)\,dt\}$ is bounded, and therefore has a convergent subsequence, say $\{\int_{U_1} G_k^1(t)\,dt\}$. The sequence $\{\int_{U_2} G_k^1(t)\,dt\}$ is bounded, and thus has a convergent subsequence, say $\{\int_{U_2} G_k^2(t)\,dt\}$. We can continue building these sub-sub- ... -sub-sequences, and by a standard diagonal procedure get a subsequence (G_i) of (G_k) such that for every U_j the sequence $\{\int_{U_j} G_i(t)\,dt\}$ for $j = 1, 2, \dots$ will converge. The uniform integrability together with the density of (U_j) implies that the $\{\int_U G_i(t)\,dt\}$ converges for every measurable $U \subset I$. Thus the sequence (G_i) is a weak Cauchy sequence of $\mathscr{A}(I, \boldsymbol{R}^n)$. Therefore by Theorem 1.7 there is $G \in \mathscr{A}(I, \boldsymbol{R}^n)$ such that $\lim_{i\to\infty} \int_U G_i(t)\,dt = \int_U G(t)\,dt$ for each measurable set $U \subset I$. ■

The following three theorems give the connection between the weak convergence of the set-valued functions in $\mathscr{A}(I, \boldsymbol{R}^n)$ and the weak convergence of their selectors in $L(I, \boldsymbol{R}^n)$.

THEOREM 1.9. *If* (G_k) *is a sequence of* $\mathscr{A}(I, \boldsymbol{R}^n)$ *weakly convergent to* $G \in \mathscr{A}(I, \boldsymbol{R}^n)$ *and* $g_k \in \mathscr{F}(G_k)$ *for* $k = 1, 2, \dots$, *then the sequence* (g_k) *has a subsequence weakly convergent in* $L(I, \boldsymbol{R}^n)$.

Proof. By Corollary 1.2, the sequence (g_k) is uniformly integrable in $L(I, \boldsymbol{R}^n)$. Thus by Dunford's theorem it is relatively weakly sequentially compact in $L(I, \boldsymbol{R}^n)$. ■

THEOREM 1.10. *If* (G_k) *is a sequence of* $\mathscr{A}(I, \boldsymbol{R}^n)$ *weakly convergent to* $G \in \mathscr{A}(I, \boldsymbol{R}^n)$ *and if* (g_k) *is a sequence of selectors of* G_k *for* $k = 1, 2, \dots$, *such that* $g_k \rightharpoonup g$ *in* $L(I, \boldsymbol{R}^n)$ *as* $k \to \infty$, *then* $g(t) \in \operatorname{co} G(t)$ *for a.e.* $t \in I$.

Proof. For every $p \in \boldsymbol{R}^n$ the inequality $p \cdot g_k(t) \leqslant s(p, G_k(t))$ holds. Since $\{s(p, G_k(\cdot))\}$ converges weakly to $s(p, G(\cdot))$ and since $\{p \cdot g_k(\cdot)\}$ converges weakly to $p \cdot g(\cdot)$ it follows that almost everywhere and for every $p \in \boldsymbol{R}^n$ the inequality $p \cdot g(t) \leqslant s(p, G(t))$ holds. Let us observe that we can deduce "almost everywhere for every $p \in \boldsymbol{R}^n$ ..." from "for every $p \in \boldsymbol{R}^n$, almost everywhere ..." since it is enough to show the inequality for a dense sequence of vectors. Since for every set A the convex hull of A is $\{a: p \cdot a \leqslant s(p, A)$ for every $p \in \boldsymbol{R}^n\}$ (see Section II.1.2), it follows that $g(t)$ belongs for a.e. $t \in I$ to the convex hull of $G(t)$. ■

THEOREM 1.11. *Let (G_k) be a sequence of $\mathscr{A}(I, \boldsymbol{R}^n)$ weakly convergent to $G \in \mathscr{A}(I, \boldsymbol{R}^n)$ and let $g \in \mathscr{F}(G)$. Then there exists a sequence (g_k) of selectors of G_k, $k = 1, 2, \ldots$, such that (g_k) converges weakly to g in $L(I, \boldsymbol{R}^n)$.*

Proof. Without loss of generality we can take $I = [0, 1]$. Let $[(j-1)2^{-m}, j2^{-m}]$ for $j = 1, \ldots, 2^m$ be the mth dyadic partition of $[0, 1]$. There are 2^m intervals in the partition, and denote them by $U_1, \ldots, U_{2m}$. Since (G_k) converges weakly to G it follows from Theorem 1.6 that for k large enough, say $k \geqslant k(m)$ we have $h(\int_{U_j} G_k(t)dt, \int_{U_j} G(t)dt) \leqslant 1/m2^m$ for every $j = 1, \ldots, 2^m$. Therefore a selector $g_{k,m}$ of G_k for $k \geqslant k(m)$ exists such that $\left|\int_{U_j} g_{k,m}(t)dt - \int_{U_j} g(t)dt\right| \leqslant 1/m2^m$ for every $j = 1, \ldots, 2^m$.

Notice that by this choice if U is a union of members of the mth partition then $\left|\int_{U_j} g_{k,m}(t)dt - \int_{U_j} g(t)dt\right| \leqslant 1/m$. Without loss of generality we can assume that the sequence $k(m)$ for $m = 1, 2, \ldots$ is strictly increasing. Define now the selector g_k of G_k by $g_k = g_{k,m}$ if $k(m) \leqslant k < k(m+1)$. It is clear that for every finite union U of dyadic intervals the sequence $\{\int_U g_k(t)dt\}$ converges to $\int_U g(t)dt$. The uniform integrability of (g_k) implies that $\{\int_U g_k(t)dt\}$ converges to $\int_U g(t)dt$ for every measurable set $U \subset I$ and hence that (g_k) converges weakly to g. ■

Now, we show that Theorems 1.9 and 1.10 hold if a sequence (G_k) of $\mathscr{A}(I, \boldsymbol{R}^n)$ is uniformly integrable and is upper weakly convergent to $G \in \mathscr{A}(I, \boldsymbol{R}^n)$, i.e. is such that $\lim_{k\to\infty} \bar{h}(\int_U G_k(t)dt, \int_U G(t)dt) = 0$ for every measurable set $U \subset I$. We have the following theorem.

THEOREM 1.12. *Let (G_k) be a uniformly integrable sequence of $\mathscr{A}(I, \boldsymbol{R}^n)$ such that $\lim_{k\to\infty} \bar{h}(\int_U G_k(t)dt, \int_U G(t)dt) = 0$ for every measurable set $U \subset I$, where $G \in \mathscr{A}(I, \boldsymbol{R}^n)$. Then every sequence (g_k) with $g_k \in \mathscr{F}(G_k)$ for $k = 1, 2, \ldots$ has a subsequence weakly convergent to any $g \in L(I, \boldsymbol{R}^n)$ such that $g(t) \in \operatorname{co} G(t)$ for a.e. $t \in I$.*

Proof. By Corollary 1.1 the collection of all selectors of a sequence (G_k) is uniformly integrable in $L(I, \boldsymbol{R}^n)$. Therefore, by virtue of Dunford's theorem it is relatively weakly sequentially compact in $L(I, \boldsymbol{R}^n)$. On the other hand, in view of Theorem 1.8 a sequence (G_k) is relatively weakly sequentially compact in $\mathscr{A}(I, \boldsymbol{R}^n)$. Thus there are $F \in \mathscr{A}(I, \boldsymbol{R}^n)$ and any subsequence, say (G_{k_n}) of (G_k) such that $\lim_{n\to\infty} h(\int_U F(t)dt, \int_U G_{k_n}(t)dt) = 0$ for every measurable set $U \subset I$. Furthermore, we have $\lim_{n\to\infty} \bar{h}(\int_U G_{k_n}(t)dt, \int_U G(t)dt) = 0$ for a measurable set $U \subset I$. Therefore, for each measurable set $U \subset I$ we get

$$\bar{h}\Big(\int_U F(t)dt, \int_U G(t)dt\Big) \leqslant h\Big(\int_U F(t)dt, \int_U G_{k_n}(t)dt\Big) + \bar{h}\Big(\int_U G_{k_n}(t)dt, \int_U G(t)dt\Big),$$

i.e., $\int_U F(t)dt \subset \int_U G(t)dt$ and therefore also $\int_U \operatorname{co} F(t)dt \subset \int_U \operatorname{co} G(t)dt$ for each measurable set $U \subset I$.

By the relative weak sequential compactness of (g_k) we can select a subsequence, say (g_{k_n}) of (g_k) weakly converging to any $g \in L(I, \boldsymbol{R}^n)$ and such that $g_{k_n} \in \mathscr{F}(G_{k_n})$. By Theorem 1.10 we have $g(t) \in \operatorname{co} F(t)$ for a.e. $t \in I$. Therefore, $\int_U g(t)dt \in \int_U \operatorname{co} F(t)dt \subset \int_U \operatorname{co} G(t)dt$ for every measurable set $U \subset I$. Hence, by Corollary II.3.4 it follows that $g(t) \in \operatorname{co} G(t)$ for a.e. $t \in I$. ■

1.3. Fundamental properties of linear mappings $\mathscr{I}$ and $\mathscr{D}$

We shall consider here mappings $\mathscr{I}$ and $\mathscr{D}$ defined on $L(I, \boldsymbol{R}^n)$ and $\mathrm{AC}(I, \boldsymbol{R}^n)$, respectively by

$$(\mathscr{I}u)(t) = \int_\sigma^t u(\tau)d\tau \quad \text{for } u \in L(I, \boldsymbol{R}^n),\ t \in I, \tag{1.2}$$

$$(\mathscr{D}x)(t) = \dot{x}(t) \quad \text{for } x \in \mathrm{AC}(I, \boldsymbol{R}^n),\ \text{a.e. } t \in I, \tag{1.3}$$

where $I := [\sigma, \sigma+a]$ with $a > 0$.

LEMMA 1.13. *The mapping $\mathscr{I}$ defined by* (1.2) *has the following properties*:

(i) *$\mathscr{I}$ is linear isometry of $L(I, \boldsymbol{R}^n)$ on $\mathrm{AC}^0(I, \boldsymbol{R}^n)$,*

(ii) *$\mathscr{I}$ is nonexpensive as a mapping of $L(I, \boldsymbol{R}^n)$ into $C^0(I, \boldsymbol{R}^n)$, i.e. $|\mathscr{I}u - \mathscr{I}v|_a \leqslant |u-v|_a$ for every $u, v \in L(I, \boldsymbol{R}^n)$,*

(iii) *the restriction of $\mathscr{I}$ to each weakly compact set $\Lambda \subset L([\sigma, \sigma+a], \boldsymbol{R}^n)$ is sequentially strong-weakly continuous on Λ as a mapping of Λ into $C(I, \boldsymbol{R}^n)$, i.e. for every $u \in \Lambda$ and every sequence (u_n) of Λ weakly converging to u one has $|\mathscr{I}u_n - \mathscr{I}u|_a \to 0$, where $|\cdot|_a$ denotes the supremum norm of $C^0([\sigma, \sigma+a], \boldsymbol{R}^n)$.*

Proof. Recall, the norms $|\cdot|_a$, $|\cdot|_a$ and $\|\cdot\|_a$ of $C^0(I, \boldsymbol{R}^n)$, $L(I, \boldsymbol{R}^n)$ and $\mathrm{AC}^0(I, \boldsymbol{R}^n)$ with $I = [\sigma, \sigma+a]$ are defined by $|x|_a := \sup_{\sigma \leqslant t \leqslant \sigma+a} |x(t)|$, $|u|_a := \int_\sigma^{\sigma+a} |u(t)|dt$ and $\|x\|_a := \int_\sigma^{\sigma+a} |\dot{x}(t)|dt$, respectively, for $x \in C^0(I, \boldsymbol{R}^n)$, $u \in L(I, \boldsymbol{R}^n)$ and $x \in \mathrm{AC}^0(I, \boldsymbol{R}^n)$. Hence it follows that $\|\mathscr{I}u - \mathscr{I}v\|_a = |u-v|_a$ for every $u, v \in L(I, \boldsymbol{R}^n)$. Then $\mathscr{I}$ is the isometry of $L(I, \boldsymbol{R}^n)$ on $\mathrm{AC}^0(I, \boldsymbol{R}^n)$. The linearity of $\mathscr{I}$ is a consequence of its definition.

Since $|\mathscr{I}u - \mathscr{I}v|_a = \sup_{\sigma \leqslant t \leqslant \sigma+a} \Big|\int_\sigma u(\tau)d\tau - \int_\sigma v(\tau)d\tau\Big| \leqslant |u-v|_a$, then $\mathscr{I}$ is nonexpensive as a mapping of $L(I, \boldsymbol{R}^n)$ into $C^0(I, \boldsymbol{R}^n)$.

Let Λ be any weakly compact subset of $L(I, \boldsymbol{R}^n)$ and let $u \in \Lambda$ and (u_n) be an arbitrary sequence of Λ weakly converging to u. Then for every $t \in I$ we have $\int_\sigma^t u_n(\tau)d\tau \to \int_\sigma^t u(\tau)d\tau$ as $n \to \infty$. Let $x_n = \mathscr{I}u_n$ and $x = \mathscr{I}u$. We have $x_n(t) \to x(t)$ for $t \in I$ as $n \to \infty$. Then by Egoroff's theorem for every $\delta > 0$ there is a measurable set $E \subset I$ with $\mu(I \setminus E) < \delta$ such that $\sup_{t \in E} |x_n(t) - x(t)| \to 0$ as $n \to \infty$. By Dunford's theorem, Λ is uniformly integrable in $L(I, R^n)$. Since, $u, u_n \in \Lambda$ then for every $\varepsilon > 0$ there is a $\delta_\varepsilon > 0$ such that for every measurable set $E_\varepsilon \subset I$ with $\mu(I \setminus E_\varepsilon) < \delta_\varepsilon$ we have $\int_{I \setminus E_\varepsilon} |u_n(\tau)| d\tau < \frac{1}{8}\varepsilon$ and $\int_{I \setminus E_\varepsilon} |u(\tau)| d\tau < \frac{1}{8}\varepsilon$ for $n = 1, 2, \ldots$ Now, for every $\varepsilon > 0$ we can find a measurable set $E_\varepsilon \subset I$ and a positive integer $N_\varepsilon \geqslant 1$ such that $\sup_{t \in E_\varepsilon} |x_n(t) - x(t)| \leqslant \frac{1}{4}\varepsilon$ for $n \geqslant N_\varepsilon$ and $\int_{I \setminus E_\varepsilon} |u_n(\tau)| d\tau < \frac{1}{8}\varepsilon$, $\int_{I \setminus E_\varepsilon} |u(\tau)| d\tau < \frac{1}{8}\varepsilon$ for $n = 1, 2, \ldots$ Thus for every $\varepsilon > 0$ and $n \geqslant N_\varepsilon$ we have

$$|x_n - x|_a \leqslant 2 \sup_{t \in E_\varepsilon} |x_n(t) - x(t)| + 2 \int_{I \setminus E_\varepsilon} |u_n(t)| dt + 2 \int_{I \setminus E_\varepsilon} |u(\tau)| d\tau \leqslant \varepsilon.$$

Then $|x_n - x|_a \to 0$ as $n \to \infty$. ∎

COROLLARY 1.5. *The mapping $\mathscr{I}$ defined by* (1.2) *is weakly-weakly continuous* (w.-w.c.) *as a mapping of $L(I, R^n)$ on* $\mathrm{AC}^0(I, R^n)$.

Indeed, by (i) of Lemma 1.13, $\mathscr{I}$ is a linear mapping of a Banach space $X := L(I, R^n)$ on a Banach space $Y := \mathrm{AC}^0(I, R^n)$. Then by the Dunford–Schwartz theorem it is continuous with respect to norm topologies in X and Y if and only if it is continuous with respect to the weak topologies. But, by (i) of Lemma 1.13, $\mathscr{I}$ is continuous with respect to norm topologies in X and Y, because it is isometry. Therefore $\mathscr{I}$ is w.-w.c. from X to Y. ∎

COROLLARY 1.6. *For every weakly compact set $\Lambda \subset L(I, R^n)$, $K := \mathscr{I}(\Lambda)$ is a compact subset of $C^0(I, R^n)$ and a weakly compact subset of* $\mathrm{AC}^0(I, R^n)$. *Furthermore, K is convex if Λ is convex.* ∎

COROLLARY 1.7. *For every nonempty weakly compact set $\Lambda \subset L(I, R^n)$ the restriction of $\mathscr{I}$ to Λ is completely continuous on Λ as a mapping from Λ into $C^0(I, R^n)$, i.e., it is continuous on Λ and $\mathscr{I}(B)$ is relatively compact in $C^0(I, R^n)$ for every $B \subset \Lambda$.* ∎

LEMMA 1.14. *For every relatively weakly compact set $\Lambda \subset L(I, R^n)$) the restriction of the mapping $\mathscr{D}$ defined by* (1.3) *to the set $K := \mathscr{I}(\Lambda)$ is sequentially weakly-strongly continuous on K as a mapping defined on a subset of $C^0(I, R^n)$ into $L(I, R^n)$, i.e. for every $x \in K$ and every sequence (x_n) of K converging in the norm topology of $C^0(I, R^n)$ to x, $\mathscr{D}x_n \rightharpoonup \mathscr{D}x$ in $L(I, R^n)$ as $n \to \infty$.*

Proof. Let Λ be a relatively weakly compact subset of $L(I, R^n)$, $x \in K$ and let (x_n) be an arbitrary sequence of K such that $|x_n - x|_a \to 0$ as $n \to \infty$. Since $K = \mathscr{I}(\Lambda)$ then $\dot{x}_n \in \Lambda$ for each $n = 1, 2, \ldots$ Thus $(\dot{x}_n)$ is uniformly integrable. Now result follows immediately from Lemma I.4.5. ∎

2. Subtrajectory and trajectory integrals of set-valued functions

We consider here topological properties of subtrajectory and trajectory integrals of set-valued functions that belong to the space $\mathscr{A}(I, \boldsymbol{R}^n)$. Furthermore, we will also be interested in the properties of subtrajectory and trajectory integrals of families of such set-valued functions. Recall, that for a given $G \in \mathscr{A}(I, \boldsymbol{R}^n)$, the family of all L-integrable, in fact of all L-measurable, selectors of G is denoted by $\mathscr{F}(G)$ and called the *subtrajectory integrals of* G and its image $\mathscr{I}\mathscr{F}(G)$ by a linear mapping $\mathscr{I}$ defined by (1.2), *trajectory integrals of* G.

2.1. Topological properties of subtrajectory integrals

We shall consider here the Banach spaces $L(I, \boldsymbol{R}^n)$ and $\mathrm{AC}^0(I, \boldsymbol{R}^n)$ as locally convex topological vector spaces with their norm and weak topologies.

Given sets $A \subset L(I, \boldsymbol{R}^n)$ and $B \subset \mathrm{AC}^0(I, \boldsymbol{R}^n)$ we will denote by $\overline{[A]}_L$, $\overline{[B]}_{\mathrm{AC}}$ and $\overline{[B]}_C$ the closure of A and B in the norm topologies of $L(I, \boldsymbol{R}^n)$, $\mathrm{AC}^0(I, \boldsymbol{R}^n)$ and $C^0(I, \boldsymbol{R}^n)$, respectively, whereas $\overline{[A]}_L^w$ and $\overline{[B]}_{AC}^w$ will denote their closures in the weak topology of $L(I, \boldsymbol{R}^n)$ and $\mathrm{AC}^0(I, \boldsymbol{R}^n)$ respectively.

THEOREM 2.1. *For every* $G \in \mathscr{A}(I, \boldsymbol{R}^n)$, $\mathscr{F}(G)$ *is a nonempty subset of* $L(I, \boldsymbol{R}^n)$ *such that*:

(i) $\mathscr{F}(G)$ *is a closed and bounded subset of* $\Lambda_G := \{u \in L(I, \boldsymbol{R}^n)\colon |u(t)| \leqslant ||G(t)||$ *for a.e.* $t \in I\}$,

(ii) $\mathscr{F}(G)$ *is relatively weakly compact in* $L(I, \boldsymbol{R}^n)$,

(iii) $\mathscr{I}\mathscr{F}(G)$ *is a nonempty closed and bounded subset of* $\mathrm{AC}^0(I, \boldsymbol{R}^n)$,

(iv) $\mathscr{I}\mathscr{F}(G)$ *is relatively weakly compact in* $\mathrm{AC}^0(I, \boldsymbol{R}^n)$,

(v) $\overline{\mathscr{I}[\mathscr{F}(G)]_L^w}$ *is compact in* $C^0(I, \boldsymbol{R}^n)$ *and weakly compact in* $\mathrm{AC}^0(I, \boldsymbol{R}^n)$.

Proof. Since G is measurable then by Kuratowski–Ryll-Nardzewski measurable selection theorem we have $\mathscr{F}(G) \neq \emptyset$. Obviously, $\mathscr{F}(G) \subset \Lambda_G$. Since Λ_G is weakly compact in $L(I, \boldsymbol{R}^n)$ then $\mathscr{F}(G)$ is bounded and relatively weakly sequentially compact. Thus by Eberlein–Šmulian's theorem (ii) holds. Suppose $u \in \overline{[\mathscr{F}(G)]}_L$ and let (u_k) be a sequence of $\mathscr{F}(G)$ such that $|u_k - u|_a \to 0$ as $k \to \infty$. Let (u_{k_i}) be a subsequence of (u_k) such that $u_{k_i}(t) \to u(t)$ for a.e. $t \in I$ as $i \to \infty$. Since $u_{k_i}(t) \in G(t)$ for $i = 1, 2, \ldots$ and a.e. $t \in I$ and G has compact values, we get $u(t) \in G(t)$ for a.e. $t \in I$, i.e. $u \in \mathscr{F}(G)$. Thus (i) holds. Now, by Eberlein–Šmulian's theorem, $\overline{[\mathscr{F}(G)]}_L^w$ is weakly compact and therefore, by (iii) of Lemma 1.13, $\overline{\mathscr{I}[\mathscr{F}(G)]_L^w}$ is a compact subset of $C^0(I, \boldsymbol{R}^n)$. Furthermore, by (i) of Lemma 1.13, $\mathscr{I}$ is a homeomorphism between $L(I, \boldsymbol{R}^n)$ and $\mathrm{AC}^0(I, \boldsymbol{R}^n)$. Hence, and (i), it follows that $\mathscr{I}\mathscr{F}(G)$ is a closed subset of $\mathrm{AC}^0(I, \boldsymbol{R}^n)$. Now, by (i) of Lemma 1.13 and weak compactness of $\overline{[\mathscr{F}(G)]}_L^w$, $\overline{\mathscr{I}[\mathscr{F}(G)]_L^w}$ is weakly compact in $\mathrm{AC}^0(I, \boldsymbol{R}^n)$. Since $\mathscr{I}\mathscr{F}(G) \subset \overline{\mathscr{I}[\mathscr{F}(G)]_L^w}$, then $\mathscr{I}\mathscr{F}(G)$ is bounded and relatively weakly sequentially compact in $\mathrm{AC}^0(I, \boldsymbol{R}^n)$. Hence, by Eberlein–Šmulian's theorem, (iii) follows. ■

THEOREM 2.2. *For every* $G \in \mathscr{A}(I, \boldsymbol{R}^n)$ *the following equalities hold:*

(i) $\mathscr{I}\overline{[\mathscr{F}(G)]}_L^w = \overline{[\mathscr{I}\mathscr{F}(G)]}_C$ *and*

(ii) $\mathscr{I}\overline{[\mathscr{F}(G)]}_L^w = \overline{[\mathscr{I}\mathscr{F}(G)]}_{AC}^w$.

Proof. We have $\mathscr{I}\mathscr{F}(G) \subset \mathscr{I}\overline{[\mathscr{F}(G)]}_L^w$. Then, by the compactness of $\mathscr{I}\overline{[\mathscr{F}(G)]}_L^w$ in $C^0(I, \boldsymbol{R}^n)$ we get $\overline{[\mathscr{I}\mathscr{F}(G)]}_C \subset \mathscr{I}\overline{[\mathscr{F}(G)]}_L^w$.

Suppose $x \in \mathscr{I}\overline{[\mathscr{F}(G)]}_L^w$. Then $\dot{x} \in \overline{[\mathscr{F}(G)]}_L^w$ and hence, since $\overline{[\mathscr{F}(G)]}_L^w$ is weakly compact, by Šmulian theorem there is a sequence (u_n) of $\mathscr{F}(G)$ weakly convergent to $\dot{x}$. Of course $\mathscr{I}u_n \in \mathscr{I}\mathscr{F}(G)$. By (iii) of Lemma 1.13, we have $|\mathscr{I}u_n - x|_a \to 0$ as $n \to \infty$. Therefore $x \in \overline{[\mathscr{I}\mathscr{F}(G)]}_C$, i.e., $\mathscr{I}\overline{[\mathscr{F}(G)]}_L^w \subset \overline{[\mathscr{I}\mathscr{F}(G)]}_C$.

Similarly as above we have $\overline{[\mathscr{I}\mathscr{F}(G)]}_{AC}^w \subset \mathscr{I}\overline{[\mathscr{F}(G)]}_L^w$. Suppose $x \in \mathscr{I}\overline{[\mathscr{F}(G)]}_L^w$ and let (u_n) be a sequence of $\mathscr{F}(G)$ weakly convergent to $\dot{x}$. In particular, hence it follows $(\mathscr{I}u_n)(t) \to x(t)$ for all $t \in I$. Furthermore, by (iv) of Theorem 2.1 and Eberlein–Šmulian theorem $\overline{[\mathscr{I}\mathscr{F}(G)]}_{AC}^w$ is weakly compact in $AC^0(I, \boldsymbol{R}^n)$. Therefore a sequence (x_n) with $x_n = \mathscr{I}u_n$ for $n = 1, 2, \ldots$ is equiabsolutely continuous. Since $x(t) = \lim_{n\to\infty} x_n(t)$ at each $t \in I$, then (x_n) is weakly converging in $AC^0(I, \boldsymbol{R}^n)$ to x. Thus $x \in \operatorname{Seq\,cl}_w(\mathscr{I}\mathscr{F}(G)) \subset \overline{[\mathscr{I}\mathscr{F}(G)]}_{AC}^w$, i.e., $\mathscr{I}\overline{[\mathscr{F}(G)]}_L^w \subset \overline{[\mathscr{I}\mathscr{F}(G)]}_{AC}^w$. ∎

THEOREM 2.3. *For every* $G \in \mathscr{A}(I, \boldsymbol{R}^n)$, $\mathscr{F}(\operatorname{co} G)$ *is a convex weakly compact subset of* $L(I, \boldsymbol{R}^n)$ *such that* $\mathscr{F}(\operatorname{co} G) = \overline{[\mathscr{F}(G)]}_L^w$.

Proof. Since $\|\operatorname{co} G(t)\| \leqslant \|G(t)\|$ for $t \in I$, then by Theorem II.3.9, for every $G \in \mathscr{A}(I, \boldsymbol{R}^n)$ we have also $\operatorname{co} G \in \mathscr{A}(I, \boldsymbol{R}^n)$. Furthermore, by (i) and (ii) of Theorem 2.1, $\mathscr{F}(\operatorname{co} G)$ is nonempty, closed bounded and relatively weakly sequentially compact in $L(I, \boldsymbol{R}^n)$. We have of course $\mathscr{F}(\operatorname{co} G) \subset \Lambda_G$. Since $\operatorname{co} G(t)$ is convex, $\mathscr{F}(\operatorname{co} G)$ is also a convex subset of $L(I, \boldsymbol{R}^n)$. Then $\mathscr{F}(\operatorname{co} G)$ is weakly closed and therefore weakly compact in $L(I, \boldsymbol{R}^n)$. Hence and $\mathscr{F}(G) \subset \mathscr{F}(\operatorname{co} G)$ it follows that $\overline{[\mathscr{F}(G)]}_L^w \subset \mathscr{F}(\operatorname{co} G)$.

Suppose now $u \in \mathscr{F}(\operatorname{co} G)$, i.e., u is a measurable selector of $\operatorname{co} G$. Hence and Theorem II.3.20 it follows that for every subinterval $I_j \subset I$ there is a measurable function $v_j \colon I_j \to \boldsymbol{R}^n$ such that $v_j(t) \in G(t)$ for a.e. $t \in I_j$ and $\int_{I_j} v_j(t)\,dt = \int_{I_j} u(t)\,dt$. Select now for each positive integer m, a family $\{I_j^m\}_{j\in\{1,\ldots,N_m\}}$ of subintervals I_j^m of I in such a way that $I_j^m \cap I_i^m = \emptyset$ for $i \neq j$, $\bigcup_{j=1}^{N_m} I_j^m = I$ and such that $\int_{I_j^m} \|G(t)\|\,dt < 1/2m$ for $j \in N_m = \{1, \ldots, N_m\}$. Let $v_j^m(t) \in G(t)$ for a.e. $t \in I_j^m$ and $\int_{I_j^m} v_j^m(t)\,dt = \int_{I_j^m} u(t)\,dt$. Put $v_m = \sum_{j=1}^{N_m} \chi_i^m v_j^m$, where χ_j^m denotes the characteristic function of I_j^m. By the decomposability of $\mathscr{F}(G)$ we get $v_m(t) \in G(t)$ for a.e. $t \in I$ and $\int_{I_j^m} v_m(t)\,dt = \int_{I_j^m} u(t)\,dt$ for each $j \in N_m$ and $m = 1, 2, \ldots$ Then $v_m \in \mathscr{F}(G)$ and $|(\mathscr{I}v_m)(t) -$

$-(\mathscr{I}u)(t)| = \left|\int_{\sigma}^{t} [v_m(\tau)-u(\tau)]\,d\tau\right| \leqslant 2\int_{I_j^k} \|G(\tau)\|\,d\tau \leqslant 1/m$ for each $m = 1, 2, \ldots,$ where $k \in N_m$ is such that $t \in I_j^k$. Therefore $(\mathscr{I}v_m)(t) \to (\mathscr{I}u)(t)$ as $m \to \infty$ for each $t \in I$. Now, similarly as in the proof of (iii) in Lemma 1.13, hence it follows that $|\mathscr{I}v_m - \mathscr{I}u|_a \to 0$ as $m \to \infty$. Since $\mathscr{I}v_m, \mathscr{I}u \in \mathscr{I}\Lambda_G$ by virtue of Lemma 1.14 it implies that $\mathscr{D}\mathscr{I}v_m \rightharpoonup \mathscr{D}\mathscr{I}u$, i.e., $v_m \rightharpoonup u$ as $m \to \infty$. Therefore, $u \in \mathrm{Seq\,cl_w}(\mathscr{F}(G)) \subset \overline{[\mathscr{F}(G)]}_L^w$, i.e. $\mathscr{F}(\mathrm{co}\,G) \subset \overline{[\mathscr{F}(G)]}_L^w$. ■

Now, as a corollary of Theorems 2.1–2.3 we obtain.

THEOREM 2.4. *For every $G \in \mathscr{A}(I, \boldsymbol{R}^n)$, $\mathscr{I}\mathscr{F}(\mathrm{co}\,G)$ is a nonempty convex compact subset of $C^0(I, \boldsymbol{R}^n)$ and a convex weakly compact subset of* $\mathrm{AC}^0(I, \boldsymbol{R}^n)$. *Moreover one has:*

(i) $\mathscr{I}\mathscr{F}(\mathrm{co}\,G) = \overline{[\mathscr{I}\mathscr{F}(G)]}_C$ *and*

(ii) $\mathscr{I}\mathscr{F}(\mathrm{co}\,G) = \overline{[\mathscr{I}\mathscr{F}(G)]}_{AC}^w$. ■

2.2. *Subtrajectory integrals of set-valued functions depending on parameters*

Let $I = [\sigma, \sigma+a]$ and let $\mathscr{K}_\lambda$ and B_λ denote closed balls of $C^0(I, \boldsymbol{R}^n)$ and $L(I, \boldsymbol{R}^n)$, respectively with centres at the origin and a radius $\lambda > 0$. We shall consider a family of all set-valued functions $G\colon I \times \mathscr{K}_\lambda \times B_\lambda \to \mathrm{Comp}(\boldsymbol{R}^n)$ such that $G(\cdot, x, z) \in \mathscr{A}(I, \boldsymbol{R}^n)$ for each fixed $(x, z) \in \mathscr{K}_\lambda \times B_\lambda$. We will deal with equivalence classes of members of the above family, where equivalence is defined by equality almost everywhere in I for each fixed $(x, z) \in \mathscr{K}_\lambda \times B_\lambda$. As usual, we will not distinguish between a function and its equivalence class. The space of all such equivalence classes is denoted by $H_\lambda(I, \boldsymbol{R}_n)$. We will also deal with a subspace $\mathscr{H}_\lambda(I, \boldsymbol{R}^n) \subset H_\lambda(I, \boldsymbol{R}^n)$ of all $G \in H_\lambda(I, \boldsymbol{R}^n)$ such that a family $\{G(\cdot, x, z)\colon (x, z) \in \mathscr{K}_\lambda \times B_\lambda\} \subset \mathscr{A}(I, \boldsymbol{R}^n)$ is uniformly integrable. We define a metric δ on $\mathscr{H}_\lambda(I, \boldsymbol{R}^n)$ by

$$\delta(F, G) = \sup\{d(F(\cdot, x, z), G(\cdot, x, z))\colon (x, z) \in \mathscr{K}_\lambda \times B_\lambda\}$$

for $F, G \in \mathscr{H}_\lambda(I, \boldsymbol{R}^n)$, where d is the metric defined on $\mathscr{A}(I, \boldsymbol{R}^n)$ by (1.1). Observe that for every $F, G \in \mathscr{H}_\lambda(I, \boldsymbol{R}^n)$ we have $\delta(F, G) < +\infty$, because by the uniform integrability of $F, G \in \mathscr{H}_\lambda(I, \boldsymbol{R}^n)$ there is a positive number M such that $\max\left(\int_\sigma^{\sigma+a} \|F(t, x, z)\|, \int_\sigma^{\sigma+a} \|G(t, x, z)\|\,dt\right) \leqslant M$ for every $(x, z) \in \mathscr{K}_\lambda \times B_\lambda$. It is easy to see that δ is a metric on $\mathscr{H}_\lambda(I, \boldsymbol{R}^n)$.

LEMMA 2.5. *$(\mathscr{H}_\lambda(I, \boldsymbol{R}^n), \delta)$ is a complete metric space.*

Proof. Let (G_n) be a Cauchy sequence of $\mathscr{H}_\lambda(I, \boldsymbol{R}^n)$. Then, $d(G_n(\cdot, x, z), G_m(\cdot, x, z)) \to 0$ as $n, m \to \infty$ uniformly with respect to $(x, z) \in \mathscr{K}_\lambda \times B_\lambda$. Thus, for each fixed $(x, z) \in \mathscr{K}_\lambda \times B_\lambda$, $\{G_n(\cdot, x, z)\}$ is a Cauchy sequence of $\mathscr{A}(I, \boldsymbol{R}^n)$. Therefore, for every $(x, z) \in \mathscr{K}_\lambda \times B_\lambda$ there exists $G(\cdot, x, z) \in \mathscr{A}(I, \boldsymbol{R}^n)$ such that $d(G_n(\cdot, x, z), G(\cdot, x, z)) \to 0$ as $n \to \infty$. It is clear that $\{G_n(\cdot, x, z)\}$ tends to $G(\cdot, x, z)$ uniformly with respect to $(x, z) \in \mathscr{K}_\lambda \times B_\lambda$ as $n \to \infty$.

Since for arbitrarily large n and a measurable set $E \subset I$

$$\sup_{(x,z)\in\mathcal{K}_\lambda\times B_\lambda}\int_E \|G(t,x,z)\|\,dt \leqslant \delta(G,G_n)+\sup_{(x,z)\in\mathcal{K}_\lambda\times B_\lambda}\int_E \|G_n(t,x,z)\|\,dt$$

then $G\in\mathcal{H}_\lambda(I,R^n)$. ■

REMARK 2.1. We can also consider a weak convergence of sequences of $\mathcal{H}_\lambda(I,R^n)$. Namely, we will say that a sequence (G_n) of $\mathcal{H}_\lambda(I,R^n)$ *weakly converges to* $G\in\mathcal{H}_\lambda(I,R^n)$ if for every measurable set $U\subset I$ one has $\lim_{n\to\infty}\sup_{(x,z)\in\mathcal{K}_\lambda\times B_\lambda} h\left(\int_U G_n(t,x,z)\,dt, \int_U G(t,x,z)\,dt\right)=0$. We can also prove that $\mathcal{H}_\lambda(I,R^n)$ is weakly sequentially complete, i.e., that each weak Cauchy sequence of $\mathcal{H}_\lambda(I,R^n)$ has a limit in $\mathcal{H}_\lambda(I,R^n)$. Indeed, if (G_n) is a weak Cauchy sequence of $\mathcal{H}_\lambda(I,R^n)$, i.e., such that $\lim_{n\to\infty}\sup_{(x,z)\in\mathcal{K}_\lambda\times B_\lambda} h\left(\int_U G_n(t,x,z)\,dt, \int_U G_m(t,x,z)\,dt\right)=0$ for each measurable set $U\subset I$, then for every fixed $(x,z)\in\mathcal{K}_\lambda\times B_\lambda$ a sequence $\{G_n(\cdot,x,z)$ is a weak Cauchy sequence of $\mathcal{A}(I,R^n)$. Thus for every $(x,z)\in\mathcal{K}_\lambda\times B_\lambda$ there exists $G(\cdot,x,z)\in\mathcal{A}(I,R^n)$ such that a sequence $\{G_n(\cdot,x,z)\}$ weakly converges to $G(\cdot,x,z)$ for fixed $(x,z)\in\mathcal{K}_\lambda\times B_\lambda$. Similarly as in the proof of Lemma 2.5 we can easily verify that $\{G_n(\cdot,x,z)\}$ weakly converges to $G(\cdot,x,z)$ uniformly with respect to $(x,z)\in\mathcal{K}_\lambda\times B_\lambda$, i.e., that (G_n) weakly converges to G in $\mathcal{H}_\lambda(I,R^n)$. Hence, similarly as in the proof of Lemma 2.5, finally it follows that $G\in\mathcal{H}_\lambda(I,R^n)$ since for a measurable set $E\subset I$ and $n\in N$ sufficiently large we have

$$\sup_{(x,z)\in\mathcal{K}_\lambda\times B_\lambda}\left\|\int_E G(t,x,z)\,dt\right\|$$

$$\leqslant \sup_{(x,z)\in\mathcal{K}_\lambda\times B_\lambda} h\left(\int_E G(t,x,z)\,dt, \int_E G_n(t,x,z)\,dt\right)+$$

$$+\sup_{(x,z)\in\mathcal{K}_\lambda\times B_\lambda}\left\|\int_E G_n(t,x,z)\,dt\right\|. \ ■$$

Given $G\in H_\lambda(I,R^n)$ and each fixed $(x,z)\in\mathcal{K}_\lambda\times B_\lambda$, by $\mathcal{F}(G)(x,z)$ we will denote the subtrajectory integrals of a set-valued function $G(\cdot,x,z)\in\mathcal{A}(I,R^n)$, and by $\mathcal{I}\mathcal{F}(G)(x,z)$ its trajectory integrals, i.e. $\mathcal{F}(G)(x,z)=\mathcal{F}[G(\cdot,x,z)]$ and $\mathcal{I}\mathcal{F}(G)(x,z)=\mathcal{I}\mathcal{F}[G(\cdot,x,z)]$. Then we arrive to new set-valued functions defined on $\mathcal{K}_\lambda\times B_\lambda$ by taking $\mathcal{F}(G)(x,z)$ and $\mathcal{I}\mathcal{F}(G)(x,z)$ for each $(x,z)\in\mathcal{K}_\lambda\times B_\lambda$. We shall denote them still by $\mathcal{F}(G)$ and $\mathcal{I}\mathcal{F}(G)$, respectively. It is quite natural, by virtue of Lemma 2.1 to consider $\mathcal{F}(G)$ and $\mathcal{I}\mathcal{F}(G)$ as set-valued functions with values in spaces $\mathrm{Cl}(L(I,R^n))$ and $\mathrm{Cl}(\mathrm{C}^0(I,R^n))$, respectively. The Hausdorff metrics in $\mathrm{Cl}(L(I,R^n))$ and $\mathrm{Cl}(\mathrm{C}^0(I,R^n))$ induced by the norms of $L(I,R^n)$ and $\mathrm{C}^0(I,R^n)$ will be denoted by H_L and H_C, respectively. They are defined by theirs subdistances $\bar{\mathrm{H}}_L$ and $\bar{\mathrm{H}}_C$, respectively (see Section II.1.1).

Given $G\in\mathcal{H}(I,R^n)$ we will be interested in sufficient conditions that imply

Lipschitz continuity, H-upper and H-lower semicontinuities, strong-weak, weak-strong and weak-weak upper and lower semicontinuities of mappings $\mathscr{F}(G)$ and $\mathscr{I}\mathscr{F}(G)$.

Recall, we say that $\mathscr{F}(G)$ is *Lipschitz continuous on* $\mathscr{K}_\lambda \times B_\lambda$ if there is a number $L > 0$ such that

$$H_L(\mathscr{F}(G)(x, z), \mathscr{F}(G)(\bar{x}, \bar{z})) \leqslant L\max(|x-\bar{x}|_a, |z-\bar{z}|_a) \tag{2.2}$$

for each $(x, z), (\bar{x}, \bar{z}) \in \quad_\lambda \times B_\lambda$.

Similarly, we say that $\mathscr{F}(G)$ is *Lipschitz continuous on* B_λ uniformly with respect to $x \in \mathscr{K}_\lambda$ if there is a number $L > 0$ such that

$$H_L(\mathscr{F}(G)(x, z), \mathscr{F}(G)(x, \bar{z})) \leqslant L|z-\bar{z}|_a \tag{2.3}$$

for each $x \in \mathscr{K}_\lambda$ and $z, \bar{z} \in B_\lambda$.

We will say that $\mathscr{F}(G)$ is *H-upper (H-lower) semicontinuous on* $\mathscr{K}_\lambda \times B_\lambda$ if for every $(x, z) \in \mathscr{K}_\lambda \times B_\lambda$ and every sequence $\{(x_n, z_n)\}$ such that $|x_n - x|_a \to 0$ and $|z_n - z|_a \to 0$ as $n \to \infty$ we have $\lim_{n\to\infty} \bar{H}_L(\mathscr{F}(G)(x_n, z_n), \mathscr{F}(G)(x, z)) = 0$ ($\lim_{n\to\infty} \bar{H}_L(\mathscr{F}(G)(x, z), \mathscr{F}(G)(x_n, z_n)) = 0$).

Finally, we say that $\mathscr{F}(G)\colon \mathscr{K}_\lambda \times B_\lambda \to \mathrm{Cl}(L(I, \boldsymbol{R}^n))$ is weakly-weakly upper (lower) semicontinuous on $\mathscr{K}_\lambda \times B_\lambda$ if for every weakly closed set $M \subset L(I, \boldsymbol{R}^n)$ the set $\mathscr{F}(G)^-(M) = \{(x, z) \in \mathscr{K}_\lambda \times B_\lambda\colon \mathscr{F}(G)(x, z) \cap M \neq \emptyset\}$ $(\mathscr{F}(G)_-(M) = \{(x, z) \in \mathscr{K}_\lambda \times B_\lambda\colon \mathscr{F}(G)(x, z) \subset M\})$ is sequentially weakly closed in the following sense: for every sequence $\{(x_n, z_n)\}$ of $\mathscr{F}(G)^-(M)$ (of $\mathscr{F}(G)_-(M)$) such that $|x_n - x|_a \to 0$ and $z_n \rightharpoonup z$ as $n \to \infty$ we have $(x, z) \in \mathscr{F}(G)^-(M)((x, z) \in \mathscr{F}(G)_-(M))$.

Similarly, we will say that $\mathscr{F}(G)$ is *weakly-strongly upper (lower) semicontinuous on* $\mathscr{K}_\lambda \times B_\lambda$ if for every closed set $M \subset L(I, \boldsymbol{R}^n)$ the set $\mathscr{F}(G)^-(M)(\mathscr{F}(G)_-(M))$ is sequentially weakly closed in the above sense.

Recall, that weak-weak, strong-weak upper or lower semicontinuities are briefly denoted by w.-w.u.s.c., s.-w.u.s.c. or w.-w.l.s.c. and s.-w.l.s.c., respectively.

Let $G \in H_\lambda(I, \boldsymbol{R}^n)$ be given. In what follows we will consider the following types of continuities of G with respect to its last two variables.

We say that G is *Lipschitz continuous with respect to its last two variables* if there is a Lebesgue integrable function $k\colon I \to \boldsymbol{R}$ such that for every $x, \bar{x} \in \mathscr{K}_\lambda$, $z, \bar{z} \in B_\lambda$ and a.e. $t \in I$ we have

$$h(G(t, x, z), G(t, \bar{x}, \bar{z})) \leqslant k(t)\max(|x-\bar{x}|_t, |z-\bar{z}|_t), \tag{2.4}$$

where $|x-\bar{x}|_t = \sup_{\sigma \leqslant \tau \leqslant t} |x(\tau) - \bar{x}(\tau)|$ and $|z-\bar{z}|_t = \int_\sigma^t |z(\tau) - \bar{z}(\tau)|\,d\tau$.

We say that G is *Lipschitz continuous with respect to its last variable uniformly with respect to* $x \in \mathscr{K}_\lambda$ if there is a Lebesgue integrable function $k\colon I \to \boldsymbol{R}$ such that

$$h(G(t, x, z), G(t, x, \bar{z})) \leqslant k(t)|z-\bar{z}|_t \tag{2.5}$$

for a.e. $t \in I$, $x \in \mathscr{K}_\lambda$ and $z, \bar{z} \in B_\lambda$.

We will say that G is *weakly-strongly upper (lower) semicontinuous with respect to its last two variables* if for every $(x, z) \in \mathscr{K}_\lambda \times B_\lambda$ and every sequence $\{(x_n, z_n)\}$ of $\mathscr{K}_\lambda \times B_\lambda$ such that $|x_n - x|_a \to 0$ and $|z - \bar{z}|_a \to 0$ as $n \to \infty$ we have

$$\lim_{n\to\infty} \bar{h}\Big(\int_U (G(t, x_n, z_n)dt, \int_U G(t, x, z)dt\Big) = 0 \tag{2.6}$$

$$\Big(\lim_{n\to\infty} \bar{h}\Big(\int_U G(t, x, z)dt, \int_U G(t, x_n, z_n)dt\Big) = 0\Big)$$

for every measurable set $U \subset I$.

We say that G is *strongly-weakly upper (lower) semicontinuous with respect to its last two variables* if for every $(x, z) \in \mathscr{K}_\lambda \times B_\lambda$ and every sequence $\{(x_n, z_n)\}$ of $\mathscr{K}_\lambda \times B_\lambda$ such that $|x_n - x|_a \to 0$ and $z_n \rightharpoonup z$ as $n \to \infty$ we have

$$\lim_{n\to\infty} \int_\sigma^{\sigma+a} \bar{h}\big(G(t, x_n, z_n), G(t, x, z)\big)dt = 0 \tag{2.7}$$

$$\Big(\lim_{n\to\infty} \int_\sigma^{\sigma+a} \bar{h}\big(G(t, x, z), G(t, x_n, z_n)\big)dt = 0\Big).$$

We will say that G is *weakly-weakly upper (lower) semicontinuous with respect to its last two variables* if for every $(x, z) \in \mathscr{K}_\lambda \times B_\lambda$ and every sequence $\{(x_n, z_n)\}$ of $\mathscr{K}_\lambda \times B_\lambda$ such that $|x_n - x|_a \to 0$ and $z_n \rightharpoonup z$ as $n \to \infty$, (2.6) holds for every measurable set $U \subset I$.

Analogously we can define strong-strong upper (lower) semicontinuity of G with respect to its last two variables. Similarly as in Section II.2.1 we have in fact above defined some types of sequential weak-weak, strong-weak, etc. upper and lower semicontinuities. They will be reduced in further investigations to upper and lower semicontinuities. Therefore, here and in Section II.2.1 we dropped "sequential" by the calling of these type semicontinuities.

If for each fixed $t \in I$ a set-valued function $G(t, \cdot, \cdot)$ is upper (lower) semicontinuous on $\mathscr{K}_\lambda \times B_\lambda$ we will write: G is u.s.c. (l.s.c.) with respect to its last two variables. If G is simultanously u.s.c. and l.s.c. with respect to its last two variables we call it *continuous* (c.) *with respect to its last two variables.* In a similar way we can also define the following types continuities of G with respect to its last two variables: w.-s.c., s.-w.c., w.-w.c. and s.-s.c.

We say that G is *upper (lower) semicontinuous* weakly, briefly u.s.c.w. (l.s.c.w.) *with respect to its last two variables* if for every $(x, z) \in \mathscr{K}_\lambda \times B_\lambda$ and every sequence $\{(x_n, z_n)\}$ such that $|x_n - x|_a \to 0$ and $z_n \rightharpoonup z$ as $n \to \infty$ we have

$$\lim_{n\to\infty} \bar{h}\big(G(t, x_n, z_n), G(t, x, z)\big) = 0 \quad \big(\lim_{n\to\infty} \bar{h}\big(G(t, x, z, G(t, x_n, z_n)\big) = 0\big)$$

for a.e. $t \in I$.

A set-valued function $G \in H_\lambda(I, \boldsymbol{R}^n)$ is said to be *$\mathscr{A}$-continuous with respect to its last two variables* ($\mathscr{A}$-c.) if a function $\mathscr{K}_\lambda \times B_\lambda \ni (x, z) \to G(\cdot, x, z) \in \mathscr{A}(I, \boldsymbol{R}^n)$ is continuous on $\mathscr{K}_\lambda \times B_\lambda$.

COROLLARY 2.1. *A set-valued function $G \in H_\lambda(I, \boldsymbol{R}^n)$ is $\mathscr{A}$-continuous with respect to its last two variables if and only if it is s.-s.u.s.c. and s.-s.l.s.c. with respect to its last two variables.* ■

A set-valued function $G \in H_\lambda(I, \boldsymbol{R}^n)$ is said to be *$\mathscr{A}$-Lipschitz continuous with respect to its last two variables* if there exists a number $L \geqslant 0$ such that

$$d(G(\cdot, x, z), G(\cdot, \bar{x}, \bar{z})) \leqslant L\max[|x-\bar{x}|_a, |z-\bar{z}|_a]$$

for $(x, z), (\bar{x}, \bar{z}) \in \mathscr{K}_\lambda \times B_\lambda$.

Similarly we define the $\mathscr{A}$-Lipschitz continuity of $G \in H_\lambda(I, \boldsymbol{R}^n)$ with respect to its last variable uniformly with respect to second one.

It is easy to see that for every $G \in \mathscr{H}_\lambda(I, \boldsymbol{R}^n)$ the following system of implications is true, where c.w. denotes the continuity weakly of G with respect to its last two variables.

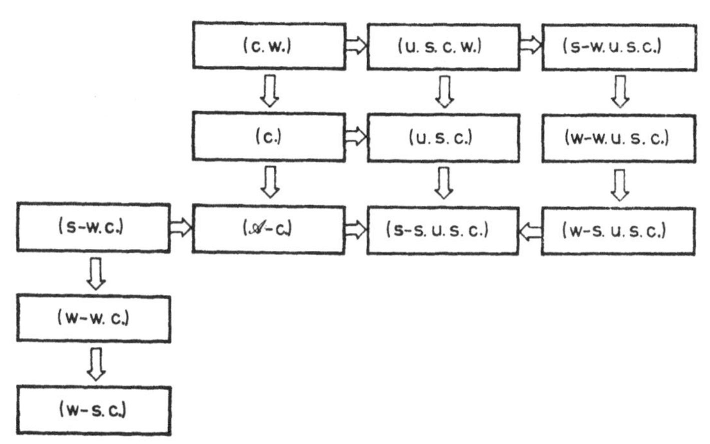

Fig. 6.

We have the following lemmas.

LEMMA 2.6. *Suppose $G \in H_\lambda(I, \boldsymbol{R}^n)$ is Lipschitz continuous with respect to its last two variables. Then for every $L > 0$ there is a norm $|\cdot|$ of $L(I, \boldsymbol{R}^n)$ equivalent to $|\cdot|_a$ and such that for every $z, \bar{z} \in B_\lambda$ and $x, y \in AC^0(I, \boldsymbol{R}^n) \cap \mathscr{K}_\lambda$ we have*

$$\mathrm{H}(\mathscr{F}(G)(x, z), \mathscr{F}(G)(y, \bar{z})) \leqslant L\max(|\dot{x}-\dot{y}|, |z-\bar{z}|), \tag{2.8}$$

where H *is the Hausdorff metric on* $\mathrm{Cl}(L(I, \boldsymbol{R}^n))$ *induced by the norm* $|\cdot|$.

Proof. Let u be an arbitrary element of $\mathscr{F}(G)(x, z)$ and select $v \in \mathscr{F}(G)(y, \bar{z})$ such that

$$\begin{aligned}|u(t)-v(t)| &= \mathrm{dist}(u(t), G(t, y, \bar{z})) \\ &\leqslant h(G(t, x, \bar{z}), G(t, y, z)) \leqslant k(t)\max(|x-y|_t, |z-\bar{z}|_t)\end{aligned}$$

for a.e. $t \in I$. Let $l = 2/L$ and let us renorm $L(I, \boldsymbol{R}^n)$ with an equivalent norm $|\cdot|$

to $|\cdot|_a$ by letting $|w| = \int_\sigma^{\sigma+a} e^{-lK(t)}|w(t)|\,dt$ for $w \in L(I, R^n)$, where $K(t) = \int_\sigma^t k(\tau)d\tau$. We have

$$|u-v| = \int_\sigma^{\sigma+a} e^{-lK(t)}|u(t)-v(t)|\,dt$$

$$\leqslant \int_\sigma^{\sigma+a} k(t)e^{-lK(t)}\max(|x-y|_t, |z-\bar{z}|_t)dt$$

$$\leqslant \int_\sigma^{\sigma+a} k(t)e^{-lK(t)} \sup_{\sigma\leqslant\tau\leqslant t} |x(\tau)-y(\tau)|\,dt+$$

$$+\int_\sigma^{\sigma+a} k(t)e^{-lK(t)}\Big(\int_\sigma^t |z(\tau)-z(\tau)|\,d\tau)dt\Big).$$

By interchanging the integration order, we obtain

$$\int_\sigma^{\sigma+a} k(t)e^{-lK(t)} \sup_{\sigma\leqslant\tau\leqslant t} |x(\tau)-y(\tau)|\,dt$$

$$\leqslant \int_\sigma^{\sigma+a} k(t)e^{-lK(t)}\Big(\int_\sigma^t |\dot{x}(\tau)-\dot{y}(\tau)d\tau\Big)dt = \int_\sigma^{\sigma+a}\int_\sigma^t [k(t)e^{-lK(t)}|\dot{x}(\tau)-\dot{y}(\tau)|]d\tau\, dt$$

$$= \int_\sigma^{\sigma+a}\int_\tau^{\sigma+a} [|\dot{x}(\tau)-\dot{y}(\tau)|k(t)e^{-lK(t)}]dt\,d\tau$$

$$= -\frac{1}{l}e^{-lK(\sigma a)}\int_\sigma^{\sigma+a} |\dot{x}(\tau)-\dot{y}(\tau)|d\tau+\frac{1}{l}\int_\sigma^{\sigma+a} k(\tau)e^{-lK(t)}|\dot{x}(\tau)-\dot{y}(\tau)|\,dt$$

$$\leqslant \tfrac{1}{2}L|\dot{x}-\dot{y}|$$

and

$$\int_\sigma^{\sigma+a} k(t)e^{-l(K)t}\Big(\int_\sigma^t |z(\tau)-\bar{z}(\tau)|d\tau\Big)\,dt \leqslant \tfrac{1}{2}L|z-\bar{z}|.$$

Then $|u-v| \leqslant L\max(|\dot{x}-\dot{y}|, |z-\bar{z}|)$ for each $u \in \mathscr{F}(G)(x, z)$. Hence it follows $\bar{\mathrm{H}}(\mathscr{F}(G)(x, z), \mathscr{F}(y, \bar{z})) \leqslant L\max(|\dot{x}-\dot{y}|, |z-\bar{z}|)$. In a similaer way we obtain $\bar{\mathrm{H}}(\mathscr{F}(G)(y, \bar{z}), \mathscr{F}(G)(x, z) \leqslant L\max(|\dot{x}-\dot{y}|, |z-\bar{z}|$. Therefore, finally we have $\mathrm{H}(\mathscr{F}(G)(x, z), \mathscr{F}(G)(y, z) \leqslant L\max(|\dot{x}-\dot{y}|, |z-\bar{z}|)$ for each $z, \bar{z} \in B$ and $x, y \in \mathrm{AC}^0(I, R^n)\cap\mathscr{K}_\lambda$. ■

In a similar way we prove the following lemma.

Lemma 2.7. *Suppose $G \in H_\lambda(I, \boldsymbol{R}^n)$ is Lipschitz continuous with respect to its last variable uniformly with respect to $x \in \mathscr{K}_\lambda$. Then for every $L > 0$ there is a norm $|\cdot|$ of $L(I, \boldsymbol{R}^n)$ equivalent to $|\cdot|_a$ and such that for every $z, \bar{z} \in B_\lambda$ and $x \in \mathscr{K}_\lambda$ we have*

$$\mathrm{H}(\mathscr{F}(G)(x, z), \mathscr{F}(G)(x, \bar{z})) \leqslant L|z-\bar{z}|, \tag{2.9}$$

where H is a Hausdorff metric in $\mathrm{Cl}(L(I, \boldsymbol{R}^n))$ induced by the norm $|\cdot|$. ■

Lemma 2.8. *If $G \in H_\lambda(I, \boldsymbol{R}^n)$ is $\mathscr{A}$-continuous with respect to its last two variables, then $\mathscr{F}(G)$ is l.s.c. and H-u.s.c. on $\mathscr{K}_\lambda \times B_\lambda$.*

Proof. Let us observe first that if $G \in \mathscr{H}_\lambda(I, \boldsymbol{R}^n)$ is continuous on $\mathscr{K}_\lambda \times B_\lambda$ then it is also $\mathscr{A}$-continuous with respect to its last two variables. Indeed, let $(x_0, z_0) \in \mathscr{K}_\lambda \times B_\lambda$ be fixed and $\{(x_n, z_n)\}$ be an arbitrary sequence of $\mathscr{K}_\lambda \times B_\lambda$ such that $\max\{|x_n - x_0|_a, |z_n - z_0|_a\} \to 0$ as $n \to \infty$. For a.e. $t \in I$ one has $h(G(t, x_n, z_n), G(t, x_0, z_0)) \to 0$ as $n \to \infty$. Furthermore, it is clear that $\lim_{\mu(E)\to 0} \int_E h(G(t, x_n, z_n), G(t, x_0, z_0))dt = 0$ uniformly with respect to $n \in N$, because the family $\{G(\cdot, x, z)\}_{(x,z)\in\mathscr{K}_\lambda \times B_\lambda}$ is uniformly integrable. Therefore, by Vitalis' convergence theorem we have $\lim_{n\to\infty} \int_I h(G(t, x_n, z_n), G(t, x_0, z_0))dt = 0$ i.e., G is $\mathscr{A}$-continuous on $\mathscr{K}_\lambda \times B_\lambda$.

Now, let $(x_0, z_0) \in \mathscr{K}_\lambda \times B_\lambda$ and a sequence $\{(x_n, z_n)\}$ of $\mathscr{K}_\lambda \times B_\lambda$ be such as above. By virtue of Theorem II.3.13 for every $u_0 \in \mathscr{F}(G)(x_0, z_0)$ and $n \in N$ there exists $u_n \in \mathscr{F}(G)(x_n, z_n)$ such that

$$\begin{aligned} |u_n(t) - u_0(t)| &= \operatorname{dist}(u_0(t), G(t, x_{,n} z_n)) \\ &\leqslant \bar{h}(G(t, x_0, z_0), G(t, x_n, z_n) \leqslant h(G(t, x_n, z_n), G(t, x_0, z_0) \end{aligned}$$

for a.e. $t \in I$. Hence, it follows $|u_n - u_0|_a \leqslant d(G(\cdot, x_n, z_n), G(\cdot, x_0, z_0))$, where d is a metric defined by (1.1). Then $\operatorname{dist}(u_0, \mathscr{F}(G)(x_n, z_n)) \leqslant d(G(\cdot, x_n, z_n), G(\cdot, x_0, z_0))$ for every $u_0 \in \mathscr{F}(G)(x_0, z_0)$ and $n \in N$. Therefore, $\bar{\mathrm{H}}_L(\mathscr{F}(G)(x_0, z_0), \mathscr{F}(G)(x_n, z_n)) \leqslant d(G(\cdot, x_n, z_n), G(\cdot, x_0, z_0))$ for $n \in N$. In a similar way one obtains $\bar{\mathrm{H}}_L(\mathscr{F}(G)(x_n, z_n), \mathscr{F}(G)(x_0, z_0)) \leqslant d(G(\cdot, x_n, z_n), G(\cdot, x_0, z_0))$ for $n \in N$. Thus $\bar{\mathrm{H}}_L(\mathscr{F}(G)(x_n, z_n), \mathscr{F}(G)(x_0, z_0)) \leqslant d(G((\cdot, x_n, z_n), G(\cdot, x_0, z_0))$ for each $n \in N$. Hence it follows, that $\mathrm{H}_L(\mathscr{F}(G)(x_n, z_n)\mathscr{F}(G)(x_0, z_0)) \to 0$ as $n \to \infty$, because $d(G(\cdot, x_n, z_n), G(\cdot, x_0, z_0)) \to 0$ as $n \to \infty$. Thus $\mathscr{F}(G)$ is H-continuous at $(x_0, z_0) \in \mathscr{K}_\lambda \times B_\lambda$ and therefore H-u.s.c. and H-l.s.c. at (x_0, z_0). Hence, in particular it follows that $\mathscr{F}(G)$ is l.s.c. on $\mathscr{K}_\lambda \times B_\lambda$. ■

In a similar way we obtain

Lemma 2.9. *If $G \in H_\lambda(I, \boldsymbol{R}^n)$ is s.-s.l.s.c.(s. -s.u.s.c.) with respect to its last two variables then $\mathscr{F}(G)$ is l.s.c. (H-u.s.c.) on $\mathscr{K}_\lambda \times B_\lambda$.* ■

Now, we will investigate the weak sequential upper and lower semicontinuities of $\mathscr{F}(G)$ on $\mathscr{K}_\lambda \times B_\lambda$. We have the following results.

LEMMA 2.10. *Suppose $G \in \mathscr{H}_\lambda(I, \boldsymbol{R}^n)$ has convex values and is w.-w.u.s.c. (w.-s.u.s.c.) with respect to its last two variables. Then $\mathscr{F}(G)$ is w.-w.u.s.c. (w.-s.u.s.c.) on $\mathscr{K}_\lambda \times B_\lambda$.*

Proof. By virtue of Theorem II.2.5, it sufficies only to show that for every sequence $\{(x_n, z_n)\}$ of $\mathscr{K}_\lambda \times B_\lambda$ such that $|x_n - x|_a \to 0$ and $z_n \rightharpoonup z$ ($|z_n - z|_a \to 0$, respectively) as $n \to \infty$ and every sequence (u_n) of $L(I, \boldsymbol{R}^n)$ such that $u_n \in \mathscr{F}(G)(x_n, z_n)$ for $n = 1, 2, \ldots$ there exists a subsequence, say (u_k) of (u_n) weakly convergent to $u \in \mathscr{F}(G)(x, z)$.

Suppose above sequences $\{(x_n, z_n)\}$ and (u_n) are given. Since G is *w.-w.u.s.c.* with respect to its last two variables, for every measurable set $U \subset I$ we have $\lim_{n\to\infty} \bar{h}(\int_U G(t, x_n, z_n)dt, \int_U G(t, x, z)dt) = 0$. Furthermore, the sequence $\{G(\cdot, x_n, z_n)\}$ of $\mathscr{A}(I, \boldsymbol{R}^n)$ is uniformly integrable, because, $G \in \mathscr{H}_\lambda(I, \boldsymbol{R}^u)$. Then, by Theorem 1.12, there exists a subsequence, say (u_k) of (u_n) weakly convergent to $u \in \mathscr{F}(\mathrm{co}\, G(\cdot, x, z)) = \mathscr{F}(G)(x, z)$. ∎

LEMMA 2.11. *Suppose $G \in \mathscr{H}_\lambda(I, \boldsymbol{R}^n)$ has convex values and is w.-w.l.s.c. (w.-s.l.s.c.) with respect to its last two variables. Then $\mathscr{F}(G)$ is w.-w.l.s.c. (w.-s.l.s.c.) on $\mathscr{K}_\lambda \times B_\lambda$.*

Proof. By the definition we must show that for every weakly closed set $M \subset L(I, \boldsymbol{R}^n)$ the set $\mathscr{F}(G)_-(M) = \{(x, z) \in \mathscr{K}_\lambda \times B_\lambda\colon \mathscr{F}(G)(x, z) \subset M\}$ is sequentially weakly closed, i.e., such that for every sequence $\{(x_n, z_n)\}$ of $\mathscr{F}(G)_-(M)$ with $|x_n - x|_a \to 0$ and $z_n \rightharpoonup z$ as $n \to \infty$ we also have $\mathscr{F}(G)(x, z) \subset M$ ($\mathscr{F}(G)_-(M)$ is closed, respectively).

Let M be an arbitrary weakly closed set in $L(I, \boldsymbol{R}^n)$, $\{(x_n, z_n)\}$ any sequence of $\mathscr{F}(G)_-(M)$ and $(x, z) \in \mathscr{K}_\lambda \times B_\lambda$ be such that $|x_n - x|_a \to 0$ and $z_n \rightharpoonup z$ as $n \to \infty$. Let u be an arbitrary point of $\mathscr{F}(G)(x, z)$. Then $u(t) \in G(t, x, z)$ for a.e. $t \in I$. Because G is w-w.l.s.c. with respect to its last two variables then for every measurable set $U \subset I$ one has $\bar{h}(\int_U G(t, x, z)dt, \int_U G(t, x_n, z_n))dt \to 0$ as $n \to \infty$. Hence in particular we obtain $\mathrm{dist}(\int_U u(t)dt, \int_U G(t, x_n, z_n)dt) \to 0$ as $n \to \infty$ for every measurable set $U \subset I$. Thus, for every measurable set $U \subset I$ there is a sequence (u_n) of measurable functions of U into $\boldsymbol{R}^n$ such that $\lim_{n\to\infty} \left|\int_U u(t)dt - \int_U u_n(t)dt\right| = 0$ and $u_n(t) \in G(t, x_n, z_n)$ for a.e. $t \in U$.

Let $U_1, U_2, \ldots$ be a sequence of Borel sets in I that are dense in Borel field of I. In particular there is a subsequence, say (U_{i_k}) of (U_i) such that $\mu(I \vartriangle U_{i_k}) \to 0$ as $k \to \infty$. Thus

$$\mu(I \setminus \bigcup_{k=1}^{\infty} U_{i_k}) = \mu\Big[\bigcap_{k=1}^{\infty} (I \setminus U_{i_k})\Big] = \lim_{k\to\infty} \mu(I \setminus U_{i_k}) \leqslant \lim_{k\to\infty} \mu(I \Delta U_{i_k}) = 0.$$

Since $\bigcup_{k=1}^{\infty} U_{i_k} \subset \bigcup_{i=1}^{\infty} U_i$, then we have also $\mu(I \setminus \bigcup_{l=1}^{\infty} U_l) = 0$. Let $u_n^k\colon H_k \to \boldsymbol{R}^n$

be for every fixed $k = 1, 2, \dots$ and $n = 1, 2, \dots$ a measurable function such that $\lim_{n\to\infty} \left|\int_{H_k} [u(t) - u_n^k(t)dt\right| = 0$, where $H_1 = U_1$ and $H_k = U_k \setminus \bigcup_{i=1}^{k-1} U_i$ for $k = 2, \dots$ Put $v_n(t) = u_n^k(t)$ for $t \in H_k$ and $k, n = 1, 2, \dots$

Now, let $\varepsilon > 0$ be an arbitrary number and U a measurable subset of I and suppose V and U_j are Borel subsets of I such that V is a measurable kernel of U and $\int_{V \triangle U_j} \|G(t, x, z)\| dt \leqslant \varepsilon/2$ for every $(x, z) \in \mathscr{K}_\lambda \times B_\lambda$. Such U_j exists because the sequence (U_i) is dense in the Borel field of I and the family $\{G(\cdot, x, z)\}_{(x,z)\in\mathscr{K}_\lambda\times B_\lambda}$ is uniformly integrable. Taking $\varepsilon > 0$ sufficiently small we can assume $V \cap U_j \neq \emptyset$. Let H_k, $k = 1, 2, \dots$ be such as above. We have $U_j = \bigcup_{k=1}^{j} H_k$ and $H_k \cap H_m = \emptyset$ for $k \neq m$. Let us observe that $V \cup (U_j \setminus V) = U_j \cup (V \setminus U_j)$, $V \cap (U_j \setminus V) = \emptyset$ and $U_j \cap (V \setminus U_j) = \emptyset$. Therefore, for every $n = 1, 2, \dots$ we get

$$\begin{aligned}
\left|\int_V [v_n(t) - u(t)]dt\right| &= \left|\int_V [v_n(t) - u(t)]dt + \int_{U_j\setminus V} [v_n(t) - u(t)]dt - \right.\\
&\quad \left. - \int_{U_j\setminus V} [v_n(t) - u(t)]dt\right| \leqslant \left|\int_{V\cup(U_j\setminus V)} [v_n(t) - u(t)]dt\right| + \\
&\quad + \left|\int_{U_j\setminus V} v_n(t) - u(t)]dt\right| \leqslant \left|\int_{U_j\cup(V\setminus U_j)} [v_n(t) - u(t)]dt\right| + \\
&\quad + \left|\int_{U_j\setminus V} \|G(t, x, z)\| dt + \int_{U_j\setminus V} \|G(t, x, z)\| dt\right. \\
&\leqslant \left|\int_{U_j} [v_n(t) - u(t)]dt\right| + 2\cdot \sup_{(x,z)\in\mathscr{K}_\lambda\times B_\lambda} \int_{V\triangle U_j} \|G(t, x, z\| dt \\
&\leqslant \sum_{k=1}^{j} \left|\int_{H_k} [u(t) - u(t)]dt\right| + \varepsilon.
\end{aligned}$$

Then $\lim_{n\to\infty} \left|\int_V [v_n(t) - u(t)]dt\right| \leqslant \varepsilon$ for every $\varepsilon > 0$. Since $\int_U [v_n(t) - u(t)]dt = \int_V [v_n(t) - u(t)]dt$ then for every measurable set $U \subset I$ we have $\lim_{n\to\infty} \int_U [v_n(t) - u(t)]dt = 0$. Thus $v_n \rightharpoonup u$ as $n \to \infty$. Since $v_n(t) \in G(t, x_n, z_n)$ for a.e. $t \in I$ and $(x_n, z_n) \in \mathscr{F}(G)_-(M)$ for $n = 1, 2, \dots$ then for every $n = 1, 2, \dots, v_n \in M$. Therefore $v \in M$. Thus $\mathscr{F}(G)(x, z) \subset M$. (The closedness $\mathscr{F}^-(G)(M)$ can be verified similarly.) ■

As a corollary from the above result we obtain the following lemma.

LEMMA 2.12. *Suppose $G \in \mathscr{H}_\lambda(I, \mathbf{R}^n)$ has convex values and is w.-w.l.s.c. (w.-s.l.s.c.) with respect to its last two variables. Then $\mathscr{I}\mathscr{F}(G)$ is s.-w.l.s.c. (s.-s.l.s.c.) on $\mathscr{K}_\lambda \times B_\lambda$.*

Proof. Let M be a closed subset of $C^0(I, \mathbf{R}^n)$ and $\{(x_n, z_n)\}$ be an arbitrary sequence of $\mathscr{I}\mathscr{F}(G)_-(M) = \{(x, z) \in \mathscr{K}_\lambda \times B_\lambda\colon\ \mathscr{I}\mathscr{F}(G)(x, z) \subset M\}$ such that $|x_n - x|_a \to 0$ and $z_n \rightharpoonup z$ as $n \to \infty$. For every $y \in \mathscr{I}\mathscr{F}(G)(x, z)$ there is $u \in \mathscr{F}(G)(x, z)$ such that $y = \mathscr{I}u$, i.e. $\dot{y} = u$. Similarly as in the proof of Lemma 2.11 for a given above $u \in \mathscr{F}(G)(x, z)$ we can define a sequence (v_n) such that $v_n \in \mathscr{F}(G)(x_n, z_n)$ and so that

$v_n \to u$ as $n \to \infty$. Put $y_n = \mathscr{I}v_n$. We have $y_n \in \mathscr{I}\mathscr{F}(G)(x_n, z_n) \subset M$ for $n = 1, 2, \ldots$ and by (iii) of Lemma 1.13 also that $|y_n - y|_a \to 0$ as $n \to \infty$. Therefore, $y \in M$, i.e $\mathscr{I}\mathscr{F}(G)(x, z) \subset M$. ∎

Finally, we prove the following lemma.

LEMMA 2.13. *Suppose* $G \in H_\lambda(I, \mathbf{R}^n)$ *is* $\mathscr{A}$*-Lipschitz continuous with respect to its last two variables* ($\mathscr{A}$*-Lipschitz continuous with respect to its last variable uniformly with respect the second one). Then the mapping* $\mathscr{F}(G)\colon \mathscr{K}_\lambda \times B_\lambda \to \mathrm{Cl}(L(I, \mathbf{R}^n))$ *is Lipschitz continuous on* $\mathscr{K}_\lambda \times B_\lambda$ *(Lipschitz continuous with respect to its last variable uniformly with respect to the second one).*

Proof. Similarly as in the proof of Lemma 2.6 we can verify that to every $u \in \mathscr{F}(G)(x, z)$ we can find any $v \in \mathscr{F}(G)(\bar{x}, \bar{z})$ such that $|u(t) - v(t)| \leqslant h(G(t, x, z), G(t, \bar{x}, \bar{z}))$ for a.e. $t \in [\sigma, \sigma + a]$ and $(x, z), (\bar{x}, \bar{z}) \in \mathscr{K}_\lambda \times B_\lambda$. Therefore,

$$\overline{\mathrm{H}}_L(\mathscr{F}(G)(x, z), \mathscr{F}(G)(\bar{x}, \bar{z})) \leqslant \int_I h(G(t, x, z), G(t, \bar{x}, \bar{z}))dt$$

for $(\bar{x}, \bar{z})$, $(x, z) \in \mathscr{K}_\lambda \times B_\lambda$. Similarly, for (x, z), $(\bar{x}, \bar{z}) \in \mathscr{K}_\lambda \times B_\lambda$ we obtain

$$\overline{\mathrm{H}}_L(\mathscr{F}(G)(\bar{x}, \bar{z}), \mathscr{F}(G)(x, z)) \leqslant \int_I h(G(t, x, z), G(t, \bar{x}, \bar{z})dt.$$

Thus,

$$\mathrm{H}_L(\mathscr{F}(G)(x, z), \mathscr{F}(G)(\bar{x}, \bar{z})) \leqslant \int_I h(G(t, x, z, G(t, \bar{x}, \bar{z}))dt$$

for (x, z), $(\bar{x}, \bar{z}) \in \mathscr{K}_\lambda \times B_\lambda$. Hence, it follows

$$\mathrm{H}_L(\mathscr{F}(G)(x, z), \mathscr{F}(G)(\bar{x}, \bar{z})) \leqslant L \max[|x - \bar{x}|_a, |z - \bar{z}|_a]$$

for every (x, z), $(\bar{x}, \bar{z}) \in \mathscr{K}_\lambda \times B_\lambda$. ∎

2.3. Continuous selection theorem

We will present here any continuous selection theorem for subtrajectory integrals of set-valued functions $G \in \mathscr{H}_\lambda(I, \mathbf{R}^n)$. Suppose $G \in \mathscr{H}_\lambda(I, \mathbf{R}^n)$ is such that $\int_\sigma^{\sigma+a} \|G(t, x, z)\| dt \leqslant \lambda$ for $(x, z) \in \mathscr{K}_\lambda \times B_\lambda$. Let us observe that for every $G \in \mathscr{H}_\lambda(I, \mathbf{R}^n)$ there exists a number $M > 0$ such that $\int_\sigma^{\sigma+a} \|G(t, x, z)\| dt \leqslant M$ for each $(x, z) \in \mathscr{K}_\lambda \times B_\lambda$. Furthermore, by the definition of $\mathscr{H}_\lambda(I, \mathbf{R}^n)$ the family $\{G(\cdot, x, z)\}_{(x, z) \in \mathscr{K}_\lambda \times B_\lambda}$ is uniformly integrable. Thus, by Corollary 1.1 hence it follows that the collection of all measurable selections of this family, i.e. the set $\Omega_\lambda = \bigcup \{\mathscr{F}(G)(x, z)\colon (x, z) \in \mathscr{K}_\lambda \times B_\lambda\}$ is relatively weakly compact in $L(I, \mathbf{R}^n)$. If furthermore $G \in \mathscr{H}_\lambda(I, \mathbf{R}^n)$ is such that $\int_\sigma^{\sigma+a} \|G(t, x, z)\| dt \leqslant \lambda$ for $(x, z) \in \mathscr{K}_\lambda \times B_\lambda$, we have $\Omega_\lambda \subset B_\lambda$. Since B_λ is convex and closed, it is also a weakly closed subset of $L(I, \boldsymbol{R}^n)$. Thus $\overline{[\Omega_\lambda]}_L^w \subset B_\lambda$. Hence it follows that $\mathrm{co}\overline{[\Omega_\lambda]}_L^w \subset B_\lambda$. Put $\boldsymbol{P}_\lambda = \mathrm{co}\overline{[\Omega\ ]}_L^w$. It is clear by Krein–Šmulian theorem that $\boldsymbol{P}_\lambda$ is a convex weakly compact subset of $L(I, \boldsymbol{R}^n)$, Of course, $\boldsymbol{P}_\lambda \neq \emptyset$.

Put now $\boldsymbol{K}_\lambda = \mathscr{I}(\boldsymbol{P}_\lambda)$, where $\mathscr{I}$ is a mapping defined by (1.2). By Corollary 1.4, $\boldsymbol{K}_\lambda$ is a compact convex subset of $C^0(I, \boldsymbol{R}^n)$. If $\boldsymbol{P}_\lambda \subset B_\lambda$ then $\boldsymbol{K}_\lambda \subset \mathscr{K}_\lambda$. Therefore, in this case we can define on $I \times \boldsymbol{K}_\lambda$ a set-valued function $G\square\mathscr{D}$ by setting $(G\square\mathscr{D}((t, x) := G(t, x, \mathscr{D}x)$ for $t \in I$ and $x \in \boldsymbol{K}_\lambda$, where $\mathscr{D}$ is a mapping defined by (1.3). We denote by $\mathscr{F}(G\square\mathscr{D})$ the subtrajectory integrals of such defined set-valued function $G\square\mathscr{D}$. In a similar we can define on $I \times B_\lambda$ a mapping $G\square\mathscr{I}$.

Let us observe that $\mathscr{F}(G\square\mathscr{D})(x) = \mathscr{F}(G)(x, \mathscr{D}x)$ and $\mathscr{F}(G\square\mathscr{I})(z) = \mathscr{F}(G)(\mathscr{I}z, z)$ for $x \in \boldsymbol{K}_\lambda$ and $z \in B_\lambda$, i.e. $\mathscr{F}(G\square\mathscr{D}) = \mathscr{F}(G) \circ (\mathrm{id}_{\boldsymbol{K}_\lambda}, \mathscr{D})$ and $\mathscr{F}(G\square\mathscr{I}) = \mathscr{F}(G) \circ (\mathscr{I}, \mathrm{id}_{B_\lambda})$, where $\mathrm{id}_{\boldsymbol{K}_\lambda}$ and id_{B_λ} denote the identity mappings defined on $\boldsymbol{K}_\lambda$ and B_λ, respectively. Therefore the properties of mappings $\mathscr{F}(G\square\mathscr{D})$ and $\mathscr{F}(G\square\mathscr{I})$ follow immediately from the properties of mappings $\mathscr{F}(G)$, $(\mathrm{id}_{\boldsymbol{K}_\lambda}, \mathscr{D})$: $\boldsymbol{K}_\lambda \in x \to (x, \dot{x}) \in \boldsymbol{K}_\lambda \times \boldsymbol{P}_\lambda$ and $(\mathscr{I}, \mathrm{id}_{B_\lambda})$: $B_\lambda \ni z \to (\mathscr{I}z, z) \in \mathscr{K}_\lambda \times B_\lambda$.

We shall prove here that for every $G \in \mathscr{H}_\lambda(I, \boldsymbol{R}^n)$ such that $\int_\sigma^{\sigma+a} \|G(t, x, z)\|\,dt \leqslant \lambda$ for $(x, z) \in \mathscr{K}_\lambda \times B_\lambda$ that is s.-w.l.s.c. with respect to its last two variables there exists a continuous function $g\colon \boldsymbol{K}_\lambda \to L(I, \boldsymbol{R}^n)$ such that $g(x) \in \mathscr{F}(G\square\mathscr{D})(x)$ for $x \in \boldsymbol{K}_\lambda$.

LEMMA 2.14. *Let $G \in \mathscr{H}_\lambda(I, \boldsymbol{R}^n)$ be s.-w.l.s.c. with respect to its last two variables. Then $\mathscr{F}(G\square\mathscr{D})$ is l.s.c. on $\boldsymbol{K}_\lambda$.*

Proof. It is clear that $(G\square\mathscr{D})(\cdot, x) \in \mathscr{A}(I, \boldsymbol{R}^n)$ for every $x \in \boldsymbol{K}_\lambda$. By Lemma 1.14 the mapping $(\mathrm{id}_{\boldsymbol{K}_\lambda}, \mathscr{D})$ sends sequences of $\boldsymbol{K}_\lambda$ strongly convergent in $C^0(I, \boldsymbol{R}^n)$ into sequences of $C^0(I \times \boldsymbol{R}^n) \times L(I, \boldsymbol{R}^n)$ which corresponding coordinates are strongly and weakly convergent in $C^0(I, \boldsymbol{R}^n)$ and $L(I, \boldsymbol{R}^n)$, respectively. Therefore similarly as in the proof of Lemma 2.8 we can easily see that $\mathscr{F}(G\square\mathscr{D})$ is H-l.s.c. on $\boldsymbol{K}_\lambda$. ■

We will deal now with decomposable subsets of $L(I, \boldsymbol{R}^n)$. Recall, we say that a set $H \subset L(I, \boldsymbol{R}^n)$ is decomposable if, for each $u, v \in H$ and any L-measurable set $A \subset I$ we have $\chi_A u + \chi_{I \setminus A} v \in H$, where χ_A is the characteristic function of $A \subset I$.

LEMMA 2.15. *Let $H \in \mathrm{Cl}(L(I, \boldsymbol{R}^n))$ be decomposable and let $e(H)(t) := \inf\{|u(t)|: u \in H\}$ for a.e. $t \in I$. Then there exists a sequence (u_n) of H such that for almost everywhere in I one has*

$$|u_1(t)| \geqslant |u_2(t)| \geqslant \dots \tag{2.10}$$

and

$$e(H)(t) = \lim_{n\to\infty} |u_n(t)| \tag{2.11}$$

for a.e. $t \in I$.

Proof. By Theorem I.4.1, there exists a sequence (v_n) of H such that $e(H)(t) = \inf_{n\geq 1} |v_n(t)|$ for a.e. $t \in I$. Let us put $u_1 = v_1$ and inductively $u_{n+1} = u_n \chi_{I_n} + v_{n+1}\chi_{I\setminus I_n}$, where $I_n = \{t: |u_n(t)| < |v_{n+1}(t)|\}$. Then (2.10) and (2.11) are implied by the inequality $|u_n(t)| \leqslant \inf\{|v_1(t)|, \ldots, |v_{n+1}(t)|\}$ for a.e. $t \in I$. ■

In what follows we will deal with a nonatomic countable additive complete n-dimensional vector measure $\vec{\mu} = (\mu_1, \ldots, \mu_n)$ defined on σ-field $\mathscr{L}(I)$ of all Lebesgue measurable subset of the compact interval $I := [\sigma, \sigma+a]$.

We shall prove now the following propositions.

PROPOSITION 2.1. *For the above measure $\vec{\mu}$ there exists a family $\{A_\alpha\}_{\alpha\in[0,1]}$ of measurable subsets of I such that*

(i) $A_\alpha \subset A_\beta$ *for* $\alpha < \beta$ *and*

(ii) $(\vec{\mu}A_\alpha) = \alpha\vec{\mu}(I)$.

Proof. By Corollary I.3.1 we may construct a family of sets A_α satisfying (i) and (ii) for $\alpha = k/2^n$ where $n \in N$ and $k = 0, \ldots, 2^n$. Indeed, immediately from Corollary I.3.1 it follows that for an arbitrary $B \in \mathscr{L}(I)$ there exists $A \in \mathscr{L}(I)$ such that $A \subset B$ and $\vec{\mu}(A) = \frac{1}{2}\vec{\mu}(B)$. Put $B = I$ and let A_1^1 be a measurable subset of B such that $\mu(A_1^1) = \frac{1}{2}\mu(B) = \frac{1}{2}\mu(I)$. Put $A_1^2 = I\setminus A_1$. We also have $\vec{\mu}(A_1^2) = \frac{1}{2}\vec{\mu}(I)$. Suppose A_m^k are defined for $m = 1, \ldots, n$ and $k = 1, \ldots, 2^m$ such that $\vec{\mu}(A_m^k) = (\frac{1}{2})^m\mu(I)$. Applying again Corollary I.3.1 to A_n^k we choose a measurable set $A_{n+1}^k \subset A_n^k$ such that $\vec{\mu}(A_{n+1}^k) = \frac{1}{2}\vec{\mu}(A_n^k)$. Let $A_{n+1}^l = A_{n+1}^k$ for $l = 2k-1$ and $A_{u+1}^l = A_n^k \setminus A_{n+1}^k$ for $l = 2k$. We have defined A_{n+1}^l for $l = 1, \ldots, 2^{n+1}$ such that $\vec{\mu}(A_{n+1}^l) = (\frac{1}{2})^{n+1}\vec{\mu}(I)$. Now for every $\alpha = k/2^n$ we define $A_{k/2^n} = \bigcup_{l=1}^{k} A_n^l$. It is not difficult to verify that (i) and (ii) for such family $\{A_\alpha\}_{\alpha\in[0,1]}$ are satisfied.

Generally, this procedure will lead to construction of a dyadic structure of I. Finally, for an arbitrary $\alpha \in [0, 1]$ we define A_α to be the union of all $A_{k/2^n}$ such that $k/2^n \leqslant \alpha$. Condition (i) holds by the definition of A_α, while (ii) follows from the continuity of the measure with respect to Lebesgue measure μ. ■

REMARK 2.1. We may additionally require in Proposition 2.1 that $\mu(A_\alpha) = \alpha$, where μ denotes the Lebesgue measure on $\mathscr{L}(I)$. Indeed, it is enough to construct the family $\{A_\alpha\}$ for the measure $\vec{\nu} = (\mu, \mu_1, \ldots, \mu_n)$. ■

Let us consider a family of nonatomic complete countable additive vector measures $\vec{\mu}_s = (\mu_s^1, \ldots, \mu_s^m)$ with $s \in S$.

PROPOSITION 2.2. *Assume that the mapping $s \to \vec{\mu}_s$ from a compact topological space S into the space M of all countable additive n-dimensional vector measures $\vec{\mu}$ with the topology induced by the norm $\|\vec{\mu}\|$ equal to the variation of $\vec{\mu}$ is continuous. Then for every $\varepsilon > 0$ there exists a family of measurable sets $\{A_\alpha\}_{\alpha\in[0,1]}$ with the properties*

(i) $A_\alpha \subset A_\beta$ *for* $\alpha < \beta$,

(ii) $|(\vec{\mu}_s(A_\alpha)-\alpha\vec{\mu}_s(I)| < \varepsilon$ *for all* $\alpha \in [0, 1]$ *and* $s \in S$ *and*
(iii) $\mu(A_\alpha) = \alpha$.

Proof. Let us take an $\varepsilon > 0$. The family of open sets $\{V_{s_0}\}_{s_0 \in S}$ given by the formulae $V_{s_0} = \{s \in S: ||\vec{\mu}_s - \vec{\mu}_{s_0}|| < \frac{1}{2}\varepsilon$ is an open covering of the compact space S. Let $s_1, \ldots, s_k \in S$ be such that $S = V_{s_1} \cup \ldots \cup V_{s_k}$. By Proposition 2.1 for the measure $\nu = (\vec{\mu}_{s_1}, \ldots, \vec{\mu}_{s_k}, \mu)$ there exists a family of measurable sets $\{A_\alpha\}_{\alpha \in [0,1]}$ such that (i) holds and $\vec{\nu}(A_\alpha) = \alpha\vec{\nu}(I)$ for all $\alpha \in [0, 1]$. By Remark 2.1 we also have $\mu(A_\alpha) = \alpha$. Then to end the proof we have to show that the family $\{A_\alpha\}$ satisfies (ii). For an arbitrary $\alpha \in [0, 1]$ and $s \in S$ we have

$$|\vec{\mu}_s(A_\alpha) - \alpha_s\vec{\mu}(I)|$$
$$\leqslant |\vec{\mu}_s(A_\alpha) - \vec{\mu}_{s_i}(A_\alpha)| + |\vec{\mu}_{s_i}(A_\alpha) - \alpha\vec{\mu}_{s_i}(I)| + |\alpha(\vec{\mu}_{s_i}(I) - \vec{\mu}_s(I))|,$$

where s_i is such that $s \in V_{s_i}$. The first and last terms of the right-hand side of the above inequality are estimated by $\frac{1}{2}\varepsilon$ because $s \in V_{s_i}$, while the middle term is equal to 0 because $\vec{\nu}(A_\alpha) = \alpha\vec{\nu}(I)$. Then (ii) holds. ■

Let S be a topological space and $\mathscr{S}^i(s) \in L(I, \boldsymbol{R}^1)$ for $s \in S$ and $i = 1, \ldots, m$. Immediately from properties of Lebesgue integrals it follows that a mapping $\vec{\mu}_s$: $\mathscr{L}(I) \to \boldsymbol{R}^n$ defined by

$$\vec{\mu}_s(E) = \int_E (\mathscr{S}^1(s)(t), \ldots, \mathscr{S}^r(s)(t))dt$$

for $E \in \mathscr{L}(I)$ and fixed $s \in S$ is a countable additive nonatomic complete vector measure. It is continuous as a mapping from S into the space M defined in Proposition 2.2 if and only if a mapping $S \ni s \to (\mathscr{S}^1(s), \ldots, \mathscr{S}^r(s)) \in L(I, \boldsymbol{R}^n)$ is continuous.

PROPOSITION 2.3. *Let* $\{A_\alpha\}_{\alpha \in [0,1]}$ *be a family of measurable subsets of* I *with the following properties*:
(i) $A_\alpha \subset A_\beta$ *for* $\alpha < \beta$,
(ii) $\mu(A_\alpha) = \alpha$
and let p: $S \to [0, 1]$ *and* k: $S \to L(I, \boldsymbol{R}^n)$, *where* S *is a topological space, be continuous maps. Then the mapping* $l(s) = k(s)\chi_{A_{p(s)}}$ *is continuous.*

Proof. The continuity of l follows from inequalities

$$|k(s)\chi_{A_{p(s)}} - k(s_0)\chi_{A_{p(s_0)}}|$$
$$\leqslant |k(s)\chi_{A_{p(s)}} - k(s_0)\chi_{A_{p(s)}}| + |k(s_0)\chi_{A_{p(s)}} - k(s_0)\chi_{A_{p(s_0)}}|$$
$$\leqslant |k(s) - k(s_0)| + \int_{A_{p(s)} \triangle A_{p(s_0)}} |k(s_0)(t)|\,dt$$

and the equality $\mu(A_{p(s)} \triangle A_{p(s_0)}) = |p(s) - p(s_0)|$, which is true for arbitrary s_0 and s from S. ■

PROPOSITION 2.4. *Let $H \in \mathrm{Cl}(L(I, \mathbf{R}^n))$ be decomposable. Then there exists an element $u_0 \in H$ such that*

$$e(H)(t) = |u_0(t)| \quad \textit{for a.e. } t \in I. \tag{2.12}$$

Proof. Let (u_n) be a sequence of H such that (2.10) and (2.11) are satisfied. Put $P(t) = \overline{\{u_n(t): n = 1, 2, \dots\}} \cap B(0, e(H)(t))$, where $B(0, \delta)$ denotes the closed ball of $\mathbf{R}^n$ with the centre at 0 and a radius $\delta > 0$. It is easy to see that $P(t) \neq \emptyset$ for a.e. $t \in I$. By Theorem II.3.4 and Corollary II.3.1, mappings $F_1: I \ni t \to \overline{\bigcup_{n=1}^{\infty} \{u_n(t)\}} \subset \mathbf{R}^n$ and $F_2: I \ni t \to B(0, e(H)(t)) \subset \mathbf{R}^n$ are measurable. Hence, by Theorem II.3.4, $F_1 \cap F_2$ is also measurable. Therefore, by Theorem II.3.10 there is a measurable selector, say u_0, of $F := F_1 \cap F_2$. We shall prove that $u_0 \in H$. Fix $i = 1, 2, \dots$ and for $n = 1, 2, \dots$ put $I_n^i = \{t \in I: |u_n(t) - u_0(t)| \leqslant 1/i\}$. Then $\mu(I \setminus \bigcup_{n=1}^{\infty} I_n^i) = 0$ and by the decomposability of H and (2.10) we see that a function v_i given by $v_i = \sum_{n=1}^{\infty} \chi_{E_n^i} u_n(t)$, where $E_1^i = I_1^i$ and $E_n^i = I_n^i \setminus \bigcup_{k<n} I_k^i$ belongs to H and inequality $|v_i(t) - u_0(t)| \leqslant 1/i$ holds a.e. in I. And so $u_0 = \lim_{i \to \infty} v_i$ belongs to H. Clearly, u_0 satisfies (2.12). ∎

We will now deal with set-valued mappings defined on subsets of a given metric space (Z, d) with values in $\mathrm{Cl}(L(I, \mathbf{R}^n))$. A set-valued function $H: X \to \mathrm{Cl}(L(I, \mathbf{R}^n))$ with $X \subset Z$ is said to be *decomposable* if $H(x)$ is a decomposable subset of $L(I, \mathbf{R}^n)$ for each $x \in X$.

PROPOSITION 2.5. *Let $H: X \to \mathrm{Cl}(L(I, \mathbf{R}^n))$ be l.s.c. and decomposable and let $P(x) := \{v \in L(I, \mathbf{R}^1): v(t) \geqslant e(H(x))(t)$ for a.e. $t \in I\}$. Then a set-valued function $P: X \ni x \to P(x) \in \mathscr{P}(L(I, \mathbf{R}^1))$ is l.s.c., decomposable and has closed and convex values.*

Proof. We shall prove only that P is l.s.c. The rest is obvious. Let M be an arbitrary closed set in $L(I, \mathbf{R}^1)$. It is enough to show that if for every sequence (x_n) of X converging to x_0 as $n \to \infty$ we have $P(x_n) \subset M$, then also $P(x_0) \subset M$. For this purpose take an arbitrary $v_0 \in P(x_0)$. By Proposition 2.4, there exists a function $u_0 \in H(x_0)$ such that $v_0(t) \geqslant |u_0(t)| = e(H(x_0))(t)$ for a.e. $t \in I$. Let (u_n) be a sequence of $L(I, \mathbf{R}^n)$ with $u_n \in H(x_n)$ for $n = 1, 2, \dots$ such that $|u_n - u_0|_a \to 0$ as $n \to \infty$. Such sequence (u_n) exists, by Theorem II.2.9, because H is l.s.c. Then the sequence (v_n) of the form $v_n(t) = |u_n(t)| + v_0(t) - |u_0(t)|$ for a.e. $t \in I$ is converging to v_0 as $n \to \infty$ and is such that $v_n \in P(x_n) \subset M$. Since M is closed we have also $v_0 \in M$. But v_0 is an arbitrary point of $P(x_0)$; then $P(x_0) \subset M$. ∎

PROPOSITION 2.6. *Let $H: X \to \mathrm{Cl}(L(I, \mathbf{R}^n))$ with $X \in \mathrm{Cl}(Z)$ be an l.s.c. and decomposable set-valued function and suppose $\mathscr{S}: X \to L(I, \mathbf{R}^1)$ and $g: X \to L(I, \mathbf{R}^n)$ are continu-*

ous mappings such that the set

$$L(x) = \{u \in H(x) \colon |u(t) - g(x)(t)| < \mathscr{L}(x)(t) \text{ for a.e. } t \in I\}$$

is nonempty for any $x \in X$. *Then the mapping* $L\colon X \ni x \to L(x) \subset L(I, R^n)$ *is decomposable and l.s.c.*

Proof. Let M be an arbitrary closed subset in $L(I, R^n)$. Similarly as in the proof of Proposition 2.5, it is enough to show that if the inclusion $L(x_n) \subset M$ holds for all $n = 1, \ldots,$ where (x_n) is a sequence of X converging to x_0 as $n \to \infty$, then $L(x_0) \subset M$. For this purpose take an arbitaryr point $u_0 \in L(x_0)$. Because of the lower semicontinuity of H there exists a sequence (u_n) of $L(I, R^n)$ with $u_n \in H(x_n)$ such that $\lim_{n\to\infty} |u_n - u_0|_a = 0$. Without any loss of generality we may assume that $u_n(t) \to u_0(t)$, $g(x_n)(t) \to g(x_0)(t)$ and $\mathscr{S}(x_n)(t) \to \mathscr{S}(x_0)(t)$ almost everywhere in I as $n \to \infty$. For each $i = 1, 2, \ldots$ let $I_i \subset I$ be such a compact set that the functions u_n, $g(x_n)$ and $\mathscr{S}(x_n)$ restricted to I_i are continuous and converge uniformly to u_0, $g(x_0)$ and $\mathscr{S}(x_0)$, respectively and that the following inequality holds:

$$\int_{I \setminus I_i} \mathscr{S}(x_0)(t)\,dt < \frac{1}{i} \tag{2.13}$$

Such sets $I \subset I$ exist by Lusin's and Egoroff's theorems.

Since for $t \in I_i$, $|u_0(t) - g(x_0)(t)| < \mathscr{S}(x_0)(t)$, and $|u_n(t) - g\ (x)| - \mathscr{S}(x_n)(t) \to |u_0(t) - g(x_0)(t)| - \mathscr{S}(x_0)(t)$ uniformly on I_i as $n \to \infty$ then there exists n_i such that for $n \geqslant n_i$ and all $t \in I_i$ we have the inequality

$$|u_n(t) - g(x_n)(t)| < \mathscr{S}(x_n)(t). \tag{2.14}$$

We may additionally assume that $n_1 < n_2 < \ldots$ Put $v_n = u_n \lambda_{I_i} + w_n \lambda_{I \setminus I_i}$ for $n_i \leqslant n < n_{i+1}$, where w_n are arbitrary but fixed elements from $L(x_n)$ for $n = 1, 2, \ldots$ Then the sequence (v_n) is converging to u_0, because for $n_i \leqslant n < n_{i+1}$ we have

$$\begin{aligned}
&|v_n - u_0|_a \\
&\leqslant \int_{I \setminus I_i} |w_n(t) - g(x_n)(t)|\,dt + \int_{I \setminus I_i} |g(x_n)(t) - g(x_0)(t)|\,dt + \\
&\quad + \int_{I \setminus I_i} |g(x_0)(t) - u_0(t)|\,dt + \int_{I_i} |u_n(t) - u_0(t)|\,dt \\
&\leqslant 2 \int_{I \setminus I_i} \mathscr{S}(x_0)(t)\,dt + |\mathscr{S}(x_n) - \mathscr{S}(x_0)|_a + |g(x_n) - g(x_0)|_a + |u_n - u_0|_a \\
&< \frac{2}{i} + |\mathscr{S}(x_n - \mathscr{S}(x_0)|_a + |g(x_n) - g(x_0)|_a + |u_n - u_0|_a.
\end{aligned}$$

By decomposability of H it is easy to check that $v_n \in L(x_n) \subset M$ for sufficiently large n because $w_n \in L(x_n)$ and (2.14) holds on I_i for n large enough. Then $u_0 \in M$. But u_0 is an arbitrary point of $L(x_0)$. Thus $L(x_0) \subset M$. ■

We shall prove now the following basic lemma.

LEMMA 2.16. *Let* $H\colon X \to \mathrm{Cl}(L(I, R^n))$ with $X \in \mathrm{Comp}(Z)$ *be l.s.c. and decomposable.*

Then for every $\varepsilon > 0$ there exist continuous functions $g: X \to L(I, R^n)$ and $\mathscr{S}: X \to L(I, R^1)$ such that

$$\int_I \mathscr{S}(x)(t)\,dt < \varepsilon \quad \text{for each } x \in X \tag{2.15}$$

and the set

$$L(x) = \{u \in H(x): |u(t)-g(x)(t)| < \mathscr{S}(x)(t) \text{ for a.e. } t \in I\} \tag{2.16}$$

is nonempty for each $x \in X$.

Proof. Fix $\varepsilon > 0$ and let $\bar{x} \in X$ and select any $\bar{u} \in H(\bar{x})$. Put $P_{\bar{x}\bar{u}}(x) := \{v \in L(I, R^1): v(t) \geqslant e(H(x)-\bar{u})(t)$ for a.e. $t \in I\}$. By Proposition 2.5 $P_{\bar{x}\bar{u}}: X \to \mathrm{Cl}(L(I, R^1))$ is l.s.c. and has closed convex values. Furthermore $0 \in P_{\bar{x}\bar{u}}(\bar{x})$ because $e(H(\bar{x})-\bar{u})(t) = 0$ for a.e. $t \in I$. Then by Michael's continuous selection theorem (see Corollary II.4.1) there exists a continuous function $\mathscr{S}_{\bar{x}\bar{u}}: X \to L(I, R^1)$ such that $\mathscr{S}_{\bar{x}\bar{u}}(\bar{x}) = 0$ and $\mathscr{S}_{\bar{x}\bar{u}}(x) \in P_{\bar{x}\bar{u}}(x)$, i.e. $\mathscr{S}_{\bar{x}\bar{u}}(x) \geqslant e(H(x)-\bar{u})(t)$ for $x \in X$ and a.e. $t \in I$. Consider the family $\{V_{\bar{x}\bar{u}}: \bar{x} \in X,\ \bar{u} \in H(\bar{x})\}$ given by the formula

$$V_{\bar{x}\bar{u}} := \left\{x \in X: \int_I \mathscr{S}_{\bar{x}\bar{u}}(x)(t)\,dt < \tfrac{1}{4}\varepsilon\right\}.$$

Since $\bar{x} \in V_{\bar{x}\bar{u}}$ and $V_{\bar{x}\bar{u}}$ are open then the family defined above is an open covering of the compact set X. We can establish a finite partition of unity $p_1(x), \ldots, p_N(x)$ subordinate to this covering. Let $V_{x_i u_i}$ be such that $p_i^{-1}(0, 1] \subset V_{x_i u_i}$ for $i = 1, 2, \ldots, N$. Then for every $x \in X$ and $i = 1, 2, \ldots, N$ the following inequalities are satisfied:

$$p_i(x) \int_I \mathscr{S}_i(x)(t)\,dt \leqslant \tfrac{1}{4}\varepsilon p_i(x), \quad \text{where } \mathscr{S}_i = \mathscr{S}_{x_i u_i}. \tag{1.17}$$

Consider a measure $\vec{\mu}_x$ with the Radon–Nikodym derivatives $(\mathscr{S}_1(x)(t), \ldots, \mathscr{S}_N(x)(t))$. Since $\mathscr{S}_i$ are continuous on X in the norm topology of $L(I, R^1)$, $\vec{\mu}_x$ is continuous in the space M of nonatomic complete vector measure $\vec{\mu} = (\mu_1, \ldots, \mu_N)$ with the topology induced by the norm $\|\vec{\mu}\|$ equal to the variation of μ (see Proposition 2.2) then by Remark 2.1 and Proposition 2.2 there exists a family $\{I_\alpha\}_{\alpha \in [0,1]}$ of measurable sets such that

$$I_\alpha \subset I_\beta \quad \text{for } \alpha < \beta \tag{2.18}$$

$$|\vec{\mu}_x(I_\alpha) - \alpha\vec{\mu}_x(I)| < \tfrac{1}{4}\varepsilon/N \quad \text{for all } x \in X \text{ and } \alpha \in [0, 1] \tag{2.19}$$

and

$$\mu(I_\alpha) = \alpha. \tag{2.20}$$

Define now functions $\mathscr{S}$ and g by setting

$$\mathscr{S}(x) = \sum_{i=1}^{N} (\mathscr{S}_i(x) + \tfrac{1}{4}\varepsilon)\chi_{I_{z_i(x)} \setminus I_{z_{i-1}(x)}}, \tag{2.21}$$

and

$$g(x) = \sum_{i=1}^{N} u_i \chi_{I_{z_i(x)} \setminus I_{z_{i-1}(x)}}, \tag{2.22}$$

where $z_0(x) = 0$, $z_i(x) = p_1(x)+ \dots +p_i(x)$ for $i = 1, \dots, N$. From Proposition 2.3, it follows that g and $\mathscr{S}$ are continuous on X. We shall prove that $\int_I \mathscr{S}(x)(t)dt < \varepsilon$ for each $x \in X$. From (2.19) we have

$$\Big|\int_{I_\alpha} \mathscr{S}_i(x)(t)dt - \alpha \int_I \mathscr{S}_i(x)(t)dt\Big| < \tfrac{1}{4}\varepsilon/N \quad \text{for all } \alpha \in [0, 1].$$

Therefore,

$$\int_{I_{z_i(x)} \setminus I_{z_{i-1}(x)}} \mathscr{S}_i(x)(t)dt = \int_{I_{z_i(x)}} \mathscr{S}_i(x)(t)dt - \int_{I_{z_{i-1}(x)}} \mathscr{S}_i(x)/(t)dt$$
$$< \big(z_i(x) - z_{i-1}(x)\big) \int_I \mathscr{S}_i(x)(t)dt + \tfrac{1}{2}\varepsilon/N.$$

Since $p_i(x) = z_i(x) - z_{i-1}(x)$, by (2.21) we have

$$\int_I \mathscr{S}(x)(t)dt < \sum_{i=1}^{N} p_i(x) \int_I \mathscr{S}_i(x)(t)dt + \tfrac{3}{4}\varepsilon$$

and by (2.17) we obtain the required estimation.

It remains to establich that $L(x) \neq \emptyset$ for $x \in X$. By Proposition 2.4 for each $x \in X$ and $i = 1, \dots, N$ there is $u_x^i \in H(x)$ such that we have

$$|u_x^i(t) - u_i(t)| = e\big(H(x) - u_i\big)(t) \quad \text{for a.e. } t \in I. \tag{2.23}$$

Then the function $u_x = \sum_{i=1}^{N} u_x^i \chi_{I_{z_i(x)} \setminus I_{z_{i-1}(x)}}$ belongs to $H(x)$, because $H(x)$ is decomposable. On the other hand, by (2.21)–(2.23) for a.e. $t \in I$ we have

$$|u_x(t) - g(x)(t)| \leqslant \sum_{i=1}^{N} |u_x^i(t) - u_i(t)| \chi_{I_{z_i(x)} \setminus I_{z_{i-1}(x)}}$$
$$\leqslant \sum_{i=1}^{N} \mathscr{S}_i(x) \chi_{I_{z_i(x)} \setminus I_{z_{i-1}(x)}} < \mathscr{S}(x)(t).$$

Therefore, $u_x \in L(x)$. ■

Now we prove the following continuous selection theorem.

THEOREM 2.17. *If* $\mathscr{G}: I \times X \to \mathrm{Comp}(\boldsymbol{R}^n)$ *is such that* $\mathscr{G}(\cdot, x) \in \mathscr{A}(I, \boldsymbol{R}^n)$ *and* $\mathscr{G}(t, \cdot)$ *is l.s.c. on* X *with* $X \in \mathrm{Comp}(Z)$ *for fixed* $x \in X$ *and* $t \in I$, *respectively then there is a continuous function* $g: X \to L(I, \boldsymbol{R}^n)$ *such that* $g(x) \in \mathscr{F}(\mathscr{G})(x)$ *for every* $x \in X$.

Proof. We shall define by induction a decreasing sequence of set-valued functions $H_n: X \to \mathrm{Cl}(L(I, \boldsymbol{R}^n))$ for $n = 0, 1, \dots,$ which are decomposable and l.s.c. and and sequences of continuous functions $g_n: X \to L(I, \boldsymbol{R}^n)$ and $\mathscr{S}_n: X \to L(I, \boldsymbol{R}^1)$ for $n = 1, 2, \dots$ with the properties

$$\int_I \mathscr{S}_n(x)(t)dt < (\tfrac{1}{2})^n \quad \text{for } x \in X \text{ and } n = 1, 2, \dots \tag{2.24}$$

and such that the sets $L_{n+1}(x)$ defined by

$$L_{n+1}(x) = \{u \in H_n(x)\colon |u(t)-g_n(x)(t)| < \mathscr{S}_n(x)(t) \text{ for a.e. } t \in I\} \tag{2.25}$$

are nonempty for $x \in X$ and $n = 0, 1 \ldots$ For $n = 0$, put $H_0 = \mathscr{F}(\mathscr{G})$. It is clear that H_0 is l.s.c. and decomposable. If for fixed $n \geqslant 0$, the set-valued function H_n is defined, then the continuous functions $g_{n+1}\colon X \to L(I, \boldsymbol{R}^n)$ and $\mathscr{S}_{n+1}\colon X \to L(I, \boldsymbol{R}^1)$ are defined by Lemma 2.16 with $\varepsilon = (\frac{1}{2})^{n+1}$, so that for $x \in X$ the set $L_{n+1}(x) = \{u \in H_n(x)\colon |u(t)-g_{n+1}(x)(t)| < \mathscr{S}_{n+1}(x)(t)$ for a.e. $t \in I\}$ is nonempty and $\int_I \mathscr{S}_{n+1}(x)(t)dt < (\frac{1}{2})^{n+1}$. Then, by Proposition 2.6, we can put for every $x \in X$, $H_{n+1}(x) = \overline{[L_{n+1}(x)]_L}$. It is clear that $H_{n+1}(x) \subset H_n(x)$. For each $x \in X$ and $n = 1, 2, \ldots$, let u_n^x be an arbitrary point of $H_n(x)$. Since $H_{n+p}(x) \subset H_n(x)$, we have $u_{n+p}^x \in H_n(x)$ for each $p \geqslant 0$. Therefore, by (2.25) we have, for each n and $p \geqslant 0$, the inequality

$$|g_n(x)(t)-u_{n+p}^x(x)(t)| \leqslant \mathscr{S}_n(x)(t) \quad \text{for a.e. } t \in I. \tag{2.26}$$

Hence, it follows $|g_n(x)(t)-g_{n+p}(x)(t)| \leqslant \mathscr{S}_n(x)(t)+\mathscr{S}_{n+p}(x)(t)$ for $n = 1, 2, \ldots$, $p \geqslant 0$, $x \in X$ and a.e. $t \in I$. Because of (2.24) the last inequality implies that (g_n) converges uniformly in the $L(I, \boldsymbol{R}^n)$-norm topology to a continuous function $g\colon X \to L(I, \boldsymbol{R}^n)$. Again from (2.24) and (2.26) it follows that $|g_n(x)-u_n^x|_a \to 0$ as $n \to \infty$; hence $g(x) \in H_0(x) = \mathscr{F}(\mathscr{G})(x)$ for $x \in X$. ∎

Now as a corollary from Lemma 2.14 and Theorem 2.17 we obtain.

THEOREM 2.18. *If* $G \in \mathscr{H}_\lambda(I, \boldsymbol{R}^n)$ *is s.-w.l.s.c. with respect to its last two variables and such that* $\int_\sigma^{\sigma+a} \|G(t, x, z)\|dt \leqslant \lambda$ *for* $(x, z) \in \mathscr{K}_\lambda \times B_\lambda$, *then there is a continuous function* $g\colon K_\lambda \to L(I, \boldsymbol{R}^n)$ *such that* $g(x) \in \mathscr{F}(G \,\square\, \mathscr{D})(x)$ *for every* $x \in K_\lambda$. ∎

3. Fixed point properties of subtrajectory and trajectory integrals depending on parameters

We investigate here the existence of fixed points of set-valued functions $\mathscr{F}(G \,\square\, \mathscr{D})$ or $\mathscr{F}(G \,\square\, \mathscr{I})$ and $\mathscr{I}\mathscr{F}(G \,\square\, \mathscr{D})$ with $G \in H_\lambda(I, \boldsymbol{R}^n)$. The existence theorems will depend on some type of upper and lower semicontinuities of G with respect to its last two variables and the type of values of G.

3.1. Existence of fixed points of subtrajectory integrals of set-valued functions with non-convex values

We begin with the case of a set-valued function G defined on $I \times C(I, \boldsymbol{R}^n) \times L(I, \boldsymbol{R}^n)$ and Lipschitz continuous with respect to its last two variables. Immediately from Covitz–Nadler fixed point theorem we obtain the following global existence theorems.

THEOREM 3.1. *Suppose* $G\colon I \times C(I, \boldsymbol{R}^n) \times L(I, \boldsymbol{R}^n) \to \mathrm{Comp}(\boldsymbol{R}^n)$ *with* $I = [\sigma, \sigma+a]$ *is such that*

(i) *$G(\cdot, x, z) \in \mathscr{A}(I, \boldsymbol{R}^n)$ for fixed (x, z),*

(ii) *G is Lipschitz continuous on $C(I, \boldsymbol{R}^n) \times L(I, \boldsymbol{R}^n)$ with respect to its last two variables.*

Then there exists $z \in L(I, \boldsymbol{R}^n)$ such that $z \in \mathscr{F}(G \Box \mathscr{I})(z)$.

Proof. Similarly as in the proof of Lemma 2.6 we can show that for every $L > 0$ there is a norm $|\cdot|$ of $L(I, \boldsymbol{R}^n)$ equivalent to the norm $|\cdot|_a$ of $L(I, \boldsymbol{R}^n)$ and such that $\mathrm{H}(\mathscr{F}(G \Box \mathscr{I})(z), \mathscr{F}(G \Box \mathscr{I})(\bar{z}) \leqslant L|z-\bar{z}|$ for every $z, \bar{z} \in L(I, \boldsymbol{R}^n)$. Take $L \in (0, 1)$ and put $X = L(I, \boldsymbol{R},)$ and $\varrho(z_1, z_2) = |z_1 - z_2|$ for every $z_1, z_2 \in X$. Obviously, (X, ϱ) is a complete metric space and $\mathscr{F}(G \Box \mathscr{I})$ is a set-valued contractive mapping of X into $\mathrm{Cl}(X)$, because by (i) of Theorem 2.1, $\mathscr{F}(G \Box \mathscr{I})(z) \in \mathrm{Cl}(L(I, \boldsymbol{R}^n))$ for every $z \in X$. Now, by Covitz–Nadler fixed point theorem $\mathscr{F}(G \Box \mathscr{I})$ has in X a fixed point. ∎

THEOREM 3.2. *Suppose $G\colon I \times C(I, \boldsymbol{R}^n) \times L(I, \boldsymbol{R}^n) \to \mathrm{Comp}(\boldsymbol{R}^n)$ with $I = [\sigma, \sigma+a]$ is such that*

(i) *$G(\cdot, x, z) \in \mathscr{A}(I, \boldsymbol{R}^n)$ for fixed (x, z),*

(ii) *G is $\mathscr{A}$-Lipschitz continuous with respect to its last two variables with Lipschitz constant $L < 1$.*

Then there is $z \in L(I, \boldsymbol{R}^n)$ such that $z \in \mathscr{F}(G \Box \mathscr{I})(z)$.

Proof. Take $X = L(I, \boldsymbol{R}^n)$ and $\varrho(z_1, z_2) = |z_1 - z_2|_a$. Similarly as in the proof of Lemma 2.13 we obtain $\mathrm{H}_L(\mathscr{F}(G)(x, z), \mathscr{F}(G)(\bar{x}, \bar{z})) \leqslant L\max[|x-\bar{x}|_a |z-\bar{z}|_a$, for $x, \bar{x} \in C(I, \boldsymbol{R}^n)$ and $z, \bar{z} \in L(I, \boldsymbol{R}^n)$. Hence, in particular it follows

$$\begin{aligned}\mathrm{H}_L\big(\mathscr{F}(G \Box \mathscr{I})(z), \mathscr{F}(G \Box \mathscr{I})(\bar{z})\big) &= \mathrm{H}_L\big(\mathscr{F}(G)(\mathscr{I}z, z), \mathscr{F}(G)(\mathscr{I}\bar{z}, \bar{z})\big)\\ &\leqslant L\max[|\mathscr{I}z - \mathscr{I}\bar{z}|_a, |z-\bar{z}|_a] \leqslant L|z-\bar{z}|_a.\end{aligned}$$

Therefore, $\mathscr{F}(G \Box \mathscr{I})$ is a set-valued contractive mapping of X into $\mathrm{Cl}(X)$. Thus, by Covitz–Nadler fixed point theorem there is $z \in X$ such that $z \in \mathscr{F}(G \Box \mathscr{I})(z)$. ∎

If condition $L < 1$ in Theorem 3.2 is not satisfied, we can obtain the following local existence theorem:

THEOREM 3.3. *Suppose $G\colon I \times C(I, \boldsymbol{R}^n) \times L(I, \boldsymbol{R}^n) \to \mathrm{Comp}(\boldsymbol{R}^n)$ is such that*

(i) *$G(\cdot, x, z) \in \mathscr{A}(I, \boldsymbol{R}^n)$ for fixed (x, z),*

(ii) *there is a continuous function $K\colon \boldsymbol{R}^+ \to \boldsymbol{R}^+$ such that*

$$\int_\sigma^t h\big(G(\tau, x, z), G(\tau, \bar{x}, \bar{z})\big)d\tau \leqslant K(t-\sigma)\max[|x-\bar{x}|_t, |z-\bar{z}|_t]$$

for $t \in I$ and $(x, z), (\bar{x}, \bar{z}) \in \mathscr{K}_\lambda \times B_\lambda$.

Then there are $\alpha \in [0, a]$ and $z_\alpha \in L(I, \boldsymbol{R}^n)$ such that $z_\alpha(t) = 0$ for a.e. $t \in (\sigma+ +\alpha, \sigma+a]$ and $z_\alpha \in \mathscr{F}(G_\alpha \Box \mathscr{I})(z_\alpha)$, where $(G_\alpha \Box \mathscr{I})(t, z_\alpha) := (G \Box \mathscr{I})(t, z_\alpha)$ for $t \in [\sigma, \sigma+\alpha]$ and $(G_\alpha \Box \mathscr{I})(t, z) = \{0\}$ for $t \in (\sigma+\alpha, \sigma+a]$.

Proof. If $K(a) < 1$ then we take $\alpha = a$ and $z_\alpha = z$, where $z \in L(I, \boldsymbol{R}^n)$ is a fixed point of $\mathscr{F}(G \Box \mathscr{I})$. It exists by Theorem 3.2. Suppose $K(a) \geqslant 1$. Since $K(0) = 0$

and K is continuous on $\boldsymbol{R}^+$ there is $\alpha \in (0, a)$ such that $K(\alpha) < 1$. Let X be a subspace of $L(I, \boldsymbol{R}^n)$ containing all functions $z_\alpha \in L(I, \boldsymbol{R}^n)$ such that $z_\alpha(t) = 0$ for a.e. $t \in (\sigma + +\alpha, \sigma + a]$. It is clear that $(X, |\cdot|_a)$ is also a Banach space. For every $z_\alpha, \bar{z}_\alpha \in X$ we have

$$\int_{\sigma+\alpha}^{\sigma+a} h[(G \square \mathscr{I})(t, z_\alpha), (G \square \mathscr{I})(t, z_\alpha)]\,dt = 0.$$

Therefore, for every $z_\alpha, \bar{z}_\alpha \in X$ one has

$$\int_{\sigma}^{\sigma+a} h[(G \square \mathscr{I})(t, z_\alpha), (G \square \mathscr{I})(t, \bar{z}_\alpha)]\,dt \leqslant K(\alpha)|z_\alpha - \bar{z}_\alpha|_a.$$

Hence, similarly as in the proof of Lemma 2.13 one obtains $\mathrm{H}_L(\mathscr{F}(G_\alpha \square \mathscr{I})(z_\alpha)$ $\mathscr{F}(G_\alpha \square \mathscr{I})(\bar{z}_\alpha)) \leqslant K(\alpha)|z_\alpha - \bar{z}_\alpha|_a$, where H_L is the Hausdorff metric in $\mathrm{Cl}(X)$ induced by the norm $|\cdot|$. Now, by Covitz–Nadler fixed point theorem there is $z_\alpha^1 \in X$ such that $z_\alpha \in \mathscr{F}(G_\alpha \square \mathscr{I})(z_\alpha)$. ■

Similarly we obtain the following existence theorems.

THEOREM 3.4. *Suppose $G \in \mathscr{H}_\lambda(I, \boldsymbol{R}^n)$ is such that* $\int_\sigma^{\sigma+a} \|G(t, x, z)\|\,dt \leqslant \lambda$ for $(x, z) \in \mathscr{K}_\lambda \times B_\lambda$. *If G is Lipschitz continuous with respect to its last two variables then there is $z \in \boldsymbol{P}_\lambda$ such that $z \in \mathscr{F}(G \square \mathscr{I})(z)$, where* $\boldsymbol{P}_\lambda = \mathrm{co}\overline{[\bigcup\{\mathscr{F}(G)(x, z)\colon (x, z) \in \boldsymbol{K}_\lambda \times S_\lambda\}]}_L^{\mathrm{w}}$.

Proof. By the definition of $\boldsymbol{P}_\lambda$ we have $\boldsymbol{P}_\lambda \subset B_\lambda$ and $\mathscr{F}(G \square \mathscr{I})(z) \subset \boldsymbol{P}_\lambda$ for each $z \in \boldsymbol{P}_\lambda$. Moreover $\boldsymbol{P}_\lambda$ is a closed subsets of $L(I, \boldsymbol{R}^n)$. Now take $L \in (0, 1)$ and let $\cdot|$ be a norm of $L(I, \boldsymbol{R}^n)$ equivalent to $|\cdot|$ and such that (2.8) is satisfied. Since for every $z \in \boldsymbol{P}_\lambda$, $(\mathscr{I}z, z) \in \mathscr{K}_\lambda \times B_\lambda$ then for every $z, \bar{z} \in \boldsymbol{P}_\lambda$ one obtains $H(\mathscr{F}(G \square \mathscr{I})(z),$ $\mathscr{F}(G \square \mathscr{I})(\bar{z})) \leqslant L|z - \bar{z}|$. Put $X = \boldsymbol{P}_\lambda$ and $\varrho(z_1, z_2) = |z_1 - z_2|$. Obviously, (X, ϱ) is a complete metric space. It is also clear that $\mathscr{F}(G \square \mathscr{I})$ is a contractive mapping of X into $\mathrm{Cl}(X)$. Therefore, by Covitz–Nadler fixed point theorem a fixed point of $\mathscr{F}(G \square \mathscr{I})$ exists in X. ■

THEOREM 3.5. *Suppose $G \in \mathscr{H}_\lambda(I, \boldsymbol{R}^n)$ is such that $\int_\sigma^{\sigma+a} \|G(t, x, z)\|\,dt \leqslant \lambda$ for $(x, z) \in \mathscr{K}_\lambda \times B_\lambda$. If G is $\mathscr{A}$-Lipschitz continuous with respect to its last two variables with the Lipschitz constant $L < 1$ then there is $z \in \boldsymbol{P}_\lambda$ such that $z \in \mathscr{F}(G \square \mathscr{I})(z)$.* ■

Now, using continuouss election Theorem 2.18 we obtain the following existenice theorem.

THEOREM 3.6. *Let $G \in \mathscr{H}_\lambda(I, \boldsymbol{R}^n)$ be s.-w.l.s.c. with respect to its last two variables and such that $\int_\sigma^{\sigma+a} \|G(t, x, z)\|\,dt \leqslant \lambda$ for $(x, z) \in \mathscr{K}_\lambda \times B_\lambda$. Then there is $x \in \boldsymbol{K}^\lambda$ such that $\dot{x} \in \mathscr{F}(G \square \mathscr{D})(x)$, where $\boldsymbol{K}_\lambda = \mathscr{I}(\boldsymbol{P}_\lambda)$.*

Proof. By Theorem 2.18 there exists a continuous function $g\colon \boldsymbol{K}_\lambda \to L(I, \boldsymbol{R}^n)$ such that $g(x) \in \mathscr{F}(G \square \mathscr{D})(x)$ for each $x \in \boldsymbol{K}_\lambda$. Let T_g be a mapping defined on $\boldsymbol{K}_\lambda$ by taking $T_g(x) := \mathscr{I}g(x)$ for $x \in \boldsymbol{K}_\lambda$. For every $x \in \boldsymbol{K}_\lambda$ we have $g(x) \in \boldsymbol{P}_\lambda$, where $\boldsymbol{P}_\lambda$ is such as in Section 2.3. Therefore, $T_g(x) \in \boldsymbol{K}_\lambda$ for each $x \in \boldsymbol{K}_\lambda$, because $\boldsymbol{K}_\lambda = \mathscr{I}(\boldsymbol{P}_\lambda)$. Clearly T_g is continuous on $\boldsymbol{K}_\lambda$ and $\boldsymbol{K}_\lambda$ is a compact and convex subset of the Banach space $C^0(I, \boldsymbol{R}^n)$. Then by Schauder–Tychonov's fixed point theorem there is $x \in \boldsymbol{K}_\lambda$ such that $x = T_g(x) := \mathscr{I}g(x)$. But $\dot{x} = g(x)$ and $g(x) \in \mathscr{F}(G \square \mathscr{D})(x)$. Then $\dot{x} \in \mathscr{F}(G \square \mathscr{D})(x)$. ■

3.2. Existence of fixed points of subtrajectory integrals of set-valued functions with convex values

Now we will deal with $G \in \mathscr{H}_\lambda(I, \boldsymbol{R}^n)$ having convex values and such that $\int_\sigma^{\sigma+a} ||G(t, x, z)||\,dt \leqslant \lambda$ for $(x, z) \in \mathscr{K}_\lambda \times B_\lambda$.

We shall prove the following fixed points theorems.

THEOREM 3.7. *Suppose $G \in \mathscr{H}_\lambda(I, \boldsymbol{R}^n)$ has convex values, is w.-w.l.s.c. with respect to its last two variables and such that $\int_\sigma^{\sigma+a} ||G(t, x, z)||\,dt \leqslant \lambda$ for $(x, z) \in \mathscr{K}_\lambda \times B_\lambda$. Then there exists $x \in \boldsymbol{K}_\lambda$ such that $x \in \mathscr{I}\mathscr{F}(G \square \mathscr{D})(x)$, where $\boldsymbol{K}_\lambda = \mathscr{I}(\boldsymbol{P}_\lambda)$.*

Proof. By virtue of Lemma 2.12, a set-valued map $\mathscr{I}\mathscr{F}(G)$ is s.-w.l.s.c. on $\mathscr{K}_\lambda \times \times B_\lambda$. Then $\mathscr{I}\mathscr{F}(G \square \mathscr{D})$ is l.s.c. on $\boldsymbol{K}_\lambda$. Furthermore, by Theorems 2.1, 2.2 and 2.4, it has compact convex values contained in $\boldsymbol{K}_\lambda$. Thus by Michael's selection theorem there is a continuous function $f\colon \boldsymbol{K}_\lambda \to C^0(I, \boldsymbol{R}^n)$ such that $f(x) \in \mathscr{I}\mathscr{F}(G \square \mathscr{D})(x)$ for $x \in \boldsymbol{K}_\lambda$. Then, f is a continuous map on compact convex subset $\boldsymbol{K}_\lambda$ of $C^0(I, \boldsymbol{R}_\lambda)$ such that $f(\boldsymbol{K}_\lambda) \subset \boldsymbol{K}_\lambda$. Therefore, by Schauder–Tychonov's fixed point theorem there is $x \in \boldsymbol{K}_\lambda$ such $x = f(x)$. Since $f(x) \in \mathscr{I}\mathscr{F}(G \square \mathscr{D})(x)$, we have $x \in \mathscr{I}\mathscr{F}(G \square \mathscr{D})(x)$. ■

THEOREM 3.8. *Let $G \in \mathscr{H}_\lambda(I, \boldsymbol{R}^n)$ be w.-w.u.s.c. with respect to its last two variables, have convex values and be such that* $\int_\sigma^{\sigma+a} ||G(t, x, z)||\,dt \leqslant \lambda$ for $(x, z) \in \mathscr{K}_\lambda \times B_\lambda$. *Then there exists $z \in \boldsymbol{P}_\lambda$ such that $z \in \mathscr{F}(G \square \mathscr{I})(z)$.*

Proof. By virtue of Lemma 2.10, $\mathscr{F}(G)$ is w.-w.u.s.c. on $\mathscr{K}_\lambda \times B_\lambda$. Hence, by (iii) of Lemma 1.13, it follows that $\mathscr{F}(G \square \mathscr{I})$ is w.-w.u.s.c. on $\boldsymbol{P}_\lambda$. Furthermore, it is easy to see that for each $z \in \boldsymbol{P}_\lambda$, $\mathscr{F}(G \square \mathscr{I})(z)$ is a weakly compact convex subset of $\boldsymbol{P}_\lambda$. Therefore, by the weak form of Kakutani–Ky Fan fixed point theorem there is $z \in \boldsymbol{P}_\lambda$ that such $z \in \mathscr{F}(G \square \mathscr{I})(z)$. ■

THEOREM 3.9. *Suppose $G \in \mathscr{H}_\lambda(I, \boldsymbol{R}^n)$ has convex values, is continuous with respect to its second variable, is Lipschitz continuous with respect to its last variable uniformly with respect to the second one and is such that $\int_\sigma^{\sigma+a} ||G(t, x, z)||\,dt \leqslant \lambda$ for $(x, z) \in \mathscr{K}_\lambda \times B_\lambda$. Then there is $z \in \boldsymbol{P}_\lambda$ such that $z \in \mathscr{F}(G \square \mathscr{I})(z)$.*

Proof. Let us observe that G is continuous with respect to its last two variables. Then by Lemma 2.8, $\mathscr{F}(G)$ is l.s.c. and H-u.s.c. on $\mathscr{K}_\lambda \times B_\lambda$. Furthermore, by Lemma 2.7, $\mathscr{F}(G)$ is Lipschitz continuous with respect to its last variable uniformly with respect to $x \in \mathscr{K}_\lambda$ and by eventual changing of the norm of $L(I, \boldsymbol{R}^n)$ its Lipschitz constant L can be taken from the interval $[0, 1)$. Since $\int_\sigma^{\sigma+a} \|G(t, x, z)\| dt \leqslant \lambda$ then there is a nonempty closed bounded convex subset $\boldsymbol{P}_\lambda \subset L(I, \boldsymbol{R}^n)$ such that for every $(x, z) \in \boldsymbol{K}_\lambda \times \boldsymbol{P}_\lambda$ with $\boldsymbol{K}_\lambda = \mathscr{J}(\boldsymbol{P}_\lambda)$, $\mathscr{F}(G)(x, z)$ is a closed convex subset of $\boldsymbol{P}_\lambda$. Therefore, the restriction of $\mathscr{F}(G)$ to $\boldsymbol{K}_\lambda \times \boldsymbol{P}_\lambda$ satisfies the assumptions of fiexd point Theorem II.4.8. Then there is $z \in \boldsymbol{P}_\lambda$ such that $z \in \mathscr{F}(G \,\square\, \mathscr{J})(z)$. ■

In a similar way we obtain the following theorem.

THEOREM 3.10. *Suppose $G \in \mathscr{H}_\lambda(I, \boldsymbol{R}^n)$ has convex values and is such that $\int_\sigma^{\sigma+a} \|G(t, x, z)\| dt \leqslant \lambda$ for $(x, z) \in \mathscr{K}_\lambda \times B_\lambda$. Suppose furthermore G is $\mathscr{A}$-continuous with respect to its second variable and $\mathscr{A}$-Lipschitz continuous with respect to its last argument uniformly with respect to the second one with Lipschitz constant $L < 1$. Then there exists $z \in \boldsymbol{P}_\lambda$ such that $z \in \mathscr{F}(G \,\square\, \mathscr{J})(z)$.*

Proof. Consider a restriction of $\mathscr{F}(G)$ to $\boldsymbol{K}_\lambda \times \boldsymbol{P}_\lambda \subset \mathscr{K}_\lambda \times B_\lambda$. By virtue of Theorems 2.1, 2.2, 2.4 and Lemmas 2.8 and 2.13 this mapping has the following properties:

(i) $\mathscr{F}(G)(x, z)$ is a nonempty, closed and convex subset of $\boldsymbol{P}_\lambda$ for each $(x, z) \in \boldsymbol{K}_\lambda \times \boldsymbol{P}_\lambda$,

(ii) $\mathscr{F}(G)(x, \cdot)$ is a contraction on $\boldsymbol{P}_\lambda$ uniformly with respect to $x \in \boldsymbol{K}_\lambda$,

(iii) $\mathscr{F}(G)$ is l.s.c. and H-u.s.c. on $\boldsymbol{K}_\lambda \times \boldsymbol{P}_\lambda$.

By Corollary 1.7, $\mathscr{J}$ is a completely continuous mapping of $\boldsymbol{P}_\lambda$ into $C^0(I, \boldsymbol{R}^n)$. Then, by Theorem II.4.8, there is $z \in \boldsymbol{P}_\lambda$ such that $z \in \mathscr{F}(G)(\mathscr{J}z, z)$, i.e., $z \in \mathscr{F}(G \,\square\, \mathscr{J})(z)$. ■

3.3. Properties of set of fixed points of subtrajectory integrals

Let $G \in \mathscr{H}_\lambda(I, \boldsymbol{R}^n)$ be such that $\int_\sigma^{\sigma+a} \|G(t, x, z)\| dt \leqslant \lambda$ for $(x, z) \in \mathscr{K}_\lambda \times B_\lambda$. Denote by $S(G)$ the set of all fixed points of $\mathscr{F}(G \,\square\, \mathscr{J})$. It is easy to see that $S(G) \subset \boldsymbol{P}_\lambda$. We shall investigate here the properties of sets $S(G)$ with $G \in \mathscr{H}^{\mathrm{C}}_\lambda(I, \boldsymbol{R}^n) \subset \mathscr{H}_\lambda(I, \boldsymbol{R}^n)$ as well as a set-valued mapping $\mathscr{H}^{\mathrm{C}}_\lambda(I, \boldsymbol{R}^n) \ni G \to S(G) \subset \boldsymbol{P}_\lambda$, where $\mathscr{H}^{\mathrm{C}}_\lambda(I, \boldsymbol{R}^n)$ is a subset of $\mathscr{H}_\lambda(I, \boldsymbol{R}^n)$ containing all $G \in \mathscr{H}_\lambda(I, \boldsymbol{R}^n)$ with convex values that are w.-w.u.s.c. with respect to its last two variables and such that $\int_\sigma^{\sigma+a} \|G(t, x, z)\| dt \leqslant \lambda$ for $(x, z) \in \mathscr{K}_\lambda \times B_\lambda$.

LEMMA 3.11. *$(\mathscr{H}^{\mathrm{C}}_\lambda(I, \boldsymbol{R}^n), \delta)$ with δ defined by* (2.1) *is a complete metric space.*

Proof. By virtue of Lemma 2.5 it sufficies only to show that $\mathscr{H}^{\mathrm{C}}_\lambda(I, \boldsymbol{R}^n)$ is a closed subset of the metric space $(\mathscr{H}_\lambda(I, \boldsymbol{R}^n), \delta)$.

Suppose (G_n) is a sequence of $\mathscr{H}^C_\lambda(I, R^n)$ converging in δ-metric topology to any $G \in \mathscr{H}_\lambda(I, R^n)$. By Corollary 1.3, G has convex values. Furthermore, we have

$$\int_\sigma^{\sigma+a} \big| \|G_n(t, x, z)\| - \|G(t, x, z)\| \big| dt$$
$$\leqslant \int_\sigma^{\sigma+a} h\big(G_n(t, x, z), G(t, x, z)\big) dt \leqslant \delta(G_n, G)$$

and

$$\int_\sigma^{\sigma+a} \|G(t, x, z)\| dt$$
$$\leqslant \int_\sigma^{\sigma+a} \big| \|G(t, x, z)\| - \|G_n(t, x, z)\| \big| dt + \int_\sigma^{\sigma+a} \|G_n(t, x, z)\| dt$$
$$\leqslant \int_\sigma^{\sigma+a} \big| \|G(t, x, z)\| - \|G_n(t, x, z)\| \big| dt + \lambda$$

for $n \in N$ and $(x, z) \in \mathscr{K}_\lambda \times B_\lambda$. Then $\int_\sigma^{\sigma+a} \|G(t, x, z)\| dt \leqslant \lambda$ for $(x, z) \in \mathscr{K}_\lambda \times B_\lambda$.

Moreover, $\lim_{n\to\infty} \delta(G_n, G) = 0$ implies that for every $\varepsilon > 0$ there is $N \in N$ such that $\sup_{(x, y)} \int_I h(G_N t, x, z), G(t, x, z)) dt < \frac{1}{2}\varepsilon$. For fixed $(x, z) \in \mathscr{K}_\lambda \times B_\lambda$ and every sequence $\{(x_k, z_k)\}$ of $\mathscr{K}_\lambda \times B_\lambda$ such that $|x_k - x|_a \to 0$ and $z_k \rightharpoonup z$ as $k \to \infty$ we have $\lim_{k\to\infty} \bar{h}\big(\int_U G_N(t, x_k, z_k) dt, \int_U G_N(t, x, z) dt\big) = 0$ for every measurable set $U \subset I$. For every measurable set $U \subset I$ one obtains

$$\bar{h}\Big(\int_U G(t, x_k, z_k) dt, \int_U G(t, x, z) dt$$
$$\leqslant \sup_{(x, z)} \int_I h\big(G(t, x, z), G_N(t, x, z)\big) dt + \bar{h}\Big(\int_U G_N(t, x_k, z_k) dt, \int_U G_N(t, x, z) dt\Big) +$$
$$+ \sup_{(x, y)} \int_I h\big(G_N(t, x, z), G(t, x, z)\big) dt$$

for each $k = 1, 2, \ldots$ Then $\lim_{k\to\infty} \bar{h}\big(\int_U G(t, x_k, z_k) dt, \int_U G(t, x, z) dt\big) \leqslant \varepsilon$ for every $\varepsilon > 0$ and each measurable set $U \subset I$. Thus $\lim_{k\to\infty} \bar{h}\big(\int_U G(t, x_k, z_k) dt, \int_U G(t, x, z) dt\big) = 0$ for each measurable set $U \subset I$ and therefore $G \in \mathscr{H}^C_\lambda(I, R^n)$. ■

Similarly as in Remark 2.1 we can define in $\mathscr{H}^C_\lambda(I, R^n)$ a weak convergence of sequence of $\mathscr{H}^C_\lambda(I, R^n)$. We will show that $\mathscr{H}^C_\lambda(I, R^n)$ is weakly sequentially complete.

LEMMA 3.12. *$\mathscr{H}^C_\lambda(I, R^n)$ is sequentially weakly complete, i.e., every weak Cauchy sequence of $\mathscr{H}^C_\lambda(I, R^n)$ has a weak limit in $\mathscr{H}^C_\lambda(I, R^n)$.*

Proof. Suppose that (G_n) is a weak Cauchy sequence of $\mathscr{H}^C_\lambda(I, \boldsymbol{R}^n) \subset \mathscr{H}_\lambda(I, \boldsymbol{R}^n)$. By virtue of Corollary 1.4, for every $(x, z) \in \mathscr{K}_\lambda \times B_\lambda$ there is $G(\cdot, x, z) \in \mathscr{A}^C(I, \boldsymbol{R}^n)$ such that $\lim_{n\to\infty} h(\int_U G_n(t, x, z)dt, \int_U G(t, x, z)dt) = 0$ for each measurable set $U \subset I$ and $(x, z) \in \mathscr{K}_\lambda \times B_\lambda$. Similarly as in Remark 2.1 it can be easily seen that the last limit exists uniformly with respect to $(x, z) \in \mathscr{K}_\lambda \times B_\lambda$ and hence that $G \in \mathscr{H}_\lambda(I, \boldsymbol{R}^n)$. Now, similarly as in the proof of Lemma 3.11 we can see that G is w.-w.u.s.c. with respect to its last two variables, because for every $n \in \boldsymbol{N}$ and $(\tilde{x}, \tilde{z}), (\bar{x}, \bar{z}) \in \mathscr{K}_\lambda \times B_\lambda$ one has

$$\bar{h}\left(\int_U G(t, \bar{x}, \bar{z})dt, \int_U G(t, \tilde{x}, \tilde{z})dt\right) \leqslant \sup_{(x,z)} h\left(\int_U G(t, x, z)dt, \int_U G_n(t, x, z)dt\right) +$$

$$+\bar{h}\left(\int_U G_n(t, \bar{x}, \bar{z})dt, \int_U G_n(t, \tilde{x}, \tilde{z})dt\right) + \sup_{(x,z)} h\left(\int_U G_n(t, x, z)dt, \int_U G(t, x, z)dt\right)$$

for each measurable set $U \subset I$.

Now, we show that $\int_\sigma^{\sigma+a} \|G(t, x, z)\|dt \leqslant \lambda$ for $(x, z) \in \mathscr{K}_\lambda \times B_\lambda$. Indeed, suppose there is $(\bar{x}, \bar{z}) \in \mathscr{K}_\lambda \times B_\lambda$ such that $\int_\sigma^{\sigma+a} \|G(t, \bar{x}, \bar{z})\|dt > \lambda$. Let $g_\lambda \in \mathscr{F}(G(\cdot, x, z)$ be such that $|g_\lambda(t)| = \|G(t, \bar{x}, \bar{z})\|$ for a.e. $t \in I := [\sigma, \sigma+a]$. Such selector exists by Theorem II.3.13 (see the proof of Theorem 1.3). By virtue of Theorem 1.11 there exists a sequence, say (g_n) of selectors of $G_n(\cdot, \bar{x}, \bar{z})$, $n = 1, 2, \dots$ such that $g_n \rightharpoonup g_\lambda$ as $n \to \infty$. Hence, it follows $|g_\lambda|_a \leqslant \liminf_{n\to\infty} |g_n|_a$. But, $G_n \in \mathscr{H}^C_\lambda(I, \boldsymbol{R}^n)$ then $|g_n|_a \leqslant \lambda$ for every $n = 1, 2, \dots$ Then we have $\lambda < |g_n|_a \leqslant \liminf_{n\to\infty} |g_n|_a \leqslant \lambda$; a contradiction. Then finally, we have $G \in \mathscr{H}^C_\lambda(I, \boldsymbol{R}^n)$. ■

In what follows for a given $G \in \mathscr{H}_\lambda(I, \boldsymbol{R}^n)$ we shall still denote by $S(G)$ the set of all fixed points of $\mathscr{F}(G \Box \mathscr{I})$.

THEOREM 3.13. *For every $G \in \mathscr{H}^C_\lambda(I, \boldsymbol{R}^n)$, the set $S(G)$ is a nonempty weakly compact subset of $\boldsymbol{P}_\lambda$.*

Proof. Let $G \in \mathscr{H}^C_\lambda(I, \boldsymbol{R}^n)$ be given. By Theorem 3.8 we have $S(G) \neq \emptyset$. It is clear that $S(G) \subset \boldsymbol{P}_\lambda$. Since $\boldsymbol{P}_\lambda$ is weakly compact, in $L(I, \boldsymbol{R}^n)$ with $I := [\sigma, \sigma+a]$, then $S(G)$ is relatively weakly sequentially compact in $L(I, \boldsymbol{R}^n)$. Then by Eberlein–Šmulian's theorem $\overline{[S(G)]}^w_L$ is weakly compact. We shall show that $S(G)$ is weakly closed. Let $u \in \overline{[S(G)]}^w_L$. By Šmulian's theorem there exists a sequence (u_n) of $S(G)$ weakly converging to u as $n \to \infty$. Then for every measurable set $U \subset I$ we have $\left|\int_U u_n(t)dt - \int_U u(t)dt\right| \to 0$ as $n \to \infty$. Furthermore, $u_n \in \mathscr{F}(G \Box \mathscr{I})(u_n)$, because $u_n \in S(G)$ for each $n = 1, 2, \dots$ Then, in particular, for each measurable set $U \subset I$, $\int_U u_n(t)dt \in \int_U G(t, \mathscr{I}u_n, u_n)dt$ for $n = 1, 2, \dots$ Therefore, for each measurable set

$U \subset I$ and $n = 1, 2, \dots$ one has

$$\operatorname{dist}\Big(\int_U u(t)\,dt, \int_U G(t, \mathscr{I}u, u)\,dt\Big)$$
$$\leqslant \Big|\int_U u(t)\,dt - \int_U u_n(t)\,dt\Big| + \operatorname{dist}\Big(\int_U u_n(t)\,dt, \int_U G(t, \mathscr{I}u_n, u_n)\,dt\Big) +$$
$$+ \bar{h}\Big(\int_U G(t, \mathscr{I}u_n, u_n)\,dt, \int_U G(t, \mathscr{I}u, u)\,dt\Big).$$

Then $\int_U u(t)\,dt \in \int_U G(t, \mathscr{I}u, u)\,dt$ for each measurable set $U \subset I$. Hence, by Corollary II.3.4 it follows that $u \in S(G)$. Thus $\overline{[S(G)]}^w_L \subset S(G)$ and therefore $S(G)$ is weakly compact in $L(I, \boldsymbol{R}^n)$. ■

We shall now consider a set-valued mapping $S(\cdot)$: $\mathscr{H}^C_\lambda(I, \boldsymbol{R}^n) \ni G \to S(G) \in \mathrm{Cl}(L(I, \boldsymbol{R}^n))$. We will say that $S(\cdot)$ is w.-w.u.s.c. on $\mathscr{H}^C_\lambda(I, \boldsymbol{R}^n)$ if for every $G \in \mathscr{H}^C_\lambda(I, \boldsymbol{R}^n)$, each sequence (G_n) of $\mathscr{H}^C_\lambda(I, \boldsymbol{R}^n)$ weakly converging in $\mathscr{H}^C_\lambda(I, \boldsymbol{R}^n)$ to G and every sequence (u_n) of $L(I, \boldsymbol{R}^n)$ such that $u_n \in S(G_n)$ for $n = 1, 2, \dots$ there exists a subsequence of (u_n) weakly converging to $u \in S(G)$.

Similarly, we say that $S(\cdot)$ is w.-s.u.s.c. on $\mathscr{H}^C_\lambda(I, \boldsymbol{R}^n)$ if for every $G \in \mathscr{H}^C_\lambda(I, \boldsymbol{R}^n)$ and each sequence (G_n) of $\mathscr{H}^C_\lambda(I, \boldsymbol{R}^n)$ converging in the δ-metric topology to G every sequence (u_n) of $L(I, \boldsymbol{R}^n)$ such that $u_n \in S(G_n)$ for $n = 1, 2, \dots$ has a subsequence weakly converging to any $u \in S(G)$.

We can also consider a set-valued mapping $\mathscr{C}(\cdot)$: $\mathscr{H}^C_\lambda(I, \boldsymbol{R}^n) \in G \to \mathscr{C}(G) \in \mathrm{Comp}(C^0(I, \boldsymbol{R}^n))$, where $\mathscr{C}(G)$ denotes for fixed $G \in \mathscr{H}^C_\lambda(I, \boldsymbol{R}^n)$ the set of all fixed points of $\mathscr{I}\mathscr{F}(G \,\square\, \mathscr{D})$. Similarly we can also define the strong-weak and strong-strong sequential upper semicontinuity (s.-w.u.s.c. and s.-s.u.s.c.) of $\mathscr{C}(\cdot)$ on $\mathscr{H}^C_\lambda(I, \boldsymbol{R}^u)$. In the case of s.-s.u.s.c. we will simply write u.s.c. instead of s.-s.u.s.c.

Finally, let us observe that for every $G \in \mathscr{H}^C_\lambda(I, \boldsymbol{R}^n)$ one has $\mathscr{C}(G) = \mathscr{I}S(G)$. Indeed, let $u \in S(G)$ and $x = \mathscr{I}u$. Since $u(t) \in G(t, \mathscr{I}u, u)$ for a.e. $t \in I$, then for a.e. $t \in I$ we also have $\dot{x}(t) \in G(t, x, \mathscr{D}x)$, i.e., $\dot{x} \in \mathscr{F}(G \,\square\, \mathscr{D})(x)$ and therefore $x \in \mathscr{I}\mathscr{F}(G \,\square\, \mathscr{D})(x)$. Then $\mathscr{I}S(G) \subset \mathscr{C}(G)$. If $x \in \mathscr{C}(G)$ then $\dot{x} \in S(G)$ and therefore $x \in \mathscr{I}S(G)$. Therefore, we also have $\mathscr{C}(G) \subset \mathscr{I}S(G)$. Now, as a Corollary from the above, Theorem 3.13 and Lemma 1.13 we obtain the following theorem.

THEOREM 3.14. *For every $G \in \mathscr{H}^C_\lambda(I, \boldsymbol{R}^n)$ the set $\mathscr{C}(G)$ is a nonempty compact subset of $\boldsymbol{K}_\lambda$ and weakly compact subset of* $\mathrm{AC}^0(I, \boldsymbol{R}^n)$. ■

Now, we prove following continuous dependence theorems.

THEOREM 3.15. *The mapping $S(\cdot)$ is w.-w.u.s.c. on $\mathscr{H}^C_\lambda(I, \boldsymbol{R}^n)$.*

Proof. Let G be an arbitrary element of $\mathscr{H}^C_\lambda(I, \boldsymbol{R}^n)$ and (G_n) a sequence weakly converging in $\mathscr{H}^C_\lambda(I, \boldsymbol{R}^n)$ to G. Suppose (u_n) is a sequence of $L(I, \boldsymbol{R}^n)$ such that $u_n \in S(G_n)$ for each $n = 1, 2, \dots$ Then $u_n(t) \in G_n(t, \mathscr{I}u_n, u_n)$ for a.e. $t \in I$ and n

$= 1, 2, \ldots$ Since $S(G_n) \subset \boldsymbol{P}_\lambda$ for $n = 1, 2, \ldots$ and $\boldsymbol{P}_\lambda$ is weakly compact then there is a subsequence, say (u_k) of (u_n), weakly converging to any $u \in \boldsymbol{P}_\lambda$. For each measurable set $U \subset I$ and $k = 1, 2, \ldots$ one has

$$\operatorname{dist}\Big(\int_U u(t)dt, \int_U (G(t, \mathscr{I}u, u)dt\Big)$$

$$\leqslant \Big|\int_U [u(t) - u_k(t)]dt\Big| + \operatorname{dist}\Big(\int_U u_k(t)dt, \int_U G_k(t, \mathscr{I}u_k, u_k)dt\Big) +$$

$$+ \sup_{(x,z)} \bar{h}\Big(\int_U G_k(t, x, z)dt, \int_U G(t, x, z)dt\Big) +$$

$$+ \bar{h}\Big(\int_U G(t, \mathscr{I}u_k, u_k)dt, \int_U G(t, \mathscr{I}u, u)dt\Big).$$

Hence, similarly as in the proof of Theorem 3.13 it follows $u \in S(G)$. ∎

REMARK 3.1. In fact, we have proved in the last theorem that for every $G \in \mathscr{H}_\lambda^{\mathrm{C}}(I, \boldsymbol{R}^n)$ and every sequence (G_n) of $\mathscr{H}_\lambda^{\mathrm{C}}(I, \boldsymbol{R}^n)$ upper weakly converging to G, i.e. such that

$$\sup_{(x,z) \in \mathscr{K}_\lambda \times S_\lambda} \bar{h}\Big(\int_U G_n(t, x, z)dt, \int_U G(t, x, z)dt\Big) \to 0 \quad \text{as } n \to \infty$$

for each measurable set $U \subset I$, every sequence (u_n) of $L(I, \boldsymbol{R}^n)$ such that $u_n \in S(u_n)$ has a subsequence weakly converging to any $u \in S(u)$. ∎

REMARK 3.2. It is clear that $S(\cdot)$ is w.-s.u.s.c. on $\mathscr{H}_\lambda^{\mathrm{C}}(I. \boldsymbol{R}^n)$, because every δ-converging sequence of $\mathscr{H}_\lambda^{\mathrm{C}}(I, \boldsymbol{R}^n)$ is also weakly converging in $\mathscr{H}_\lambda^{\mathrm{C}}(I, \boldsymbol{R}^n)$. ∎

THEOREM 3.16. *The mapping $\mathscr{C}(\cdot)$ is s.-w.u.s.c. on $\mathscr{H}_\lambda^{\mathrm{C}}(I, \boldsymbol{R}^n)$ as a mapping of $\mathscr{H}_\lambda^{\mathrm{C}}(I, \boldsymbol{R}^n)$ into* Comp($C^0(I, \boldsymbol{R}^n)$) *and is w.-w.u.s.c. as a mapping of $\mathscr{H}_\lambda^{\mathrm{C}}(I, \boldsymbol{R}^n)$ into $\mathscr{P}(\mathrm{AC}^0(I, \boldsymbol{R}^n))$. In particular, $\mathscr{C}(\cdot)$ is u.s.c. as a mapping of $\mathscr{H}_\lambda^{\mathrm{C}}(I, \boldsymbol{R}^n)$ into* Comp($C^0(I, \boldsymbol{R})$). ∎

COROLLARY 3.1. *For every $G \in \mathscr{H}_\lambda^{\mathrm{C}}(I, \boldsymbol{R}^n)$ and each sequence (G_n) of $\mathscr{H}_\lambda^{\mathrm{C}}(I, \boldsymbol{R}^n)$ upper weakly converging to G (see Remark* 3.1) *every sequence (x_n) of $C^0(I, \boldsymbol{R}^n)$ such that $x_n \in \mathscr{C}(G_n)$ for $n = 1, 2, \ldots$ has a subsequence converging in the norm topology of $C^0(I, \boldsymbol{R}^n)$ to any $x \in \mathscr{C}(G)$.* ∎

We now consider a more complicated situation. Let $a > 0$ be given and suppose I and $J := [m, M]$ are compact intervals of $\boldsymbol{R}^1$ such that for every $\sigma \in I$ we have $I_\sigma := [\sigma, \sigma + a] \subset J$. Given $z \in L(I_\sigma, \boldsymbol{R}^m)$ denote by z^J the extension of z on the interval J defined by

$$z^J(t) = \begin{cases} z(t) & \text{for a.e. } t \in I_\sigma, \\ 0 & \text{for a.e. } t \in J \setminus I_\sigma. \end{cases} \tag{3.1}$$

Given $w \in L(J, \boldsymbol{R}^n)$ by $w|_\sigma$ we shall denote the restriction of w to the interval I_σ. Finally, given $\sigma \in I$ by $\mathscr{I}_\sigma$ we shall denote a mapping defined on $L(I_\sigma, \boldsymbol{R}^n)$ by (1.2) and B_λ^σ a closed ball of $L(I_\sigma, \boldsymbol{R}^n)$ with a radius $\lambda > 0$ and centred at the origin.

We shall consider here the space $\mathscr{H}_I^C$ of all set-valued functions $G \in \mathscr{H}_\lambda^C(I_\sigma, R^n)$ with $\sigma \in I$, i.e., $\mathscr{H}_I^C := \bigcup_{\sigma\in I} \mathscr{H}_\lambda^C(I_\sigma, R^n)$. Given $G \in \mathscr{H}_\lambda^C(I_\sigma, R^n)$ with $\sigma \in I$ we define an extension $\hat{G}$ of $G \square \mathscr{I}_\sigma$ on the whole set $J \times B_\lambda^J$ by setting

$$\hat{G}(t, z) = \begin{cases} (G \square \mathscr{I}_\sigma)(t, z|_\sigma) & \text{for } t \in I_\sigma, z \in B_\lambda^J, \\ \{0\} & \text{for } t \in J \setminus I_\sigma, z \in B_\lambda^J, \end{cases} \tag{3.2}$$

where B_λ^J denotes the closed ball of $L(J, R^n)$ with the centre at the origin and a radius $\lambda > 0$.

Denote by $S_\sigma(G)$ the set of all fixed points of $\mathscr{F}(G \square \mathscr{I}_\sigma)$ with $\sigma \in I$ and $G \in \mathscr{H}_\lambda^C(I_\sigma, R^n)$. By $S(\hat{G})$ we will denote the set of all fixed points of $\mathscr{F}(\hat{G})$ where $\hat{G}$ is the extension of $G \square \mathscr{I}_\sigma$ defined by (3.2). It is evident that $S(\hat{G})|_\sigma = S_\sigma(G)$ and $S_\sigma(G)^J = S(\hat{G})$ where $S(\hat{G})|_\sigma := \{z|_\sigma\colon z \in S(G)\}$ and $S_\sigma(G)^J := \{z^J\colon z \in S_\sigma(G)\}$. We shall consider sets $S_\sigma(G)$ with $\sigma \in I$ as subsets of the metric space $(\mathscr{L}(J), l)$ defined in Section I.4, i.e., by $\mathscr{L}(J) := \bigcup_{\sigma\in I} L(I_\sigma, R^n)$ and $l(u_1, u_2) = \max\big(|\sigma_1 - \sigma_2|, \int |u_1^J(t) - u_2^J(t)| dt\big)$, where $\sigma_1, \sigma_2 \in I$ are such that $u_1 \in L(I_{\sigma_1}, R^m)$ and $u_2 \in L(I_{\sigma_2}, R^m)$.

Given sequence (G_n) of $\mathscr{H}_I^C$ with $G_n \in \mathscr{H}_\lambda^C(I_{\sigma_n}, R^n)$, we will say that (G_n) is l-weakly converging to $G \in \mathscr{H}_\lambda^C(I_\sigma, R^n)$ if

$$\lim_{n\to\infty} \max\Big(|\sigma_n - \sigma|, \sup_{z\in B_\lambda^J} h\Big(\int_U \hat{G}_n(t, z)\,dt, \int_U \hat{G}(t, z)\,dt\Big)\Big)$$

for every measurable set $U \subset J$, where $\hat{G}_n$ and $\hat{G}$ are extensions of G_n and G, respectively defined by (3.2).

We shall show now that a mapping $\mathscr{Z}\colon \mathscr{H}_I^C \ni G \to S_\sigma(G) \subset \mathscr{L}(J)$ is l-w.-w.u.s.c. on $\mathscr{H}_I^C$ in the sense defined similarly as above, i.e., that for each $G \in \mathscr{H}_I^C$ and every sequence (G_k) of $\mathscr{H}_I^C$ l-weakly converging to $G \in \mathscr{H}_I^C$ every sequence (u_k) of $\mathscr{L}(J)$ such that $u_k \in S_{\sigma_k}(G_k)$ for $k = 1, 2, \ldots$, where $\sigma_k \in I$ is such that $G_k \in \mathscr{H}_\lambda^C(I_{\sigma_k}, R^n)$ has a subsequence l-weakly converging in $\mathscr{L}(J)$ to any $u \in S_\sigma(G)$, where $\sigma \in I$ is such that $G \in \mathscr{H}_\lambda^C(I_\sigma, R^n)$.

We begin with the following lemma.

LEMMA 3.17. *For every $G \in \mathscr{H}_I^C$ an extension $\hat{G}$ of G defined by* (3.2) *belongs to $\mathscr{H}_\lambda^C(J, R^n)$, i.e., it is such that*

(i) *$\hat{G}(\cdot, z) \in \mathscr{A}^C(J, R^n)$ for $z \in B_\lambda^J$,*

(ii) *$\hat{G}$ is w.-w.u.s.c. with respect to its second variable,*

(iii) *the family $\{G(\cdot, z)\}_{z\in B_\lambda^J}$ is uniformly integrable on J and $\int_J ||\hat{G}(t, z)|| dt \leqslant \lambda$ for every $z \in B_\lambda^J$.*

Proof. (i) follows immediately from the properties of $G \in \mathscr{H}_\lambda^C(I_\sigma, R^n)$ with any $\sigma \in I$ and (3.2). Furthermore, for every sequence (z_n) of B_λ^J weakly converging to $z \in B_\lambda^J$ we have $z_n|_\sigma \in B_\lambda^\sigma$ for every $n = 1, 2, \ldots$ and $z_n|_\sigma \rightharpoonup z|_\sigma$ in $L(I_\sigma, R^n)$ as $n \to \infty$.

Then, by the properties of G, for every measurable set $E \subset I_\sigma$ we have

$$\bar{h}\left(\int_E (G \square \mathcal{I}_\sigma)(t, z|_n)dt, \int_E (G \square \mathcal{I}_\sigma)(t, z|_\sigma)dt\right) \to 0 \quad \text{as } n \to \infty.$$

Therefore, for every open set $U \subset J$ we obtain

$$\lim_{n\to\infty} \bar{h}\left(\int_U \hat{G}(t, z_n)dt, \int_U G(t, z)dt\right)$$

$$= \lim_{n\to\infty} \bar{h}\left(\int_{U\cap I_\sigma} (G \square \mathcal{I}_\sigma)(t, z_n|_\sigma)dt, \int_{U\cap I_\sigma} (G \square \mathcal{I}_\sigma)(t, z|_\sigma)dt\right) = 0.$$

In a similar way we can verify that (iii) holds. ■

COROLLARY 3.2. *For every sequence (G_n) of $\mathcal{H}_I^C$ l-weakly converging to $G \in \mathcal{H}_I^C$, every sequence (v_n) of $L(J, R^n)$ such that $v_n \in S(\hat{G}_n)$ for every $n = 1, 2, \ldots$ has a subsequence weakly converging in $L(J, R^m)$ to $v \in S(\hat{G})$.*

Indeed, it follows immediately from Lemma 3.17 and Theorem 3.15. ■

Now we obtain the following theorem.

THEOREM 3.18. *A mapping $\mathcal{Z}: \mathcal{H}_I^C \ni G \to S_\sigma(G) \subset \mathcal{L}(J)$ is l-w.-w.u.s.c. on $\mathcal{H}_I^C$.*

Proof. Let G be arbitrarily fixed in $\mathcal{H}_I^C$ and let (G_k) be a sequence of $\mathcal{H}_I^C$ l-weakly converging to G. Suppose (u_k) is a sequence of $\mathcal{L}(J)$ such that $u_k \in S_{\sigma_k}(G_k)$, with $\sigma_k \in I$ such that $G_k \in \mathcal{H}_\lambda^C(I_{\sigma_k}, R^n)$ for $k = 1, 2, \ldots$ Put $v_k = u_k^J$ with u_k^J defined by (3.1) for $k = 1, 2, \ldots$ It is clear that $v_k \in S(\hat{G}_k)$, where $\hat{G}_k$ denotes the extension of $G_k \square \mathcal{I}_{\sigma_k}$ defined by (3.2). Since (G_k) l-weakly converges to G then $|\sigma_k - \sigma| \to 0$ as $k \to \infty$, where $\sigma \in I$ is such that $G \in \mathcal{H}_\lambda^C(I_\sigma, R^n)$. By Corollary 3.2, there is a subsequence of (v_k), say again (v_k) weakly converging in $L(J, R^n)$ to any $v \in S(\hat{G})$. Let $u = v|_\sigma$. It is clear that $v = u^J$ and $u \in S_\sigma(G)$. Let $\tilde{u}_k = v_k|_{\sigma_k}$. It is easy to see that $(\tilde{u}_k)$ is a subsequence of (u_k), l-weakly converging to u. Thus $\mathcal{Z}$ is l-w.-w.u.s.c. on $\mathcal{H}_I^C$. ■

REMARK 3.3. The above result is also true if in the proof of Theorem 3.18 we take upper l-weakly converging sequence (G_k) of $\mathcal{H}_I^C$ instead of l-weakly converging one. ■

4. SUBTRAJECTORY AND TRAJECTORY INTEGRALS σ-SELECTIONABLE SET-VALUED FUNCTIONS

We will consider here set-valued functions $G \in \mathcal{H}_\lambda(I, R^n)$ that are u.s.c.w. on $I \times \times \mathcal{K}_\lambda \times B_\lambda$, i.e., such that for every $(t, x, z) \in I \times \mathcal{K}_\lambda \times B_\lambda$ and every sequence $\{(t_n, x_n, z_n)\}$ of $I \times \mathcal{K}_\lambda \times B_\lambda$ with $|t_n - t| + |x_n - x|_a \to 0$ and $z_n \rightharpoonup z$ as $n \to \infty$ one has $\lim_{n\to\infty} \bar{h}(G(t_n, x_n, z_n), G(t, x, z) = 0$. Furthermore we shall assume that G has convex values and is such that $\int_{\sigma}^{\sigma+a} ||G(t, x, z)|| dt \leqslant \lambda$ for $(x, z) \in \mathcal{K}_\lambda \times B_\lambda$. Similarly, as in 2.3 we define sets $\boldsymbol{P}_\lambda$, $\boldsymbol{K}_\lambda$ and a set-valued function $G \,\square\, \mathcal{D}$. We have of course $\boldsymbol{K}_\lambda \times \boldsymbol{P}_\lambda \subset \mathcal{K}_\lambda \times B_\lambda$ and $G \,\square\, \mathcal{D}$ is u.s.c. on $I \times \boldsymbol{K}_\lambda$. We will show that the mapping $\mathcal{I}\mathcal{F}(G \,\square\, \mathcal{D})$ is σ-selectionable and then that the set of all its fixed points is acyclic. Recall (see Section II.4.2) that a mapping $F: X \to \mathcal{P}(Y)$ where X and Y are linear normed space is said to be σ-selectionable if there exists a decreasing sequenes (F_n) of u.s.c. mappings $F_n: X \to \mathrm{Comp}(Y)$ such that F_n has for each $n = 1, 2, \ldots$ a continuous selection and for each $x \in X$ we have $F(x) = \bigcap_{n=1}^{\infty} F_n(x)$.

4.1. σ-selectionable approximation theorem

We begin with the following lemmas.

LEMMA 4.1. *Let C_1 and C_2 be nonempty bounded subsets of $\boldsymbol{R}$ and $C^0(I, \boldsymbol{R}^n)$, respectively and let $k_1: I \to \boldsymbol{R}^+$ and $k_2: I \times \boldsymbol{K}_\lambda \to \boldsymbol{R}^+$ be defined by*

$$k_1(t) = \inf\{|t-\tau|: \tau \in C_1\} \tag{4.1}$$

and

$$k_2(t, x) = \inf\{|x-u|_t: u \in C_2\}, \text{ with } |x-u|_t = \sup_{\sigma \leqslant \tau \leqslant t} |x(\tau)-u(\tau)|. \tag{4.2}$$

Then

(i) $|k_1(t_1)-k_2(t_2)| \leqslant |t_1-t_2|$ *for* $t_1, t_2 \in I$,

(ii) $k_2(\cdot, x)$ *in nondecreasing for fixed* $x \in \boldsymbol{K}_\lambda$,

(iii) $|k_2(t, x_1)-k_2(t, x_2)| \leqslant |x_1-x_2|_t$ *for every* $t \in I$ *and* $x_1, x_2 \in \boldsymbol{K}_\lambda$.

Proof. Clearly, for arbitrary $\tau \in C_1$ we have $k_1(t_1) \leqslant |t_1-\tau| \leqslant |t_1-t_2|+|t_2-\tau|$. Then $k_1(t_1) \leqslant |t_1-t|_2+k_2(t_2)$. Similary, we get $k_2(t_2) \leqslant |t_1-t_2|+k(t_1)$. Thus (i) holds. Condition (ii) follows immediately from the definition of k_2; (iii) can be obtained similarly as (i). ■

We shall consider here a compact set $I \times \boldsymbol{K}_\lambda \subset \boldsymbol{R} \times C^0(I, \boldsymbol{R}^n)$ as a compact metric space with a reletive topology generated by the norm $||\cdot||$ of $\boldsymbol{R} \times C(I, \boldsymbol{R}^n)$ defined by $||(t, x)|| = \max(|t|, |x|_a)$ for $(t, x) \in \boldsymbol{R} \times C(I, \boldsymbol{R}^n)$.

LEMMA 4.2. *To any finite open covering $\{\Omega_i\}_{i \in N}$ of the compact metric space $(I \times \boldsymbol{K}_\lambda, d)$ we can associate a partition of unity $\{p_i\}_{i \in N}$ subordinate to it and such that*

(i) $p_i(\cdot, x)$ *is measurable for fixed* $x \in \boldsymbol{K}_\lambda$,

(ii) *for every $(t, x) \in I \times \boldsymbol{K}_\lambda$ there exists a neighbourhood U_{tx} of (t, x) and a number $L_{tx} > 0$ such that $|p_i(\tau, x_1)-p_i(\tau, x_2)| \leqslant L_{tx}|x_1-x_2|_\tau$ for (τ, x_1), $(\tau, x_2) \in U_{tx}$ and $i \in N$,*

(iii) $\sum_{i \in N} p\,(t, x) = 1$ *for* $(t, x) \in I \times \boldsymbol{K}_\lambda$.

Proof. Let a covering $\{\Omega_i\}_{i \in N}$ of $I \times \boldsymbol{K}_\lambda$ be given and let us define for each $i \in N$ a function $k_i: I \times \boldsymbol{K}_\lambda \to \boldsymbol{R}^+$ by setting

$$k_i(t, x) = \begin{cases} 0 & \text{for } (t, x) \in I \times \mathcal{K}_\lambda \setminus \Omega_i, \\ k_1^i(t)+k_2^i(t, x) & \text{for } (t, x) \in \Omega_i, \end{cases} \tag{4.4}$$

where $k_1^i(t) = \inf\{|t-\tau|: \tau \in I \setminus C_1^i\}$ and $k_2^i(t, x) = \inf\{|x-u|_t: u \in \boldsymbol{K}_\lambda \setminus C_2^i\}$; $C_1^i = \Pi_R(\Omega_i)$ and $C_2^i = \Pi_C(\Omega_i)$. Here, Π_R and Π_C denote projection onto $\boldsymbol{R}$ and $C^0(I, \boldsymbol{R}^n)$, respectively.

By virtue of Lemma 4.1, for every $i \in N$ functions k_1^i and k_2^i are such that k_1^i is continuous on I, $k_2^i(\sigma, x) = 0$ for $x \in \boldsymbol{K}_\lambda$, $|k_2^i(t, x_1) - k_2^i(t, x_2)| \leqslant |x_1 - x_2|_t$ for $t \in I$, $x_1, x_2 \in \boldsymbol{K}_\lambda$ and $k_2^i(\cdot, x)$ is measurable for fixed $x \in \boldsymbol{K}_\lambda$. Furthermore, $\sum_{i \in N} k_1^i(t) > 0$ for each $t \in I$. Hence, by continuity of $\sum_{i \in N} k_1^i(\cdot)$ on I it follows the existence of a number $m > 0$ such that $m \leqslant \sum_{i \in N} k_1^i(t)$ for each $t \in I$. Therefore, for each $i \in N$ $k_i(\cdot, x)$ is measurable for fixed $x \in \boldsymbol{K}_\lambda$ and such that $|k_i(t, x_1) - k_i(t, x_2)| \leqslant |x_1 - x_2|_t$ for $t \in I$ and $x_1, x_2 \in \boldsymbol{K}_\lambda$. Furthermore, $\sum_{i \in N} k_i(t, x)$ is bounded above and $m \leqslant \sum_{i \in N} k_i(t, x)$ for each $(t, x) \in I \times \boldsymbol{K}_\lambda$. Hence it follows that a function $w: I \times \boldsymbol{K}_\lambda \to \boldsymbol{R}^+$ defined by $w(t, x) = \left[\sum_{i \in N} k_i(t, x)\right]^{-1}$ is well defined on $I \times \boldsymbol{K}_\lambda$ and has the following properties: $w(\cdot, x)$ is measurable for fixed $x \in \boldsymbol{K}_\lambda$, $w(t, \cdot)$ is continuous and such that $w(t, x) \leqslant 1/m$ for fixed $(t, x) \in I \times \boldsymbol{K}_\lambda$.

Put now, $p_i(t, x) = k_i(t, x) \cdot w(t, x)$ for $i \in N$ and $(t, x) \in I \times \boldsymbol{K}_\lambda$. It is clear that $\sum_{i \in N} p_i(t, x) = 1$ for $(t, x) \in I \times \boldsymbol{K}_\lambda$, $p_i(\cdot, x)$ is measurable and $p_i(t, \cdot)$ is continuous for each $i \in N$ and fixed $x \in \boldsymbol{K}$ and $t \in I$, respectively.

Now, we shall show that each p_i is locally Lipschitzean on $\boldsymbol{K}_\lambda$. Indeed, let $(t, x) \in I \times \boldsymbol{K}_\lambda$ be fixed. Let U_{tx} be a neigbourhood of (t, x) and $N^{(t,x)} = \{1, 2, \ldots, N_{tx}\}$ be a subset of N such that $\Omega_i \cap U_{tx} = \emptyset$ for $i \notin N^{(t,x)}$. For every $(\tau, x_1), (\tau, x_2) \in U_{tx}$ we have

$$\begin{aligned}
&|p_i(\tau, x_1) - p_i(\tau, x_2)| \\
&\leqslant w(\tau, x_1)|k_i(\tau, x_1) - k_i(\tau, x_2)| + k_i(\tau, x_2)|w(\tau, x_1) - w(t, x_2)| \\
&\leqslant \frac{1}{m}|x_1 - x_2| + p_i(\tau, x_2) w(\tau, x_1) \sum_{i \in N_{tx}} |k_i(\tau, x_1) - k_i(\tau, x_2)| \\
&\leqslant \frac{1}{m}(1 + N_{tx})|x_1 - x_2|_\tau .
\end{aligned}$$

Thus conditions (i)–(iii) are satisfied. ■

Now we prove the following approximation theorem.

THEOREM 4.3. *Let* $I := [\sigma, \sigma + a]$ *and assume* $G \in \mathscr{H}_\lambda(I, \boldsymbol{R}^n)$ *is u.s.s.w. on* $I \times \mathscr{K}_\lambda \times B_\lambda$, *has convex values and is such that* $\int_\sigma^{\sigma+a} \|G(t, x, z)\| dt \leqslant \lambda$ *for* $(x, z) \in \mathscr{K}_\lambda \times B_\lambda$. *Then there exists a sequence* (H_n) *of set-valued functions* $H_n = \sum_{i \in N_n} p_i^n(t, x) S_i^n$, *where* S_i^n *are closed and convex and* p_i^n *satisfy for each* $i \in N_n$ *and* $n = 1, 2, \ldots$ *conditions* (i)–(iii) *of Lemma* 4.2. *Furthermore, set-valued functions* H_n *have the following properties*

(i) $H_n(\cdot, x) \in \mathscr{A}(I, \boldsymbol{R}^n)$ *for fixed* $x \in \boldsymbol{K}_\lambda$ *and* $H_n(t, \cdot)$ *is continuous for fixed* $t \in I$,

(ii) $(G \,\square\, \mathscr{D})(t, x) \subset \ldots \subset H_{n+1}(t, x) \subset H(t, x) \subset \ldots \subset H_0(t, x)$ *for* $n = 0, 1, \ldots$, $t \in I$ *and* $x \in \boldsymbol{K}_\lambda$,

(iii) *for every* $\varepsilon > 0$ *and* $(\bar{t}, \bar{x}) \in I \times \boldsymbol{K}_\lambda$ *there is a number* $N(\varepsilon, \bar{t}, \bar{x}) \geqslant 0$ *such that* $H_n(t, x) \subset (G \,\square\, \mathscr{D})(t, x) + \varepsilon B$ *for* $n \geqslant N(\varepsilon, \bar{t}, \bar{x})$, $t \in I$ *and* $x \in \boldsymbol{K}_\lambda$.

Proof. Similarly as above we shall consider $I \times \boldsymbol{K}_\lambda$ as a metric space with the relative topology generated by the norm $\|\cdot\|$ defined by $\|(t, x)\| = \max(|t|, |x|_a)$ for $(t, x) \in \boldsymbol{R} \times C^0(I, \boldsymbol{R}^n)$. Let S denote the the closed convex hull of $(G \,\square\, \mathscr{D})(I \times \boldsymbol{K}_\lambda)$. By Proposition II.2.3 and Mazur theorem S is compact in $\boldsymbol{R}^n$, because $G \,\square\, \mathscr{D}$ is u.s.c. on $I \times \boldsymbol{K}_\lambda$ and $I \times \boldsymbol{K}_\lambda$ is compact. Fix $\varrho > 0$ and let us cover $I \times \boldsymbol{K}_\lambda$ with the open balls of $B^\circ[(t_i^0, x_i^0), \varrho]$ of $(I \times \boldsymbol{K}_\lambda, d)$ with the centre at $(t_i^0, x_i^0) \in I \times \boldsymbol{K}_\lambda$, a radius $\varrho > 0$ and $i \in N_0$, where N_0 is a finite subset of N. Put $S_i^{(0)} := \overline{\text{co}}(G \,\square\, \mathscr{D})(B^\circ[(t_i^0, x_i^0), 2\varrho])$ and $\Omega_i^{(0)} := B^\circ[(t_i^0, x_i^0), \varrho]$ for $i \in N_0$.

Let us define now a set-valued function H_0 by setting $H_0(t, x) := \sum_{i \in N_0} p_i^0(t, x) S_i^{(0)}$ for $(t, x) \in I \times \boldsymbol{K}_\lambda$, where $\{p_i^0\}_{i \in N_0}$ is a partition of unity defined in Lemma 4.2 corresponding to a covering $\{\Omega_i^{(0)}\}_{i \in N_0}$. It is clear that $H_0(t, x) \in \text{Conv}(\boldsymbol{R}^n)$ for $(t, x) \in I \times \boldsymbol{K}_\lambda$. By Corollary II.3.1 for every $i \in N_0$, $p_i^0(\cdot, x) S_i^{(0)}$ is measurable for fixed $x \in \boldsymbol{K}_\lambda$. Therefore also $H_0(\cdot, x)$ is measurable for fixed $x \in \boldsymbol{K}_\lambda$. Hence in particular it follows that $H_0(\cdot, x) \in \mathscr{A}(I, \boldsymbol{R}^n)$ because $H_0(t, x) \subset \overline{\text{co}}(G \,\square\, \mathscr{D})(I \times \boldsymbol{K}_\lambda)$ and the latter set is compact in $\boldsymbol{R}^n$. Immediately from Lemma II.1.5, Corollary II.1.2 and Lemma 4.2 it follows that $H_0(t, \cdot)$ is H-continuous and therefore also continuous on $\boldsymbol{K}_\lambda$ for fixed $t \in I$.

In order to define H_1 we do the same as before with the finite open covering $\{B^\circ[(t_i^1, x_i^1), \frac{1}{3}\varrho]\}_{i \in N_1}$ and an associated locally Lipschitzean with respect to $x \in \boldsymbol{K}_\lambda$ partition of unity $\{p_i^1\}_{i \in N_1}$. As before we set for all $i \in N_1$: $S_i^1 = \overline{\text{co}}(G \,\square\, \mathscr{D})(B^\circ[(t_i^1, x_i^1), \frac{2}{3}\varrho]) \subset S$ and then we can also define $H_1(t, x) = \sum_{i \in N_1} p_i^1(t, x) S_i^1$ for $(t, x) \in I \times \boldsymbol{K}_\lambda$. The set-valued function H_1 enjoys the same properties as H_0.

We shall now prove that for any $(t, x) \in I \times \boldsymbol{K}_\lambda$, $H_1(t, x) \subset H_0(t, x)$. Let us fix $(t, x) \in I \times \boldsymbol{K}_\lambda$. We define $N_0^{(t, x)} = \{i \in N_0 \colon (t, x) \in B^\circ[(t_i^0, x_i^0), \varrho]\}$ and $N_1^{(t, x)} = \{i \in N_1 \colon (t, x) \in B^\circ[(t_i^0, x_i^0), \frac{1}{3}\varrho]\}$. Let $i_0 \in N_0^{(t, x)}$ and $i_1 \in N_1^{(t, x)}$ be given. Then if $(\tau, y) \in B^\circ[(t_{i_1}^1, x_{i_1}^1), \frac{2}{3}\varrho]$ we get $|\tau - t_{i_1}^1| + |y - x_{i_1}^1|_a \leqslant \frac{2}{3}\varrho$ with $|t - t_{i_0}^0| + |x - x_{i_0}^0|_a \leqslant \varrho$ and $|t - t_{i_1}^1| + |x - x_{i_1}^1|_a \leqslant \frac{1}{3}\varrho$. Thus we have $|\tau - t_{i_0}^0| + |y - x_{i_0}^0|_a \leqslant \frac{2}{3}\varrho + \frac{1}{3}\varrho + \varrho = 2\varrho$. Then $B^\circ[(t_{i_1}^1, x_{i_1}^1), \frac{2}{3}\varrho] \subset B^\circ[(t_{i_0}^0, x_{i_0}^0), 2\varrho]$ for all $i_0 \in N_0^{(t, x)}$ and all $i_1 \in N_1^{(t, x)}$. This leads to $S_{i_1}^1 \subset S_{i_0}^{(0)}$ for such indexes. Then for all $i_1 \in N_1^{(t, x)}$ we have:

$$S_{i_1}^1 \subset \sum_{i \in N_0^{(t, x)}} p_i^0(t, x) S_i^{(0)} = \sum_{i \in N_0} p_i^0(t, x) S_i^{(0)} = H_0(t, x),$$

this being true by convexity arguments and since $\{p_i^0\}_{i \in N_0}$ is a partition of unity associated to $\{B^\circ[(t_i^0, x_i^0), \varrho]\}_{i \in N_0}$ which in particular says that $p_i^0(t, x) = 0$ if i

$\notin N_0^{(t,x)}$. And then for the same reasons we get:

$$H_1(t,x) = \sum_{i \in N_0} p_i^1(t,x) S_i^1 = \sum_{i \in N_0^{(t,x)}} p_i^1(t,x) S_i^1 \subset H_0(t,x).$$

We shall now prove for example that for any $(t,x) \in I \times \boldsymbol{K}_\lambda$, $(G \,\square\, \mathscr{D})(t,x) \subset H_0(t,x)$. Indeed from the construction of $S^{(0)}$ we have $(G \,\square\, \mathscr{D})(t,x) \subset S_i^{(0)}$ for all $i \in N_0^{(t,x)}$. Then always for convexity reasons and since $\{p_i^0\}_{i \in N_0}$ is locally finite partition of unity associated to $\{B^o[(t_i^0, x_i^0), \varrho\}_{i \in N_0}$, we get

$$(G \,\square\, \mathscr{D})(t,x) \subset \sum_{i \in N_0^{(t,x)}} p_i^0(t,x) S_i^{(0)} = \sum_{i \in N_0} p_i^0(t,x) S_i^{(0)} = H_0(t.\,x).$$

Now let us define $\varrho_n = (\frac{1}{3})^n \varrho$ for any $n \in N$. Then as for H_0 associated to $\varrho_0 = \varrho$ and for H_1 associated to $\varrho_1 = \frac{1}{3}\varrho$ we can build by induction a sequence of set-valued functions H_n: $I \times \boldsymbol{K}_\lambda \to S$, $n \in N$, each of them being continuous with respect to $x \in \boldsymbol{K}_\lambda$, measurable on I for fixed $x \in \boldsymbol{K}_\lambda$ nonempty convex compact valued for a.e. $t \in I$ and verifying properties (ii). To end the proof we have to show that (iii) is satisfied.

Let $(t,x) \in I \times \boldsymbol{K}_\lambda$ be given. Since $G \,\square\, \mathscr{D}$ is u.s.c. on $I \times \boldsymbol{K}_\lambda$, for any $\varepsilon > 0$, there exists $\eta_\varepsilon^{(t,x)} > 0$ such that $|\tau - t| + |y - x|_a \leqslant \eta_\varepsilon^{(t,x)}$ implies $(G \,\square\, \mathscr{D})(\tau, y) \subset (G \,\square\, \mathscr{D})(t,x) + \varepsilon B$. Then there obviously exists a number $N(\varepsilon, t, x)$ such that for $n \geqslant N(\varepsilon, t, x)$ we have $\varrho_n \leqslant \frac{1}{3}\eta_\varepsilon^{(t,x)}$. Let us define as before $N_n^{(t,x)} = \{i \in N_n$: $(t,x) \in B^o[(t_i^n, x_i^n), \varrho_n]\}$. For the same reasons as for H_0 and H_1 we can write $H_n(t,x) = \sum_{i \in N_n^{(t,x)}} p_i^n(t,x) S_i^n$, where $S_i^n = \overline{\text{co}}(G \,\square\, \mathscr{D})(B^o[(t_i^n, x_i^n), 2\varrho_n]) \subset S$. Then for all $(\tau, y) \in B^o[(t_i^n, x_i^n), 2\varrho]_n$ with $i \in N_n^{(t,x)}$ we have $|\tau - t| + |y - x|_a \leqslant |\tau - t_i^n| + |y - x_i^n| + |t_i^n - t| + |x_i^n - x| \leqslant 2\varrho_n + \varrho_n = 3\varrho_n < \eta_\varepsilon^{(t,x)}$ if we take $n \geqslant N(\varepsilon, t, x)$. Thus for all $n \geqslant N(\varepsilon, t, x)$ we have $(G \,\square\, \mathscr{D})(\tau, y) \subset (G \,\square\, \mathscr{D})(t,x) + \varepsilon B$ for all $(\tau, y) \in B^o[(t_i^n, x_i^n), 2\varrho_n]$ with $i \in N_n^{(t,x)}$. But since $(G \,\square\, \mathscr{D})(t,x) + \varepsilon B$ is closed convex we get $S_i^n \subset (G \,\square\, \mathscr{D})(t,x) + \varepsilon B$ for all $i \in N_n^{(t,x)}$. And then always by convexity we get: $H_n(t,x) = \sum_{i \in N_n^{(t,x)}} p_i^n(t,x) S_i^n \subset (G \,\square\, \mathscr{D})(t,x) + \varepsilon B$ for $n \geqslant N(\varepsilon, t, x)$. ■

Remark 4.1. If $G \in \mathscr{H}_\lambda(I, \boldsymbol{R}^n)$ satisfies conditions of Theorem 4.3 then for every $n = 1, 2, \ldots$ the family $\bigcup_{n-1}^{\infty} \{H_n(\cdot, x)\}_{x \in K_\lambda}$ with H_n defined in Theorem 4.3 is uniformly integrable on I.

Indeed, since $S = \overline{\text{co}}(G \,\square\, \mathscr{D})(I \times \boldsymbol{K}_\lambda)$ is compact then there is $M > 0$ such that $||H_n(t,x)|| \leqslant M$ for $n = 1, 2, \ldots$ and every $(t,x) \in I \times \boldsymbol{K}_\lambda$. Hence it follows that the family $\bigcup_{n=1}^{\infty} \{H_n(\cdot, x)\}_{x \in \boldsymbol{K}_\lambda}$ is uniformly integrable, because for every measurable set $E \subset I$ we have $\int_E ||H_n(t,x)||\, dt \leqslant M\mu(E)$ for each $n = 1, 2, \ldots$ and $x \in \boldsymbol{K}_\lambda$. ■

Remark 4.2. Suppose G: $I \times \mathscr{K}_\lambda \times B_\lambda \to \text{Comp}(\boldsymbol{R}^n)$ has convex values, is u.s.c.w. and such that $||G(I \times \mathscr{K}_\lambda \times B_\lambda)|| \leqslant \lambda/a$. Then there is a sequence (H_n) of set-valued functions H_n: $I \times \boldsymbol{K}_\lambda \to \overline{\text{co}}(G \,\square\, \mathscr{D})(I \times \boldsymbol{K}_\lambda)$ satisfying conditions (i)–(iii) of Theorem

4.3 and such that $\int_{\sigma}^{\sigma+a} ||H_n(t, x)||\,dt \leqslant \lambda$ for each $n = 1, 2, \ldots$ and $x \in \boldsymbol{K}_\lambda$.

Indeed, let us observe that $G(\cdot, x, z)$ is measurable on $I = [\sigma, \sigma+a]$ (see Theorem II.3.2) and $||G(t, x, z)|| \leqslant \lambda/a$ for $(t, x, z) \in I\times\mathscr{K}_\lambda\times B_\lambda$. Therefore, $G \in \mathscr{H}_\lambda(I, \boldsymbol{R}^n)$ and $\int_{\sigma}^{\sigma+a} ||G(t, x, z)||\,dt \leqslant \lambda$ for $(x, z) \in \mathscr{K}_\lambda\times B_\lambda$. Then, by Theorem 4.3 for each $n = 1, 2, \ldots$ a set-valued functions $H_n(t, x) = \sum p_i^n(t, x)S_i^n$, whit $S_i \subset \overline{\text{co}}\,G(I\times \boldsymbol{K}_\lambda\times \boldsymbol{P}_\lambda) \subset \overline{\text{co}}\,G(I\times\mathscr{K}_\lambda\times B_\lambda)$ can be defined. But $\overline{\text{co}}||G(I\times\mathscr{K}_\lambda\times B_\lambda)|| \leqslant \lambda/a$. Then $||H_n(t, x)|| \leqslant \sum_{i\in N_n} p_i^n(t, x)||S_i^n|| \leqslant \lambda/a$ and therefore, $\int_{\sigma}^{\sigma+a} ||H_n(t, x)||\,dt \leqslant \lambda$ for $x \in \boldsymbol{K}_\lambda$ and $n = 1, 2, \ldots$ ■

As a corollary of the above results we obtain the following theorem.

THEOREM 4.4. *Let* $G \in \mathscr{K}_\lambda\,(I, \boldsymbol{R}^n)$ *having convex values, be u.s.c.w. on* $I\times\mathscr{K}_\lambda\times B_\lambda$ *and such that* $\int_{\sigma}^{\sigma+a} ||G(t, x, z)||\,dt \leqslant \lambda$ *for* $(x, z) \in \mathscr{K}_\lambda\times B_\lambda$. *Then there exists a sequence* (H_n) *of set-valued functions* H_n: $I\times \boldsymbol{K}_\lambda \to \text{Conv}(\boldsymbol{R}^n)$ *satisfying conditions* (i), (ii) *of Theorem* 4.3 *and a sequence* (g_n) *of functions* g_n: $I\times \boldsymbol{K}_\lambda \to \boldsymbol{R}^n$ *such that*

(i) $(G\square\mathscr{D})\,(t, x) = \bigcap_{n=1}^{\infty} H_n(t, x)$ *for* $(t, x) \in I\times \boldsymbol{K}_\lambda$,

(ii) $g_n(t, x) \in H_n(t, x)$ *for* $(t, x) \in I\times \boldsymbol{K}_\lambda$ *and* $n = 1, 2, \ldots,$

(iii) $g_n(\cdot, x) \in L(I, \boldsymbol{R}^n)$ *and a mapping* $\boldsymbol{K}_\lambda \ni x \to g_n(\cdot, x) \in L(I, \boldsymbol{R}^n)$ *is continuous on* $\boldsymbol{K}_\lambda$ *for* $n = 1, 2, \ldots$

Moreover, if $||G\,(I\times\mathscr{K}_\lambda\times B_\lambda)|| \leqslant \lambda/a$ *then*

(iv) $\int_{\sigma}^{\sigma+a} ||g_n(t, x)||\,dt \leqslant \lambda$ *for* $x \in \boldsymbol{K}_\lambda$ *and* $n = 1, 2, \ldots$

Proof. By Theorem 4.3 there exists a sequence (H_n) of set-valued functions H_n: $I\times \boldsymbol{K}_\lambda \to \text{Conv}(\boldsymbol{R}^n)$ such that (i)–(iii) of this theorem are satisfied. By (ii) we have $(G\square\mathscr{D})(t, x) \subset \bigcap_{n\geqslant 1} H_n(t, x)$ for $(t, x) \in I\times \boldsymbol{K}_\lambda$, whereas (iii) implie $\bigcap_{n\geqslant 1} H_n(t, x) \subset (G\square\mathscr{D})(t, x)$ for $(t, x) \in I\times \boldsymbol{K}_\lambda$. Then (i) holds. By Lemma 4.2 it follows that a function g_n: $I\times \boldsymbol{K}_\lambda \to \boldsymbol{R}^n$ defined by $g_n(t, x) = \sum_{i\in N_n} p_i^0(t, x)s_i^n$ for $(t, x) \in I\times \boldsymbol{K}_\lambda$, with $s_i^n \in S_i^n$ for $i \in N_n$ is such that $g_n(t, x) \in H_n(t, x)$ for $(t, x) \in I\times \boldsymbol{K}_\lambda$, $g_n(\cdot, x) \in L(I, \boldsymbol{R}^n)$ for fixed $x \in \boldsymbol{R}^n$ and $g_n(t, \cdot)$ is continuous for fixed $t \in I$. Since $S = \overline{\text{co}}(G\square\mathscr{D})(I\times \boldsymbol{K}_\lambda)$ is compact in $\boldsymbol{R}^n$ and $g_n(t, x) \in S$ then there is M such $|g_n(t, x)| \leqslant M$ for every $(t, x) \in I\times \boldsymbol{K}_\lambda$. Therefore, $g_n(\cdot, x) \in L(I, \boldsymbol{R}^n)$ and for each $x \in \boldsymbol{K}_\lambda$ and every sequence (x_k) of $\boldsymbol{K}_\lambda$ converging to x we have $\int_{\sigma}^{\sigma+a} |g_n(t, x_k) - g_n(t, x)|\,dt \to 0$ as $k \to \infty$. Thus a function $\boldsymbol{K}_\lambda \in x \to g_n(\cdot, x) \in L(I, \boldsymbol{R}^n)$ is continuous. Finally, by Remark 4.2 we have $\int_{\sigma}^{\sigma+a} ||g_n(t, x)||\,dt \leqslant \lambda$ for $x \in \boldsymbol{K}_n$ and $n = 1, 2, \ldots$ ■

4.2. Properties of trajectory integrals of σ-selectionable set-valued functions

We shall prove the following theorem.

THEOREM 4.5. *Let* $G \in \mathscr{K}_\lambda(I, \boldsymbol{R}^n)$ *have convex values, be u.s.c.w. on* $I \times \mathscr{K}_\lambda \times B_\lambda$ *and such that* $\int_\sigma^{\sigma+a} \|G(t, x, z)\| dt \leqslant \lambda$ *for* $(x, z) \in \mathscr{K}_\lambda \times B_\lambda$. *Then the set-valued mapping* $\mathscr{I}\mathscr{F}(G\Box\mathscr{D})\colon \boldsymbol{K}_\lambda \to \mathrm{Conv}(C(I, \boldsymbol{R}^n))$ *is σ-selectionable and has a fixed point in* $\boldsymbol{K}_\lambda$.

Proof. By Theorem 4.4 we have $(G\Box\mathscr{D})(t, x) = \bigcap_{n \geqslant 1} H_n(t, x)$ for $(t, x) \in I \times \boldsymbol{K}_\lambda$, where $H_n\colon I \times \boldsymbol{K}_\lambda \to \mathrm{Conv}(\boldsymbol{R}^n)$ have for each $n \geqslant 1$ a selector $g_n\colon I \times \boldsymbol{K}_\lambda \to \boldsymbol{R}^n$ satisfying (iii) of Theorem 4.4. We have $\mathscr{F}(G\Box\mathscr{D})(x) \subset \boldsymbol{P}_\lambda$, $\mathscr{F}(G\Box\mathscr{D})(x) = \bigcap_{n \geqslant 1} \mathscr{F}(H_n)(x)$ and $g_n(\cdot, x) \in \mathscr{F}(H_n)(x)$ for each $x \in \boldsymbol{K}_\lambda$ and $n = 1, 2, \ldots$ Hence in particular it follows that $\mathscr{I}\mathscr{F}(G\Box\mathscr{D})(x) = \bigcap_{n \geqslant 1} \mathscr{I}\mathscr{F}(H_n)(x) \subset \boldsymbol{K}_\lambda$ and $f_n(x) := \mathscr{I}g_n(\cdot, x) \in \mathscr{I}\mathscr{F}(H_n)(x)$ for each $n \geqslant 1$ and $x \in \boldsymbol{K}_\lambda$. By Lemma 2.8 and (iii) of Lemma 1.13, $\mathscr{I}\mathscr{F}(H_n)$ is u.s.c. for every $n = 1, 2, \ldots$ Thus a set-valued mapping $\mathscr{I}\mathscr{F}(G\Box\mathscr{D})\colon \boldsymbol{K}_\lambda \ni x \to \mathscr{I}\mathscr{F}(G\Box\mathscr{D})(x) \subset \boldsymbol{K}_\lambda$ is σ selectionable. By Theorem II.4.10 it follows that $\mathscr{I}\mathscr{F}(G\Box\mathscr{D})$ has a fixed point in $\boldsymbol{K}_\lambda$. ■

Now we shall prove the following approximation theorem for set $\mathscr{C}(G)$ of all fixed points of the trajectory integral $\mathscr{I}\mathscr{F}(G\Box\mathscr{D})$ of $(G\Box\mathscr{D})$.

THEOREM 4.7. *Let* $G\colon I \times \mathscr{K}_\lambda \times B_\lambda \to \mathrm{Conv}(\boldsymbol{R}^n)$ *be u.s.c.w. on* $I \times \mathscr{K}_\lambda \times B_\lambda$ *and such that* $\|G(I \times \mathscr{K}_\lambda \times B_\lambda)\| \leqslant \lambda/a$. *Then there exists a sequence of convex compact subsets* U^m *and a sequence of continuous closed functions* $s^m\colon U_m \to C(I, \boldsymbol{R}^n)$ *such that*

(i) $\mathscr{C}(G) \subset \ldots \subset s_{m+1}(U^{m+1}) \subset s_m(U^m) \subset \ldots$ *for* $m = 0, 1, \ldots,$

(ii) *for every* $\varepsilon > 0$ *there is a number* $N_\varepsilon \geqslant 0$ *such that* $s_m(U^m) \subset \mathscr{C}(G) + \varepsilon B$ *for* $m \geqslant N_\varepsilon$, *where* B *denotes the closed unit ball of* $C^0(I, \boldsymbol{R}^n)$.

Furhermore, for every $x \in \mathscr{C}(G)$, *the set* $\{u \in U^m\colon s_m(u) = x\}$ *is convex.*

Proof. By Theorem 4.4, $G\Box\mathscr{D}$ can be approximated by a decreasing sequence of set-valued functions $H_m\colon I \times \boldsymbol{K}_\lambda \to \mathrm{Conv}(\boldsymbol{R}^n)$ of the form

$$H_m(t, x) = \sum_{i \in N_m} p_i^m(t, x) S_i^m \subset (\lambda/a) B,$$

where N_m is a finite subset of N, $p_i^m\colon I \times \boldsymbol{K}_\lambda \to \boldsymbol{R}^+$ satisfy conditions of Lemma 4.2 and where S_i^m are convex compact subsets of $\boldsymbol{R}^n$ contained in the ball of radius $\lambda/a > 0$.

Let us denote by $\mathscr{C}(H_m)$ the set of all fixed points of $\mathscr{I}\mathscr{F}(H_m)$ for each $m = 0, 1, \ldots$ Since $(G\Box\mathscr{D})(t, x) \subset \ldots \subset H_{m+1}(t, x) \subset H_m(t, x) \subset \ldots$ for $(t, x) \in I \times \boldsymbol{K}_\lambda$, we have also $\mathscr{C}(G) \subset \ldots \subset \mathscr{C}(H_{m+1}) \subset \mathscr{C}(H_m) \subset \ldots$ We shall prove that for all $\varepsilon > 0$ there is $N_\varepsilon \geqslant 0$ such that for all $m \geqslant N_\varepsilon$, we have $\mathscr{C}(H_m) \subset \mathscr{C}(G) + \varepsilon B$.

Assume the contrary: There exists $\bar{\varepsilon} > 0$ and a seqeunce (x_m) of $C^0(I, \boldsymbol{R}^n)$ with $x_m \in \mathscr{C}(H_m)$ satisfying $\mathrm{dist}(x_m, \mathscr{C}(G)) > \bar{\varepsilon}$. We have $x_m \in \mathscr{I}\mathscr{F}(H_m)(x_m)$ for each $m = 0, 1, \ldots,$ then $x_m \in \boldsymbol{K}_\lambda$ and therefore, there is a subsequence, say again (x_m)

of (x_m) and $\bar{x} \in C^0(I, R^n)$ such that $|x_m - \bar{x}|_a \to 0$ as $m \to \infty$. Hence in particular it follows $\dot{x}_m \to \dot{\bar{x}}$ as $m \to \infty$.

On the other hand, by (iii) of Theorem 4.3 for all $(t, x) \in I \times K_\lambda$ there exists an $N_\varepsilon^{tx} \geq 0$ such that $H_m(t, x) \subset (G \square \mathscr{D})(t, x) + \varepsilon B$ when $m \geq N_\varepsilon^{tx}$, i.e., $\lim_{m \to \infty} \bar{h}(H_m(t, x), (G \square \mathscr{D})(t, x)) = 0$ for $(t, x) \in I \times K_\lambda$. In particular, we have $\lim_{m \to \infty} \bar{h}(H_m(t, \bar{x}), (G \square \mathscr{D})(t, \bar{x})) = 0$ for $t \in I$. Since $\dot{x}_m(t) \in H_m(t, x_m)$ then, for $i \in N_m$, there is $s_i^m \in S_i^m$ such that $\dot{x}(t) = \sum_{i \in N_m} p_i^m(t, x_m) s_i^m$ for a.e. $t \in I$ and $m = 0, 1, \ldots$ Let $f_m(t, x) := \sum_{i \in N_m} p_i^m(t, x) s_i^m$ for $(t, x) \in I \times K_\lambda$ and $m = 0, 1, \ldots$ We have $f_m(t, x) \in H_m(t, x)$ for $(t, x) \in I \times K_\lambda$ and $m = 0, 1, \ldots$ Furthermore, each $f_m(t, \cdot)$ is continuous on K_λ and by compactness of K_λ also uniformly continuous. Thus for each fixed m and $x_m, \bar{x} \in K_\lambda$ with $|x_m - \bar{x}|$ sufficiently small we have $|f_m(t, x_m) - f_m(t, \bar{x})| \leq \frac{1}{2}\varepsilon$ for fixed $t \in I$ and a given $\varepsilon > 0$. Therefore, for m sufficiently large we have

$$\begin{aligned}
\operatorname{dist}(\dot{x}_m(t), (G \square \mathscr{D})(t, \bar{x})) &= \operatorname{dist}(f_m(t, x_m), (G \square \mathscr{D})(t, \bar{x})) \\
&\leq |f_m(t, x_m) - f_m(t, \bar{x})| + \operatorname{dist}(f_m(t, \bar{x})(G \square \mathscr{D})(t, \bar{x})) \\
&\leq \tfrac{1}{2}\varepsilon + \bar{h}(H_m(t, \bar{x}), (G \square \mathscr{D})(t, \bar{x})) \leq \varepsilon
\end{aligned}$$

for a.e. $t \in I$ and m sufficiently large such that $\bar{h}(H_m(t, \bar{x}), (G \square \mathscr{D})(t, \bar{x})) \leq \frac{1}{2}\varepsilon$.

Then $\dot{x}_m(t) \in (G \square \mathscr{D})(t, \bar{x}) + \varepsilon B$ for a.e. $t \in I$ and m sufficiently large. Since $\dot{x}_m \rightharpoonup \dot{\bar{x}}$, $|x_m - \bar{x}|_a \to 0$ as $m \to \infty$ and $(G \square \mathscr{D})(t, \bar{x}) + \varepsilon B$ is a convex subset of R^n, then $\bar{x} \in \mathscr{C}(G)$ and at the same time $\operatorname{dist}(\bar{x}, \mathscr{C}(G)) \geq \bar{\varepsilon} > 0$. This contradiction proves that $\mathscr{C}(H_m) \subset \mathscr{C}(G) + \varepsilon B$ for $m \geq N_\varepsilon$.

We set $U^m = \prod_{i \in N_m} L^\infty(I, S_i^m)$, which is a convex compact subset of $L^\infty(I, R^n)^{N_m}$ supplied with the weak-*topology. Let $u^m(\cdot) = \{u_i^m(\cdot)\}_{i \in N_m}$ be a function of U^m. Since the mapping $I \times K_\lambda \in (t, x) \to \sum_{i \in N_m} p_i^m(t, x) u_i^m(t)$ is measurable with respect to $t \in I$, locally Lipschitzean with respect to $x \in K_\lambda$ and is a selector of H_m, then there exists the unique solution $x \in K_\lambda$ of an initial value problem

$$\begin{aligned}
&\dot{x}(t) = \sum_{i \in N_m} p_i^m(t, x) u_i^m(t) \quad \text{for} \quad \text{a.e. } t \in I, \\
&x(\sigma) = 0.
\end{aligned} \tag{4.5}$$

Indeed, since for every $m = 1, 2, \ldots$ the function $g_m \colon I \times K_\lambda \to R^n$ defined by $g_m(t, x) := \sum p_i^m(t, x) u_i^m(t)$ for $(t, x) \in I \times K_\lambda$ is a selector of H_m, then by Remarks 4.1 and 4.2 the family $\{g_m(\cdot, x)\}_{x \in K_\lambda}$ is uniformly integrable and $g_m(\cdot, x) \in P_\lambda$, i.e., $\mathscr{I} g_m(\cdot, x) \in K_\lambda$ for $x \in K_\lambda$. Furthermore, by the properties of functions p_i^m for every $(t, x) \in I \times K_\lambda$ there is an open set $\theta_{tx} \subset I \times K_\lambda$ and a number $L_{tx} > 0$ such that $|g_m(\tau, x_1) - g_m(\tau, x_2)| \leq L_{tx} |x_1 - x_2|_\tau$ for each (τ, x_1), $(\tau, x_2) \in \theta_{tx}$. Hence in particular it follows that a mapping $K_\lambda \ni x \to \mathscr{I} g_m(\cdot, x) \in K_\lambda$ is continuous on K_λ. Since K_λ is a compact convex subset of $C^0(I, R^n)$ then by Schauder–Tikhonov's fixed point theorem there is a $x \in K_\lambda$ such that $x = \mathscr{I} g_m(\cdot, x)$. Suppose there are two such fixed points $x_1, x_2 \in K_\lambda$ that are different on any subinterval $[t, t+\alpha] \subset I$.

Suppose (τ, x_1), $(\tau, x_2) \in \theta_{tx}$ for $\tau \in [t, t+\alpha]$ and any $x \in \boldsymbol{K}_\lambda$. We assume that $x_1(\tau) = x_2(\tau)$ for $\tau \in [\sigma, t]$. Therefore such $x \in \boldsymbol{K}_\lambda$ exists. Now, for $\tau \in [t, t+\alpha]$ and every $\eta > 0$ we have

$$\sup_{t \leqslant s \leqslant \tau} |x_1(s) - x_2(s)| \leqslant \eta + L_{tx} \int_t^\tau \sup_{t \leqslant s \leqslant p} |x_1(s) - x_2(s)| dp.$$

Therefore, by Gronwall's inequality we get

$$\sup_{t \leqslant s \leqslant \tau} |x_1(s) - x_2(s)| \leqslant \eta \exp(L_{tx}\alpha)$$

for $\tau \in [t, t+\alpha]$ and every $\eta > 0$. Hence it follows that

$$\sup_{t \leqslant s \leqslant +\alpha} |x_1(s) - x_2(s)| = 0.$$

This contradiction proves that (4.5) has only one solution in $\boldsymbol{K}_\lambda$.

We shall denote by $s_m(u^m)$ this solution. Hence, we have defined a mapping s_m: $U^m \to C^0(I, \boldsymbol{R}^n)$. Since the solution of (4.5) can be written by

$$x(t) = \sum_{i \in N_m} \int_\sigma^t p_i^m(\tau, x) u_i^m(\tau) d\tau$$

we see at once that s_m is a continuous mapping.

Furthermore, for any closed set $A \subset U^m$ and any $x \in \overline{[s_m(A)]_C}$ there exists a sequence (u_k) of A such that $|s_m(u_k) - x|_a \to 0$ as $k \to \infty$. Since U^m is compact then there is a subsequence, say again (u_k) of (u_k) and $u \in A \subset U^m$ such that (u_k) converges to u as $k \to \infty$. Hence it follows that $x = s_m(u)$, i.e. that $x \in s_m(A)$. Therefore s_m is a closed mapping.

We have of course $s_m(u^m) \in \mathscr{C}(H_m)$ for each $u^m \in U^m$. Therefore, $s_m(U^m) \subset \mathscr{C}(H_m) + \varepsilon B$. Suppose $x \in \mathscr{C}(H_m)$. Then $\dot{x}(t) \in \sum_{i \in N_m} p_i^m(t, x) S_i^m$ for a.e. $t \in I$. Therefore, by Theorem II.3.12, for each $i \in N_m$ there exists a function $u_i^m \in L^\infty(I, S_i^m)$ such that $\dot{x}(t) = \sum_{i \in N_m} p_i^m(t, x) u_i^m(t)$ for a.e. $t \in I$. Thus $\mathscr{C}(H_m) \subset s_m(U^m)$. Hence, finally it follows that $\mathscr{C}(G) \subset s_m(U^{m+1}) \subset s_m(U^m) \subset \ldots$ for $m \geqslant N_\varepsilon$. The subset of controls u^m mentioned above is obviously convex. ∎

Now as a corollary from the above theorem we obtain the following result.

THEOREM 4.8. *Let* G: $I \times \mathscr{K}_\lambda \times B_\lambda \to \mathrm{Conv}(\boldsymbol{R}^m)$ *with* $I := [\sigma, \sigma + a]$ *be u.s.c.w. and such that* $\|G(I \times \mathscr{K}_\lambda \times B_\lambda)\| \leqslant \lambda/a$. *Then the set* $\mathscr{C}(G)$ *is acyclic.*

Proof. By Theorem 4.4 we have $\mathscr{C}(G) = \bigcap_{m \geqslant 0} \mathscr{C}(H_m)$ and hence, by Theorem 4.7, that $\mathscr{C}(G) = \bigcap_{m \geqslant 0} s_m(U^m)$. By Theorem 4.7, s_m satisfies the assumptions of Vietoris–Begle's theorem. Since each U^m is compact and convex, then $s_m(U^m)$ is compact and by Vietoris–Begle's theorem acyclic. Since $\mathscr{C}(G) = \bigcap_{m \geqslant 0} s_m(U^m)$ this implies that $\mathscr{C}(G)$ is also acyclic. ∎

5. Relaxation theorems

We give here some sufficient conditions under which the set of fixed points of the trajectory integrals of a set-valued function $G \in \mathscr{H}_\lambda(I, \boldsymbol{R}^n)$ is dense in the set of all fixed points of the trajectory integrals of $\operatorname{co} G \in \mathscr{H}_\lambda(I, \boldsymbol{R}^n)$. This problem is particularly important in the control theory; solutions to the convexified problem are often called *relaxed solutions*, and the problem we have mentioned, is called the *problem of relaxation*. We shall prove that the relaxation property for a giern $G \in \mathscr{H}_\lambda(I, \boldsymbol{R}^n)$ holds when some Nagumo's types conditions for G are satisfied. It can be proved that the relaxation property does not necessarily hold when G is only continuous.

5.1. *Some properties of Stieltjes integrals*

Let $I = [\sigma, \sigma+a]$ and $G \in \mathscr{H}_\lambda(I, \boldsymbol{R}^n)$ be such that $\int_\sigma^{\sigma+a} \|G(t, x, z)\| dt \leqslant \lambda$ for $(x, z) \in \mathscr{K}_\lambda \times B_\lambda$. Let $\boldsymbol{K}_\lambda$ and $\boldsymbol{P}_\lambda$ be such as in Section 2.3. For a given $\mathscr{S} \in \mathrm{AC}_{0r}$, we shall denote by $\boldsymbol{K}_\lambda^{\mathscr{S}}$ and $\boldsymbol{P}_\lambda^{\mathscr{S}}$ subsets of $\mathrm{AC}([\sigma-r, \sigma+a], \boldsymbol{R}^n)$ and $L([\sigma-r, \sigma+a], \boldsymbol{R}^n)$ defined by $\boldsymbol{K}_\lambda^{\mathscr{S}} = \{\mathscr{S} \oplus x\colon x \in \boldsymbol{K}_\lambda\}$ and $\boldsymbol{P}_\lambda^{\mathscr{S}} = \{\dot{\mathscr{S}} \oplus z\colon z \in \boldsymbol{P}_\lambda\}$, where $\mathscr{S} \oplus x$ and $\dot{\mathscr{S}} \oplus z$ are equal to $\mathscr{S}$ and $\dot{\mathscr{S}}$, respectively on $[\sigma-r, \sigma]$ and to $\mathscr{S}(0)+x$ and z on $(\sigma, \sigma+a]$. Given $(x, z) \in C([\alpha-r, \beta], \boldsymbol{R}^n) \times L([\alpha-r, \beta], \boldsymbol{R}^n)$ and fixed $t \in I$ by x_t and z_t we shall denote mappings defined on $[-r, 0]$ by setting $x_t(s) := x(t+s)$ and $z_t(s) := z(t+s)$ for $s \in [-r, 0]$.

We have the following lemma.

Lemma 5.1. *For every $x \in C([\alpha-r, \beta], \boldsymbol{R}^n)$ and $z \in L([\alpha-r, \beta], \boldsymbol{R}^n)$ with $r \geqslant 0$, mappings $v(\cdot, x)\colon [\alpha, \beta] \to C_{0r}$ and $w(\cdot, x)\colon [\alpha, \beta] \to L_{0r}$ defined on $[\alpha, \beta]$ by setting $v(t, x) := x_t$ and $w(t, x) := z_t$ for each $t \in [\alpha, \beta]$ are continuous on $[\alpha, \beta]$.*

Proof. It is clear that $x_t \in C_{0r}$ and $z \in L_{0r}$ for each $x \in C([\alpha-r, \beta], \boldsymbol{R}^n)$, $z \in L([\alpha-r, \beta], \boldsymbol{R}^n)$ and $t \in [\alpha, \beta]$.

Let $|\cdot|$ and $\left|\cdot\right|$ be norms of $C([\alpha-r, \beta], \boldsymbol{R}^n)$ and $L([\alpha-r, \beta], \boldsymbol{R}^n)$, respectively. For every $t, \tau \in [\alpha, \beta]$ one has

$$|x_t - x_\tau|_0 = \sup_{-r \leqslant s \leqslant 0} |x(t+s) - x(\tau+s)| \leqslant \omega_x(|t-\tau|),$$

where ω_x is the modulus of continuity of x. Then $\lim_{|t-\tau| \to 0} |x_t - x_\tau|_0 = 0$.

Let $\varepsilon > 0$ be arbitrarily fixed and let $x_\varepsilon \in C([\alpha-r, \beta], \boldsymbol{R}^n)$ be such that $\left|x^\varepsilon - z\right| \leqslant \frac{1}{2}\varepsilon$. Such x^ε exists because $C([\alpha-r, \beta], \boldsymbol{R}^n)$ is dense in $L([\alpha-r, \beta], \boldsymbol{R}^n)$. Denote by ω_ε a modulus of continuity of x^ε. Now, for $t, \tau \in [\alpha, \beta]$ one obtains

$$\begin{aligned}
\left|z_t - z_\tau\right|_0 &\leqslant \left|z_t - x_t^\varepsilon\right|_0 + \left|x_t^\varepsilon - x_\tau^\varepsilon\right|_0 + \left|x_\tau^\varepsilon - z_\tau\right|_0 \\
&\leqslant \int_{t-r}^{t} |z(p) - x^\varepsilon(p)| dp + r\omega_\varepsilon(|t-\tau|) + \int_{\tau-r}^{\tau} |(x^\varepsilon(p) - z(p)| dp \\
&\leqslant 2\left|x^\varepsilon - z\right| + r\omega_\varepsilon(|t-\tau|).
\end{aligned}$$

Hence it follows $\lim_{|t-\tau|\to 0} |z_t - z_\tau|_0 \leqslant \varepsilon$ for every $\varepsilon > 0$. Thus, we also have $\lim_{|t-\tau|\to 0} |z_t - z_\tau|_0 = 0$. ■

Let $\nu \in \mathrm{BV}(I, \boldsymbol{R}^1)$ and $z \in \boldsymbol{P}_\lambda^{\mathscr{S}}$ be given and denote by V_ν total variation of ν. Let $v(\cdot, z)$: $I \ni t \to z_t \in L_{0r}$. By virtue of Lemma 5.1, $v(\cdot, z)$ is continuous on I. Then, by Theorem I.3.3, the Stieltjes integral $\int_\sigma^t v(\tau, z)d\nu(\tau)$ exists for each fixed $t \in I$ and $|\int_\sigma^t v(\tau, z)d\nu(\tau)|_0 \leqslant \sup_{\sigma\leqslant\tau\leqslant t} |v(\tau, z)|_0 V_\nu$. This integral will be also denoted by $\int_\sigma^t z_\tau d\nu(\tau)$. We shall show that a family $\{\int_\sigma^t v(\tau, \cdot)d\nu(\tau)\}_{t\in I}$ of functions from $\boldsymbol{P}_\lambda^{\mathscr{S}}$ into L_{0r} is equi-w.-w.s.c. on $\boldsymbol{P}_\lambda^{\mathscr{S}}$, i.e., is such that for every $z \in \boldsymbol{P}_\lambda^{\mathscr{S}}$ and every sequence (z_n) of $\boldsymbol{P}_\lambda^{\mathscr{S}}$ weakly converging to z one has

$$\lim_{n\to\infty} \sup_{t\in I} \left| \int_U \left[\int_\sigma^t v(\tau, z_n)d\nu(\tau) - \int_\sigma^t v(\tau, z)d\nu(\tau) \right](s)ds \right| = 0 \tag{5.1}$$

for every measurable set $U \subset [-r, 0]$. We begin with the following lemmas.

LEMMA 5.2. *The family $\{v(t, \cdot)\}_{t\in I}$ of mappings from $\boldsymbol{P}_\lambda^{\mathscr{S}}$ into L_{0r} is equi-w.-w.s.c. on $\boldsymbol{P}_\lambda^{\mathscr{S}}$.*

Proof. Let $z \in \boldsymbol{P}_\lambda^{\mathscr{S}}$ be fixed and suppose (z_n) is any sequence of $\boldsymbol{P}_\lambda^{\mathscr{S}}$ weakly converging to z. Let $x = \mathscr{S} \oplus \mathscr{I}z$ and $x_n = \mathscr{S} \oplus \mathscr{I}z_n$. By Lemma 1.13 we have $|x_n - x|_a \to 0$ as $n \to \infty$. Put $\psi_n(t, s) = v(t, z_n)(s) - v(t, z)(s)$. Now, for every $s_1, s_2 \in [-r, 0]$ we have

$$\left|\int_{s_1}^{s_2} \psi_n(t, s)ds\right| = \left|\int_{s_1}^{s_2} [(\mathscr{D}x_n)(t+s) - (\mathscr{D}x)(t+s)]ds\right| =$$

$$= |[x_n(t+s_2) - x(t+s_2)] - [x_n(t+s_1) - x(t+s_1)]| \leqslant 2|x_n - x|_a$$

for $n = 1, 2, \ldots$ Therefore, $\lim_{n\to\infty} |\int_{s_1}^{s_2} \psi_n(t, s)ds| = 0$ uniformly with respect to $t \in I$.

Since $\boldsymbol{P}_\lambda^{\mathscr{S}}$ is weakly compact it is uniformly integrable. Then for every $\varepsilon > 0$ there is $\delta(\varepsilon) > 0$ such that for each measurable set $E \subset [\sigma - r, \sigma + a]$ with $\mu(E) < \delta(\varepsilon)$ and each $z \in \boldsymbol{P}_\lambda^{\mathscr{S}}$ we have $\int_E |z(\tau)d\tau| < \frac{1}{2}\varepsilon$. Let U be an arbitrary measurable subset of $[-r, 0]$ with $\mu(U) > 0$. Select a family $\{(s_i, s_i'): 1 \leqslant i \leqslant N_\varepsilon\}$ of disjoint open intervals of $\boldsymbol{R}^1$ such that $\mu(U \triangle H) < \delta(\varepsilon)$, where $H = \bigcup_{i=1}^{N_\varepsilon} (s_i, s_i') \cap [-r, 0]$. Similary as above we can show that for every $i = 1, 2, \ldots, N_\varepsilon$, $\lim_{n\to\infty} \left| \int_{(s_i, s_i')\cap[-r,0]} \psi_n(t, s)ds \right| = 0$ uniformly with respect to $t \in I$. Therefore we have also $|\int_H \psi_n(t, s)ds| \to 0$ uniformly with respect to $t \in I$ as $n \to \infty$. Now, for every $n = 1, 2, \ldots$ and $t \in I$ we have

$$\left|\int_U \psi_n(t,s)ds\right|$$
$$\leqslant \left|\int_H \psi_n(t,s)ds\right| + \left|\int_{U\setminus H} \psi_n(t,s)ds\right| + \left|\int_{H\setminus U} \psi_n(t,s)ds\right|$$
$$\leqslant \left|\int_H \psi_n(t,s)ds\right| + \int_{U\triangle H} |(z_n)_t(s)|ds + \int_{U\triangle H} |z_t(s)|ds$$
$$\leqslant \left|\int_H \psi_n(t,s)ds\right| + \int_{U\triangle H+t} |z_n(\tau)|d\tau + \int_{U\triangle H+t} |z(\tau)|d\tau$$
$$\leqslant \varepsilon + \left|\int_H \psi_n(t,s)ds\right|$$

because $\mu(U\triangle H+t) = \mu(U\triangle H) < \delta(\varepsilon)$ for each $t\in I$. Thus $\lim_{n\to\infty}\left|\int_U \psi_n(t,s)ds\right| \leqslant \varepsilon$ uniformly with respect to $t\in I$ for every $\varepsilon > 0$. ■

LEMMA 5.3. *Let* $\nu\in \mathrm{BV}(I, \boldsymbol{R}^1)$ and $z\in \boldsymbol{P}_\lambda^{\mathscr{S}}$ *be given. Then for every* $t\in I$ *and every measurable set* $U\in[-r,0]$ *one has*

$$\int_U\left[\int_\sigma^t v(\tau,z)d\nu(\tau)\right](s)ds = \int_\sigma^t\left[\int_U v(\tau,z)(s)ds\right]d\nu(\tau).$$

Proof. Let $t\in I$ and $U\subset[-r,0]$ be fixed. Observe first that a function $I\ni\tau \to \int_U v(\tau,z)(s)ds\in \boldsymbol{R}^n$ is continuous on I because it is a superposition of continuous mappings $I\in\tau\to v(\tau,z)\in L_{0r}$ and $L_{0r}\ni u\to \int_U u(s)ds\in \boldsymbol{R}^u$. Let $A[v(\cdot,z)](\tau) := \int_U v(\tau,z)(s)ds$. Since $A[v(\cdot,z)]$ is continuous and $\nu\in \mathrm{BV}(I,\boldsymbol{R})$ then by Theorem I.3.3, the Stieltjes integral $\int_\sigma^t A[v(\cdot,z)](\tau)d\nu(\tau)$ exists. Let (Δ_n) be any sequence of partitions $\Delta_n = \{t_0^n, t_1^n, \ldots, t_{N_n}^n\}\in\pi(\sigma,t)$ and let $S_n := \sum_{i=1}^{N_n} v(\tau_i,z)\,[\nu(t_i^n)-\nu(t_{i-1}^n)]$ with $\tau_i\in[t_{i-1}^n, t_i^n]$. For every $n=1,2,\ldots$ and a measurable set $E\subset[-r,0]$ we have $\int_E |S_n(s)|ds \leqslant V_\nu \max_{\tau\in I}\int_E |v(\tau,z)(s)|ds$. Hence it follows that a sequence (S_n) is uniformly integrable, because a family $\{v(\tau,z)\}_{\tau\in I}$ is uniformly integrable. Then by Vitali's theorem, $\lim_{n\to\infty}\int_U S_n(s)ds = \int_U \lim_{n\to\infty} S_n(s)ds = \int_U\left(\int_\tau^t v(\tau,z)d\nu(\tau)\right)(s)ds$. On the other hand,

$$\lim_{n\to\infty}\int_U S_n(s)ds = \lim_{n\to\infty}\sum_{i=1}^{N_n} A[v(\cdot,z)]\,(\tau_i)\,[\nu(t_i^n)-\nu(t_{i-1}^n)]$$
$$= \int_\sigma^t A[v(\cdot,z)]\,(\tau)d\nu(\tau). \;■$$

COROLLARY 5.1. *For every $x \in K_\lambda^{\mathscr{S}}$ and $v \in \mathrm{BV}(I, R^1)$ we have*

$$\int_{-r}^{0} \Big(\int_{\sigma}^{t} \dot{x}_\tau dv(\tau)\Big) ds = \int_{\sigma}^{t} [x(\tau) - x(\tau - r)] dv(\tau)$$

for each $t \in I$.

Indeed, we have

$$\int_{-r}^{0} \Big(\int_{\sigma}^{t} \dot{x}_\tau dv(\tau)\Big) ds = \Big(\int_{-r}^{0} \dot{x}(\tau + s) ds\Big) dv(\tau) = \int_{\sigma}^{t} [x(\tau) - x(\tau - r)] dv(\tau). \ \blacksquare$$

Now we can prove our main lemma.

LEMMA 5.4. *Let $v \in \mathrm{BV}(I, R^1)$. The family $\{\int_{\sigma}^{t} v(\tau, \cdot) dv(\tau)\}_{t \in I}$ of mappings from $P_\lambda^{\mathscr{S}}$ into L_{0r} is equi-w.-w.s.c. on $P_\lambda^{\mathscr{S}}$.*

Proof. By virtue of Lemma 5.2, the family $(v(t, \cdot)\}_{t \in I}$ is equi-w.-w.s.c. on $P_\lambda^{\mathscr{S}}$. Then for every fixed $z \in P_\lambda^{\mathscr{S}}$ and every sequence (z_n) of $P_\lambda^{\mathscr{S}}$ weakly converging to z one has

$$\lim_{n \to \infty} \sup_{\sigma \leqslant \tau \leqslant t} \Big| \int_U [v(\tau, z_n) - v(\tau, z)] (s) ds \Big| = 0$$

for every measurable set $U \subset [-r, 0]$. Hence, by Lemma 5.3 and the properties of the Stieltjes integrals, the result follows, because

$$\Big| \int_U \Big(\int_{\sigma}^{t} v(\tau, z_n) dv(\tau) - \int_{\sigma}^{t} v(\tau, z) dv(\tau) \Big] (s) ds \Big|$$

$$= \Big| \int_{\sigma}^{t} \Big(\int_U [v(\tau, z_n) - v(\tau, z)] (s) ds \Big) dv(\tau)$$

$$\leqslant V_v \sup_{\sigma \leqslant \tau \leqslant t} \Big| \int_U [v(\tau, z_n) - v(\tau, z)] (s) ds \Big|$$

for every measurable set $U \subset [-r, 0]$. $\blacksquare$

Let us denote by $\mathscr{L}(L_{0r}, R^n)$ the Banach space of all linear bounded operators from L_{0r} into R^n with the norm $\|\cdot\|_{\mathscr{L}}$ defined in the usual way by $\|A\|_{\mathscr{L}} = \sup\{|A(u)|: |u|_0 = 1\}$ for $A \in \mathscr{L}(L_{0r}, R^n)$. Let us consider a mapping $L: I \to \mathscr{L}(L_{0r}, R^n)$. By Riesz representation theorem there exists a map $\eta^L: I \times [-r, 0] \to \mathscr{L}(R^n, R^n)$ such that for every $u \in L_{0r}$, fixed $t \in I$ and $i, j = 1, \ldots, n$ we have $\eta_{ij}(t, \cdot) \in L^\infty([-r, 0], R)$ and $L(t)u = \int_{-r}^{0} \eta\ (t, s) u(s) ds$, where $\eta^L = [\eta_{ij}]_{m \times n}$ is a $(n \times n)$-matrix.

We shall consider now a family $\{\Psi(t, \cdot)\}_{t \in I}$ of mappings from $P_\lambda^{\mathscr{S}}$ into R^ndefined by $\Psi(t, z) := \int_{-r}^{0} \eta^L(t, s) z_t(s) ds$ for $z \in P_\lambda^{\mathscr{S}}$ and $t \in I$. It will be called equi-s.-w.s.c.

on $P_\lambda^{\mathscr{S}}$ if for every $z \in P_\lambda^{\mathscr{S}}$ and every sequence (z_n) of $P_\lambda^{\mathscr{S}}$ weakly converging to z one has $\lim_{n\to\infty} \sup_{t\in I}|\Psi(t, z_n)-\Psi(t, z)| = 0$.

LEMMA 5.5. *If a mapping $L: I \to \mathscr{L}(L_{0r}, R^n)$ is continuous, then the family $\{\Psi(t, \cdot)\}_{t\in I}$ is equi-s.-w.s.c. on $P_\lambda^{\mathscr{S}}$.*

Proof. Let us observe that $\Psi(t, \cdot) = L(t) \circ v(t, \cdot)$ and let $z \in P_\lambda^{\mathscr{S}}$ be arbitrarily fixed and (z_n) any sequence of $P_\lambda^{\mathscr{S}}$ weakly converging to z, Since $\|L(t)[v(t, z_n)-v(t, z)]\| \leqslant \sup_{t\in I} \|L(t)\|_{\mathscr{L}}|v(t, z_n)-v(t, z)|$ for $t \in I$ and $n = 1, 2, \ldots$ then the result follows immediately from Lemma 5.2. ∎

5.2. Continuous selection theorems

We shall consider here a set-valued function $G \in \mathscr{H}_\lambda(I, R^n)$ satisfying Kamke conditions. Recall, the function ω: $I\times R^+ \to R^+$ is said to be *Kamke function* if the following conditions are satisfied:

(i) $\omega(\cdot, u)$ is measurable for fixed $u \in R^+$,

(ii) a family $\{\omega(\cdot, u)\}_{u\in R^+}$ is uniformly integrable and $\omega(t, 0) = 0$ for a.e. $t \in I$,

(iii) $\omega(t, \cdot)$ is continuous for fixed $t \in I$,

(iv) the unique absolutely continuous solution of $0 \leqslant u(t) \leqslant \int_0^t \omega(s, u(s))ds$

is $u(t) \equiv 0$.

Let $\nu \in \mathrm{BV}(I, R^1)$ and L: $I \to \mathscr{L}(L_0, R^n)$ be given. Without loss of generality we shall assume that $V_\nu \leqslant 1$. For the sake of simplicity, for fixed $t \in I$ and $x, y \in K_\lambda^{\mathscr{S}}$ with $K_\lambda^{\mathscr{S}}$ such as in Section 5.1, we denote

$$R_1(t, x, y) = \max\Big\{|x-y|_t, \sup_{\sigma\leqslant p\leqslant t}\Big|\int_{-r}^{0}\Big(\int_{\sigma}^{p}[\dot{x}_\tau-\dot{y}_\tau]d\nu(\tau)(s)\Big)ds\Big|\Big\} \tag{5.2}$$

and

$$R_2(t, x, y) = \max\{\|L(t)\|_{\mathscr{L}}|x-y|_t, \sup_{0\leqslant\tau\leqslant t}|L(\tau)[\dot{x}_\tau-\dot{y}_\tau]|\}. \tag{5.3}$$

Let $\psi \in \mathrm{AC}_{0r}$ be given. We shall consider here set-valued functions $G \in \mathscr{H}_\lambda(I, R^n)$ with $\int_\sigma^{\sigma+a} \|G(t, x, z)\|dt \leqslant \lambda$ for $(x, z) \in \mathscr{K}_\lambda\times B_\lambda$ and such that there are $\nu \in \mathrm{BV}(I, R^1)$ and a Kamke function ω: $I\times R^+ \to R^+$ such that

$$h\big((G\Box\mathscr{D})(t, x), (G\Box\mathscr{D})(t, y)\big) \leqslant \omega\big(t, R_1(t, \psi\oplus x, \psi\oplus y)\big) \tag{5.4}$$

for a.e. $t \in I$ and $x, y \in K_\lambda$. It will be proved that for every $\varepsilon > 0$, $\Phi \in K_\lambda$ and $\xi \in \mathscr{F}(\mathrm{co}(G\Box\mathscr{D}))(\Phi)$ there exists a continuous function g: $K_\lambda \to L(I, R^n)$ such that $g(x) \in \mathscr{F}(G\Box\mathscr{D})(x)$ for each $x \in K_\lambda$ and

$$\max\Big\{\Big|\int_\sigma^t [g(x)(\tau)-\xi(\tau)]d\tau\Big|, \Big|\int_{-r}^{0}\Big(\int_\sigma^t[(\dot\psi\oplus g(x))_\tau-(\dot\psi\oplus\xi)_\tau]d\nu(\tau)\Big)(s)ds\Big|\Big\} \tag{5.5}$$

$$\leqslant \varepsilon+\int_\sigma^t\omega\big(s, R_1(s, \psi\oplus x, \psi\oplus\Phi)\big)ds$$

for $x \in K_\lambda$ and $t \in I$.

We begin with the following lemmas.

LEMMA 5.6. *Let* $\psi \in \mathrm{AC}_{0r}$ *be given. Suppose* $G \in \mathscr{H}_\lambda(I, R^n)$ *is such that* $\int_\sigma^{\sigma+a} \|G(t, x, z)\| dt \leqslant \lambda$ *for* $(x, z) \in \mathscr{K}_\lambda \times B_\lambda$ *and there are* $\nu \in \mathrm{BV}(I, R^1)$ *and a Kamke function* $\omega\colon I \times R^+ \to R^+$ *such that* (5.4) *is satisfied for* $x, y \in K$, *and a.e.* $t \in I$. *Then* $(G\Box\mathscr{D})(t, \cdot)$ *is continuous on* K *for a.e. fixed* $t \in I$.

Proof. The continuity of $(G\Box\mathscr{D})(t, \cdot)$ follows immediately from (5.4) and properties of Kamke functions, because for every $x \in K_\lambda$ and every sequence (x_n) of K_λ converging to x we have $R_1(t, \psi\oplus x_n, \psi\oplus x) \to 0$ as $n \to \infty$. ■

REMARK 5.1. *For every* $x, y \in K_\lambda$, $u \in \mathscr{F}(G\Box\mathscr{D})(y)$ *and a.e.* $t \in I$ *we have* $e(\mathscr{F}(G\Box\mathscr{D})(x) - u)(t) \leqslant h((G\Box\mathscr{D})(t, x), (G\Box\mathscr{D})(t, y))$.

Indeed, let $x, y \in K_\lambda$ be fixed and u an arbitrary point of $\mathscr{F}(G\Box\mathscr{D})(y)$.Since $(G\Box\mathscr{D})(\cdot, x)$ is measurable then by Theorem II.3.13, there is $v \in \mathscr{F}(G\Box\mathscr{D})(x)$ such that $|v(t) - u(t)| = \mathrm{dist}(u(t), (G\Box\mathscr{D})(t, x))$ for a.e. $t \in I$. Since $e(\mathscr{F}(G\Box\mathscr{D})(x) - u)(t) \leqslant |v(t) - u(t)|$ for a.e. $t \in I$, then for a.e. $t \in I$ we also have $e(\mathscr{F}(G\Box\mathscr{D})(x) - u)(t) \leqslant h((G\Box\mathscr{D})(t, x), (G\Box\mathscr{D})(t, y))$. ■

LEMMA 5.7. *Let* $\psi \in \mathrm{AC}_{0r}$ *be given. Suppose* $G \in \mathscr{H}_\lambda(I, R^n)$ *satisfies the assumptions of Lemma 5.6 and is such that a family* $\{(G\Box\mathscr{D})(t, \cdot)\}_{t\in I}$ *is equicontinuous on* K_λ. *Then for every* $\varepsilon > 0, \Phi \in K_\lambda$ *and* $\xi \in \mathscr{F}(\mathrm{co}(G\Box\mathscr{D}))(\Phi)$ *there exist continuous functions* $g\colon K_\lambda \to L(I, R^n)$ *and* $\mathscr{S}\colon K_\lambda \to L(I, R^1)$ *such that*

$$\int_I \mathscr{S}(x)(t)dt < \varepsilon \quad \text{for each } x \in K_\lambda, \tag{5.6}$$

and

$$\max\left\{\left|\int_\sigma^t [g(x)(\tau) - \xi(\tau)]d\tau, \left|\int_{-r}^0 \left(\int_\sigma^t [(\dot\psi\oplus g(x))_\tau - (\dot\psi\oplus\xi)_\tau]d\nu(\tau)\right)(s)ds\right|\right\}\right. \tag{5.7}$$

$$\leqslant \varepsilon + \int_\sigma^t \omega(s, R_1(s, \psi\oplus x, \psi\oplus\Phi))ds$$

for $t \in I$ *and every* $x \in K_\lambda$. *Moreover the set*

$$L(x) = \{u \in \mathscr{F}(G\Box\mathscr{D})(x)\colon |u(t) - g(x)(t)| < \mathscr{S}(x)(t) \text{ for a.e. } t \in I\} \tag{5.8}$$

is nonempty for each $x \in K_\lambda$.

Proof. Let $\varepsilon > 0$, $\Phi \in K_\lambda$, $\xi \in \mathscr{F}(\mathrm{co}(G\Box\mathscr{D}))(\Phi)$ and $\psi \in \mathrm{AC}_{0r}$ be given. By virtue of Theorem 2.3 there exists a sequence (ξ_n) of $\mathscr{F}(G\Box\mathscr{D})(\Phi)$ weakly converging to ξ. Hence, by Lemma 5.4, it follows that $\lim_{n\to\infty}\left|\int_{-r}^0 \left(\int_\sigma^t [(\dot\psi\oplus\xi_n)(\tau+s) - (\dot\psi\oplus\xi)(\tau+s)]d\nu(\tau)\right)ds\right| = 0$ uniformly with respect to $t \in I$. Furthermore, we have $\lim_{n\to\infty}\sup_{\sigma\leqslant t\leqslant\sigma+a}\left|\int_\sigma^t |[\xi_n(\tau) - \xi(\tau)]d\tau\right| = 0$. Indeed, for every $t \in I$ we have $\left|\int_0^t [\xi_n(\tau) -\right.$

$-\xi(\tau)]d\tau| \leqslant |\mathscr{I}\xi_n - \mathscr{I}\xi|_a$ and by Lemma 1.13, $\lim_{n\to\infty} |\mathscr{I}\xi_n - \mathscr{I}\xi|_a = 0$. Thus, there is $\bar{N} \geqslant 1$ such that for every $t \in I$ and $n \geqslant \bar{N}$ one has $\left|\int_{-r}^{0} \left(\int_{\sigma}^{t} [(\dot{\psi}\oplus\xi_n)(\tau+s) - (\dot{\psi}\oplus\xi)(\tau+s)]d\nu(\tau)\right)ds\right| < \frac{1}{2}\varepsilon$ and $\left|\int_{\sigma}^{t} [\xi_n(\tau)-\xi(\tau)]d\tau\right| < \frac{1}{2}\varepsilon$. Put $\bar{\xi} = \xi_{\bar{N}}$ and select for every $\bar{x} \in \boldsymbol{K}_\lambda$, $\bar{u} \in \mathscr{F}(G\Box\mathscr{D})(\bar{x})$ such that $|\bar{u}(t) - \bar{\xi}(t)| = e(\mathscr{F}(G\Box\mathscr{D})(\bar{x}) - \bar{\xi})(t)$ for a.e. $t \in I$. It is posible because of Proposition 2.4. Similarly as in the proof of Lemma 2.16, select for $\bar{x} \in \boldsymbol{K}_\lambda$ and above taken $\bar{u} \in \mathscr{F}(G\Box\mathscr{D})(\bar{x})$ a continuous function $\mathscr{S}_{\bar{x}\bar{u}}\colon \boldsymbol{K}_\lambda \to L(I, \boldsymbol{R}^1)$ such that $\mathscr{S}_{\bar{x}\bar{u}}(\bar{x}) = 0$ and $\mathscr{S}_{\bar{x}\bar{u}}(x)(t) \geqslant e(\mathscr{F}(G\Box\mathscr{D})(x) - \bar{u})(t)$ for $x \in \boldsymbol{K}_\lambda$ and a.e. $t \in I$.

It is possible, because by Lemma 2.8, $\mathscr{F}(G\Box\mathscr{D})$ is l.s.c. on $\boldsymbol{K}_\lambda$. By the uniform equicontinuity of $\{(G\Box\mathscr{D})(t, \cdot)\}_{t\in I}$ on $\boldsymbol{K}_\lambda$, there exists a number $\delta_\varepsilon > 0$ such that for every $x \in B^\circ(\bar{x}, \delta_\varepsilon)\cap \boldsymbol{K}_\lambda$ and $t \in I$ we have $h((G\Box\mathscr{D})(t, \bar{x}), (G\Box\mathscr{D})(t, x)) < \frac{1}{2}\varepsilon a$, where $B^\circ(\bar{x}, \delta_\varepsilon)$ is an open ball of $C^0(I, \boldsymbol{R}^n)$ with the centre at $\bar{x} \in \boldsymbol{K}_\lambda$ and a radius $\delta_\varepsilon > 0$. To taken above $\bar{x}$ and $\bar{u}$, let $V_{\bar{x}\bar{u}} = \{x \in B(\bar{x}, \delta_\varepsilon)\colon \int_I \mathscr{S}_{\bar{x}\bar{u}}(x)(t)dt < \frac{1}{4}\varepsilon\}$ and consider the family $\{V_{\bar{x}\bar{u}}\colon \bar{x} \in \boldsymbol{K}_\lambda, \bar{u} \in \mathscr{F}(G\Box\mathscr{D})(\bar{x})\}$ of open subsets of $C^0(I, \boldsymbol{R}^n)$.

Similarly as in the proof of Lemma 2.16 we can establish a finite partition of unity $p_1(x), \ldots, p_N(x)$ subordinate to the above given open covering of a compact set $\boldsymbol{K}_\lambda$. Let $V_{x_i u_i}$ be such that $p^{-1}(0, 1] \subset V_{x_i u_i}$ for $i = 1, 2, \ldots, N$. Now, similarly as in the proof of Lemma 2.16, we define continuous functions $\mathscr{S}\colon \boldsymbol{K}_\lambda \to L(I, \boldsymbol{R}^1)$ and $g\colon \boldsymbol{K}_\lambda \to L(I, \boldsymbol{R}^n)$ by (2.21) and (2.22), respectively. They are such that (5.6) holds for each $x \in \boldsymbol{K}_\lambda$ and the set $L(x)$ defined by (5.8) is nonempty for each $x \in \boldsymbol{K}_\lambda$. For every $x \in \boldsymbol{K}_\lambda$ and $t \in I$ one has

$$
\begin{aligned}
&\left|\int_{\sigma}^{t} [g(x)(\tau) - \xi(\tau)]d\tau\right| \\
&\leqslant \sum_{i=1}^{A} \int_{\sigma}^{t} |u_i(\tau) - \bar{\xi}(\tau)| \chi_{I_{z_i(x)} I_{z_{i-1}(x)}}(\tau)d\tau + \left|\int_{\sigma}^{t} [\xi(\tau) - \bar{\xi}(\tau)]d\tau\right| \\
&\leqslant \sum_{i=1}^{N} \int_{\sigma}^{t} |e(\mathscr{F}(G\Box\mathscr{D})(x_i) - \xi)(\tau) \cdot \chi_{I_{z_i(x)}\setminus I_{z_{i-1}(x)}}(\tau)|d\tau + \tfrac{1}{2}\varepsilon \\
&\leqslant \sum_{i=1}^{N} \int_{\sigma}^{t} h((G\Box\mathscr{D})(\tau, x_i), (G\Box\mathscr{D})(\tau, \Phi))(\tau) \cdot \chi_{I_{z_i(x)}\setminus I_{z_{i-1}(x)}}(\tau)d\tau + \tfrac{1}{2}\varepsilon \\
&\leqslant \sum_{i=1}^{N} \int_{\sigma}^{t} h((G\Box\mathscr{D})(\tau, x_i), (G\Box\mathscr{D})(\tau, x))(\tau) \cdot \chi_{I_{z_i(x)}\setminus I_{z_{i-1}(x)}}(\tau)d\tau + \\
&+ \int_{\sigma}^{t} \omega(\tau, \boldsymbol{R}_1(\tau, \psi\oplus x, \psi\oplus\Phi))d\tau + \tfrac{1}{2}\varepsilon \\
&\leqslant \tfrac{1}{2}\varepsilon/a \cdot \sum_{i=1}^{N} \int_{\tau}^{t} \chi_{I_{z_i(x)}\setminus I_{z_{i-1}(x)}}(\tau)d\tau + \tfrac{1}{2}\varepsilon + \int_{\sigma}^{t} \omega(\tau, \boldsymbol{R}_1(\tau, \psi\oplus x, \psi\oplus\Phi))d\tau
\end{aligned}
$$

$$= \varepsilon + \int_{\sigma}^{t} \omega\big(\tau, R_1(\tau, \psi\oplus x, \psi\oplus\Phi)\big)d\tau.$$

Similarly, for $x \in K_\lambda$ and $t \in I$ we obtain

$$\Big|\int_{-r}^{0}\Big(\int_{\sigma}^{t}[(\dot\psi\oplus g(x))_\tau-(\dot\psi\oplus\xi)_\tau]d\nu(\tau)\Big)(s)ds\Big|$$

$$\leqslant \Big|\int_{-r}^{0}\Big(\int_{\sigma}^{t}[(\dot\psi\oplus g(x))(\tau+s)-(\dot\psi\oplus\bar\xi)(\tau+s)]d\nu(\tau)\Big)ds\Big|+$$

$$+\Big|\int_{-r}^{0}\Big(\int_{\sigma}^{t}[(\dot\psi\oplus\xi)(\tau+s)-(\dot\psi\oplus\bar\xi)(\tau+s)]d\nu(\tau)\Big)ds\Big|$$

$$\leqslant V_\nu \sup_{\sigma\leqslant\tau\leqslant t}\int_{-r}^{0}|(\dot\psi\oplus\bar\xi)(\tau+s)-(\dot\psi\oplus g(x))(\tau+s)|ds+\tfrac{1}{2}\varepsilon \leqslant \int_{\sigma}^{t}|\bar\xi(\tau)-$$

$$-g(x)(\tau)|d\tau+\tfrac{1}{2}\varepsilon \leqslant \varepsilon+\int_{\sigma}^{t}\omega\big(\tau, R_1(\tau,\psi\oplus x,\psi\oplus\Phi)\big)d\tau.$$

Therefore, (5.7) holds. ■

LEMMA 5.8. *Let $\psi \in AC_{0r}$ be given. Suppose $G \in \mathscr{H}_\lambda(I, R^n)$ satisfies the assumptions of Lemma 5.6 and is such that a family $\{(G\Box\mathscr{D})(t, \cdot)\}_{t\in I}$ is equicontinuous on K . Then for every $\varepsilon > 0$, $\Phi \in K_\lambda$ and $\xi \in \mathscr{F}(\mathrm{co}(G\Box\mathscr{D}))(\Phi)$ there exists a continuous function $g\colon K_\lambda \to L(I, R^n)$ such that $g(x) \in \mathscr{F}(G\Box\mathscr{D})(x)$ for $x \in K_\lambda$ and so that* (5.7) *is satisfied for $t \in I$ and $x \in K_\lambda$.*

Proof. Let $\varepsilon > 0$ be arbitrarily fixed and let (ε_k) be a sequence of positive numbers such that $\sum_{k=1}^{\infty}\varepsilon_k < \frac{1}{2}\varepsilon$ and let $g_1\colon K_\lambda \to L(I, R^n)$ and $\mathscr{S}_1\colon K_\lambda \to L(I, R^n)$ be defined by virtue of Lemma 5.7 corresponding to ε_1, $\Phi \in K_\lambda$, $\xi \in \mathscr{F}(\mathrm{co}(G\Box\mathscr{D}))(\Phi)$ and $\psi \in AC_{0r}$ such that the set $L(x)$ defined by (5.8) is nonempty for every $x \in K_\lambda$ and (5.6) and (5.7) are satisfied. Put now $L_1(x) = \{u \in \mathscr{F}(G\Box\mathscr{D})(x)\colon |u(t)-g_1(x)(t)| < \mathscr{S}_1(x)(t)$ for a.e. $t \in I\}$ for $x \in K_\lambda$. By Lemma 5.7 we have $L_1(x) \neq \emptyset$ for each $x \in K_\lambda$ and therefore by Proposition 2.6, a set-valued function $L_1\colon K_\lambda \ni x \to L_1(x) \subset L(I, R^n)$ is decomposable and l.s.c. Thus, a set-valued function $H_1\colon K_\lambda \to \mathrm{Cl}(L(I, R^n))$ defined by $H_1(x) = \overline{[L_1(x)]_L}$ is decomposable and l.s.c., too. Now, by virtue of Lemma 2.16 there exist continuous functions $g_2\colon K_\lambda \to L(I, R^n)$ and $\mathscr{S}_2\colon K_\lambda \to L(I, R^n)$ such that $\int_I \mathscr{S}_2(x)(t)dt < \varepsilon_1$ and the set $L_2(x) = \{u \in H_1(x)\colon |u(t)-g_2(x)(t)| < \mathscr{S}_2(x)(t)$ for a.e. $t \in I\}$ is nonempty for each $x \in K_\lambda$.

In a similar way we can see that for each $k = 1, 2, \ldots$ there exist continuous functions $g_k\colon K_\lambda \to L(I, R^n)$ and $\mathscr{S}_k\colon K_\lambda \to L(I, R^n)$ such that $\int_I \mathscr{S}_k(x)(t)dt < \varepsilon_k$

and the set $L_k(x) = \{u \in H_{k-1}(x) \colon |u(t)-g_k(x)(t)| < \mathscr{S}_k(x)(t)$ for a.e. $t \in I\}$ is nonempty for every $x \in \boldsymbol{K}_\lambda$, where $H_{k-1}(x) = \overline{[L_{k-1}(x)]_L}$.

Now, similarly as in the proof of Theorem 2.17 we can get $|g_n(x)(t)-g_{n+p}(x)(t)| \leqslant \mathscr{S}_n(x)(t)+\mathscr{S}_{n+p}(x)(t)$ for $n = 1, 2, \ldots$, $p = 0, 1, 2, \ldots$, $x \in \boldsymbol{K}_\lambda$ and a.e. $t \in I$. Hence, in particular it follows that (g_n) converges uniformly in the $L(I, \boldsymbol{R}^n)$-norm to a continuous function $g\colon \boldsymbol{K}_\lambda \to L(I, \boldsymbol{R}^n)$ such that $g(x) \in \mathscr{F}(G\square\mathscr{D})(x)$ for $x \in \boldsymbol{K}_\lambda$. Furthermore, $|g_{k+1}(x)-g_k(x)|_a \leqslant \varepsilon_{k+1}+\varepsilon_k$ for $k = 1, 2, \ldots$ and $x \in \boldsymbol{K}_\lambda$. Therefore, for $x \in \boldsymbol{K}_\lambda$ and $t \in I$ one has

$$\begin{aligned}
&\left|\int_\sigma^t [g(x)(\tau)-\xi(\tau)]d\tau\right| \\
&= \left|\int_\sigma^t \big([g_0(x)(\tau)-\xi(\tau)]+[g(x)(\tau)-g_0(x)(\tau)]\big)d\tau\right| \\
&\leqslant \varepsilon_1+\int_\sigma^t \omega\big(\tau, \boldsymbol{R}_1(\tau, \psi\oplus x, \psi\oplus\Phi)\big)d\tau+\sum_{k=1}^{\infty}|g_{k+1}(x)-g_k(x)|_a \\
&\leqslant \varepsilon_1+\int_\sigma^t \omega\big(\tau, \boldsymbol{R}_1(\tau, \psi\oplus x, \psi\oplus\Phi)\big)d\tau+\sum_{k=1}^{\infty}(\varepsilon_{k+1}+\varepsilon_k) \\
&\leqslant \varepsilon+\int_\sigma^t \omega\big(\tau, \boldsymbol{R}_1(\tau, \psi\oplus x, \psi\oplus\Phi)\big)d\tau.
\end{aligned}$$

Similarly for $t \in I$ and $x \in \boldsymbol{K}_\lambda$ we obtain

$$\begin{aligned}
&\left|\int_{-r}^{0}\Big(\int_\sigma^t [(\dot\psi\oplus g(x))_\tau-(\dot\psi\oplus\xi)_\tau]dv(\tau)\Big)(s)ds\right| \\
&\leqslant \varepsilon+\int_\sigma^t \omega\big(\tau, \boldsymbol{R}_1(\tau, \psi\oplus x, \psi\oplus\Phi)\big)d\tau.
\end{aligned}$$

Thus, (5.7) holds. ■

Now we can prove the following selection theorem.

THEOREM 5.9. *Let $\psi \in \mathrm{AC}_{0r}$ be given and suppose $G \in \mathscr{H}_\lambda(I, \boldsymbol{R}^n)$ is such that $\int_\sigma^{\sigma+a} ||G(t, x, z)||\,dt \leqslant \lambda$ for $(x, z) \in \mathscr{K}_\lambda^{\psi}\times B_\lambda$ and there exist $v \in \mathrm{BV}(I, \boldsymbol{R}^1)$ and a Kamke function $\omega\colon I\times \boldsymbol{R}^+ \to \boldsymbol{R}^+$ such that (5.4) is satisfied for every $x, y \in \boldsymbol{K}_\lambda$ and a.e. $t \in I$. Then for every $\varepsilon > 0$, $\Phi \in \boldsymbol{K}_\lambda$ and $\xi \in \mathscr{F}(\mathrm{co}(G\square\mathscr{D}))(\Phi)$ there exists a continuous function $g\colon \boldsymbol{K}_\lambda \to L(I, \boldsymbol{R}^n)$ such that $g(x) \in \mathscr{F}(G\square\mathscr{D})(x)$ for $x \in \boldsymbol{K}_\lambda$ and so that (5.7) is satisfied for $t \in I$ and $x \in \boldsymbol{K}_\lambda$.*

Proof. Let us observe that $(G\square\mathscr{D})(\cdot, x)$ is measurable and has compact values for a.e. $t \in I$ and fixed $x \in \boldsymbol{K}_\lambda$. Moreover by Lemma 5.6, $(G\square\mathscr{D}(t, \cdot)$ is continuous on $\boldsymbol{K}_\lambda$ for a.e. fixed $t \in I$. Therefore, by Theorem II.3.7 for every $n \in N$ there exists a closed set $E_n \subset I$ with $\mu(I \setminus E) < (\frac{1}{2})^n$ and such that $(G\Box\mathscr{D})$ restricted to each $E_n \times \boldsymbol{K}_\lambda$ for $n = 1, 2, \dots$ is uniformly continuous. Put $F_1 = E_1$ and $F_n = E_n \setminus \bigcup_{k=1}^{n-1} E_k$ for $n = 2, 3, \dots$ We have $\bigcup_{n=1}^{\infty} E_n = \bigcup_{n=1}^{\infty} F_n$, $F_n \cap F_m = \emptyset$ for $n \neq m$ and $\mu(I \setminus \bigcup_{n=1}^{\infty} F_n) = 0$. Define now for every $n = 1, 2, \dots$ mappings G_n: $I \times \mathscr{K}_\lambda \times B_\lambda \to \mathrm{Cl}(\boldsymbol{R}^n)$, ξ_n: $I \to \boldsymbol{R}^n$ and ω_n: $I \times \boldsymbol{R}^+ \to \boldsymbol{R}^+$ by setting $G_n := \chi_{F_n} G$, $\xi_n := \chi_{F_n} \xi$ and $\omega_n := \chi_{F_n} \omega$. It is easy to see that $\xi_n \in \mathscr{F}(\mathrm{co}(G_n\Box\mathscr{D}))(\Phi)$ and $h((G_n\Box\mathscr{D})(t, x), (G_n\Box\mathscr{D})(t, y)) \leqslant \omega_n(t, \boldsymbol{R}_1(t, \psi\oplus x, \psi\oplus y)$ for $x, y \in \boldsymbol{K}_\lambda$, a.e. $t \in I$ and $n = 1, 2, \dots$ Furthermore, for every $n = 1, 2, \dots$ the family $\{(G_n\Box\mathscr{D})(t, \cdot)\}_{t\in I}$ is uniformly equicontinuous on $\boldsymbol{K}_\lambda$. Therefore, for every $\Phi \in \boldsymbol{K}_\lambda$, $\xi_n \in \mathscr{F}(\mathrm{co}(G_n\Box\mathscr{D}))(\Phi)$, $n = 1, 2, \dots$ and $\varepsilon_n > 0$ there exists a continuous function g_n: $\boldsymbol{K}_\lambda \to L(I, \boldsymbol{R}^n)$ such that $g_n(x) \in \mathscr{F}(G_n\Box\mathscr{D})(x)$ for $x \in \boldsymbol{K}_\lambda$ and such that

$$\max\left\{\left|\int_\sigma^t [g_n(x)(\tau) - \xi_n(\tau)]d\tau\right|, \left|\int_{-r}^0 \left(\int_\sigma^t [(\dot{\psi}\oplus g_n(x))_\tau - (\dot{\psi}\oplus\xi_n)_\tau] d\nu(\tau)\right)(s)ds\right|\right\}$$

$$\leqslant \varepsilon_n + \int_\sigma^t \omega_n\big(\tau, \boldsymbol{R}_1(\tau, \psi\oplus x, \psi\oplus\Phi)\big)d\tau$$

for $x \in \boldsymbol{K}_\lambda$, $t \in I$ and $n = 1, 2, \dots$ Let g: $\boldsymbol{K}_\lambda \to L(I, \boldsymbol{R}^n)$ be defined by $g(x) := \chi_{\bigcup_{n=1}^{\infty} F_n} \cdot g_n(x)$ for $x \in \boldsymbol{K}_\lambda$. Let us observe that for every positive integer $N \geqslant 1$ we have $E_N = \bigcup_{i=0}^{N} F_i$. Furthermore, by the uniform integrability of G for every $\eta > 0$ there is $N_\eta \geqslant 1$ such that $\int_{I\setminus E_{N_\eta}} \|G(t, x, z)\|dt \leqslant \frac{1}{2}\eta$ for $(x, z) \in \mathscr{K}_\lambda \times B_\lambda$. Hence it follows that the function g: $\boldsymbol{K}_\lambda \to L(I, \boldsymbol{R}^n)$ defined above is continuous on $\boldsymbol{K}_\lambda$. Indeed, let $x \in \boldsymbol{K}_\lambda$ be arbitrarily taken and suppose (x_k) is a sequence of $\boldsymbol{K}_\lambda$ converging to x. For every $\eta > 0$ and $k = 1, 2, \dots$ we have

$$\int_\sigma^{\sigma+a} |g(x_k)(t) - g(x)(t)|dt$$

$$= \int_{E_{N_\eta}} |g(x_k)(t) - g(x)(t)|dt + 2 \sup_{(x,z)\in\mathscr{K}_\lambda\times B_\lambda} \int_{I\setminus E_{N_\eta}} \|G(t, x, z)\|dt$$

$$\leqslant \sum_{i=1}^{N_\eta} \int_{F_i} |g_i(x_k)(t) - g_i(x)(t)|dt + \eta.$$

Then $\lim_{k\to\infty} |g(x_k) - g(x)|_a \leqslant \eta$ for every $\eta > 0$, i.e., $\lim_{k\to\infty} |g(x_k) - g(x)|_a = 0$.

Let $\varepsilon > 0$, $\Phi \in \boldsymbol{K}_\lambda$ and $\xi \in \mathscr{F}(\mathrm{co}(G\Box\mathscr{D}))(\Phi)$ be given and suppose (ε_n) is a sequence of positive numbers such that $\sum_{n=1}^{\infty} \varepsilon_n \leqslant \varepsilon$. Let g: $\boldsymbol{K}_\lambda \to L(I, \boldsymbol{R}^n)$ be a continuous function defined above to given $\Phi \in \boldsymbol{K}_\lambda$, $\xi \in \mathscr{F}(\mathrm{co}(G\Box\mathscr{D}))(\Phi)$ and a sequence (ε_n). We have of course $g(x) \in \mathscr{F}(G\Box\mathscr{D})(x)$ for $x \in \boldsymbol{K}_\lambda$. Furthermore, for $x, y \in \boldsymbol{K}_\lambda$

and $t \in I$ one has

$$\max\Big\{\Big|\int_\sigma^t [g(x)(\tau)-\xi(\tau)]d\tau\Big| \int_{-r}^0 \Big(\int_\sigma^t [(\dot\psi \oplus g(x))_\tau - (\dot\psi\oplus\xi_\tau)]dv(\tau)\Big)(s)ds\Big|\Big\}$$

$$\leqslant \sum_{n=1}^\infty \max\Big\{\Big|\int_\xi^t [g_n(x)(\tau)-\xi_n(\tau)]d\tau\Big|, \Big|\int_{-r}^0 \Big(\int_\sigma^t [(\dot\psi\oplus g_n(x))_\tau - (\dot\psi\oplus\xi_n)_\tau]dv(\tau)\Big)(s)ds\Big|\Big\}$$

$$\leqslant \sum_{n=1}^\infty \varepsilon_n + \sum_{n=1}^\infty \int_\sigma^t \omega_n\big(\tau, R_1(\tau, \psi\oplus x, \psi\oplus\Phi)\big)d\tau \leqslant \varepsilon$$

$$+\sum_{n=1}^\infty \int_{[\sigma,t]\cap F_n} \omega\big(\tau, R_1(\tau, \psi\oplus x, \psi\oplus\Phi)\big)d\tau$$

$$= \varepsilon + \int_\sigma^t \omega\big(\tau, R_1(\tau, \psi\oplus x, \psi\oplus\Phi)\big)d\tau. \quad \blacksquare$$

Let X be a nonempty compact subset of the Banach space $C(I, C_{0r})$ such that $\{u(t)\colon u\in X\} \subset AC_{0r}$.

In a similar way we obtain the following theorem.

THEOREM 5.10. *Let $\mathscr{G}\colon I\times X \to \mathrm{Comp}(R^n)$ be such that $\mathscr{G}(\cdot, x)$ is measurable on I and such that there exists a continuous operator $L\colon I \to \mathscr{L}(L_{0r}, R^n)$ and a Kamke function $\omega\colon I\times R^+ \to R^+$ such that $\sup_{t\in I}\{\|L(t)\|_{\mathscr{L}} h(\mathscr{G}(t,u), \mathscr{G}(t,v)) \leqslant \omega(t, R(t,u,v))$ for a.e. $t\in I$ and $u, v\in X$, where $R(t,u,v) := \max\{\|L(t)\|_{\mathscr{L}} |u(t)-v(t)|_0, \sup_{\sigma\leqslant\tau\leqslant t} |L(\tau)[\dot u(\tau)-\dot v(\tau)]|\}$. Then for every $\psi\in L_{0r}$, $\varepsilon>0$, $\Phi\in X$ and $\xi\in\mathscr{F}(\mathrm{co}\,\mathscr{G})(\Phi)$ there exists a continuous function $g\colon X\to L(I, R^n)$ such that $g(u)\in\mathscr{F}(\mathscr{G})(u)$ for $u\in X$ and such that for $t\in I$ and $u\in X$ one has*

$$\max\Big\{\|L(t)\|_{\mathscr{L}} \sup_{\sigma\leqslant s\leqslant t}\Big|\int_\sigma^t [g(u)(\tau)-\xi(\tau)]d\tau\Big|, L(t)[(\psi\oplus g(u))_t - (\psi\oplus\xi)_t]\Big|\Big\}$$

$$\leqslant \varepsilon + \int_\sigma^t \omega\big(s, R(s,u,\Phi)\big)ds. \tag{5.9}$$

Proof. Without loss of the generality we may assume that $|L|_a := \sup_{\sigma\leqslant t\leqslant\sigma+a} \|L(t)\|_{\mathscr{L}} \leqslant 1$. Now we can repeat the proofs of Lemmas 5.7, 5.8 and Theorem 5.9. Here we only verify that the function $g_1\colon X\to L(I, R^n)$ defined similarly as in the proof of Lemma 5.7 for any $\varepsilon_1>0$, by the additional condition dealing with the family $\{\mathscr{G}(t,\cdot)\}_{t\in I}$, satisfies (5.9). The rest of the proof is the same as above.

For every $x\in X$ and $t\in I$ we have

$$\|L(t)\|_{\mathscr{L}} \sup_{\sigma\leqslant s\leqslant t}\Big|\int_\sigma^s [g_1(x)(\tau)-\xi(\tau)]d\tau\Big|$$

$$\leqslant |L|_a \sup_{\sigma \leqslant t \leqslant t} \left| \int_\sigma^s |[\xi(\tau) - \bar{\xi}(\tau)] d\tau \right| +$$

$$+ \sup_{\sigma \leqslant s \leqslant t} \left(\sum_{i=1}^N |L|_a \int_{a_\sigma}^s |u_i(\tau) - \bar{\xi}(\tau)| \cdot \chi_{I_{z_i(x)} \setminus I_{z_{i-1}(x)}}(\tau) d\tau \right)$$

$$\leqslant \tfrac{1}{2}\varepsilon + \sup_{\sigma \leqslant s \leqslant t} \left(\sum_{i=1}^N \int_\sigma^s |L|_a e(\mathscr{F}(\mathscr{G})(x_i) - \bar{\xi})(\tau) \cdot \chi_{I_{z_i(x)} \setminus I_{z_{i-1}(x)}}(\tau) d\tau \right)$$

$$\leqslant \varepsilon + \int_\sigma^t \omega\big(s, R(s, x, \Phi)\big) ds$$

and

$$|L(t)\,[(\psi \oplus g_1(x))_t - (\psi \oplus \xi)_t]|$$

$$\leqslant ||L(t)||_{\mathscr{L}} |(\psi \oplus g_1(x))_t - (\psi \oplus \xi)_t|_0 = ||L(t)||_{\mathscr{L}} \left| \int_{t-r}^t [(\psi \oplus g_1(x))(\tau) \right.$$

$$\left. - (\psi \oplus \xi)(\tau)] dt \right| \leqslant ||L(t)||_{\mathscr{L}} \sup_{\sigma \leqslant s \leqslant t} \left| \int_\sigma^t |[g_1(x)(\tau) - \xi(\tau)] d\tau \right|$$

$$\leqslant \varepsilon + \int_\sigma^t \omega\big(s, R(s, x, \Phi)\big) ds.$$

Hence (5.9) follows. ∎

In a similar way we obtain

THEOREM 5.11. *Let $\psi \in \mathrm{AC}_{0r}$ be given and suppose $G \in \mathscr{H}_\lambda(I, \boldsymbol{R}^n)$ is such that $\int_\sigma^{\sigma+a} ||G(t, x, z)|| dt \leqslant \lambda$ for $(x, z) \in \mathscr{K}_\lambda \times B_\lambda$ and there exist a continuous operator $L\colon I \to \mathscr{L}(L_{0r}, \boldsymbol{R}^n)$ and a Kamke function $\omega\colon I \times \boldsymbol{R}^+ \to \boldsymbol{R}^+$ such that*

$$\sup_{\sigma \leqslant t \leqslant \sigma + a} \big(||L(t)||_{\mathscr{L}}\big) h\big((G \square \mathscr{D})(t, x), (G \square \mathscr{D})(t, y)\big) \leqslant \omega\big(t, \boldsymbol{R}_2(t, \psi \oplus x, \psi \oplus \Phi)\big) \tag{5.10}$$

for $x, y \in \boldsymbol{K}_\lambda$ and a.e. $t \in I$, where $\boldsymbol{R}_2(t, \psi \oplus x, \psi \oplus \Phi)$ is defined by (5.3). Then for every $\varepsilon > 0$, $\Phi \in \boldsymbol{K}_\lambda$ and $\xi \in \mathscr{F}(\mathrm{co}(G \square \mathscr{D}))(\Phi)$ there exists a continuous function $g\colon \boldsymbol{K}_\lambda \to L(I, \boldsymbol{R}^n)$ such that $g(x) \in \mathscr{F}(G \square \mathscr{D})(x)$ for $x \in \boldsymbol{K}$ and

$$\max\left\{ ||L(t)||_{\mathscr{L}} \sup_{\sigma \leqslant s \leqslant t} \left| \int_\sigma^t [g(x)(\tau) - \xi(\tau)] d\tau \right|, \right. \tag{5.11}$$

$$\left. |L(t)[(\dot{\psi} \oplus g(x))_t - (\dot{\psi} \oplus \xi)_t]| \right\} \leqslant \varepsilon + \int_\sigma^t \omega\big(s, \boldsymbol{R}_2(s, \psi \oplus x, \psi \oplus \Phi)\big) ds$$

for $t \in I$, $x \in \boldsymbol{K}_\lambda$. ∎

5.3. Relaxation theorems

We shall present here some relaxation theorems for the trajectory integrals of set-valued functions G satisfying some special conditions mentioned in Theorems 5.9 and 5.11.

THEOREM 5.12. *Let* $G \in \mathcal{H}_\lambda(I, \boldsymbol{R}^n)$ *be such that* $\int_\sigma^{\sigma+a} ||G(t, x, z)|| dt \leqslant \lambda$ *for* $(x, z) \in \mathcal{K}_\lambda \times B_\lambda$ *and suppose there are* $\mathcal{S} \in \mathrm{AC}_{0r}$, $\nu \in \mathrm{BV}(I, \boldsymbol{R}^n)$ *and a Kamke function* ω: $I \times \boldsymbol{R}^+ \to \boldsymbol{R}^+$ *such that* (5.4) *is satisfied for each* $x, y \in \boldsymbol{K}_\lambda$ *and a.e.* $t \in I$. *Then*

(i) $\mathcal{I}S(\operatorname{co} G) = \overline{[\mathcal{I}S(G)]_C}$ *and*

(ii) $\mathcal{I}S(\operatorname{co} G) = \overline{[\mathcal{I}S(G)]^w_{AC}}$, *where* $S(\operatorname{co} G)$ *and* $S(G)$ *denote sets of all fixed points of* $\mathcal{F}(\operatorname{co}(G \Box \mathcal{I}))$ *and* $\mathcal{F}(G \Box \mathcal{I})$, *respectively.*

Proof. Let us observe that for the proof it sufficies only to show that $\mathcal{I}\overline{[S(G)]^w_L} = \overline{[\mathcal{I}S(G)]_C} = \overline{[\mathcal{I}S(G)]^w_{AC}}$ and that $S(\operatorname{co} G) = \overline{[S(G)]^w_L}$. To see that $\mathcal{I}\overline{[S(G)]^w_L} = \overline{[\mathcal{I}S(G)]_C} = \overline{[\mathcal{I}S(G)]^w_{AC}}$, let us observe that $\mathcal{I}S(G) \subset \mathcal{I}\overline{[S(G)]^w_L}$. Hence, by the compactness of $\mathcal{I}\overline{[S(G)]^w_L}$ in $C^0(I, \boldsymbol{R}^n)$ and its weak compactness in $\mathrm{AC}^0(I, \boldsymbol{R}^n)$ (see Theorems 2.1 and 2.2) we obtain $\overline{[\mathcal{I}S(G)]_C} \subset \mathcal{I}\overline{[S(G)]^w_L}$ and $\overline{[\mathcal{I}S(G)]^w_{AC}} \subset \mathcal{I}\overline{[S(G)]^w_L}$. We shall prove that $\mathcal{I}\overline{[S(G)]^w_L} \subset \overline{[\mathcal{I}S(G)]_C}$ and $\mathcal{I}\overline{[S(G)]^w_L} \subset \overline{[\mathcal{I}S(G)]^w_{AC}}$. Indeed, suppose $x \in \mathcal{I}\overline{[S(G)]^w_L}$. Then $\dot{x} \in \overline{[S(G)]^w_L} \subset \boldsymbol{P}_\lambda$. Since $\boldsymbol{P}_\lambda$ is weakly compact then $S(G)$ is relatively weakly compact in $L(I, \boldsymbol{R}^n)$. Thus, by Šmulian's theorem, there exists a sequence (z_n) of $S(G)$ weakly convergent to $\dot{x}$. By Lemma 1.13, we have $|\mathcal{I}z_n - x|_a \to 0$ as $n \to \infty$. Since $\mathcal{I}z_n \in \mathcal{I}S(G)$, then $x \in \overline{[\mathcal{I}S(G)]_C}$, i.e. $\mathcal{I}\overline{[S(G)]^w_L} \subset \overline{[\mathcal{I}S(G)]_C}$. On the other hand the weak convergence of (z_n) to $\dot{x}$ implies that (z_n) is uniformly integrable and therefore, the sequence $(\mathcal{I}z_n)$ is equiabsolutely continuous on I. Furthermore, $(\mathcal{I}z_n)(t) \to x(t)$ for $t \in I$ as $n \to \infty$. Therefore, $(\mathcal{I}z_n)$ converges weakly in $\mathrm{AC}^0(I, \boldsymbol{R}^n)$ to x. Then $\mathcal{I}\overline{[S(G)]^w_L} \subset \operatorname{Seq} \operatorname{cl}_w(\mathcal{I}S(G)) \subset \overline{[\mathcal{I}S(G)]^w_{AC}}$.

We shall prove now that $S(\operatorname{co} G) = \overline{[S(G)]^w_L}$. Observe first that $S(G) \subset S(\operatorname{co} G)$ and that $S(\operatorname{co} G)$ is relatively weakly compact. Furthermore, $S(\operatorname{co} G)$ is a weakly sequentially closed subset of $L(I, \boldsymbol{R}^n)$ and therefore by Eberlein–Šmulian theorem also weakly closed. Indeed, let (z_n) be a sequence of $S(\operatorname{co} G)$ convergent weakly to any $z \in L(I, \boldsymbol{R}^n)$. There is a sequence (x_n) of $\boldsymbol{K}_\lambda$ such that $\dot{x}_n = z_n$. Furthermore by Lemma 1.13, $|x_n - \mathcal{I}z|_a \to 0$ as $n \to \infty$. Put $x = \mathcal{I}z$. Since $(G \Box \mathcal{D})(t, \cdot)$ is continuous on $\boldsymbol{K}_\lambda$ than $h((G \Box \mathcal{D})(t, x_n), (G \Box \mathcal{D})(t, x)) \to 0$ for fixed $t \in I$ as $n \to \infty$. Moreover, the family $\{(G \Box \mathcal{D})(\cdot, x)\}_{x \in \boldsymbol{K}_\lambda}$ is uniformly integrable. Then by Vitali's theorem for every $p \in \boldsymbol{R}^n$ and every measurable set $U \subset I$ the sequence $\{\int_U s(p, \operatorname{co}(G \Box \mathcal{D})(t, x_n) dt\}$ converges to $\int_U s(p, \operatorname{co}(G \Box \mathcal{D})(t, x)) dt$ as $n \to \infty$. Thus, by the definition, the sequence $\{\operatorname{co}(G \Box \mathcal{D})(\cdot, x_n)\}$ of $\mathcal{A}(I, \boldsymbol{R}^n)$ converges weakly to $\operatorname{co}(G \Box \mathcal{D})(\cdot, x)$. Since $\dot{x}_n$ is for each $n = 1, 2, \ldots$ a selector of $\operatorname{co}(G \Box \mathcal{D})(\cdot, x_n)$ and $\dot{x}_n \rightharpoonup \dot{x}$ as $n \to \infty$, then by Theorem 1.10 it follows that $\dot{x}(t) \in \operatorname{co}(G \Box \mathcal{D})(t, x)$ for a.e. $t \in I$, i.e. $z \in S(\operatorname{co} G)$. Now, we obtain $\overline{[S(G)]^w_L} \subset S(\operatorname{co} G)$. To see that $S(\operatorname{co} G)$

$\subset \overline{[S(G)]^{\mathrm{w}}_{L}}$ suppose $z \in S(\mathrm{co}\, G)$, i.e., there exists $\Phi \in \boldsymbol{K}_\lambda$ such that $\dot{\Phi} \in \mathscr{F}(\mathrm{co}(G\Box\mathscr{D})(\Phi))$ and $z = \dot{\Phi}$. By Theorem 5.9, for every $\varepsilon > 0$ there exists a continuous function $g_\varepsilon\colon \boldsymbol{K}_\lambda \to L(I, \boldsymbol{R}^n)$ such that $g_\varepsilon(x)(t) \in (G\Box\mathscr{D})(t, x)$ for $x \in \boldsymbol{K}_\lambda$ and a.e. $t \in I$, and such that the following inequality is satisfied:

$$\max\left\{\left|\int_{\sigma}^{t} [g_\varepsilon(x)(\tau) - \dot{\Phi}(\tau)]d\tau\right|, \left|\int_{-r}^{0}\left(\int_{0}^{t}[(\dot{\mathscr{S}} \oplus g_\varepsilon(x))_\tau - (\dot{\mathscr{S}} \oplus \dot{\Phi})_\tau]d\nu(\tau)\right)ds\right|\right\}$$

$$\leqslant \varepsilon + \int_{\sigma}^{t} \omega\big(s, \boldsymbol{R}_1(s, \mathscr{S}\oplus x, \mathscr{S}\oplus\Phi)\big)ds \tag{5.12}$$

for $t \in I$, $x \in \boldsymbol{K}_\lambda$ and $\varepsilon > 0$.

Similarly as in the proof of Theorem 3.6, we can see that for every $n \geqslant 1$ there exists $x_n \in \boldsymbol{K}_\lambda$ such that $\dot{x}_n = g_{1/n}(x_n)$. By (5.12) for $\tau \in [\sigma, t]$, $t \in I$ and $n \geqslant 1$ we have

$$|x_n(\tau) - \Phi(\tau)| \leqslant \frac{1}{n} + \int_{\sigma}^{\tau} \omega\big(s, \boldsymbol{R}_1(s, \mathscr{S}\oplus x, \mathscr{S}\oplus\Phi)\big)ds$$

and

$$\left|\int_{-r}^{0}\left(\int_{0}^{p} [(\mathscr{S}\oplus x_n)_\tau - \mathscr{S}\oplus\Phi_\tau]d\nu(\tau)\right)ds\right| \leqslant \frac{1}{n} + \int_{\sigma}^{p} \omega\big(s, \boldsymbol{R}_1(s, \mathscr{S}\oplus x_n, \mathscr{S}\oplus\Phi)\big)ds.$$

Hence it follows

$$\boldsymbol{R}_1(t, \mathscr{S}\oplus x_n, \mathscr{S}\oplus\Phi) \leqslant \frac{1}{n} + \int_{\sigma}^{t} \omega\big(s, \boldsymbol{R}_1(t, \mathscr{S}\oplus x_n, \mathscr{S}\oplus\Phi)\big)ds$$

for $t \in I$ and $n \geqslant 1$. Thus, by the properties of the Kamke function ω, we have $\boldsymbol{R}_1(t, x_n^{\mathscr{S}}, \Phi^{\mathscr{S}}) \leqslant 1/n$ for $t \in I$ and $n \geqslant 1$. In particular, for $n \geqslant 1$ one obtains $|x_n - \Phi|_a \leqslant 1/n$. Therefore $|x_n - \Phi|_a \to a$ as $n \to \infty$. This also implies $\dot{x}_n \rightharpoonup \dot{\Phi} = z$ as $n \to \infty$. Since $g_{1/n}(x_n) \in \mathscr{F}(G\Box\mathscr{D})(x_n)$ for each $n = 1, 2, \ldots$ then for every $n \geqslant 1$ one has $\dot{x}_n \in S(G)$. Thus $z \in \mathrm{Seq\,cl_w}(S(G)) \subset \overline{[S(G)]^{\mathrm{w}}_{L}}$. Then, $S(\mathrm{co}\, G) \subset \overline{[S(G)]^{\mathrm{w}}_{L}}$. Finally, one obtains $\mathscr{I}S(\mathrm{co}\, G) = \mathscr{I}\overline{[S(G)]^{\mathrm{w}}_{L}} = \overline{[\mathscr{I}S(G)]_{C}} = \overline{[\mathscr{I}S(G)]^{\mathrm{w}}_{AC}}$. ∎

THEOREM 5.13. *Let $G \in \mathscr{H}_\lambda(I, \boldsymbol{R}^n)$ be such that $\int_{\sigma}^{\sigma+a} \|G(t, x, z)\|dt \leqslant \lambda$ for $(x, z) \in \mathscr{K}_\lambda \times B_\lambda$ and suppose there exist $\mathscr{S} \in \mathrm{AC}_{0r}$, a continuous operator $L\colon I \to \mathscr{L}(L_0, \boldsymbol{R}^n)$ and a Kamke function $\omega\colon I \times \boldsymbol{R}^+ \to \boldsymbol{R}^+$ such that* (5.10) *is satisfied for each $x, y \in \boldsymbol{K}_\lambda$ and $t \in I$. Then*

(i) *$\mathscr{I}S(\mathrm{co}\, G) = \overline{[\mathscr{I}S(G)]_{C}}$ and*

(ii) *$\mathscr{I}S(\mathrm{co}\, G) = \overline{[\mathscr{I}S(G)]^{\mathrm{w}}_{AC}}$, where $S(\mathrm{co}\, G)$ and $S(G)$ denote sets of all fixed points of $\mathscr{F}(\mathrm{co}(G\Box\mathscr{I}))$ and $\mathscr{F}(G\Box\mathscr{I})$, respectively.*

Proof. Similarly as in the proof of Theorem 5.12, we obtain $\mathscr{I}\overline{[S(G)]_L^w} = \overline{[\mathscr{I}S(G)]_C} = \overline{[\mathscr{I}S(G)]_{AC}^w}$ and $\overline{[S(G)]_L^w} \subset S(\mathrm{co}\, G)$. To see that $S(\mathrm{co}\, G) \subset \overline{[S(G)]_L^w}$ suppose, $z \in S(\mathrm{co}\, G)$ and let $\Phi \in \boldsymbol{K}_\lambda$ be such that $\dot{\Phi} \in \mathscr{F}(\mathrm{co}(G \square \mathscr{D}))(\Phi)$ and $z = \dot{\Phi}$. Similarly as in the proof of Theorem 5.13 we can find a subset $\{x_\varepsilon\}_{\varepsilon>0}$ of $\boldsymbol{K}_\lambda$ such that $\dot{x}_\varepsilon \in S(G)$ and $\boldsymbol{R}_2(t, \mathscr{S} \oplus x_\varepsilon, \mathscr{S} \oplus \Phi) \leqslant \varepsilon + \int_\sigma^t \omega(s, \boldsymbol{R}_2(s, \mathscr{S} \oplus x_\varepsilon, \mathscr{S} \oplus \Phi))ds$ for $t \in I$ and $\varepsilon > 0$. Hence, in particular it follows the existence of a sequence (x_k) of $\boldsymbol{K}_\lambda$ such that $\dot{x}_k \in S(G)$ and $\|L(t)\|_{\mathscr{L}} |x_k - \Phi|_t \leqslant 1/k$ for each $k = 1, 2, \ldots$ and $t \in I$. Thus, $|x_k - \Phi|_a \to 0$ as $k \to \infty$ and therefore $\Phi \in \overline{[S(G)]_L^w}$. Then $S(\mathrm{co}\, G) \subset \overline{[S(G)]_L^w}$. ■

6. Viability theorems

Given $G \in \mathscr{H}_\lambda(I, \boldsymbol{R}^n)$ and a nonempty set $\Lambda \in L(I, \boldsymbol{R}^n)$; a point $u \in S(G)$, i.e., a fixed point of $\mathscr{F}(G \square \mathscr{I})$ is said to be Λ-*viable* if $u \in \Lambda$. We give here some sufficient conditions for the existence of Λ-viable fixed points of $\mathscr{F}(G \square \mathscr{I})$.

Given $\Phi \in \mathrm{AC}_{0r}$ and a nonempty set $H \subset \boldsymbol{R}^n$ such that $\Phi(0) \in H$, a fixed point u of $\mathscr{F}(G \square \mathscr{I})$ is said to be a *fixed point of* $\mathscr{F}(G \square \mathscr{I})$ with (Φ, H)-viable trajectory if $(\mathscr{I}u)(t) \in H - \Phi(0)$ for every $t \in I$.

We also give here some sufficient conditions for the existence of Λ-vaible fixed points of $\mathscr{F}(G \square \mathscr{I})$ with (Φ, H)-viable trajectory.

6.1. *Bouligand's contingent cone*

Let H be a nonempty subset of $\boldsymbol{R}^n$. We shall define the *Bouligand's contingent cone* $T_H(x)$ to H at $x \in H$ by setting

$$T_H(x) = \bigcap_{\varepsilon>0} \bigcap_{\alpha>0} \bigcup_{0<h<\alpha} \left[\frac{1}{h}(H - x) + \varepsilon B\right], \tag{6.1}$$

where B is a closed unit ball of $\boldsymbol{R}^n$.

Immediately from the above definition it follows that for a given $x \in H$, $v \in T_H(x)$ if and only if for every $\varepsilon > 0$ and every $\alpha > 0$ there are $u \in v + \varepsilon B$ and $h \in (0, \alpha)$ such that $x + hu \in H$.

We characterize the contingent cone $T_H(x)$ by using the distant function of a given point of $\boldsymbol{R}^n$ to the set H. We shall denote it by $d_H(\cdot)$, i.e., $d_H(z) = \mathrm{dist}(z, H)$ for $z \in \boldsymbol{R}^n$.

LEMMA 6.1. *Given* $x \in H$, $v \in T_H(x)$ *if and only if*

$$\liminf_{h \to 0^+} \frac{1}{h} d_H(x + hv) = 0. \tag{6.2}$$

Proof. Let $x \in H$ be given and suppose $v \in T_H(x)$. By the definition of $T_H(x)$, for every $\varepsilon > 0$ and every $\alpha > 0$ there are $h \in (0, \alpha)$ and $u \in v + \varepsilon B$ such that $x + hu \in H$. Hence,

$$\frac{1}{h} d_H(x+hv) = \frac{1}{h} \operatorname{dist}(x+hv, H) \leqslant \frac{1}{h} |(x+hv)-(x+hu)| +$$

$$+\frac{1}{h} \operatorname{dist}(x+hu, H) = |v-u| \leqslant \varepsilon.$$

Therefore, for every $\varepsilon > 0$ we have $0 \leqslant \sup_{a>0} \inf_{h \leqslant \alpha} \frac{1}{h} d(x+hv) \leqslant \varepsilon$, i.e. $\liminf_{h \to 0^+} \frac{1}{h} d_H(x+hv) = 0$. Conversely, if $\liminf_{h \to 0^+} d_H(x+hv) = 0$, then for every $\varepsilon > 0$ and every $\alpha > 0$ there exists $h \in (0, \alpha)$ such that $d_H(x+hv)/h \leqslant \frac{1}{2}\varepsilon$. Therefore, there exists $y \in H$ such that

$$\frac{1}{h} |y-(x+hv)| \leqslant \frac{1}{h} d_H(x+hv) + \frac{\varepsilon}{2} \leqslant \varepsilon.$$

Hence, it follows that for every $\varepsilon > 0$ and every $\alpha > 0$ there exist $h \in (0, \alpha)$ and $u := (y-x)/h \in v+\varepsilon B$ such that $x+hu \in H$. Therefore, by the definition of $T_H(x)$ we have $v \in T_H(x)$. ■

Finally, let us observe that for every $v \in T_H(\tilde{x})$ with $\tilde{x} \in H$ a function $\boldsymbol{R}^n \ni x \to d_H(x+hv) \in \boldsymbol{R}^+$ is continuous on $\boldsymbol{R}^n$ for every $h > 0$.

6.2. *Viability theorem for subtrajectory integrals of set-valued functions with convex values*

Let $I := [\sigma, \sigma+a]$, $G \in \mathscr{H}_\lambda(I, \boldsymbol{R}^n)$, $\Lambda \subset (I, \boldsymbol{R}^n)$, $H \subset \boldsymbol{R}^n$ and $\Phi \in \mathrm{AC}_{0r}$ be given such that $\Phi(0) \in H$ and G has Volterra's property with respect to its last two variables, i.e., for every (u, v), $(\tilde{u}, \tilde{v}) \in \mathscr{K}_\lambda \times B_\lambda$ and $t \in I$ such that $u(\tau) = \tilde{u}(\tau)$ for $\tau \in [\sigma, t]$ and $v(\tau) = \tilde{v}(\tau)$ for a.e. $\tau \in [\sigma, t]$ we have $G(t, u, v) = G(t, \tilde{u}, \tilde{v})$.

For a given $\alpha > 0$ we define $\Lambda_\alpha \subset L(I, \boldsymbol{R}^n)$ by $\Lambda_\alpha = \{u \in L(I, \boldsymbol{R}^n)\colon |u(t)| \leqslant \alpha$ for a.e. $t \in I\}$. We shall prove the following theorem.

THEOREM 6.2. *Suppose $G \in \mathscr{H}_\lambda(I, \boldsymbol{R}^n)$ has convex values, is u.s.c.w. on $I \times \mathscr{K}_\lambda \times B_\lambda$ and has Volterra's property with respect to its last two variables. Let $\Lambda \subset L(I, \boldsymbol{R}^n)$, $\Phi \in \mathrm{AC}_{0r}$ and a compact set $H \subset \boldsymbol{R}^n$ be given. Suppose $\Phi(0) \in H$. If there exist numbers $l > 0$, $r > 0$ and $M > 0$ such that $l+M \leqslant \lambda/a$, $\Lambda_\alpha \subset \Lambda$, with $\alpha = l+M$, $(\Phi \oplus \mathscr{I}u)_t(0) \in H$, $G(t, \mathscr{I}u, u) \subset MB$ and $G(t, \mathscr{I}u, u) \cap T_H[(\Phi \oplus \mathscr{I}u)_t(0)] \neq \emptyset$ for $t \in I$ and $u \in \Lambda_\alpha$, then there is $u \in S(G) \cap \Lambda$ such that $(\mathscr{I}u)(t) \in H - \Phi(0)$ for $t \in I$.*

Proof. Let N be a positive integer such that $1/k \leqslant l$ for every $k \geqslant N_l$. Put $K_\alpha := \mathscr{I}\Lambda_\alpha$. Let us observe that for every $k \geqslant N_l$ the tangential condition $(G \square \mathscr{D})(t, x) \cap T_H[(\Phi \oplus x)_t(0)] \neq \emptyset$ for $(t, x) \in I \times K_n$, implies, by Lemma 6.1 that for every $\tau \in I$ and $y \in K_\alpha$ there are $v_y^\tau \in (G \square \mathscr{D})(\tau, y)$, $h_y^\tau \in (0, 1/k)$ such that $d_H[(\Phi \oplus y)_\tau(0) + h_y^\tau v_y^\tau] \leqslant h_y^\tau/3k$. We introduce the subsets $N(\tau, y) = \{x \in \boldsymbol{R}^n\colon d_H(x+h_y^\tau v_y^\tau) < h_y^\tau/2k\}$. Since a function $\boldsymbol{R}^n \ni x \to d_H(x+h_y^\tau v_y^\tau) \in \boldsymbol{R}^+$ is continuous, then $N(\tau, y)$ is for every $(\tau, y) \in I \times K_\alpha$ an open subset of $\boldsymbol{R}^n$. Thus, for every $(\tau, y) \in I \times K_\alpha$ there is an open ball $B(\tau, y)$ of $\boldsymbol{R}^n$ centred at $(\Phi \oplus y)_\tau(0)$ and with a radius $\eta_y^\tau \in (0, 1/k)$ such that $B(\tau, y) \subset N(\tau, y)$. Let $I(\tau, y) := (\tau-\eta_y^\tau, \tau+\eta_y^\tau)$ and $S(\tau, y) := \{z \in C^0(I, \boldsymbol{R}^n)\colon |z-y|_a < \eta_y^\tau\}$, where $|\cdot|_a$ denotes the supremum norm of $C^0(I, \boldsymbol{R}^n)$. Let $g\colon I \times K_\alpha \to \boldsymbol{R}^n$ be defined by $g(t, x) := (\Phi \oplus x)_t(0)$ for $(t, x) \in I \times K_\alpha$. Since $I \times K_\alpha$

is compact in $\boldsymbol{R}\times C^0(I, \boldsymbol{R}^n)$ and g is continuous, the set $\operatorname{Graph}(g) := \{(t, x, (\Phi\oplus\oplus x)_t(0)): (t, x) \in I\times K_\alpha\}$ is compact in $\boldsymbol{R}\times C^0(I, \boldsymbol{R}^n)\times\boldsymbol{R}^n$. It is clear that $\operatorname{Graph}(g) \subset \bigcup_{(\tau, y)\in I\times K_\alpha} I(\tau, y)\times S(\tau, y)\times B(\tau, y)$. Therefore, there are $(\tau_j, y_j) \in I\times K_\alpha$ with $j = 1, \ldots, q$ such that $\operatorname{Graph}(g) \subset \bigcup_{j=1}^{q} I(\tau_j, y_j)\times S(\tau_j, y_j)\times B(\tau_j, y_j)$. For simplicity, we set $\eta_j := \eta_{y_j}^{\tau_j}$, $h_j := h_{y_j}^{\tau_j}$, $v_j := v_{y_j}^{\tau_j}$ and $h_0^k := \min_{1\leqslant j\leqslant q} h_j$.

Let $(t, x) \in I\times K_\alpha$ be fixed. It belongs to any $I(\tau_j, y_j)\times S(\tau_j, y_j)$ and $(\Phi\oplus x)_t(0) \in B(\tau_j, y_j) \subset N(\tau_j, y_j)$, because $(t, x, (\Phi\oplus x)_t(0)) \in \operatorname{Graph}(g)$. Therefore, $|t-\tau_j| < \eta_j \leqslant 1/k$, $|x-y_j|_a < \eta_j \leqslant 1/k$ and there is $x_j \in H$ such that $|v_j - (x_j - (\Phi\oplus x)_t(0))/h_j| \leqslant \frac{1}{h_j} d_H[(\Phi\oplus x)_t(0)+h_j v_j] + \frac{1}{2}k \leqslant 1/k$. Let $u_j := [x_j - (\Phi\oplus x)_t(0)]/h_j$. We have $u_j \in v_j + \frac{1}{k} B$ and $(\Phi\oplus x)_t(0)+h_j u_j \in H$.

Finally, we see that for every $(t, x) \in I\times K_\alpha$ there are $(\tau_j, y_J) \in I\times K_\alpha$, $v_j \in (G\square\mathscr{D})(\tau_j, y_j)$, $u_j \in v_j + \frac{1}{k} B$ and $h_j \in [h_0^k, 1/k]$ such that $|t-\tau_j| \leqslant 1/k$, $|x-y_j|_a \leqslant 1/k$ and $(\Phi\oplus x)_t(0)+h_j u_j \in H$. Hence, in particular it follows that $(\tau_j, y_j, v_j) \in \operatorname{Graph}(G\square\mathscr{D})$ and $v_j \in (G\square\mathscr{D})(I\times K_\alpha)$. Therefore, for every $k \geqslant N_l$ and $(t, x) \in I\times K_\alpha$ there are $h_j \in [h_0^k, 1/k]$ and $u_j \in (G\square\mathscr{D})(I\times K_\alpha)+lB$ such that $(\Phi\oplus x)_t(0)+ +h_j u_j \in H$ and $(t, x, u_j) \in \operatorname{Graph}(G\square\mathscr{D})+\frac{1}{k} S$, where S denotes a closed unit ball of a normed space $(\boldsymbol{R}\times C^0(I, \boldsymbol{R}^n)\times\boldsymbol{R}^n, \|\cdot\|)$ with $\|\cdot\|$ defined by $\|(t, x, u)\| = \max(|t|, |x|_a, |u|)$ for $(t, x, u) \in \boldsymbol{R}\times C^0(I, \boldsymbol{R}^n)\times\boldsymbol{R}^n$.

Let $k \geqslant N_l$ and $x^0 \in K_\alpha$ be fixed. Select for $(\sigma, x^0) \in I\times K_\alpha$, $h_0 \in [h_0^k, 1/k]$ and $u_0 \in (G\square\mathscr{D})(I\times K_\alpha)+lB$ such that $(\Phi\oplus x^0)_\sigma(0)+h_0 u_0 = \Phi(0)+h_0 u_0 \in H$ and $(\sigma, x^0, u_0) \in \operatorname{Graph}(G\square\mathscr{D})+\frac{1}{k}B$. Put $t_1 = \sigma+h_0$ and define $l_0(t) = (t-\sigma)u_0$ for $t \in [\sigma, t_1]$. Let $z^1\colon [\sigma-r, \sigma+a] \to \boldsymbol{R}^n$ be defined by

$$z^1(\tau) = \begin{cases} \Phi(\tau-\sigma) & \text{for } \tau \in [\sigma-r, \sigma), \\ \Phi(0)+l_0(\tau) & \text{for } \tau \in [\sigma, t_1], \\ z^1(t_1) & \text{for } \tau \in (t_1, \sigma+a]. \end{cases}$$

Put $x^1 = \mathscr{I}(z^1|_{[\sigma,\sigma+a]})$. We have $|\dot{x}^1(t)| = |u_0| \leqslant l+M$ for $t \in (\sigma, t_1)$ and $|\dot{x}^1(t)| = 0 \leqslant l+M$ for $t \in (t_1, \sigma+a)$. Therefore $x^1 \in K_\alpha$.

Now, for $(t_1, x^1) \in I\times K_\alpha$ we can find $h_1 \in [h_0^k, 1/k]$ and $u_1 \in (G\square\mathscr{D})(I\times K_\alpha)+lB$ such that $(\Phi\oplus x^1)_{t_1}(0)+h_1 u_1 \in H$ and $(t_1, x^1, u_1) \in \operatorname{Graph}(G\square\mathscr{D})+\frac{1}{k} S$. Let $t_2 = t_1+h_1$ and define $l_1(t) = (t-t_1)u_1$ for $t \in [t_1, t_2]$. Let $z^2\colon [\sigma-r, \sigma+a] \to \boldsymbol{R}^n$ be defined by

$$z^2(t) = \begin{cases} z^1(t) & \text{for } t \in [\sigma-r, t_1), \\ z^1(t_1)+l_1(t) & \text{for } t \in [t_1, t_2], \\ z^2(t_2) & \text{for } t \in (t_2, \sigma+a]. \end{cases}$$

Put, $x^2 = \mathscr{I}(\dot{z}^2|_{[\tau,\sigma+a]})$. We have $|\dot{x}^2(t)| = |\dot{x}^1(t)| \leqslant l+M$ for $t \in (\sigma, t_1)$, $|\dot{x}^2(t)| = 0 < l+M$ for $t \in (t_2, \sigma+a)$. Therefore, $x^2 \in K_\alpha$.

Continuing this procedure we can find numbers $h_0, h_1, \ldots, h_{m_k+1} \in [h_0^k, 1/k]$ and points $u_0, u_1, \ldots, u_{m_k+1} \in (G\Box\mathscr{D})(I\times K_\alpha)+lB$ such that $h_0+h_1+\ldots+h_{m_k} \leqslant a < h_0+h_1+\ldots+h_{m_k}+h_{m_k+1}$. We can also define functions z^i: $[\sigma-r, \sigma+a] \to \boldsymbol{R}^n$ and $x^i = \mathscr{I}(z|_{[\sigma,\sigma+a]})$ such that $x^i \in K_\alpha$, $(t_i, x^i, u_i) \in \text{Graph}(G\Box\mathscr{D})+\frac{1}{k}S$ and $(\Phi\oplus x^i)_{t_i}(0)+h_i u_i \in H$ for $i = 1, 2, \ldots, m_k$. Moreover, $z^{i+1}(t) = z^i(t)$ for $t \in [\sigma-r, t_i]$ for $i = 1, 2, \ldots, m_k-1$.

If $\sum_{i=0}^{m_k} h_i = a$ we define for every $k \geqslant N_l$ an absolutely continuous function ξ^k: $[\sigma, \sigma+a] \to \boldsymbol{R}^n$ by taking $\xi^k = x^{m_k}$. If $\sum_{i=0}^{m_k} h_i < a$ we define $z^{m_k+1}(t) = z^{m_k}(t)$ for $t \in [\sigma-r, t_{m_k}]$ and $z^{m_k+1}(t) = z^m(t_{m_k})+(t-t_{m_k})u_{m_k}$ for $t \in (t_{m_k}, \sigma+a]$. Then we take $\xi^k = x^{m_k+1} = \mathscr{I}(\dot{z}^{m_k+1}|_{\sigma,\sigma+a})$.

Let us observe that $\dot{\xi}^k(t) = u_i$ for $t \in (t_i, t_{i+1})$ and $i = 0, 1, \ldots, m_{k-1}$ with $t_0 := \sigma$. If $\sum_{i=0}^{m_k} h_i < a$ we also have $\dot{\xi}^k(t) = u_{m_u}$ for $t \in (t_{m_k}, \sigma+a)$. Therefore, in the both cases we have $\xi^k \in K_\alpha$ for $k \geqslant N_l$.

Since $z^i(t) = x^i(t)+\Phi(0)$ for $t \in I$, $i = 1, \ldots, m_k$ and $z^{i+1}(t) = z^i(t)$ for $t \in [\sigma-$ $-r, t_i]$ and $i = 1, \ldots, m_k-1$ then $x^{i+1}(t) = x^i(t)$ for $t \in [\sigma, t_i]$ for $i = 1, \ldots, m_k-1$.

On the other hand we have $(\Phi\oplus x^i)_{t_i}(0) := x^i(t_i)+\Phi(0)$ for $i = 1, \ldots, m_k$, $h_0 u_0 = z^1(t_1)-\Phi(0) = x_1(t_1)$, $h_i u_i = z^{i+1}(t_{i+1})-z^i(t_i) = z^{i+1}(t_i)-x^i(t_i)-\Phi(0)$ for $i = 1, \ldots, m_k-1$. Then, $x^1(t_1)+\Phi(0) = h_0 u_0+\Phi(0) \in H$ and $x^{i+1}(t_{i+1})+\Phi(0) = z^{i+1}(t_{i+1}) = (\Phi\oplus x^i)_{t_i}+h_i u_i \in H$ for $i = 1, 2, \ldots, m_k-1$. Therefore, $x^i(t_i) \in H-$ $-\Phi(0)$ for $i = 1, 2, \ldots, m_k$. Since $x^{m_k}(t_i) = x^i(t_i)$ for $i = 1, \ldots, m_k$ then for every $i = 1, \ldots, m_k$ we have $\xi^k(t_i) \in H-\Phi(0)$. By Volterra's property of G we have $(G\Box\mathscr{D})(t_i, x^{m_k}) = (G\Box\mathscr{D})(t_i, x^i)$ for every $i = 1, \ldots, m_k$, because $x^{m_k}(t) = x^i(t)$ for $t \in [\sigma, t_i]$ for every $i = 1, \ldots, m_k$. Therefore, for every $i = 1, \ldots, m_k$ and $k \geqslant N_l$ we also have $(t_i, \xi^k, \dot{\xi}^k(t)) \in \text{Graph}(G\Box\mathscr{D})+\frac{1}{k}S$ for $t \in (t_{i-1}, t_i)$, $i = 1, \ldots, m_k$ and $k \geqslant N_l$ where $t_0 = \sigma$. If $\sum_{i=0}^{m_k} h_i < a$ we also have $(t_{m_k}, \xi^k, \dot{\xi}^k(t)) \in \text{Graph}(G\Box\mathscr{D})+$ $+\frac{1}{k}S$ for $t \in (t_{m_k}, \sigma+a)$ and $k \geqslant N_l$. In both cases we have $|\dot{\xi}^k(t)| \leqslant l+M$ for a.e. $t \in I$ and $k \geqslant N_l$. Therefore, there are an absolutely continuous function ξ: $I \to \boldsymbol{R}^n$ and a subsequence, say again (ξ^k) of (ξ^k) such that $\xi \in K_\alpha$, $|\xi^k-\xi|_a \to 0$ and $\dot{\xi}^k \rightharpoonup \dot{\xi}$ in $L(I, \boldsymbol{R}^n)$ as $k \to \infty$.

By the definition we have $\xi^k(t_i) \in H-\Phi(0)$, $(t_i, \xi^k, \dot{\xi}^k(t)) \in \text{Graph}(G\Box\mathscr{D})+\frac{1}{k}S$ for $t \in (t_{i-1}, t_i)$, $i = 1, \ldots, m_k$ and $k = 1, 2, \ldots$

Since $\dot{\xi}^k \rightharpoonup \dot{\xi}$ as $k \to \infty$ then by Banach–Mazur theorem there is a sequence $\{\sum_{i=j}^{\infty} \lambda_i^j \dot{\xi}^{k_i}\}$ of convex combinations converging a.e. in I to $\dot{\xi}$ as $j \to \infty$.

For every $\eta > 0$ there is $N_\eta \geq 1$ such that for $k \geq N_\eta$, $\xi(t) \in (H-\Phi(0))+$ $+\eta B$ for $t \in [t_{i-1}, t_i]$ and $(t, \xi, \dot{\xi}^k(t)) \in \text{Graph}(G\Box\mathcal{D})+(1/k+\eta)S$ for $t \in (t_{i-1}, t_i)$ with $i = 1, \ldots, m_k$. Indeed, for every $t \in [t_{i-1}, t_i]$ we have $t_i - t \leqslant h_i \leqslant 1/k$. Since $|\xi^k - \xi|_a \to 0$ and $|\xi(t_i)-\xi(t))| \to 0$ as $k \to \infty$, then for every $\eta > 0$ there is $N_\eta \geq 1$ such that $|\xi^k - \xi|_a < \frac{1}{2}\eta$ and $|\xi(t_i)-\xi(t)| < \frac{1}{2}\eta$ for $k \geq N_\eta$. Then for $k \geq N_\eta$ we have $|\xi(t)-\xi^k(t_i)| \leqslant |\xi(t)-\xi(t_i)|+|\xi(t_i)+\xi^k(t_i)| < \eta$, i.e. $\xi(t) \in \xi^n(t_i)+\eta B$ for $t \in [t_{i-1}, t_i]$ and $k \geq N_\eta$.

Similarly, taking above N_η such that $1/k \leqslant \frac{1}{2}\eta$ for $t \in (t_{i-1}, t_i)$ and $k \geq N_\eta$ we obtain $(t, \xi, \dot{\xi}^k(t)) \in (t_i, \xi^k, \dot{\xi}^k(t))+\eta S$. Therefore, $(t, \xi, \dot{\xi}^k(t)) \in \text{Graph}(G\Box\mathcal{D})+$ $+(1/k+\eta)S$ for $t \in (t_{i-1}, t_i)$ and $k \geq N_\eta$.

We shall show now that $\xi(t) \in H-\Phi(0)$ for $t \in I$ and $\dot{\xi}(t) \in (G\Box\mathcal{D})(t, \xi)$ for a.e. $t \in I$. Indeed, for each fixed $t \in I$ we have $\xi(t) \in (H-\Phi(0))+\eta B$. Therefore, $\xi(t)$ $\in H-\Phi(0)$. Furthermore, for a.e. $t \in I$ we have $\dot{\xi}(t) = \lim \sum_{i=j}^{\infty} \lambda_i^j \dot{\xi}^{k_i}(t)$ and $(t, \xi, \dot{\xi}^k(t))$ $\in \text{Graph}(G\Box\mathcal{D})+(1/k+\eta)S$ for k sufficiently large, say for $k \geq N_\eta$. Suppose $t \in I$ is such that the last two conditions are satisfied for $k \geq N_k$. Let $p \in \boldsymbol{R}^n$ and λ $> s(p, (G\Box\mathcal{D})(t, \xi))$ be fixed, where $s(\cdot, (G\Box\mathcal{D})(t, \xi))$ denotes a support function of $(G\Box\mathcal{D})(t, \xi)$. Since $G\Box\mathcal{D}$ is u.s.c. at (t, ξ) then there is a neighborhood M_0 of 0 in $\boldsymbol{R}\times C^0(I, \boldsymbol{R}^n)$ such that when $(\tau, u) \in (t, \xi)+M_0$ we have $s(p, (G\Box\mathcal{D})(\tau, u))$ $\leqslant \lambda$. Since $(t, \xi, \dot{\xi}^k(t)) \in \text{Graph}(G\Box\mathcal{D})+\eta(k)S$ with $\eta(k) := 1/k+\eta$ for $k \geq N_\eta$ then for every $k \geq N_\eta$ there exists $(\tau_k, u_k, w_k(t)) \in \text{Graph}(G\Box\mathcal{D})$ such that $(\tau_k, u_k, w_k(t)) \in (t, \xi, \dot{\xi}^k(t))+\eta(k)S$. Suppose N_ε is such that $(\tau_k, u_k) \in (t, \xi)+M_0$ and $\eta(k) \leqslant \varepsilon$ for $k \geq N_\varepsilon$. Now, for $k \geq N_\varepsilon$ one obtains

$$\begin{aligned} p\cdot\dot{\xi}^k(t) &= p\cdot[\dot{\xi}^k(t)-w_k(t)]+p\cdot w_k(t) \leqslant p\cdot w_k(t)+\eta(k)|p| \\ &\leqslant s\big(p, (G\Box\mathcal{D})(\tau_k, u_k)\big)+\varepsilon|p| \leqslant \lambda+\varepsilon|p|. \end{aligned}$$

Hence, for $j \geq N_\varepsilon$ we have

$$p\cdot\sum_{i=j}^{\infty} \lambda_i^j \dot{\xi}^{k_i}(t) \leqslant \lambda+\varepsilon|p|.$$

Thus, by letting $j \to \infty$ one obtains

$$p\cdot\dot{\xi}(t) \leqslant \lambda+\varepsilon|p|.$$

By letting λ converge to $s(p, (G\Box\mathcal{D})(t, \xi))$ and ε to 0, we deduce that

$$p\cdot\dot{\xi}(t) \leqslant s\big(p,(G\Box\mathcal{D})(t, \xi)\big).$$

Since $(G\Box\mathcal{D})(t, \xi)$ is closed and convex we deduce that $\dot{\xi}(t) \in (G\Box\mathcal{D})(t, \xi)$.

Put $u := \dot{\xi}$. We have $u \in \Lambda_\alpha \subset \Lambda$, $u \in \mathcal{F}(G\Box\mathcal{D})(u)$ and $(\mathcal{I}u)(t) = \xi(t) \in H-$ $-\Phi(0)$ for $t \in I$. ■

REMARK 6.1. We can replace in the above theorem a compact set H by an upper semicontinuous set-valued function H: $[\sigma, \sigma+a] \to \text{Comp}(\boldsymbol{R}^n)$ such that $\Phi(0)$ $\in H(\sigma)$, $(\Phi\oplus x)_t(0) \in H(t)$ for $t \in [\sigma, \sigma+a]$ and $x \in K_\alpha$ and such that $G(t, x, \dot{x})\cap$ $\cap T_{H(t)}[(\Phi\oplus x)_t(0)] \neq \emptyset$ for $t \in [\sigma, \sigma+a]$ and $x \in K_\alpha$. ■

6.3. *Viability theorem for subtrajectory integrals of set-valued functions with nonconvex values*

We will here deal with set-valued mappings $G \in \mathscr{H}_\lambda(I, \boldsymbol{R}^n)$ with $I = [\sigma, \sigma+a]$ having Volterra's property with respect to its last two variables. We show that such mappings addmit some continuous selectors having also Volterra's property. Then we shall prove any viability theorem. We begin with some sufficient conditions for the existence of a Carathéodory selector a given set-valued function defined on the Cartesian product of given metric spaces.

Let T be a compact metric space with a positive Radon measure μ and X a metric space. A function $f\colon T \times X \to \boldsymbol{R}^n$ is said to be a *Carathéodory selector of a set-valued function* $F\colon T\times X \to \mathrm{Comp}(\boldsymbol{R}^n)$ if f is a selector for F and f is such that $f(\cdot, x)$ is μ-measurable for fixed $x \in X$ and $f(t, \cdot)$ is continuous for fixed $t \in T$. A set-valued function $F\colon T\times X \to \mathrm{Comp}(\boldsymbol{R}^n)$ is called an *M-mapping* if every its lower semicontinuous restriction has a continuous selector.

LEMMA 6.3. *Let T be a compact metric space with a positive Radon measure μ, X a separable complete metric space and (T, Σ^*, μ^*) a Lebesgue extension of $(T, \beta(T), \mu)$. Suppose $F\colon T\times X \to \mathrm{Comp}(\boldsymbol{R}\,)$ is an M-mapping such that F is $(\Sigma^* \otimes \beta(X)$, $\beta(\mathrm{Comp}(\boldsymbol{R}^n))$-measurable and $F(t, \cdot)$ is l.s.c. for fixed $t \in T$. Then F has a Carathéodory selector.*

Proof. Let E_j be compact subset of T such that $\mu(T\setminus E_j) < 2^{-j}$ and such that F restricted to $E_j \times X$ is l.s.c. Such E_j exist by Scorza-Dragoni-Castaing theorem. Since F is an M-mapping it follows that for each index j a continuous function $f_j\colon E_j\times X \to \boldsymbol{R}^n$ exists such that $f_j(t, x) \in F(t, x)$ for $(t, x) \in E_j \times X$. We define $f\colon T\times X \to \boldsymbol{R}^n$ to be equal f_j if $t \in E_j$ and $t \notin E_i$ for $i < j$. It is easy to see that such defined function f is a Carathéodory selector for F. ■

LEMMA 6.4. *Suppose $G \in \mathscr{H}_\lambda(I, \boldsymbol{R}^n)$ is s.-w.l.s.c. with respect to its last two variables and let $l > 0$, $M > 0$ be such that $l+M \leqslant \lambda/a$. Suppose furthermore G has Volterra's property with respect to its last two variables, $(G\square\mathscr{D})$ is an M-mapping $(\mathscr{L}(I)\otimes \otimes\beta(K_\alpha)$, $\beta(\mathrm{Comp}(\boldsymbol{R}^n))$-measurable with $\alpha = l+M$, where $K_\alpha = \mathscr{I}(\Lambda_\alpha)$. Then there is a Carathéodory selector $g\colon I\times K_\alpha \to \boldsymbol{R}^n$ of $G\square\mathscr{D}$ having Volterra's property with respect to its last variable.*

Proof. It is clear that $G\square\mathscr{D}\colon I\times K_\alpha \to \mathrm{Comp}(\boldsymbol{R}^n)$ satisfies conditions of Lemma 6.3 and has Volterra's property with respect to its last variable. Then, by virtue of Lemma 6.3 there is a Carathéodory selector $\tilde{g}\colon I\times K_\alpha \to \boldsymbol{R}^n$ of $G\square\mathscr{D}$. Fix $(t, x) \in I\times K_\alpha$ and define $\mathscr{S}_x(t) := \mathscr{I}(\chi_t \dot{x})$, where χ_t denotes the characteristic function of $[\sigma, t]$. It is clear that mappings $I \ni t \to \mathscr{S}_x(t) \in K_\alpha$ and $K_\alpha \ni x \to \mathscr{S}_x(t) \in K_\alpha$ are continuous for fixed x and t, respectively. Furthermore $\mathscr{S}_x(t)(\tau) = x(\tau)$ for $\tau \in [\sigma, t]$. Therefore, $(G\square\mathscr{D})(t, \mathscr{S}_x(t)) = (G\square\mathscr{D})(t, x)$. Define now on $I\times K_\alpha$ a mapping g by taking $g(t, x) = \tilde{g}(t, \mathscr{S}_x(t))$ for each $(t, x) \in I\times K_\alpha$. It is clear that $g(\cdot, x)$ is measurable for fixed $t \in I$. Moreover, $g(t, x) = \tilde{g}(t, \mathscr{S}_x(t)) \in (G\square\mathscr{D})(t, \mathscr{S}_x(t)) = (G\square\mathscr{D})(t, x)$ and $g(t, x_1) = \tilde{g}(t, \mathscr{S}_{x_1}(t)) = g(t, \mathscr{S}_{x_2}(t)) = g(t, x_2)$ for $t \in I$ and $x, x_1, x_2 \in K_\alpha$ such that $x_1(\tau) = x_2(\tau)$ for $\tau \in [\sigma, t]$. ■

Now we prove the following viability theorem.

THEOREM 6.5. *Let* $\Lambda \in L(I, \boldsymbol{R}^n)$, $\Phi \in AC_{0r}$ *and a compact set* $H \subset \boldsymbol{R}^n$ *be given and let* $G \in \mathscr{H}_\lambda(I, \boldsymbol{R}^n_{\mathcal{D}})$. *Assume* Λ, Φ, H *and* G *are such that* $\Phi(0) \in H$ *and there are numbers* $l, r, M > 0$ *such that* $\Lambda_\alpha \subset \Lambda$ *with* $\alpha = l+M \leqslant \lambda/a$, $(\Phi \oplus x)_t(0) \in H$ *and* $(G\Box\mathscr{D})(t, x) \subset MB$ *for all* $(t, x) \in I\times K_\alpha$ *with* $K_\alpha = \mathscr{I}(\Lambda_\alpha)$. *Assume* G *is such that* $\mathscr{F}(G\Box\mathscr{D})$ *has a continuous selector* q: $K_\alpha \to L(I, \boldsymbol{R}^n)$ *such that a function* g: $I\times K_\alpha \to \boldsymbol{R}^n$ *defined by* $g(t, x) = q(x)(t)$ *for* $(t, x) \in I\times K_\alpha$ *has Volterra's property with respect to its last variable. If furthermore* G *is such that* $(G\Box\mathscr{D})(t, x) \subset T_H[(\Phi \oplus x)_t(0)]$ *for* $x \in K_\alpha$ *and a.e.* $t \in I$ *then there is* $u \in S(G)\cap\Lambda$ *such that* $(\mathscr{I}u)(t) \in H-\Phi(0)$ *for* $t \in I$.

Proof. Let g: $I\times K_\alpha \to \boldsymbol{R}^n_\lambda$ be such as above. Let us observe we can assume that $g(t, x) \in T_H[(\Phi \oplus x)_t(0)]$ for every $t \in I$ and $x \in K_\alpha$. Indeed, if for fixed $x \in K_\alpha$ we define $E_x = \{t \in I\colon g(t, x) \notin T_H[(\Phi\oplus x)_t(0)]\}$, then we can define a function f: $I\times K_\alpha \to \boldsymbol{R}^n$ by $f(t, x) = g(t, x)$ for $t \in I\setminus E_x$, $x \in K_\alpha$ and $f(t, x) = 0$ for $t \in E_x$, $x \in K_\alpha$. It is clear that $f(t, x) \in T_H[(\Phi \oplus x)_t(0)]$ for every $t \in I$ and $x \in K_\alpha$. Furthermore, we have $\int_I |f(t, x)-g(t, x)|\,dt = 0$ for $x \in K_\alpha$ and therefore a mapping $K_\alpha \ni x \to f(\cdot, x) \in L(I, \boldsymbol{R}^n)$ is also continuous on K_α. Let $I := [\sigma, \sigma+a]$, fix $x \in K_\alpha$ and denote by E^x a set of all $\tau \in I$ such that for every $\tau \in I\setminus E^x$ we have

$$\lim_{\eta\to\infty}\left|g(\tau, x)-\frac{1}{n}\int_{\tau-\eta}^{\tau}\tilde{g}(s, x)\,ds\right| = 0, \tag{6.3}$$

where $\tilde{g}(t, x) = \dot{\Phi}(t-\sigma)$ for a.e. $t \in [\sigma-r, \sigma]$, $x \in K_\alpha$ and $\tilde{g}(t, x) = g(t, x)$ for $(t, x) \in I\times K_\alpha$.

Let $\varepsilon > 0$ be arbitrarily fixed. By (6.3) for every $x \in K_\alpha$ and $\tau \in I\setminus E^x$ we can select $\eta(\tau, x) > 0$ such that for every $\eta \in (0, \min(a, \eta(\tau, x))$ we have

$$\left|g(\tau, x)-\frac{1}{n}\int_{\tau-\eta}^{\tau}\tilde{g}(s, x)\,ds\right| < \tfrac{1}{9}\varepsilon. \tag{6.4}$$

Observe that a mapping $K_\alpha \in x \to g(\cdot, x) \in L(I, \boldsymbol{R}^n)$ is continuous on a compact set $K_\alpha \subset C^0(I, \boldsymbol{R}^n)$. Therefore, for fixed $\eta \in (0, \min(a, \eta(\tau, x))$ and a given above $\varepsilon > 0$ there is $\delta(\tau, x) > 0$ such that for every $x^1, x^2, \tilde{x} \in K_\alpha$ and every measurable set $E \subset [\sigma-r, \sigma+a]$ we have

$$\int_{\sigma}^{\sigma+\varrho} |g(s, x^1)-g(s, x^2)|\,ds < \tfrac{1}{9}\eta\varepsilon \tag{6.5}$$

and

$$\int_E |\tilde{g}(s, \tilde{x})|\,ds < \tfrac{1}{9}\eta\varepsilon \tag{6.6}$$

whenever $|x^1-x^2|_a < \delta(\tau, x)$ and $\mu(E) < \delta(\tau, x)$.

Finally, let us observe that by the tangential condition $g(\tau, x) \in T_H[(\Phi \oplus x)_\tau(0)]$, by Lemma 6.1, for every $(\tau, x) \in I \times K_\alpha$ there is $h_x^\tau \in (0, 1/k)$ such that

$$d_H[(\Phi \oplus x)_\tau(0) + h_x^\tau g(\tau, x)] \leqslant h_x^\tau/3k.$$

We introduce the subsets

$$N(\tau, x) = \{z \in \boldsymbol{R}^n\colon d_H\big(z + h_x^\tau g(\tau, x)\big) < h_x^\tau/2k\}.$$

Similarly as in the proof of Theorem 6.2 we see that $N(\tau, x)$ is open for every $(\tau, x) \in I \times K_\alpha$. Thus, for every $(\tau, x) \in I \times K_\alpha$ there is an open ball $\boldsymbol{K}(\tau, x)$ of $\boldsymbol{R}^n$ centred at $(\Phi \oplus x)_\tau(0)$ and a radius $\varrho_x^\tau \in (0, 1/k)$ such that $\boldsymbol{K}(\tau, x) \subset N(\tau, x)$. Let $x \in K_\alpha$, $\tau \in I \setminus E^x$ and $\eta \in (0, \min(a, \eta(\tau, x)))$ be fixed and put $\lambda_x^\tau = \min(\varrho_x^\tau, \frac{1}{2}\delta(\tau, x))$. Denote by $S(\tau, x)$ an open ball of $C^0(I, \boldsymbol{R}^n)$ centred at x and with a radius λ_x^τ and put $I(\tau, x) = (\tau - \lambda_x^\tau, \tau + \lambda_x^\tau)$. Finally, let $B(\tau, x)$ be an open ball of $\boldsymbol{R}^n$ with the centre at $(\Phi \oplus x)_\tau(0)$ and a radius λ_x^τ. We have $\lambda_x^\tau \in (0, 1/k)$, $B(\tau, x) \subset \boldsymbol{K}(\tau, x) \subset N(\tau, x)$ and (6.5) is satisfied for every $x^1, x^2 \in S(\tau, x)$. Let $\mathscr{R}\colon I \times K_\alpha \to \boldsymbol{R}^n$ be defined by $\mathscr{R}(t, x) := (\Phi \oplus x)_t(0)$ for $t \in I$ and $x \in K_\alpha$. Since $(\tau, x, (\Phi \oplus x)_\tau(0)) \in I(\tau, x) \times S(\tau, x) \times B(\tau, x)$ for every $(\tau, x) \in I \times K_\alpha$, then $\operatorname{Graph} \mathscr{R} \subset \bigcup_{x \in K_\alpha} \bigcup_{\tau \in I \setminus E^x} I(\tau, x) \times S(\tau, x) \times B(\tau, x)$. Indeed, if $(s, x) \in (I \setminus E^x) \times K_\alpha$ it is clear that $(s, x, \mathscr{R}(s, x)) \in I(s, x) \times S(s, x) \times B(s, x) \subset \bigcup_{x \in X_\alpha} \bigcup_{\tau \in I \setminus E^x} I(\tau, x) \times S(\tau, x) \times B(\tau, x)$. Suppose $(s, x) \in E^x \times K_\alpha$. By the uniform continuity of $\mathscr{R}(\cdot, x)$ on I for every $\tau \in I \setminus E^x$ there is $\nu(\tau, x) \in (0, \lambda_x^\tau)$ such that for every $s_1, s_2 \in I$ with $|s_1 - s_2| \leqslant \nu(\tau, x)$ we have $|\mathscr{R}(s_1, x) - \mathscr{R}(s_2, x)| < \lambda_x^\tau$. Since $s \in E^x$ and $\mu(E^x) = 0$ then there is $\tau \in I \setminus E^x$ such that $|s - \tau| \leqslant \nu(\tau, x)$. Therefore, $s \in I(\tau, x)$ and $|\mathscr{R}(s, x) - \mathscr{R}(\tau, x)| < \lambda_x^\tau$, i.e. $\mathscr{R}(s, x) \in B(\tau, x)$. Of course $x \in S(\tau, x)$. Then, for every $(s, x) \in I \times K_\alpha$ we have $(s, x, \mathscr{R}(s, x)) \in I(\tau, x) \times S(\tau, x) \times B(\tau, x)$ for any $\tau \in I \setminus E^x$.

Let $H_\alpha = \mathscr{R}(I \times K_\alpha)$. By the compactness of $I \times K_\alpha \times H_\alpha$ in $\boldsymbol{R} \times C^0(I, \boldsymbol{R}^n) \times \boldsymbol{R}^n$ and continuity of $\mathscr{R}$ it follows that $\operatorname{Graph}(\mathscr{R})$ is also compact in $\boldsymbol{R} \times C^0(I, \boldsymbol{R}^n) \times \boldsymbol{R}^n$. Therefore there are $x_j \in K_\alpha$ and $\tau_j \in I \setminus E^{x_j}$, $j = 1, \ldots, q$ such that $\operatorname{Graph}(\mathscr{R}) \subset \bigcup_{j=i}^{q} I_j \times S_j \times B_j$, where $I_j := I(\tau_j, x_j)$, $S_j := S(\tau_j, x_j)$ and $B_j := B(\tau_j, x_j)$. For simplicity we set $\eta_j := \min(r, \eta(\tau_j, x_j))$, $\delta_k := \delta(\tau_k, x_k)$, $\lambda_j := \lambda_{x_j}^{\tau_j}$, $h_j := h_{x_j}^{\tau_j}$, $h_0^k = \min_{i \leqslant j \leqslant q} h_j$ and $\eta_0^k = \min_{i \leqslant j \leqslant q} \eta_j$.

Let $(t, x, v) \in \operatorname{Graph}(\mathscr{R})$. There is $j \in \{1, \ldots, q\}$ such that $(t, x) \in I_j \times S_j$ and $v = \mathscr{R}(t, x) \in B_j \subset N(\tau_j, x_j)$. Therefore, $|t - \tau_j| < \lambda_j \leqslant 1/k$, $|x - x_j|_a < \lambda_j \leqslant 1/k$ and there is $y_j \in H$ such that $|g(\tau_j, x_j) - (y_j - \mathscr{R}(t, x))/h_j| \leqslant \dfrac{1}{h_j} d_H[\mathscr{R}(t, x) + h_j g(\tau_j, x_j)] + 1/2k \leqslant 1/k$. Let $u_j = [y_j - \mathscr{R}(t, x)]/h_j$. We have $|u_j - g(\tau_j, x_j)| \leqslant 1/k$ and $\mathscr{R}(t, x) + h_j u_j \in H$. We also have $\left|g(\tau_j, x_j) - \dfrac{1}{\eta} \displaystyle\int_{\tau_j - \eta}^{\tau_j} \tilde{g}(s, x_j)\, ds\right| < \frac{1}{9}\varepsilon$ for every $\eta \in (0, \eta_0^k)$. Finally, we see that for every $(t, x) \in I \times K_\alpha$ there are $x_j \in K_\alpha$, $\tau_j \in I \setminus E^{x_j}$, $u_j \in g(\tau_j, x_j) + \dfrac{1}{k} B$ and $h_j \in [h_0^k, 1/k]$ such that $|t - \tau_j| < \lambda_j \leqslant 1/k$,

$|x-x_j|_a < \lambda_j \leqslant 1/k$, $\mathscr{R}(t,x)+h_ju_j \in H$ and $|u_j - \frac{1}{\eta}\int_{\tau_j-\eta}^{\tau_j} \tilde{g}(s,x_j)ds| \leqslant 1/k+\frac{1}{9}\varepsilon$ for every $\eta \in (0,\eta_0^k)$. Hence, in particular it follows $|u_j| \leqslant 1/k+M$. Futrhermore, for $t \in I_j$ and $\eta \in (0,\eta_0^k)$ we have

$$\left|\int_{\tau-\eta}^{\tau_j} \tilde{g}(s,x_j)ds - \int_{t-\eta}^{t} \tilde{g}(s,x_j)ds\right| \leqslant \int_{(\tau_j,\tau_j+\eta)\triangle(t,t+\eta)} |\tilde{g}(s,x_j)|ds \leqslant \tfrac{1}{9}\eta\varepsilon,$$

because $\mu[(\tau_j,\tau_j+\eta)\triangle(t,t+\eta)] \leqslant 2\lambda_j \leqslant \delta_j$. Therefore for every $(t,x) \in I\times K_\alpha$ there are $u_j \in (1/k+M)B$ and $h_j \in [h_0^k, 1/k]$ such that $\mathscr{R}(t,x)+h_ju_j \in H$ and $|u_j - \frac{1}{\eta}\int_{t-\eta}^{t} \tilde{g}(s,x)ds| \leqslant 1/k+\frac{1}{3}\varepsilon$ for $\eta \in (0,\eta_0^k)$.

Finally, select $N_l \geqslant 1$ such that for $k \geqslant N_l$ we have $1/k \leqslant l$ and let $k \geqslant N_l$, $x^0 \in K_\alpha$ be fixed. Select for $(\sigma,x^0) \in I\times K_\alpha$, $h_0 \in [h_0^k, 1/k]$ and $u_0 \in (l+M)B$ such that $\mathscr{R}(\sigma,x^0)+h_0u_0 \in H$ and $\left|u_0 - \frac{1}{\eta}\int_{\sigma-\eta}^{\sigma} \dot{\Phi}(s-\sigma)ds\right| \leqslant 1/k+\frac{1}{3}\varepsilon$ for $\eta \in (0,\eta_0^k)$. Put $t_1 = \sigma+h_0$ and define $l_0(t) = (t-\sigma)u_0$ for $t \in [\sigma,t_1]$. Define $z^1\colon [\sigma-r,\sigma+a] \to \boldsymbol{R}^n$ by

$$z^1(\tau) = \begin{cases} \Phi(\tau-\sigma) & \text{for } \tau \in [\sigma-r,\sigma] \\ \Phi(0)+l_0(t) & \text{for } \tau \in [\sigma,t_1] \\ z^1(t_1) & \text{for } \tau \in [t_1,\sigma+a]. \end{cases}$$

Put $x^1 = \mathscr{I}(z^1|_{[\sigma,\sigma+a]})$. Similarly, as in the proof of Theorem 6.2 we obtain $x^1 \in K_\alpha$.

Now, for $(t_1,x^1) \in I\times K_\alpha$ we find $h_1 \in [h_0^k, 1/k]$ and $u_1 \in (l+M)B$ such that $\mathscr{R}(t_1,x^1)+h_1u_1 \in H$ and $\left|u_1 - \frac{1}{\eta}\int_{t_1-\eta}^{t_1} g(s,x^1)ds\right| \leqslant 1/k+\frac{1}{3}\varepsilon$ for $\eta \in (0,\min(h_0^k,\eta_0^k))$.

Continuing this procedure, similarly as in the proof of Theorem 6.2 we can find numbers $h_0,h_1,\ldots,h_{m_k+1} \in [h_0^k,1/k]$ and points $u_0,u_1,\ldots,u_{m_k+1} \in (l+M)B$ such that $h_0+h_1+\ldots+h_{m_k} \leqslant a < h_0+h_1+\ldots+h_{m_k}+h_{m_k+1}$. We can also define functions $z\colon [\sigma-r,\sigma+a] \to \boldsymbol{R}^n$ and $x^i = \mathscr{I}(\dot{z}^i|_{[\sigma,\sigma+a]})$ such that $x^i \in K_\alpha$, $\mathscr{R}(t_i,x^i)+h_iu_i \in H$ and $\left|u_i - \frac{1}{\eta}\int_{t_i-\eta}^{t_i} g(s,x^i)ds\right| \leqslant 1/k+\frac{1}{3}\varepsilon$ for $\eta \in (0,\min(h_0^k,\eta_0^k))$ and $i = 1,2,\ldots,m_k$, where $t_1 = \sigma+h_0$. Moreover, we have $x^{i+1}(t) = x^i(t)$ for $t \in [\sigma,t_i]$ for $i = 1,2,\ldots,m_k-1$.

Similarly as in the proof of Theorem 6.2, for every $k \geqslant N_l$ we define an absolutely continuous function $\xi^k \in K_\alpha$ such that $\xi^k(t_i) \in H-\Phi(0)$ for $i = 1,\ldots,m_k$, $\left|\dot{\xi}^k(t) - \frac{1}{\eta}\int_{t_i-\eta}^{t_i} g(s,\xi^k)ds\right| \leqslant 1/k+\frac{1}{3}\varepsilon$ for $t \in (t_i,t_{i+1})$ with $i = 1,\ldots,m_k-1$, $\eta \in$

$\in (0, \min(h_0^k, \eta_0^k))$ and $|\dot{\xi}^k(t)| \leqslant l+M$ for a.e. $t \in I$, because $\xi^k(t) = x^{i+1}(t)$ for $t \in [\sigma, t_{i+1}]$, $\dot{x}^{i+1}(t) = u_i$ for $t \in (t_i, t_{i+1})$ and by Volterra's property of g we have

$$\left|\dot{x}^{i+1}(t) - \frac{1}{\eta}\int_{t_i-\eta}^{t_i} g(s, x^{i+1})ds\right| = \left|u_i - \frac{1}{\eta}\int_{t_i-\eta}^{t_i} g(s, x^i)ds\right| \leqslant 1/k + \tfrac{1}{3}\varepsilon \quad \text{for} \quad t \in (t_i, t_{i+1})$$

with $i = 1, \dots, m_k - 1$ and $\eta \in (0, \min(h_0^k, \eta_0^k))$.

Hence, in particular it follows the existence of an absolutely continuous function $\xi \in K_\alpha$ and a subsequence, say again (ξ^k) of (ξ^k) such that $|\xi^k - \xi|_a \to 0$ and $\dot{\xi}^k \rightharpoonup \dot{\xi}$ in $L(I, \boldsymbol{R}^n)$ as $k \to \infty$.

Similarly as in the proof of Theorem 6.2 we obtain $\xi(t) \in H - \Phi(0)$ for $t \in I$. Furthermore, for $t \in (t_i, t_{i+1})$, $\eta \in (0, \min(h_0^k, \eta_0^k))$, $i = 1, \dots, m_k - 1$ and k sufficiently large, say for $k \geqslant N$, we have

$$\begin{aligned}
&\left|\dot{\xi}^k(t) - \frac{1}{\eta}\int_{t-\eta}^{t} g(s, \xi)ds\right| \\
&\leqslant \left|\dot{\xi}^k(t) - \frac{1}{\eta}\int_{t_i-\eta}^{t_i} g(s, \xi^k)ds\right| + \frac{1}{\eta}\int_{(t_i-\eta, t_i)\Delta(t-\eta, t)} |g(s, \xi^k)|\,ds \\
&\quad + \frac{1}{\eta}\int_{\sigma+a}^{\sigma} |g(s, \xi^k) - g(s, \xi)|\,ds \leqslant \frac{1}{k} + \varepsilon.
\end{aligned}$$

Since $\dot{\xi}^k \rightharpoonup \dot{\xi}$ as $k \to \infty$, then by Banach–Mazur theorem there is a sequence $\{\sum_{i=j}^{\infty} \lambda_i^j \dot{\xi}^{k_i}\}$ of convex combinations with $k_i \geqslant N$ converging a.e. in I to $\dot{\xi}$ as $j \to \infty$. We have of course

$$\begin{aligned}
&\left|\sum_{i=j}^{\infty} \lambda_i^j \dot{\xi}^{k_i}(t) - \frac{1}{\eta}\int_{t-\eta}^{t} g(s, \xi)ds\right| \\
&\leqslant \sum_{i=j}^{\infty} \lambda_i^j \left|\dot{\xi}^{k_i}(t) - \frac{1}{\eta}\int_{t-\eta}^{t} g(s, \xi)ds\right| \leqslant \sum_{i=j}^{\infty} \lambda_i^j \frac{1}{k_i} + \varepsilon
\end{aligned}$$

for $t \in (t_i t_{i+1})$, $\eta \in (0, \min(h_0^k, \eta_0^k))$, $i = 1, 2, \dots, m_k - 1$, $k \geqslant N$ and $j = 1, 2, \dots$ Therefore, for a.e. $t \in [t_i, t_{i+1}]$, $\eta \in (0, \min(h_0^k, \eta_0^k))$, $i = 1, 2, \dots, m_k - 1$ and $k \geqslant N$ we obtain $\left|\dot{\xi}(t) - \frac{1}{\eta}\int_{t-\eta}^{t} g(s, \xi)ds\right| \leqslant \varepsilon$.

Now, if $\sum_{i=0}^{m_k} h_i = a$, we obtain $\xi(t) \in H - \Phi(0)$ for $t \in I$ and $|\dot{\xi}(t) - g(t, \xi)| \leqslant \varepsilon$ for a.e. $t \in [t_1, \sigma + a]$. Hence it follows that $\dot{\xi}(t) = g(t, \xi)$ for a.e. $t \in I$.

Similarly as in the proof of Theorem 6.2, for every $t \in [t_{m_k}, \sigma + a]$ we also obtain $\xi(t) \in H - \Phi(0)$ if $\sum_{i=0}^{m_u} h_i < a$. Furthermore, by the definition of ξ^k we have $\dot{\xi}^k(t)$

$= u_{m_k}$ for $t \in (t_{m_k}, \sigma+a)$ and $\xi^k(t) = x^{m_k}(t)$ for $t \in [\sigma, t_{m_k}]$. Therefore, for $t \in (t_{m_k}, \sigma+$ $+a)$ we also have

$$\left|\dot{\xi}^k(t) - \frac{1}{\eta}\int_{t_{m_k}-\eta}^{t_{m_k}} g(s, \xi^k)ds\right| = \left|u_{m_k} - \frac{1}{\eta}\int_{t_{m_k}-\eta}^{t_{m_k}} g(s, x^k)ds\right| \leqslant \frac{1}{\eta} + \varepsilon$$

for $\eta \in (0, \min(h_0^k, \eta_0^k))$ and $k = 1, 2, \ldots$ Hence, similarly as above it follows that in the case $\sum_{i=0}^{m_k} h_i < a$ we also have $\dot{\xi}(t) = g(t, \xi)$ for a.e. $t \in I$.

Then we have proved that there exists $\xi \in K_\alpha$ such that $\xi(t) \in H - \Phi(0)$ for $t \in I$ and $\dot{\xi}(t) = g(t, \xi) \in (G\Box\mathscr{D})(t, \xi)$ for a.e. $t \in I$. Taking $u := \dot{\xi}$ we see that $u \in \Lambda_\alpha \subset \Lambda$, $u \in \mathscr{F}(G\Box\mathscr{I})(u)$ and $(\mathscr{I}u)(t) = \xi(t) \in H - \Phi(0)$ for $t \in I$. ■

Immediately from Lemma 6.4 and Theorem 6.5 we obtain the following theorem.

THEOREM 6.6. *Let $\Lambda \subset L(I, \mathbf{R}^n)$, $\Phi \in \mathrm{AC}_{0r}$ and a compact set $H \subset \mathbf{R}^n$ be given. Suppose $G \in \mathscr{H}_\lambda(I, \mathbf{R}^n)$ is s.-w.l.s.c. and has Volterra's property with respect to its last two variables. Assume Λ, Φ, H and G are such that $\Phi(0) \in H$ and there are numbers l, r, $M > 0$ such that $\Lambda_\alpha \subset \Lambda$ with $\alpha = l+M \leqslant \lambda/a$, $(\Phi\oplus x)_t(0) \in H$, $(G\Box\mathscr{D})(t, x) \subset MB \cap T_H[(\Phi\oplus x)(0)]$ for a.e. $t \in I$ and $x \in K_\alpha$ with $K_\alpha = \mathscr{J}(\Lambda_\alpha)$. If furthermore G is such that $(G\Box\mathscr{D})$ is on M-mapping, $(\mathscr{L}(I)\otimes\beta(K_\alpha), \beta(\mathrm{Comp}(\mathbf{R}^n))$ measurable then there is $u \in S(G)\cap\Lambda$ such that $(\mathscr{J}u)(t) \in H - \Phi(0)$ for $t \in I$.* ■

REMARK 6.2. If we take in Theorem 6.5, G such that instead of $(G\Box\mathscr{D})(t, x) \subset T_H[\Phi\oplus x)_t(0)]$ for $x \in K_\alpha$ and a.e. $t \in I$ only $(G\Box\mathscr{D})(t, x) \subset [T_H(\Phi\oplus x)_t(0)] + \varepsilon B$ for given $\varepsilon > 0$, a.e. $t \in I$ with $\alpha = l+M \leqslant \lambda/a$ and $x \in K_\alpha$, then for a given $\varepsilon > 0$ there exists $x^\varepsilon \in K_\alpha$ such that $x^\varepsilon(t) \in H - \Phi(0)$ for $t \in I$ and $\dot{x}^\varepsilon(t) \in (G\Box\mathscr{D})(t, x^\varepsilon) + \varepsilon B$ for a.e. $t \in I$. ■

REMARK 6.3. For every $u \in S(G)$ a function $x = \mathscr{I}u$ satisfies $\dot{x}(t) \in G(t, x, \dot{x})$ for a.e. $t \in I$. Therefore Theorems 6.2 and 6.5 can be also seen as viability theorems for functional-differential inclusions of the form $\dot{x}(t) \in G(t, x, \dot{x})$. ■

7. NOTES AND REMARKS

The trajectory integrals have been investigated systematically in Bridgland [1] and Hermes [2], Hiai and Unegaki [1] and Papageorgiou [1]. The first studies of the trajectory integrals have been connected with the investigations of Aumann integrals. Further motivation for the study of the trajectory integral has arisen in connection with the existence theory of differential inclusions and some related topics of optimal control. This approach has permited to provide the method for constructive proofs of existence, along with classical lines, thereby providing at same time a method of approximation to solutions. Systematical investigations of the subtrajectory integrals have been in Kisielewicz [4] taken up. The definition and properties of the weak convergence in the space $\mathscr{A}(I, \boldsymbol{R}^n)$ have been taken from Artstein [1], where mainly Borel measurable set-valued functions are investigated. It is not difficult to show (see for example Kisielewicz [4]) that for L-measurable set-valued function F: $I \to \text{Comp}(\boldsymbol{R}^n)$ there is a Borel measurable set-valued function equivalent to F. The ideas of the proofs of continuous selection theorems contained in this chapter arise from Fryszkowski [1]. The first result of this type has been obtained in Antosiewicz and Cellina [1]. Latter on, it was extended in Pianigiani [1] to the form applicable in the relaxation theory of differential inclusions $\dot{x}(t) \in F(t, x(t))$. Some extensions of Pianigiani's relaxation theorem on the case of differential inclusions in Banach spaces are given in Tolstonogov [1] and Papageorgiou [3]. The Antosiewicz–Cellina's and Pianigiani's results deal with set-valued functions F defined on a subsets of $\boldsymbol{R}^{n+1}$ such that $F(\cdot, x)$ is measurable for fixed x and $F(t, \cdot)$ is continuous for fixed t. The values of F are of course compact and not necessarily convex. The idea of the proof of the approximation Theorem 4.3 arise from Aubin and Cellina [1]. Also the idea of the proofs of viability Theorems contained in this chapter are related to the proofs of the theorems of such types given in Aubin and Cellina [1]. The proof of Lemma 6.3 is taken from Artstein and Prikry [1].

Chapter IV

Neutral functional-differential inclusions

There is a large, constantly growing literature on the subject of optimal processess governed by functional-differential equations. Such problems bring us in the natural way to fundamental problems of neutral functional-differential inclusions. The object of this chapter is to discuss and investigate fundamental problems of the theory of neutral functional-differential inclusions. We begin with the introducing of basic notion of the theory of nonlinear functional-differential equations and the theory of optimal control of systems described by such equations. Beside the properties of sets of solutions of neutral functional-differential inclusions also some viability and controllability problems for such systems are investigated here.

1. Neutral functional-differential equations and fundamental problems in optimal control theory of systems described by NFDEs

The base to the successful analysis of systems described by neutral functional-differential equations (NFDEs) is the proper choice of a topological space for solutions and this choice is depended upon the particular problem. The selection is usually motivated by a desire to give the solutions as much structure as possible. The most basic concern here is the continuity of solutions with respect to initial conditions and parameters. It turns out that the Sobolev space $\mathscr{W}^p([\alpha, \beta], \boldsymbol{R}^n)$ with $p \geqslant 1$ and the norm $||x||_{\alpha\beta} = |x(\alpha)| + [\int_\alpha^\beta |\dot{x}(\tau)|^p d\tau]^{1/p}$ are natural choice. The reason is quite simple that the number and location of discontinuities of $\dot{x}$ are not important in determining convergence.

From the point of view of optimal control problems it is natural to consider the space $\mathscr{X}_r(\Omega)$ defined in Section I.4 as solution space of NFDEs. The right-hand sides of NFDEs are defined on subsets of the metric space $(\boldsymbol{R} \times C_{0r} \times L_{0r}, d)$ with $r \geqslant 0$ and a metric d defined by $d[(t, x, z), (\bar{t}, \bar{x}, \bar{z})] := \max(|t-\bar{t}|, |x-\bar{x}|_0, |z-\bar{z}|_0)$ for $(t, x, z), (\bar{t}, \bar{x}, \bar{z}) \in \boldsymbol{R} \times C_{0r} \times L_{0r}$.

1.1. Neutral functional-differential equations

Let D be a nonempty open subset of $\boldsymbol{R} \times C_{0r} \times L_{0r}$ and f: $D \to \boldsymbol{R}^n$ be a function that is assumed to satisfy the following *local measurability condition* $(\mathscr{M})$:

$(\mathscr{M})$ *For every interval $I \subset \boldsymbol{R}$ and sets $\boldsymbol{K} \subset C_{0r}$ and $S \subset L_{0r}$ with $I \times \boldsymbol{K} \times S \subset D$ and every $\varepsilon > 0$ there exists a compact set $E \subset I$ with $\mu(I \setminus E) \leqslant \varepsilon$ such that the restriction of f to $E \times \boldsymbol{K} \times S$ is Borel measurable.*

In what follows by $\mathcal{M}(D, \boldsymbol{R}^n)$ we shall denote the space of all functions $f\colon D \to \boldsymbol{R}^n$ satisfying the local measurability condition $(\mathcal{M})$.

Similarly as in Section III.5.1, for a given $x \in \mathrm{AC}([t_1 - r, t_2], \boldsymbol{R}^n)$ we shall consider here a mapping $[t_1, t_2] \ni t \to x_t \in \mathrm{AC}_{0r} \subset C_{0r}$ defined by setting $x_t(s) := x(t+s)$ for each $s \in [-r, 0]$.

LEMMA 1.1. *For every* $x \in \mathrm{AC}([t_1 - r, t_2], \boldsymbol{R}^n)$ *a mapping* $[t_1, t_2] \ni t \to x_t \in \mathrm{AC}_{0r}$ *is weakly sequentially continuous on* $[t_1, t_2]$, *i.e., for each* $t \in [t_1, t_2]$ *and every sequence* (t_n) *of* $[t_1, t_2]$ *converging to* t, *the sequence* (x_{t_n}) *converges weakly in* AC_{0r} *to* x_t *as* $n \to \infty$.

Proof. Let $x \in \mathrm{AC}([t_1 - r, t_2], \boldsymbol{R}^n)$, $t \in [t_1, t_2]$, and a sequence (t_n) converging to t be given. Put $z_n := x_{t_n}$, $z := x_t$, $\dot{z}_n := \dot{x}_{t_n}$ and $\dot{z} := \dot{x}_t$ for $n = 1, 2, \ldots$ By Lemma III.5.1 we have $|z_n - z|_0 \to 0$ and $|\dot{z}_n - \dot{z}|_0 \to 0$ as $n \to \infty$. Then, in particular, $|z_n(s) - z(s)| \to 0$ for $s \in [-r, 0]$ and $\dot{z}_n \to \dot{z}$ in L_{0r} as $n \to \infty$. Therefore, by Dunford's theorem a sequence $(\dot{z}_n)$ is uniformly integrable and bounded. By Lemma I.4.3, hence it follows that a sequence (z_n) is bounded and equiabsolutely continuous. Thus, (z_n) converges weakly to z in AC_{0r}. ■

LEMMA 1.2. *For every* $f \in \mathcal{M}(D, \boldsymbol{R}^n)$, $I \times \boldsymbol{K} \times S \subset D$ *with* $I := [t_1, t_2]$, *each* $x \in C([t_1 - r, t_2], \boldsymbol{R}^n)$ *and* $u \in L([t_1 - r, t_2], \boldsymbol{R}^n)$ *such that* $(x_t, u_t) \in \boldsymbol{K} \times S$ *for* $t \in I$, *the mapping* $I \ni t \to f(t, x_t, u_t) \in \boldsymbol{R}^n$ *is L-measurable on* I.

Proof. By virtue of Lemma III.5.1, a mapping $I \ni t \to (t, x_t, u_t) \in \boldsymbol{R} \times C_{0r} \times L_{0r}$ is continuous. By the definition of $\mathcal{M}(D, \boldsymbol{R}^n)$, for every $\varepsilon > 0$ there exists a closed set $E_\varepsilon \subset I$ with $\mu(I \setminus E_\varepsilon) < \varepsilon$ such that the restriction $f|_{E_\varepsilon \times \boldsymbol{K} \times S}$ is Borel measurable. Then, by Remark II.3.3, a mapping $E_\varepsilon \ni t \to f(t, x_t, u_t) \in \boldsymbol{R}^n$ is measurable for every $\varepsilon > 0$. Now, the result follows immediately from Lusin's theorem. ■

Given $f \in \mathcal{M}(D, \boldsymbol{R}^n)$, $I \times \boldsymbol{K} \times S \subset D$ and $I = [t_1, t_2]$, a *neutral functional-differential equation* NFDE(D, f) we define as the relation

$$\dot{x}(t) = f(t, x_t, \dot{x}_t) \tag{1.1}$$

that is to be satisfied a.e. on I by any absolutely continuous function $x\colon [t_1 - r, t_2] \to \boldsymbol{R}^n$ such that $(x_t, \dot{x}_t) \in \boldsymbol{K} \times S$ for $t \in I$. Each such absolutely continuous function x is said to be a *local solution* of NFDE(D, f) on $[t_1 - r, t_2]$.

Given $(\sigma, \Phi) \in \boldsymbol{R} \times \mathrm{AC}_{0r}$ we say that $x\colon [t_1 - r, t_2] \to \boldsymbol{R}^n$ is a solution of NFDE(D, f) on $[t_1 - r, t_2]$ through (σ, Φ) if x is a solution of NFDE(D, f) on $[t_1 - r, t_2]$, $t_1 = \sigma$ and $x_\sigma = \Phi$.

Similarly as in the theory of ordinary differential equations we define for NFDEs the unique and continuable solutions through a given $(\sigma, \Phi) \in \boldsymbol{R} \times \mathrm{AC}_{0r}$.

Very often we are interested in solutions of NFDE(D, f) satisfying for each $t \in [t_1, t_2]$ the following additional condition $(t, x_t) \in A$, where A is a nonempty subset of $\boldsymbol{R} \times \mathrm{AC}_{0r}$. It is necessary to assume in such case that D and A are such that for every (t, x), $(t, y) \in A$ we also have $(t, x, \dot{y}) \in D$. Such problems for NFDE(D, f) is denoted by NFDE(D, A, f).

Let us observe that NFDE(D,f) can be expressed in the form

$$\dot{x}(t) = f(t, A(t)x, B(t)\dot{x}),$$

where for any $I \times \boldsymbol{K} \times S \subset D$ and $t \in I$, $A(t)$ and $B(t)$ denote the operators which to any continuous function x and any Lebesgue integrable function u, defined at least on the interval $[-r+t, t]$, associate the functions $A(t)x \in \boldsymbol{K}$ and $B(t)u \in S$ such that $(A(t)x)(s) = x(t+s)$ and $(B(t)u)(s) = u(t+s)$ for all or almost all $s \in [-r, 0]$.

A neutral functional-differential equation (1.1) is a very general type of functional-differential equations; particularly ordinary differential equations are obtained when $r = 0$. Other cases are considered in the following examples.

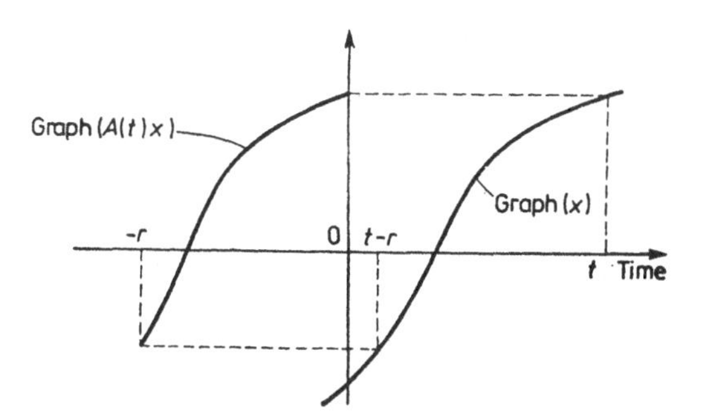

Fig. 7. Graphs of x and $A(t)x$.

EXAMPLE 1.1. Let G be an open connected subset of $\boldsymbol{R} \times \boldsymbol{R}^{2np}$ and denote by I_G and Π_G ranges of projections of G onto t-axis and onto the space $\boldsymbol{R}^{2np}$, respectively. Let ϱ_i, r_i: $I_G \to \boldsymbol{R}$ be for $i = 1, 2, \ldots, p$ nonnegative functions. Suppose g: $G \to \boldsymbol{R}^n$ is such that for every $(\bar{x}_1, \ldots, \bar{x}_{2p}) \in \Pi_G$ with $\bar{x}_i \in \boldsymbol{R}^n$, $g(\cdot, \bar{x}_1, \ldots, \bar{x}_{2p})$ is L-measurable and for every $\bar{t} \in I_G$, $g(\bar{t}, \cdot \ldots \cdot)$ is continuous. Consider a neutral difference-differential equation:

$$\begin{aligned} \dot{x}(t) = g\big(t, x(t-r_1(t)), \ldots, x(t-r_p(t)),\\ \dot{x}(t-\varrho_1(t)), \ldots, \dot{x}(t-\varrho_p(t))\big) \quad \text{for a.e. } t \in [t_1, t_2] \text{ with } [t_1, t_2] \subset I_G. \end{aligned} \tag{1.2}$$

Let $f(t, \mathscr{S}, \psi) = g(t, \mathscr{S}(-r_1(t)), \ldots, \mathscr{S}(-r_p(t)), \ \psi(-\varrho_1(t)), \ldots, \psi(-\varrho_p(t)))$ for each $(t, \mathscr{S}, \psi) \in D \subset \boldsymbol{R} \times C_{0r} \times L_{0r}$, where D is so taken that $(t, \mathscr{S}, \psi) \in D$ if and only if $(t, \mathscr{S}(-r_1(t)), \ldots, \mathscr{S}(-r_p(t)), \psi(-\varrho_1(t)), \psi(-\varrho_p(t))) \in G, 0 \leqslant r_i(t) \leqslant r$, and $0 \leqslant \varrho_i(t) \leqslant r$ for $i = 1, 2, \ldots, p$ and $t \in I_G$. Since, for every absolutely continuous function x: $I_G \to \boldsymbol{R}^n$ we have $f(t, x_t, \dot{x}_t) = g(t, x(t-r_1(t)), \ldots, x(t-r_p(t)), \dot{x}_t(t--\varrho_1(t)), \ldots, \dot{x}(t-\varrho_p(t)))$ for a.e. $t \in I_G$, then (1.2) can be written in the form of (1.1). ■

EXAMPLE 1.2. Let $G \subset \boldsymbol{R} \times \boldsymbol{R}^n$ be an open connected set and let k: $\boldsymbol{R} \times \boldsymbol{R} \times \boldsymbol{R}^{2n} \to \boldsymbol{R}^n$ be such that for every interval $[t_1, t_2] \subset I_G$ and every absolutely continuous functionn x: $[t_1-r, t_2] \to \boldsymbol{R}^n$ a function $[t_1-r, t] \ni s \to k(t, s, x(s), \dot{x}(s)) \in \boldsymbol{R}^n$ is

L-integrable on $[t-r, t]$ for fixed $t \in [t_1, t_2]$ and such that $(t, \int_{t-r}^{r} k(t, s, x(s), \dot{x}(s))ds) \in G$ for $t \in [t_1, t_2]$. Suppose $g\colon G \to \mathbf{R}^n$ is like in Example 1.1 and let us consider a neutral Volterra equation:

$$\dot{x}(t) = g\Big(t, \int_{t-r}^{t} k\big(t, s, x(s), \dot{x}(s)\big)ds\Big) \quad \text{for a.e. } t \in [t_1, t_2]. \tag{1.3}$$

Let $D \subset \mathbf{R} \times C_{0t} \times L_{0r}$ be such that $(t, \mathscr{S}, \psi) \in D$ if and only if $(t, \int_{-r}^{0} k(t, t+\tau, \mathscr{S}(\tau), \psi(\tau))d\tau) \in G$ and then put $f(t, \mathscr{S}, \psi) = g(t, \int_{-r}^{0} k(t, t+\tau, \mathscr{S}(\tau), \psi(\tau))d\tau)$ for $(t, \mathscr{S}, \psi) \in D$. Since for every absolutely continuous function $x\colon I_G \to \mathbf{R}^n$ and each $t \in [t_1, t_2] \subset I_G$ we have $f(t, x_r, \dot{x}_r) = g(t, \int_{-r}^{0} k(t, t+\tau, x(\tau), x(\tau))d\tau) = g(t, \int_{t-r}^{t} k(t, s, x(s), x(s))ds)$, then (1.3) can be written in the form of (1.1). ■

EXAMPLE 1.3. Let for fixed $t \geqslant 0$ $P(t)$ and $Q(t)$ be operators from $C([-r, \infty), \mathbf{R}^n)$ and $L([-r, \infty), \mathbf{R}^n)$, respectively into given spaces X and Y such that $P(t)x = P(t)y$ and $Q(t)u = Q(t)v$ whenever $x = y$ and $u = v$ on $[t-r, t]$, respectively. For $G \subset \mathbf{R}^+ \times X \times Y$ and $g\colon G \to \mathbf{R}^n$ satisfying conditions mentioned in Example 1.1, let us consider the following neutral differential trajectory processing equation:

$$\dot{x}(t) = g\big(t, P(t)x, Q(t)\dot{x}\big) \quad \text{for a.e. } t \in I_G. \tag{1.4}$$

For every $\mathscr{S} \in C_{0r}$, $\psi \in L_{0r}$ and any $t \geqslant 0$ we define $R(t)\mathscr{S} \in C([-r, \infty), \mathbf{R}^n)$ and $S(t)\psi \in L([-r, \infty), \mathbf{R}^n)$ by setting

$$R(t)\mathscr{L}(\tau) = \begin{cases} \mathscr{S}(\tau-t) & \text{for } t-r \leqslant \tau \leqslant t, \\ \mathscr{S}(0) & \text{for } \tau \geqslant t \end{cases}$$

and

$$S(t)\psi(\tau) = \begin{cases} \psi(\tau-t) & \text{for } t-r \leqslant \tau \leqslant t, \\ 0 & \text{for } \tau \geqslant t. \end{cases}$$

Let $D \subset \mathbf{R} \times C_{0r} \times L_{0r}$ be such that $(t, \mathscr{S}, \psi) \in D$ if and only if $(t, P(t)R(t)\mathscr{S}, Q(t)S(t)\psi) \in G$ and define $f\colon D \to \mathbf{R}^n$ by setting $f(t, \mathscr{S}, \psi) = g(t, P(t)R(t)\mathscr{S}, Q(t)S(t)\psi)$ for $(t, \mathscr{S}, \psi) \in D$. Since for a given absolutely continuous function $x\colon [t_1-r, t_2] \to \mathbf{R}^n$ with $[t_1, t_2] \subset I_G$ we have $(R(t)x_t)(\tau) = x(\tau)$ for fixed $t \in [t_1, t_2]$ and $\tau \in [t-r, t]$ then by properties of a mapping $P(t)$ we have $P(t)R(t)x_t = P(t)x$ for each fixed $t \in [t_1, t_2]$. Analogously, for a.e. $t \in [t_1, t_2]$ we obtain $Q(t)S(t)\dot{x}_t = Q(t)\dot{x}$. Therefore, for a.e. $t \in [t_1, t_2]$ we have $f(t, x_t, \dot{x}_t) = g(t, P(t)x, Q(t)\dot{x})$ whenever $x\colon [t_1-r, t_2] \to \mathbf{R}^n$ is such that $(t, x_t, \dot{x}_t) \in D$. Thus (1.4) can be written in the form of (1.1). ■

There is a lot number of applications in which the delayed argument occurs in the derivative of the state variable as well as in the independent variable. They turn out to neutral differential-difference equations. Such problems are more difficult to motivate but often arise in the study of two or more simply oscillatory systems with some interconnections between them. We will consider now the problem of a transmision line when a nonlinear capacitance is placed in shunt in the line.

EXAMPLE 1.4. Consider the network shown in Fig. 8. The equations governing the current $i(x, t)$ and voltage $v(x, t)$ are

$$\frac{\partial v}{\partial x} = -L\frac{\partial i}{\partial t}, \quad \frac{\partial i}{\partial x} = -C\frac{\partial v}{\partial t}, \quad -l_1 < x < l_2,$$

where L and C denote the specific inductance and capacitance in the line; the boundary conditions are:

$$\frac{\partial v}{\partial x} + Cz_1\frac{\partial v}{\partial t} = 0 \quad \text{at } x = l_2, \tag{1.5}$$

$$0 = E(t) - v - R_0 i \quad \text{at } x = -l_1. \tag{1.6}$$

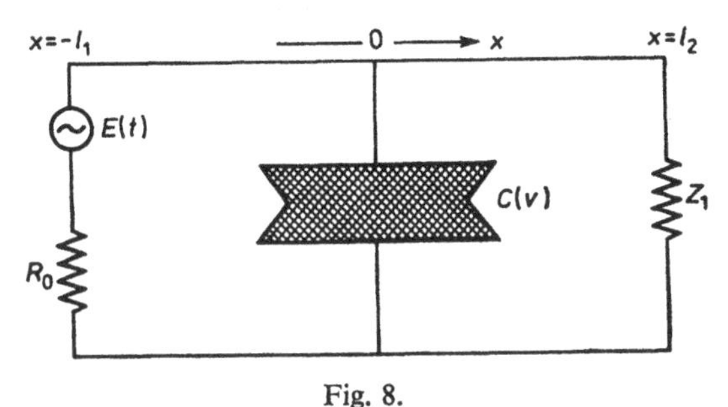

Fig. 8.

Let v_l and i_l denote the voltage and current for $x < 0$ and v_r and i_r the same things for $x > 0$. Then we have the interface condition

$$i_r - i_l = -C(v)\dot{v}(t), \tag{1.7}$$

where $v(t) = v(0, t)$ and $C(v) = -dq/dv$, with q equal to the charge on the capacitor. If we let $\sigma = 1/\sqrt{LC}$, $z = \sqrt{C/L}$, then we have

$$v_l = \frac{1}{2}[\Phi_l(x-\sigma t) + \psi_l(x+\sigma t)],$$

$$i_l = \frac{1}{2z}[\Phi_l(x-\sigma t) - \psi_l(x+\sigma t)],$$

and analogous formulas for v_r and i_r. Substituting into (1.6) and solving for Φ_l, we get

$$\Phi_l(-l_1-\sigma t) = \tilde{E}(t) + q\psi_l(-l_1+\sigma t), \tag{1.8}$$

where

$$\tilde{E}(t) = \frac{2zE(t)}{z+R_0}, \quad q = \frac{z-R_0}{z+R_0}.$$

From (1.5) we see that $-L(\partial i/\partial t) + Cz_1(\partial v/\partial t) = 0$ at $x = l_2$ and then $-Li+$

$+Cz_1 v = \text{const}$ at $x = l_2$. Replacing i and v by their general expression, we get

$$\mu\Phi_r(l_2-\sigma t)+\psi_r(l_2+\sigma t) = k_1, \tag{1.9}$$

where k_1 is a constant and $\mu = (z_1-z)/(z_1+z)$. The continuity of the voltage at the interface gives

$$\Phi_l(-\sigma t)+\psi_l(\sigma t) = \Phi_r(-\sigma t)+\psi_r(\sigma t) = 2v(t), \tag{1.10}$$

and equation (1.7) yields

$$\Phi_r(-\sigma t)-\psi_r(\sigma t)-\Phi_l(-\sigma t)+\psi_l(\sigma t) = -2zC(v)\dot{v}(t). \tag{1.11}$$

If we replace t by $t-2l_1/\sigma$ in (1.11), multiply the resulting equation by q and add it to (1.11), we obtain

$$\begin{aligned}&2z[C(v)\dot{v}(t)+qC(v(t-2l_1/\sigma))\dot{v}(t-2l_1/\sigma)]\\&= -\ [\Phi_r(-\sigma t)-\psi_r(\sigma t)-\Phi_l(-\sigma t)+\psi_l(\sigma t)+q\Phi_r(-\sigma t+2l_1)-\\&\quad -q\psi_r(\sigma t-2l_1)-q\Phi_l(-\sigma t+2l_1)+q\psi_l(\sigma t-2l_1)].\end{aligned}$$

If we call A the left-hand side of this last equality, then, from (1.8) we see that $\Phi_l(-\sigma t) = \tilde{E}(t-l_1/\sigma)+q\psi_l(-2l_1+\sigma t)$, and hence

$$\begin{aligned}A = &-[\Phi_r(-\sigma t)-\psi_r(\sigma t)-\tilde{E}(t-l_1/\sigma)+\psi_l(\sigma t)+\\&+q\Phi_r(-\sigma t+2l_1)-q\psi_r(\sigma t-2l_1)-q\Phi_l(-\sigma t+2l_1),\end{aligned}$$

and from (1.10), we can write

$$\begin{aligned}&2z[C(v)(t)\dot{v}+qC(v(t-2l_1/\sigma))\dot{v}(t-2l_1/\sigma)]\\&= -[4v(t)-2\psi_r(\sigma t)-2\tilde{E}(t-l_1/\sigma)-2q\psi_r(\sigma t-2l_1)].\end{aligned} \tag{1.12}$$

Now, from (1.9) and (1.10) we have

$$\mu[2v(t-l_2/\sigma)-\psi_r(\sigma t-l_2)]+\psi_r(l_2+\sigma t) = k_1,$$

or, replacing t by $t-l_2/\sigma$,

$$\psi_r(\sigma t)-\mu\psi_r(\sigma t-2l_2) = 2\mu v(t-2l_2/\sigma)+k_1. \tag{1.13}$$

If we multiply (1.12) by μ, replace t by $t-2l_2/\sigma$, subtract the resulting equality from (1.12) and use (1.13), we obtain

$$\begin{aligned}&2z[C(v)\dot{v}(t)+qC(v(t-2l_1/\sigma))\dot{v}(t-2l_1/\sigma)]-\\&\quad -2z\mu[C(v(t-2l_2/\sigma))\dot{v}(t-2l_2/\sigma)+qC(v(t-2l_1/\sigma-2l_2/\sigma))\\&\quad \times \dot{v}(t-2l_1/\sigma-2l_2/\sigma)\\&= [4v(t)-2\psi_r(\sigma t)-2\tilde{E}(t-l_1/\sigma)-2q\psi_r(\sigma t-2l_1)]+\\&\quad +\mu[4v(t-2l_2/\sigma)-2\psi_r(\sigma t-2l_2)-2\tilde{E}(t-l_1/\sigma-2l_2/\sigma)-2q\psi_r(\sigma t-2l_1-2l_2)]\\&= -4v(t)+2\tilde{E}(t-l_1/\sigma)+4\mu v(t-2l_2/\sigma)-2\mu\tilde{E}(t-1_1/\sigma-2l_2/\sigma)-\\&\quad -4\mu v(t-2l_2/\sigma)+2k_1-4q\mu v(t-2l_1/\sigma-2l_2/\sigma)+2qk_1\\&= -4v(t)+2\tilde{E}(t-l_1/\sigma)-2\mu\tilde{E}(t-l_1/\sigma-2l_2/\sigma)+2k_1-\\&\quad -4q\mu v(t-2l_1/\sigma-2l_2/\sigma)+2qk_1.\end{aligned}$$

Define the function $\tilde{C}(v) = \int_0^v C(s)ds$ and put $x(t) = \tilde{C}(v(t))$. Assuming that $d\tilde{C}/dv$

$= C(v)$ is nonnegative and is not identically zero on any interval, we see that there is an inverse function $v(t) = h(x(t))$. If we set

$$r_1 = 2l_1/\sigma, \quad r_2 = 2l_2/\sigma, \quad g(x) = 2h(x)/z,$$

$$p(t) = \frac{1}{2z}[2\tilde{E}(t-l_1/\sigma)-2\mu\tilde{E}(t-l_1/\sigma-2l_2/\sigma)+2k_1+2qk_1],$$

then $p(t)$ is ω-periodic if $E(t)$ is, and $x(t)$ satisfies

$$\frac{d}{dt}[x(t)+qx(t-r_1)-\mu x(t-r_2)-\mu qx(t-r_1-r_2)]$$
$$= -g(x(t))+q\mu g(x(t-r_1-r_2))+p(t). \blacksquare \tag{1.14}$$

Finally, we shall illustrate some of the basic ideas of the calculation of solutions of linear neutral difference-differential equations.

EXAMPLE 1.5. Use the continuation process to calculate the solution of

$$\begin{aligned} &\dot{x}(t) = x(t-1)+2\dot{x}(t-1) \quad &&\text{for a.e. } t \in [1, \infty) \\ &x(t) = 1 &&\text{for } t \in [0, 1]. \end{aligned} \tag{1.15}$$

We shall consider (1.15) in turn in intervals $(n, n+1)$ for $n = 1, 2, \ldots$

If the equation $x(t) = \dot{x}(t-1)+2\dot{x}(t-1)$ is to be satisfied for $t > 1$, we must have $\dot{x}(t) = 1$ for $t \in (1, 2)$, and therefore $x(t) = t$ for $t \in [1, 2]$. If equation of (1.15) is to be satisfied for $t \in (2, 3)$, we must have $\dot{x}(t) = t+1$ for $t \in (2, 3)$, and $x(t) = \frac{1}{2}t^2+t-2$ for $t \in [2, 3]$. By continuing this process, we can continue x as far forwards as we like, though it seems not to be easy to find a general formula for the solution obtained in this way. Moreover, the solution x obtained above has a derivative discontinuous at every positive integer value t. Equation of (1.15) is satisfied at such values only in the sense of left-hand limits and of right-hand limits. In fact, though there are exceptions, it is in general not true that a solution of (1.15) is continuous and has a continuous first derivative for all $t > 1$. We may say that equation (1.15) fails to "smooth out" a discontinuity of $\dot{x}(t)$ at $t = 1$. $\blacksquare$

EXAMPLE 1.6. Let $L(x)(t) \quad a_0\dot{x}(t)+a_1\dot{x}(t-r)+b_0x(t)+b_1x(t-r)$, where $a_0, a_1, b_0, b_1 \in \boldsymbol{R}$ are such that $a_0 \cdot a_1 \neq 0$. Suppose $f \in C^0([0, \infty), \boldsymbol{R})$ and $g \in C^1([0, r], \boldsymbol{R})$ are given.

Using the Laplace transform method we define a solution of the initial value problem

$$\begin{aligned} &L(x)(t) = f(t) \quad &&\text{for a.e. } t \in [r, \infty), \\ &x(t) = g(t) &&\text{for } t \in [0, r]. \end{aligned} \tag{1.16}$$

To do this let us assume that there are positive numbers C_1 and C_2 such that $|f(t)| \leqslant C_1e^{C_2t}$ for $t \geqslant 0$ and let $h(s) = a_0s+a_1se^{-rs}+b_0+b_1e^{-rs}$ be the characteristic function of the linear operator L defined above. Using the relations

$$\int_r^\infty \dot{x}(t)e^{-st}dt = -g(r)e^{-rs}+s\int_r^\infty x(t)e^{-st}dt,$$

$$\int_r^\infty x(t-r)e^{-st}dt = e^{-rs}\int_r^\infty x(t)e^{-st}dt + e^{-rs}\int_0^r g(t)e^{-st}dt, \tag{1.17}$$

$$\int_r^\infty \dot{x}(t-r)e^{-st}dt = -g(0)e^{-rs} + s\int_r^\infty x(t-s)e^{-st}dt$$

we obtain

$$h(s)\int_r^\infty x(t)e^{-st}dt = p_0(s) + q(s),$$

where

$$p_0(s) = a_0 g(r)e^{-rs} + a_1 g(0)e^{-rs} - (a_1 s + b_1)e^{-rs}\int_0^r g(t)e^{-st}dt$$

and

$$q(s) = \int_r^\infty f(t)e^{-st}dt.$$

Or, instead of the first of the relations in (1.17), we can use

$$\int_r^\infty \dot{x}(t)e^{-st}dt = -g(r)e^{-rs} + s\int_0^\infty x(t)e^{-st}dt - s\int_0^\infty g(t)e^{-st}dt$$

$$\int_r^\infty x(t-r)e^{-st}dt = e^{-rs}\int_0^\infty x(t)e^{-st}dt,$$

and obtain

$$h(s)\int_0^\infty x(t)e^{-st}dt = p(s) + q(s),$$

where

$$p(s) = a_0 g(r)e^{-rs} + a_1 g(0)e^{-rs} + (a_0 s + b_0)\int_0^r g(t)e^{-st}dt.$$

Integration by parts shows that

$$p_0(s) = a_0 g(r)e^{-rs} + a_1 g(r)e^{-2rs} - e^{-rs}\int_0^r [a_1\dot{g}(t) + b_1 g(t)]e^{-st}dt$$

and

$$p(s) = a_0 g(0) + a_1 g(0)e^{-rs} + \int_0^r [a_0\dot{g}(t) + b_0 g(t)]e^{-rt}dt.$$

We now deduce that for any sufficiently large real number C, we have

$$x(t) = \int_{(C)} e^{ts}h^{-1}(s)[p_0(s) + q(s)]ds \quad \text{for } t > r$$

or

$$x(t) = \int_{(C)} e^{ts} h^{-1}(s)[p(s)+q(s)]ds \quad \text{for } t > 0,$$

where p_0, p and q are defined above and

$$\int_{(C)} F(s)e^{st}ds = \lim_{T\to\infty} \frac{1}{2\pi i} \int_{C-iT}^{C+iT} F(s)e^{st}ds$$

for every F such that the right-hand side of the above equality exists. ■

1.2. Mayer problem of optimal control theory of systems described by NFDEs

Here we are concerned with the Mayer problem of optimal control theory of systems described by NFDE(D,f) of the form

$$\dot{x}(t) = f(t, x_t, \dot{x}_t, u(t)), \tag{1.18}$$

where u is called a *control function.*

Let D and A be nonempty open subsets of $\boldsymbol{R}\times C_{0r}\times L_{0r}$ and $\boldsymbol{R}\times \mathrm{AC}_{0r}$, respectively such that for every (t, x), $(t, y) \in A$ we have $(t, x, \dot{y}) \in D$. For every $(t, x, z) \in D$ let $U(t, x, z)$ be a given subset of the space $\boldsymbol{R}^m$ and let $M_D = \{(t, x, z, w)\colon (t, x, z) \in D,\ w \in U(t, x, z)\}$. Let C be a nonempty subset of $\boldsymbol{R}\times C_{0r}\times\boldsymbol{R}\times C_{0r}$ and Ω a bounded connected subset of A. Suppose $f\colon M_D \to \boldsymbol{R}^n$ and $g\colon C \to \boldsymbol{R}$ are given. Put $I[x, u] = g(t_1, x_{t_1}, t_2, x_{t_2})$ for every pair (x, u) of functions $(x, [t_1-r, t_2]) \in \mathscr{X}(\Omega)$ and $(u\colon [t_1, t_2]) \in \bigcup_{[\tau_1,\tau_2]\subset I_\Omega} \mathfrak{M}([\tau_1, \tau_2], \boldsymbol{R}^m)$ with $t_2-t_1 \geqslant r > 0$ satisfying

$$\begin{aligned} &\dot{x}(t) = f(t, x_t, \dot{x}_t, u(t)) && \text{for a.e. } t \in [t_1, t_2],\\ &(t, x_t) \in \Omega && \text{for } t \in [t_1, t_2],\\ &u(t) \in U(t, x_t, \dot{x}_t) && \text{for a.e. } t \in [t_1, t_2],\\ &(t_1, x_t, t_2, x_t) \in C. \end{aligned} \tag{1.19}$$

where $\mathfrak{M}([\tau_1, \tau_2], \boldsymbol{R}^m)$ denotes the space of all L-measurable functions $u\colon [\tau_1, \tau_2] \to \boldsymbol{R}^m$.

The problem of the minimization or maximization of the functional $I[x, u]$ for pairs $(x, u) \in \mathrm{AC}([t_1-r, t_2], \boldsymbol{R}^n)\times\mathfrak{M}([t_1, t_2], \boldsymbol{R}^m)$ with $t_2-t_1 \geqslant r > 0$ satisfying conditions (1.19) is said to be the *Mayer problem of optimal control* for systems described by NFDE (1.18). It is denoted by $\mathscr{M}(D,f, \Omega, C, g)$.

A pair $(x, u) \in \mathrm{AC}([t_1-r, t_2], \boldsymbol{R}^n)\times\mathfrak{M}([t_1, t_2], \boldsymbol{R}^m)$ with $t_2-t_1 \geqslant r > 0$ satisfying conditions (1.19) is said to be *admissible for the Mayer problem* $\mathscr{M}(D,f, \Omega, C, g)$ and x is said to be an *admissible trajectory*, and u an *admissibly strategy*, or *control function.*

The sets of all admissible pairs and all admissible trajectories for the Mayer problem $\mathscr{M}(D,f, \Omega, C, g)$ will be denoted by $\mathscr{A}(D,f, \Omega, C)$ and $\mathscr{T}(D,f, \Omega, C)$, respectively.

REMARK 1.1. Sometimes it is interesting to consider Euclidean form of the Mayer problem of optimal control. To get it, we must change $I[x, u]$ by $\tilde{I}[x, u] = \tilde{g}(t_1, x_{t_1}, t_2, x(t_2))$ defined for all pairs $(x, u) \in \mathrm{AC}([t_1-r, t_2], \mathbf{R}^n) \times \mathfrak{M}([t_1, t_2], \mathbf{R}^m)$ with $t_2 - t_1 > 0$ satisfying

$$\begin{aligned} &\dot{x}(t) = f(t, x_t, \dot{x}_t, u(t)) && \text{for a.e. } t \in [t_1, t_2], \\ &(t, x_t) \in \Omega && \text{for } t \in [t_1, t_2], \\ &u(t) \in U(t, x_t, \dot{x}_t) && \text{for a.e. } t \in [t_1, t_2] \\ &(t_1, x_{t_1}, t_2, x(t_2)) \in E \end{aligned}$$

instead of (1.19), where $E \subset \mathbf{R} \times \mathrm{AC}_{0r} \times \mathbf{R}^{n+1}$ and $\tilde{g}: E \to \mathbf{R}$. ■

The system described by NFDE, (1.18) is called *compatible* or *controllable with respect to the Mayer problem* $\mathcal{M}(D, f, \Omega, C, g)$ if $\mathcal{A}(D, f, \Omega, C) \neq \emptyset$.

Graphic illustrations of some attainable trajectories of the controllable Mayer problem and its Euclidean form are given in Fig. 9 and Fig. 10, respectively, where $C = \{(t_1, \Phi, t_2, \psi)\}$ and $E = \{(t_1, \Phi, t_2, x_2)\}$ for given $t_1, t_2 \in \mathbf{R}$, $\Phi, \psi \in \mathrm{AC}_{0r}$ and $x_2 \in \mathbf{R}$.

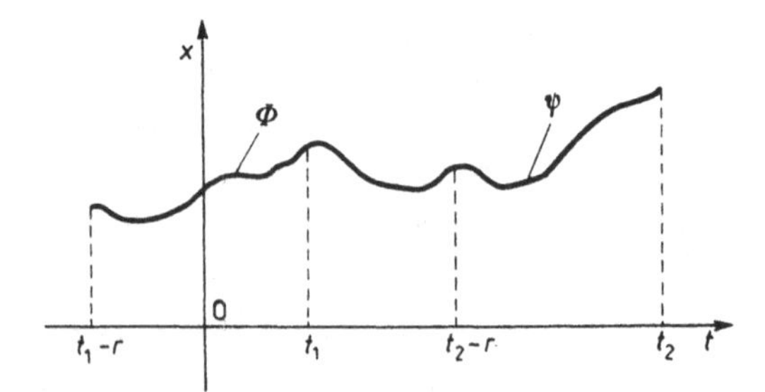

Fig. 9.

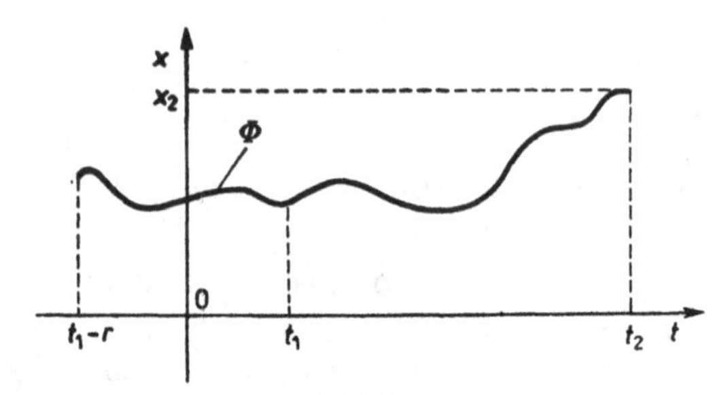

Fig. 10.

Let us observe that the controllability concept can be in some cases expressed by the nonemptness of the set $\mathcal{T}(D, f, \Omega, C)$ of all attainable trajectories. To do that let us denote by $F(t, x, z)$ for $(t, x, z) \in D$ the set of the form $\{v \in \mathbf{R}^n: v = f(t, x, z, w),\ w \in U(t, x, z)\}$, i.e.,

$$F(t, x, z) = f(t, x, z, U(t, x, z)) \quad \text{for } (t, x, z) \in D. \tag{1.20}$$

It is easily seen that for every $(x, u) \in \mathcal{A}(D, f, \Omega, C)$, an attainable trajectory

x of $\mathcal{M}(D, f, \Omega, C, g)$ satisfies

$$\begin{aligned} &\dot{x}(t) \in F(t, x_t, \dot{x}_t) && \text{for a.e. } t \in [t_1, t_2], \\ &(t, x_t) \in \Omega && \text{for } t \in [t_1, t_2], \\ &(t_1, x_{t_1}, t_2, x_{t_2}) \in C. \end{aligned} \tag{1.21}$$

Furthermore, we have the following lemma.

LEMMA 1.1. *Let* U: $D \to \mathrm{Comp}(\boldsymbol{R}\ )$ *be a set-valued function such that for every* $x \in \mathrm{AC}([t_1 - r, t_2], \boldsymbol{R}^n)$ *satisfying* $(t, x_t) \in \Omega$ *for* $t \in [t_1, t_2]$ *a set-valued function* Γ: $[t_1, t_2] \ni t \to U(t, x_t, \dot{x}_t) \in \mathrm{Comp}(\boldsymbol{R}^m)$ *is measurable. Suppose* f: $D \times \boldsymbol{R}^m \to \boldsymbol{R}^n$ *is such that*

(i) $f(t, x, z, \cdot)$ *is continuous on* $\boldsymbol{R}^m$ *for fixed* $(t, x, z) \in D$,

(ii) *for every* $x \in \mathrm{AC}([t_1 - r, t_2], \boldsymbol{R}^n)$ *satisfying* $(t, x_t) \in \Omega$ *for* $t \in [t_1, t_2]$ *a function* $[t_1, t_2] \ni t \to f(t, x_t, \dot{x}_t, w) \in \boldsymbol{R}^m$ *is L-integrable on* $[t_1, t_2]$ *for each fixed* $w \in \boldsymbol{R}^m$. *Then for every* $x \in \mathrm{AC}([[t_1 - r, t_2], \boldsymbol{R}^n)$ *satisfying* (1.21) *with* $t_2 - t_1 \geqslant r > 0$ *and* F *defined by* (1.20) *there exists* $u \in \mathfrak{M}([t_1, t_2], \boldsymbol{R}^m)$ *such that* $(x, u) \in \mathcal{A}(D, f, \Omega, C)$.

Proof. Suppose $x \in \mathrm{AC}([[t_1 - r, t_2], \boldsymbol{R}^n)$ satisfies (1.21) with $t_2 - t_1 \geqslant r > 0$ and let h: $[t_1, t_2] \times \boldsymbol{R}^m \to \boldsymbol{R}^n$ and Γ: $[t_1, t_2] \to \mathrm{Comp}(\boldsymbol{R}^m)$ be defined by

$$h(t, w) = f(t, x_t, \dot{x}_t, w) \quad \text{for } (t, w) \in [t_1, t_2] \times \boldsymbol{R}^m$$

and

$$\Gamma(t) = U(t, x_t, \dot{x}_t) \quad \text{for a.e. } t \in [t_1, t_2].$$

It is easily seen that $h(\cdot, w)$ is measurable on $[t_1, t_2]$ for fixed $w \in \boldsymbol{R}^m$, Γ is measurable on $[t_1, t_2]$ and $h(t, \cdot)$ is continuous on $\boldsymbol{R}^m$ for fixed $t \in [t_1, t_2]$ and $\dot{x}(t) \in h(t, \Gamma(t))$ for a.e. $t \in [t_1, t_2]$. Then by Theorem II.3.12, there exists a measurable selection u: $[t_1, t_2] \to \boldsymbol{R}^m$ of Γ such that $\dot{x}(t) = h(t, u(t))$ for a.e. $t \in [t_1, t_2]$, i.e., such that $\dot{x}(t) = f(t, x_t, \dot{x}_t, u(t))$ for a.e. $t \in [t_1, t_2]$. Furthermore, by the definition of Γ we have $u(t) \in U(t, x_t, \dot{x}_t)$ for a.e. $t \in [t_1, t_2]$. Then $(x, u) \in \mathcal{A}(D, f, \Omega, C)$. ∎

Now we can formulate the following equivalence theorem.

THEOREM 1.2. *Let* f: $D \times \boldsymbol{R}^m \to \boldsymbol{R}^n$ *and* U: $D \to \mathrm{Comp}(\boldsymbol{R}^m)$ *be such as in Lemma* 1.1. *Then the Mayer problem* $\mathcal{M}(D, f, \Omega, C, g)$ *is equivalent to the problem of minimization of the functional* $H[x] = g(t_1, x_{t_1}, t_2, x_{t_2})$ *for* $(x;\ [t_1 - r, t_2]) \in \mathcal{X}_r(\Omega)$ *satisfying* (1.21) *with* $t_2 - t_1 \geqslant r > 0$ *and* F *defined by* (1.20). ∎

Often a given problem has no optimal solution, but the mathematical problem and the corresponding set of solutions can be modified in such a way that an optimal solution exists, and yet neither the system of trajectories nor the corresponding values of the cost functional are essentially modified. The modified (or relaxed) problem and its solutions are of interest in themselves, and often have relevant physical interpretations.

Now, we introduce relaxed solutions as usual problems involving a finite number of ordinary strategies, which are thought of as being used at the same time according to some probability distribution. Briefly, instead of considering the usual cost functional, functional-differential systems, boundary conditions, and constrains of the form (1.19) we shall consider new cost functional, functional-differential system, boundary conditions, and constrains

$$\begin{aligned} &J[x,p,v] = g(t_1, x_{t_1}, t_2, x_{t_2}), \\ &\dot{x}(t) = h(t, x_t, \dot{x}_t, p(t), v(t)) && \text{for a.e. } t \in [t_1, t_2], \\ &(t, x_t) \in \Omega && \text{for } t \in [t_1, t_2], \\ &v(t) \in V(t, x_t, \dot{x}_t), p(t) \in \Gamma && \text{for a.e. } t \in [t_1, t_2], \\ &(t_1, x_{t_1}, t_2, x_{t_2}) \in C, \end{aligned} \tag{1.22}$$

where $h = (h_1, \ldots, h_m)$, $x = (x_1, \ldots, x_n)$, $p = (p_1, \ldots, p_\gamma)$, $v = (u^1, \ldots, u^\gamma)$ and

$$h(t, x, z, p, v) = \sum_{j=1}^{\gamma} p f(t, x, z, u^j), \; V(t, x, z) = \underbrace{U(t, x, z) \times \ldots \times U(t, x, z)}_{\gamma} \subset \boldsymbol{R}^{m\gamma}$$

and $\Gamma = \{p\colon p_j \geqslant 0, j = 1, \ldots, \gamma, p_1 + \ldots + p_\gamma = 1\}$.

Relations (1.22) can be written in the equivalent form

$$\begin{aligned} &J_0[x] = g(t_1, x_{t_1}, t_2, x_{t_2}), \\ &\dot{x}(t) \in \overline{\text{co}}\, F(t, x_t, \dot{x}_t) && \text{for a.e. } t \in [t_1, t_2], \\ &(t, x_t) \in \Omega && \text{for } t \in [t_1, t_2], \\ &(t_1, x_{t_1}, t_2, x_{t_2}) \in C, \end{aligned} \tag{1.23}$$

where F is defined by (1.20) and $\overline{\text{co}}\, F(t, x_t, \dot{x}_t)$ denotes a closed convex hull of $F(t, x_t, \dot{x}_t)$.

1.3. Lagrange problem of optimal control theory of systems described by NFDEs

Let sets D, A, $\Omega \subset A$, $U(t, x, z)$, M_D and C be such as in 1.2 and suppose $f\colon M_D \to \boldsymbol{R}^n$, $f_0\colon M_D \to \boldsymbol{R}$ and $g\colon C \to \boldsymbol{R}$ are given. Assume f and f_0 are such that there exists $(x, u) \in \text{AC}([t_1 - r, t_2], \boldsymbol{R}^n) \times \mathfrak{M}([t_1, t_2], \boldsymbol{R}^m)$ with $t_2 - t_1 \geqslant r > 0$ and $(t, x_t, \dot{x}_t, u(t)) \in M_D$ for $t \in [t_1, t_2]$ such that functions $[t_1, t_2] \ni t \to f(t, x_t, \dot{x}_t, u(t)) \in \boldsymbol{R}^n$ and $[t_1, t_2] \ni t \to f_0(t, x_t, \dot{x}_t, u(t)) \in \boldsymbol{R}$ are L-integrable on $[t_1, t_2]$. Put

$$I[x, u] = \int_{t_1}^{t_2} f_0(t, x_t, \dot{x}_t, u(t))dt \tag{1.24}$$

for all $(x, u) \in \text{AC}(t_1 - r, t_2], \boldsymbol{R}^u) \times \mathfrak{M}([t_1, t_2], \boldsymbol{R}^m)$ such that $(t, x_t, \dot{x}_t, u(t)) \in M_D$ for $t \in [t_1, t_2]$ and such that the last integral exists.

The problem of the minimization or maximization of the functional $I[x, u]$ defined by (1.24) for pairs $(x, u) \in \text{AC}([t_1 - r, t_2], \boldsymbol{R}^n) \times \mathfrak{M}([t_1, t_2], \boldsymbol{R}^m)$ with $t_2 - t_1 \geqslant r > 0$ satisfying conditions (1.19) and such that the integral in (1.24) exists is said to be the Lagrange problem of optimal control for system described by NFDE (1.18). It will be denoted by $\mathscr{L}(D, f, \Omega, C, f_0)$.

The problem of minimization or maximization of the functional

$$I[x, u] = g(t_1, x_{t_1}, t_2, x_{t_2}) + \int_{t_2}^{t_1} f_0(t, x_t, \dot{x}_t, u(t))dt \tag{1.25}$$

for pairs (x, u) satisfying conditions mentioned above in the definition of $\mathscr{L}(D, f, \Omega, C, f_0)$ is said to be the *Bolza problem of optimal control* for system described by NFDE (1.18). It will be also denoted by $\mathscr{L}(D, f, \Omega, C, g, f_0)$ and also called the *Lagrange optimal control problem.* Sometimes it is comfortable to consider an equivalent form of the Lagrange or Bolza problem instead of them. To obtain it we introduce for every $(t, x, z) \in D$ sets $F(t, x, z)$ and $\tilde{F}(t, x, z)$ by setting $F(t, x, z) = f(t, x, z, U(t, x, z))$ and

$$\tilde{F}(t, x, z) = \{(v, v_0)\colon v_0 \geqslant f_0(t, x, z, w),\ v \in F(t, x, z)\}.$$

Furthermore, for every $(t, x, z) \in D$ we put $T(t, x, z, v) = \inf\{v_0\colon (v, v_0) \in \tilde{F}(t, x, z)\}$. In general we can have $-\infty \leqslant T(t, x, z, v) \leqslant +\infty$. But here $T(t, x, z, v) = +\infty$ for $v \notin F(t, x, z)$, since the inf is taken on an empty class of real numbers. On the other hand, $-\infty \leqslant T(t, x, z, v) < +\infty$ for $v \in F(t, x, z)$ and $(t, x, z) \in D$.

Finally, for any admissible pair $(x, u) \in \mathrm{AC}([t_1 - r, t_2], \boldsymbol{R}^n) \times \mathfrak{M}([t_1, t_2], \boldsymbol{R}^m)$ with $t_2 - t_1 \geqslant r > 0$ for the Lagrange problem $\mathscr{L}(D, f, \Omega, C, f_0)$ we have

$$\begin{aligned} &I_0[x] = \int_{t_1}^{t_2} T(t, x_t, \dot{x}_t, \dot{x}(t))dt \leqslant I[x, u] \\ &\dot{x}(t) \in \tilde{F}(t, x_t, \dot{x}_t) \qquad \text{for a.e. } t \in [t_1, t_2] \\ &(t, x_t) \in \Omega \qquad \text{for } t \in [t_1, t_2] \\ &(t_1, x_{t_1}, t_2, x_{t_2}) \in C \end{aligned} \tag{1.26}$$

provided, that a function $[t_1, t_2] \ni t \to T(t, x_t, \dot{x}_t, \dot{x}(t)) \in \boldsymbol{R}$ is L-integrable on $[t_1, t_2]$.

It is interesting to note that any Lagrange minimization problem with $f_0 > 0$ can be reduced to a problem of minimum time. Indeed, if τ is a new time variable related to t, by the relation $d\tau/dt = f_0(t, x_t, \dot{x}_t, u(t))$, then the relation $t = t(\tau)$ can be inverted into a relation $\tau = \tau(t)$; the initial and terminal times t_1, t_2 become new times $\tau_1, \tau_2, \tau_1 = \tau(t_1)$, $\tau_2 = \tau(t_2)$; functional-differential systems $\dot{x}(t) = f(t, x_t, \dot{x}_t, u(t))$ become $dx/d\tau = g(\tau, x_\tau, \dot{x}_\tau, u(\tau))$, where g is defined by f and f_0; and the functional (1.24) becomes $I[x, u] = \int_{t_1}^{t_2} f_0 dt = \int_{t_1}^{t_2} d\tau = \tau_2 - \tau_1$.

Finally, let us note that there exists a theoretical equivalence of Mayer, Lagrange and Bolza problems of optimal control. It is readily seen by introducing some additional variables and new state vectors. We will give here a few examples showing how to derive in an ordinary case to the Mayer problem. Our case can be obtained similarly.

EXAMPLE 1.7. Find the minimum of the functional $I[x, u] = \int_0^1 u^2(t)dt$ with $\dot{x} = x + u$, $x(0) = x_1 = 1$, $x(1) = x_2 = 0$. This is a Lagrange problem with $n = m = 1$,

$U = \boldsymbol{R}$. It is immediately reduced to the Mayer problem $I[x, y, u] = y(1)$ with $\dot{x} = x+u$, $\dot{y} = u^2$, $n = 2$, $m = 1$, $x(0) = 1$, $x(1) = 0$, $y(0) = y_1 = 0$, and in this Mayer problem $C = \{(0, 1, 0, 1, 0, y_2)\}$ is a straight line in $\boldsymbol{R}^6$, since $y(1) = y_2 \in \boldsymbol{R}$ is undetermined. ■

EXAMPLE 1.8. Find the minimum of the functional $I[x, u] = \int_0^1 u^2(t)\,dt + [x(1)]^2$ with $\dot{x} = x+u$, $x(0) = x_1 = 1$. This is a Bolza problem with $n = m = 1$, $g = x_2^2 = [x_2(1)]^2$, $U = \boldsymbol{R}$. It is immediately reduced to the Mayer problem $I[x, y, u] = y(1)+[x(1)]^2$ with $\dot{x} = x+u$, $\dot{y} = u^2$, $n = 2$, $m = 1$, $g = y_2+x_2^2 = y(1)+ +[x(1)]^2$, $x(0) = x_1 = 1$, $y(0) = y_1 = 0$. ■

EXAMPLE 1.9. A point P moves along the x-axis monitored by the equation $\ddot{x} = u$ with $|u| \leqslant 1$. Take P from any given state $(x = a, \dot{x} = b)$ to rest at the origin $(x = 0, \dot{x} = 0)$ in the shortest possible time (a brachistochrone problem). By using phase coordinates $x = x$, $\dot{y} = x$ we have a Mayer problem of minimum time with $n = 2$, $m = 1$, differential system $\dot{x} = y$, $\dot{y} = u$, control space $U = \{u\colon -1 \leqslant u \leqslant 1\} \subset \boldsymbol{R}$, $A = \boldsymbol{R}^3$, initial data $t_1 = 0$, $x(t_1) = a$, $y(t_1) = b$, terminal data $x(t_2) = 0$, $y(t_2) = 0$, $t_2 \geqslant 0$ undetermined, and functional $I[x, y, u] = g = t_2$. ■

EXAMPLE 1.10. A point P moves along the x-axis monitored by the equation $\ddot{x}+ +x = u$ with $|u| \leqslant 1$. Take P from any given state $(x = a, \dot{x} = b)$ to rest at the origin $(x = 0, \dot{x} = 0)$ in the shortest possible time. As in Example 1.9 the problem is reduced to the Mayer problem with $n = 2$, $m = 1$ differential systems $\dot{x} = y$. $\dot{y} = -x+u$, control space $U = \{u\colon -1 \leqslant \leqslant u \leqslant 1\} \subset \boldsymbol{R}$, $A = \boldsymbol{R}^3$, initial data $t_1 = 0$, $x(t_1) = a$, $y(t_1) = b$, terminal data $x(t_2) = 0$, $y(t_2) = 0$, $t_2 \geqslant 0$ and a functional $I[x, y, u] = g = t_2$. ■

2. NEUTRAL FUNCTIONAL DIFFERENTIAL INCLUSIONS-EXISTENCE THEOREMS

It was pointed out in Sections 1.2 and 1.3 how some problems of optimal control theory for systems described by NFDEs lead to functional-differential relations of the form $\dot{x}(t) \in F(t, x_t, \dot{x}_t)$. These relations are called *neutral functional differential inclusions.* We shall present here some existence theorems for such type differential relations.

Let D be a nonempty open subset of $\boldsymbol{R}\times C_{0r}\times L_{0r}$ and $F\colon D \to \mathrm{Comp}(\boldsymbol{R}^n)$ be a set-valued function. Similarly as in Section 1.1 it will be assumed that F is such that the following local measurability condition (M) is satisfied:

(M) *For every interval $I \subset \boldsymbol{R}$ and sets $\boldsymbol{K} \subset C_{0r}$ and $S \subset L_{0r}$ with $I\times \boldsymbol{K}\times S \subset D$ and every $\varepsilon > 0$ there exists a compact set $E \subset I$ with $\mu(I\setminus E) \leqslant \varepsilon$ such that the restriction of F to $E\times \boldsymbol{K}\times S$ is Borel measurable.*

The space of all set-valued functions $F\colon D \to \mathrm{Comp}(\boldsymbol{R}^n)$ satisfying a local measurability condition (M) will be denoted by $\mathcal{M}(D, \mathrm{Comp}(\boldsymbol{R}^n))$. We shall consider here $\boldsymbol{R}\times C_{0r}\times L_{0r}$ with the metric d defined in Section IV.1.

2.1. Neutral functional differential inclusions—basic concepts

Let D be a nonempty open subset of $\boldsymbol{R}\times C_{0r}\times L_{0r}$ and let $F\in\mathcal{M}(D, \mathrm{Comp}(\boldsymbol{R}^n))$ be given.

LEMMA 2.1. *For every $F\in\mathcal{M}(D, \mathrm{Comp}(\boldsymbol{R}^n))$, $I\times \boldsymbol{K}\times S\subset D$ with $I := [t_1, t_2]$ and each $x\in C([t_1-r, t_2], \boldsymbol{R}^n)$ and $u\in L([t_1-r, t_2], \boldsymbol{R}^n)$ such that $(x_t, u_t)\in \boldsymbol{K}\times S$ for $t\in I$, the set-valued function $I\ni t\to F(t, x_t, u_t)\in \mathrm{Comp}(\boldsymbol{R}^n)$ is L-measurable on I.*

Proof. By virtue of Lemma III.5.1, a mapping $I\in t\to (t, x_t, u_t)\in \boldsymbol{R}\times C_{0r}\times L_{0r}$ is continuous. Since $F\in\mathcal{M}(D, \mathrm{Comp}(\boldsymbol{R}^n))$ then for every $\varepsilon>0$ there exists closed set $E\subset I$ with $\mu(I\setminus E)\leqslant\varepsilon$ such that the restriction of F to $E\times\boldsymbol{K}\times S$ is Borel measurable. Then, by Theorem II.3.2, F is measurable as a mapping defined on a subsets of a measurable space $(\boldsymbol{R}\times C_{0r}\times L_{0r}, \beta(\boldsymbol{R}\times C_{0r}\times L_{0r}))$ into a metric space $(\mathrm{Comp}(\boldsymbol{R}^n), h)$. Then by Remark II.3.3 a mapping $E\ni t\to F(t, x_t, u_t)\in \mathrm{Comp}(\boldsymbol{R}^n)$ is L-measurable for every $\varepsilon>0$. Thus, by Lusin–Pliś' theorem, a set-valued function $I\ni t\to F(t, x_t, u_t)\in\mathrm{Comp}(\boldsymbol{R}^n)$ is L-measurable on I. ■

Given $F\in\mathcal{M}(D, \mathrm{Comp}(\boldsymbol{R}^n))$, $I\times\boldsymbol{K}\times S\subset D$ and $I=[t_1, t_2]$ a *neutral functional-differential inclusion* NFDI(D, F) we define as the relation

$$\dot{x}(t)\in F(t, x_t, \dot{x}_t) \tag{2.1}$$

that is to be satisfied a.e. on I by an absolutely continuous function $x\colon [t_1-r, t_2]\to \boldsymbol{R}^n$ such that $(x_t, \dot{x}_t)\in\boldsymbol{K}\times S$ for $t\in I$. Each such absolutely continuous function is said to be a *local solution* of NFDI(D, F) on $[t_1-r, t_2]$.

Given $(\sigma, \Phi)\in\boldsymbol{R}\times \mathrm{AC}_{0r}$ such that $(\sigma, \Phi, \dot{\Phi})\in D$ we say that $x\colon [\sigma-r, \sigma+a]\to\boldsymbol{R}^n$ is a *solution* of NFDI(D, F) on $[\sigma-r, \sigma+a]$ through (σ, Φ) if x is a solution of NFDI(D, F) on $[\sigma-r, \sigma+a]$ and $x_\sigma=\Phi$.

Given a solution $x\colon [t_1-r, t_2]\to\boldsymbol{R}^n$ of NFDI(D, F) on $[t_1-r, t_2]$ a solution $\hat{x}\colon [\tau_1-r, \tau_2]\to\boldsymbol{R}^n$ of NFDI(D, F) on $[\tau_1-r, \tau_2]$ is said to be a *continuation of x over* $[\tau_1-r, \tau_2]$ if $\tau_1<t_1$, $t_2<\tau_2$ and $x(t)=\hat{x}(t)$ for $t\in[t_1-r, t_2]$. In this case x is called *continuable*.

A solution $x\colon [t_1-r, t_2]\to\boldsymbol{R}^n$ of NFDI(D, F) on $[t_1-r, t_2)$ is said to be *non-continuable* if no solution of NFDE(D, F) is continuation of x. In this case the interval $[t_1-r, t_2)$ is called the *maximal interval of the existence of the solution x.*

Similarly as in the case of NFDEs we are interested in solutions $x\colon [t_1-r, t_2]\to\boldsymbol{R}^n$ of NFDI(D, F) satisfying at each $t\in[t_1, t_2]$ the following additional condition: $(t, x_t)\in A$, where A is a nonempty subset of $\boldsymbol{R}\times\mathrm{AC}_{0r}$. Such problems for NFDI$(D, F)$ will be denoted by NFDI(D, A, F). It is clear that in the last case we must assume that sets D and A are such that for every (t, x), $(t, y)\in A$ we have $(t, x, \dot{y})\in D$. In what follows a pair (D, A) of sets $D\subset\boldsymbol{R}\times C_{0r}\times L_{0r}$ and $A\subset\boldsymbol{R}\times\mathrm{AC}_{0r}$ will be called *conformable* if for every (t, x), $(t, y)\in A$ one has $(t, x, y)\in D$.

Given $(\sigma, \Phi, \psi) \in \boldsymbol{R} \times C_{0r} \times L_{0r}$ and $\varrho > 0$ we denote by $I_\rho(\sigma)$, $\mathscr{K}_\rho(\Phi)$ and $B_\rho(\psi)$ closed balls of $\boldsymbol{R}$, C_{0r} and L_{0r}, respectively with centres at σ, Φ and ψ and a radius $\varrho > 0$. It is clear that $I_\rho(\sigma) \times \mathscr{K}_\rho(\Phi) \times B_\rho(\psi)$ is a closed ball of the metric space $(\boldsymbol{R} \times C_{0r} \times L_{0r}, d)$ centred at (σ, Φ, ψ) and a radius $\varrho > 0$.

A set-valued function $F \in \mathscr{M}(D, \mathrm{Comp}(\boldsymbol{R}^n))$ is said to be *locally integrable* if for every $(\sigma, \Phi, \psi) \in D$ there exists $\varrho > 0$ such that $I_\rho(\sigma) \times \mathscr{K}_\rho(\Phi) \times B_\rho(\psi) \subset D$ and $F(\cdot, u, v) \in \mathscr{A}(I_\rho(\sigma), \boldsymbol{R}^n)$ for every $(u, v) \in \mathscr{K}_\rho(\Phi) \times B_\rho(\psi)$. If furthermore, a family $\{F(\cdot, u, v)\}_{(u,v) \in \mathscr{K}_\rho(\Phi) \times B_\rho(\psi)}$ is uniformly integrable, F is called *locally uniformly integrable*. Finally, a set-valued function $F \in \mathscr{M}(D, \mathrm{Comp}(\boldsymbol{R}^n))$ is said to be a *locally integrably bounded* if for every $(\sigma, \Phi, \psi) \in D$ there exist $\varrho > 0$ and $M > 0$ such that $I_\rho(\sigma) \times \mathscr{K}_\rho(\Phi) \times B_\rho(\psi) \subset D$ and $\int_{I_\rho(\sigma)} \|F(t, u, v)\| dt \leqslant M$ for every $(u, v) \in \mathscr{K}_\rho(\Phi) \times B_\rho(\psi)$. It is clear that if $F \in \mathscr{M}(D, \mathrm{Comp}(\boldsymbol{R}^n))$ is locally integrably bounded, it is also locally integrable. We have also the following results.

LEMMA 2.2. *If $F \in \mathscr{M}(D, \mathrm{Comp}(\boldsymbol{R}^n))$ is locally uniformly integrable, then it is also locally integrably bounded.*

Proof. Let $(\sigma, \Phi, \psi) \in D$ be fixed and let $\varrho > 0$ be such that $I_\rho(\sigma) \times \mathscr{K}_\rho(\Phi) \times B_\rho(\psi) \subset D$ and so that a family $\{F(\cdot, u, v)\}_{(u,v) \in \mathscr{K}_\rho(\Phi) \times B_\rho(\psi)}$ is uniformly integrable on $I_\rho(\sigma)$. Hence, by Theorem III.1.3 one has $\lim_{\mu(E) \to 0} \int_E \|F(t, u, v)\| dt = 0$ uniformly, with respect to $(u, v) \in \mathscr{K}_\rho(\Phi) \times B_\rho(\psi)$. Then, in particular, for $\varepsilon = 1$ there exists a $\delta_1 > 0$ such that for each measurable set $E \subset I_\rho(\sigma)$ with $\mu(E) < \delta_1$ we have $\int_E \|F(t, u, v)\| dt \leqslant 1$ for every $(u, v) \in \mathscr{K}_\rho(\Phi) \times B_\rho(\psi)$. Select now a family $\{I_i;\ i = 1, \ldots, N\}$ of disjoint subintervals of $I_\rho(\sigma)$ such that $I_\rho(\sigma) = \bigcup_{i=1}^N I_i$ and $\mu(I_i) < \delta_1$ for each $i = 1, \ldots, N$. For every $(u, v) \in \mathscr{K}_\rho(\Phi) \times B_\rho(\psi)$ one obtains

$$\int_{I_\rho(\sigma)} \|F(t, u, v)\| dt \leqslant \sum_{i=1}^N \int_{I_i} \|F(t, u, v)\| dt \leqslant N. \blacksquare$$

Given $\sigma \in \boldsymbol{R}$ and $a > 0$ we define now a mapping $\mathscr{R}$: $C_{0r} \times C^0([\sigma, \sigma+a], \boldsymbol{R}^n) \to C([\sigma-r, \sigma+a], \boldsymbol{R}^n)$ by setting

$$\mathscr{R}(\Phi, x)(t) = \begin{cases} \Phi(t-\sigma) & \text{for } t \in [\sigma-r, \sigma), \\ \Phi(0)+x(t) & \text{for } t \in [\sigma, \sigma+a], \end{cases} \tag{2.2}$$

where, as usual, $C^0([\sigma, \sigma+a], \boldsymbol{R}^n)$ denotes a Banach space of all continuous functions x: $[\sigma, \sigma+a] \to \boldsymbol{R}^n$ such that $x(\sigma) = 0$.

LEMMA 2.3. *For every $\sigma \in \boldsymbol{R}$ and $a > 0$, $\mathscr{R}$ is a biaffine mapping, i.e., $\mathscr{R}(\Phi, \cdot)$ and $\mathscr{R}(\cdot, x)$ are affine mappings for fixed Φ and x, respectively.*

Proof. Indeed, by the definitions for fixed $\Phi \in C_{0r}$, $x_1, x_2 \in C^0([\sigma, \sigma+a], \boldsymbol{R}^n)$ and $\alpha, \beta \in [0, 1]$ satisfying $\alpha+\beta = 1$ one has $\mathscr{R}(\Phi, \alpha x_1+\beta x_2) = \alpha\mathscr{R}(\Phi, x_1)+$

$+\beta\mathscr{R}(\Phi, x_2)$. Similarly, for $\Phi_1, \Phi_2 \in C_{0r}$ and $x \in C^0([\sigma, \sigma+a], \boldsymbol{R}^n)$ one obtains $\mathscr{R}(\alpha\Phi_1+\beta\Phi_2, x) = \alpha\mathscr{R}(\Phi_1, x)+\beta\mathscr{R}(\Phi_2, x)$. ■

In what follows, we shall denote by $\mathscr{R}^-$ a mapping defined on $C_{0r}\times C([\sigma-r, \sigma+a], \boldsymbol{R}^n)$ by setting

$$\mathscr{R}^-(\Phi, y) = y|_{[\sigma,\sigma+a]}-\Phi(0) \quad \text{for } \Phi \in C_{0r} \tag{2.3}$$

and $y \in C([\sigma-r, \sigma+a], \boldsymbol{R}^n)$, where $y|_{[\sigma,\sigma+a]}$ denotes the restriction of y to $[\sigma, \sigma+a]$.

COROLLARY 2.1. *For every* $\sigma \in \boldsymbol{R}$, $a > 0$, $\Phi \in C_{0r}$, $x \in C^0([\sigma, \sigma+a], \boldsymbol{R}^n)$ *and* $y \in C([\sigma-r, \sigma+a], \boldsymbol{R}^n)$ *one has*

$$\mathscr{R}\big(\Phi, \mathscr{R}^-(\Phi, y)\big) = y \quad \textit{and} \quad \mathscr{R}^-\big(\Phi, \mathscr{R}(\Phi, x)\big) = x. \tag{2.4}$$

In a similar way as above, we can also define for each fixed $\sigma \in \boldsymbol{R}$ and $a > 0$ a biaffine mapping $\mathscr{L}$: $L_{0r}\times L([\sigma, \sigma+a], \boldsymbol{R}^n) \to L([\sigma-r, \sigma+a], \boldsymbol{R}^n)$ by taking

$$\mathscr{L}(\psi, z)(t) = \begin{cases} \psi(t-\sigma) & \text{for } t \in [\sigma-r, \sigma), \\ z(t) & \text{for } t \in [\sigma, \sigma+a]. \end{cases} \tag{2.5}$$

COROLLARY 2.2. *For every* $\sigma \in \boldsymbol{R}$, $a > 0$, $\Phi \in \mathrm{AC}_{0r}$ *and* $x \in \mathrm{AC}^0([\sigma, \sigma+a], \boldsymbol{R}^n)$ *one has*

$$\dot{\mathscr{R}}(\Phi, x)(t) = \mathscr{L}(\dot{\Phi}, \dot{x})(t) \quad \textit{for aa e. } t \in [\sigma-r, \sigma+a],$$

where as usual $\mathrm{AC}^0([\sigma, \sigma+a], \boldsymbol{R}^n)$ *denotes the Banach space of all absolutely continuous functions* x: $[\sigma, \sigma+a] \to \boldsymbol{R}^n$ *such that* $x(\sigma) = 0$. ■

In what follows for fixed $\sigma \in \boldsymbol{R}$, $a > 0$, $\Phi \in C_{0r}$, $\psi \in L_{0r}$, $x \in C^0([\sigma, \sigma+a], \boldsymbol{R}^n)$ and $z \in L([\sigma, \sigma+a], \boldsymbol{R}^n)$ we will denote $\mathscr{R}(\Phi, x)$ and $\mathscr{L}(\psi, z)$ by $\Phi\oplus x$ and $\psi\oplus z$, respectively.

Suppose now, for given $D \subset \boldsymbol{R}\times C_{0r}\times L_{0r}$ and $(\sigma, \Phi) \in \boldsymbol{R}\times \mathrm{AC}_{0r}$ there are numbers $a > 0$ and $\lambda > 0$ such that $(t, (\Phi\oplus x)_t(\dot{\Phi}\oplus z)_t) \in D$ for every $t \in [\sigma, \sigma+a]$, $x \in \mathscr{K}_\lambda$ and $z \in B_\lambda$, where $\mathscr{K}_\lambda$ and B_λ denote closed balls of $C^0([\sigma, \sigma+a], \boldsymbol{R}^n)$ and $L([\sigma, \sigma+a], \boldsymbol{R}^n)$ with centres at origins and a radius $\lambda > 0$.

Given F: $D \to \mathrm{Comp}(\boldsymbol{R}^n)$ a set-valued function G^F: $[\sigma, \sigma+a]\times\mathscr{K}_\lambda\times B_\lambda \to \mathrm{Comp}(\boldsymbol{R}^n)$ defined by

$$\dot{G}^F(t, x, z) := F\big(t, (\Phi\oplus x)_t, (\dot{\Phi}\oplus z)_t\big) \tag{2.6}$$

for $t \in [\sigma, \sigma+a]$, $x \in \mathscr{K}_\lambda$ and $z \in B_\lambda$ is said to be a *locally associated set-valued function* to F corresponding to $(\sigma, \Phi) \in \boldsymbol{R}\times \mathrm{AC}_{0r}$, $a > 0$ and $\lambda > 0$.

LEMMA 2.5. *Let* (D, Ω) *be a conformable pair of subsets of* $\boldsymbol{R}\times C_{0r}\times L_{0r}$ *and* $\boldsymbol{R}\times \mathrm{AC}_{0r}$, *respectively such that* D *is open in* $\boldsymbol{R}\times C_{0r}\times L_{0r}$ *and* Ω *is closed in* $\boldsymbol{R}\times C_{0r}$ *and relatively weakly compact in* $\boldsymbol{R}\times \mathrm{AC}_{0r}$. *If* $F \in \mathscr{M}(D, \mathrm{Comp}(\boldsymbol{R}\ ))$ *is locally uniformly integrable then there are numbers* $a > 0$ *and* $\lambda > 0$ *such that*

(i) *for every* $(\sigma, \Phi) \in \Omega$, $t \in [\sigma, \sigma+a]$, $x \in \mathscr{K}_\lambda$ *and* $z \in B_\lambda$ *one has* $(t, (\Phi\oplus x)_t, (\dot{\Phi}\oplus z)_t) \in D$,

(ii) *a locally associated function* G^F *to* F *defined on* $[\sigma, \sigma+a]\times\mathcal{K}_\lambda\times B_\lambda$ *corresponding to each* $(\sigma, \Phi)\in\Omega$, $a>0$ *and* $\lambda>0$ *is such that* $\int_\sigma^{\sigma+a}\|G^F(t,x,z)\|\,dt\leq\lambda$ *for* $(x,z)\in\mathcal{K}_\lambda\times B_\lambda$ *and for every* $\varepsilon>0$ *there is* $\delta>0$ *such that for every* $(\sigma,\Phi)\in\Omega$ *and every measurable set* $E\subset[\sigma,\sigma+a]$ *with* $\mu(E)<\delta$ *one has* $\int_E\|G^F(t,x,z)\|\,dt<\varepsilon$ *for every* $(x,z)\in\mathcal{K}_\lambda\times B_\lambda$, i.e. $G^F\in\mathcal{H}_\lambda(I, R^n)$ *with* $I=[\sigma,\sigma+a]$ *for every* $(\sigma,\Phi)\in\Omega$, *where* $\mathcal{H}_\lambda(I,R^n)$ *is such as in* Section III.2.2.

Proof. Let us observe first that each relatively weakly compact set of AC_{0r} closed in C_{0r} is compact in C_{0r}. Indeed, such set is equiabsolutely continuous and bounded in AC_{0r}. Therefore, it is also equicontinuous and bounded subset of C_{0r}. Thus, by Arzela–Ascoli's theorem it is relatively compact, and therefore by the closedness, also compact in C_{0r}.

Given $(\sigma,\Phi)\in\Omega$ and $\delta>0$, let $B^o((\sigma,\Phi),\delta)$ denote an open ball of $(R\times C_{0r}, d)$ with the centre (σ,Φ) and a radius $\delta>0$, where d is a metric defined by $d[(\bar{t},\bar{x}),(t,x)]:=\max(|t-\bar{t}|, |x-\bar{x}|_0)$ for $(t,x),(\bar{t},\bar{x})\in R\times C_{0r}$.

Now, lest us observe that for every $(\sigma,\Phi)\in\Omega$ we have $(\sigma,\Phi,\dot{\Phi})\in D$. Since D is open in $R\times C_{0r}\times L_{0r}$ and F is locally uniformly integrable, then for every $(\sigma,\Phi)\in\Omega$ there is $\varrho(\sigma,\Phi)>0$ such that $I_{\rho(\sigma,\Phi)}(\sigma)\times\mathcal{K}_{\rho(\sigma,\Phi)}(\Phi)\times B_{\rho(\sigma,\Phi)}(\dot{\Phi})\subset D$ and $\lim_{\mu(E)\to 0}\int_E\|F(t,u,v)\|\,dt=0$ uniformly with respect to $(u,v)\in\mathcal{K}_{\rho(\sigma,\Phi)}(\Phi)\times B_{\rho(\sigma,\Phi)}(\dot{\Phi})$ with $E\subset I_{\rho(\sigma,\Phi)}(\sigma)$. Since $\Omega\subset\bigcup_{(\sigma,\Phi)\in\Omega}B^o[(\sigma,\Phi),\frac{1}{2}\varrho(\sigma,\Phi)]$ and Ω is compact in $R\times C_{0r}$ then we can select a finite set $\{(\sigma_1,\Phi_1),\dots,(\sigma_N,\Phi_N)\}\subset\Omega$ such that $\Omega\subset\bigcup_{i=1}^N B^o((\sigma_i,\Phi_i),\delta_i)$, where $\delta_i=\frac{1}{2}\varrho(\sigma_i,\Phi_i)$ for $i=1,\dots,N$.

Let $\lambda=\frac{1}{5}\min_{1\leq i\leq N}\delta_i$ and select for every $\dot{\Phi}_i\in L_{0r}$ any $x_i\in C_{0r}$ such that $|\dot{\Phi}_i-x_i|_0\leq\lambda$ for $i=1,\dots,N$. Such x_i exists because C_{0r} is dense in L_{0r}. Denote by ω_i the modulus of continuity of x_i for $i=1,\dots,N$ and let $\omega(p)=\max_{1\leq i\leq N}\omega_i(p)$ for $p\geq 0$. By the compactness of Ω in $R\times C_{0r}$ and its relative weak compactness in $R\times AC_{0r}$ there are modulus of continuities ω_C and ω_L such that for every $(\sigma,\Phi)\in\Omega$, $s_1,s_2\in[-r,0]$ and each measurable set $E\subset[-r,0]$ one has $|\Phi(s_1)-\Phi(s_2)|\leq\omega_C(|s_1-s_2|)$ and $\int_E|\dot{\Phi}(s)|\,ds\leq\omega_L(\mu(E))$.

We can also define for every $i=1,\dots,N$ a modulus of continuity ω_i^F such that $\int_E\|F(t,u,v)\|\,dt\leq\omega_i^F(\mu(E))$ for every measurable set $E\subset I_{2\delta_i}(\sigma_i)$ and $(u,v)\in\mathcal{K}_{2\delta_i}(\Phi_i)\times B_{2\delta_i}(\dot{\Phi}_i)$, where as above $\delta_i=\frac{1}{2}\varrho(\sigma_i,\Phi_i)$. Let $\omega^F(p):=\max_{1\leq i\leq N}\omega_i^F(p)$ for $p\geq 0$.

Select now $a\in(0,\min(r,\min_{1\leq i\leq N}\delta_i)]$ such that $\omega_C(a)\leq\lambda$, $\omega_L(a)\leq\lambda$, $r\omega(a)\leq\lambda$ and $\omega\ (a)\leq\lambda$. Since $a\leq\min_{1\leq i\leq N}\delta_i$ then for every $(\sigma,\Phi)\in\Omega$ there is $i\in\{1,2,\dots,N\}$ such that for every $t\in[\sigma,\sigma+a]$, $x\in\mathcal{K}_\lambda$ and $z\in B_\lambda$ we have $|t-\sigma_i|\leq|t-$

$-\sigma|+|\sigma-\sigma_i| \leqslant a+\delta_i \leqslant 2\delta_i \leqslant \varrho(\sigma_i, \Phi_i)$, and

$$|(\Phi \oplus x)_t-\Phi_i|_0 \leqslant |(\Phi \oplus x)_t-\Phi|_0+|\Phi-\Phi_i|_0 < \sup_{r\leqslant s\leqslant 0} |(\Phi \oplus x)(t+s)-\Phi(s)|$$
$$+\delta_i = \sup_{t-r\leqslant\tau\leqslant t} |(\Phi \oplus x)(\tau)-\Phi(\tau-t)|+\delta_i \leqslant \sup_{t-r\leqslant\tau\leqslant\sigma} |\Phi(\tau-\sigma)$$
$$-\Phi(\tau-t)|+\sup_{\sigma\leqslant\tau\leqslant t} |\Phi(0)+x(\tau)-\Phi(\tau-t)|+\delta_i \leqslant 2\omega_C(a)$$
$$+|x|_a+\delta_i \leqslant 3\lambda+\delta_i \leqslant \tfrac{5}{8}\delta_i < \varrho(\sigma_i, \Phi_i)$$

and finally,

$$|(\dot{\Phi}\oplus z)_t-\dot{\Phi}_i|_0 = \int_{-r}^{0} |(\dot{\Phi}\oplus z)(t+s)-\dot{\Phi}_i(s)|ds = \int_{t-r}^{t} |(\dot{\Phi}\oplus z)(\tau)-\dot{\Phi}_i(\tau-t)|d\tau$$
$$\leqslant \int_{\sigma-r}^{\sigma} |\dot{\Phi}(\tau-\sigma)-\dot{\Phi}_i(\tau-\sigma)|\,d\tau+$$
$$+\int_{\sigma-r}^{\sigma} |\Phi_i(\tau-\sigma)-x_i(\tau-\sigma)|\,d\tau+\int_{t-r}^{\sigma} |x_i(\tau-\sigma)-x_i(\tau-t)|\,d\tau+$$
$$+\int_{t-r}^{t} |x_i(\tau-t)-\dot{\Phi}_i(\tau-t)|d\tau+\int_{\sigma}^{t} |z(\tau)-\dot{\Phi}_i(\tau-t)|\,d\tau$$
$$\leqslant |\dot{\Phi}-\dot{\Phi}_i|_0+2|\dot{\Phi}_i-x_i|_0+\int_{\sigma-t}^{\sigma} \omega(a)\,d\tau+|z|_a+\int_{\sigma-t}^{a} |\dot{\Phi}_i(s)|\,ds$$
$$\leqslant \delta_i+2\lambda+r\omega(a)+\lambda+\omega_L(a) \leqslant \delta_i+5\lambda \leqslant 2\delta_i \leqslant \varrho(\sigma_i, \Phi_i).$$

Therefore,

$$\big(t, (\Phi\oplus x)_t, (\dot{\Phi}\oplus z)_t\big) \in I_{\varrho(\sigma_i,\Phi_i)}(\sigma_i)\times \mathscr{K}_{\varrho(\sigma_i,\Phi_i)}(\Phi_i)\times B_{\varrho(\sigma_i,\Phi_i)}(\dot{\Phi}_i) \subset D$$

for every $t \in [\sigma, \sigma+a]$, $x \in \mathscr{K}_\lambda$ and $z \in B_\lambda$. Hence, by Lemma 2.1 it follows that $G^F(\cdot, x, z)$ is measurable on $[\sigma, \sigma+a]$ for fixed $(x, z) \in \mathscr{K}_\lambda \times B_\lambda$. Let $\varepsilon > 0$ be given and let $\delta > 0$ be such that $\omega^F(\delta) < \varepsilon$. For every fixed $(\sigma, \Phi) \in \Omega$ and a measurable set $E \subset [\sigma, \sigma+a]$ we have $\int_E ||F(t, (\Phi \oplus x)_t, (\dot{\Phi} \oplus z)_t)||\,dt \leqslant \omega^F(\mu(E))$ for every $(x, z) \in \mathscr{K}_\lambda \times B$ because to every $(\sigma, \Phi) \in \Omega$ there is $i \in \{1, \dots, N\}$ such that $E \subset I_{2\delta_i}(\sigma_i)$ and $(t, (\Phi \oplus x)_i, (\Phi \oplus z)_i) \in I_{2\delta_i}(\sigma_i) \times \mathscr{K}_{2\delta_i}(\Phi_i) \times B_{2\delta_i}(\dot{\Phi}_i)$ for $(t, x, z) \in E \times \mathscr{K}_\lambda \times B_\delta$. Thus for every $(x, z) \in \mathscr{K}_\lambda \times B_\lambda$ we have $\int_E ||G\,(t, x, z)||\,dt < \varepsilon$ whenever $\mu(E) < \delta$. Hence, in particular it follows that $\lim_{\mu(E)\to 0} \int_E ||G^F(t, x, z)||\,dt = 0$ uniformly with respect to $x \in \mathscr{K}_\lambda$ and $z \in B_\lambda$. Finally, for each measurable set $E \subset [\sigma, \sigma+a]$ one has $\int_E ||G^F(t, x, z)||\,dt \leqslant \omega_i^F(\mu(E)) \leqslant \omega_i^F(a) \leqslant \lambda$ for every $(x, z) \in \mathscr{K}_\delta \times B_\delta$, because $[\sigma, \sigma+a] \subset I_{2\delta_i}(\sigma_i)$. ■

2.2. Sufficient conditions for existence of local solutions of NFDIs

Let D be a nonempty open subset of $\boldsymbol{R} \times C_{0r} \times L_{0r}$ and let F: $D \to \text{Comp}(\boldsymbol{R}^n)$ be given.

A set-valued function F is said to be *locally Lipschitz continuous with respect to its last two variables* if for every $(\sigma, \Phi, \psi) \in D$ there are a number $\varrho > 0$ and a L-integrable function k: $I_\varrho(\sigma) \to \boldsymbol{R}^+$ such that $I_\varrho(\sigma) \times \mathscr{K}_\varrho(\Phi) \times B_\varrho(\psi) \subset D$ and $h(F(t, u, v), F(t, \bar{u}, \bar{v})) \leqslant k(t) \max(|u-\bar{u}|_0, |v-\bar{v}|_0)$ for a.e. $t \in I_\sigma(\sigma)$, $u, \bar{u} \in \mathscr{K}_\varrho(\Phi)$ and $\bar{v}, v \in B_\varrho(\psi)$.

A set-valued function F is said to be *locally Lipschitz continuous with respect to its last variable uniformly with respect to the second one* if for every $(\sigma, \Phi, \psi) \in D$ there are a number $\varrho > 0$ and an L-integrable function k: $I_\varrho(\sigma) \to \boldsymbol{R}^+$ such that $I_\varrho(\sigma) \times \mathscr{K}_\varrho(\Phi) \times B(_\varrho\psi) \subset D$ and $h(F(t, u, v), F(t, u, \bar{v})) \leqslant k(t)|v-\bar{v}|_0$ for a.e. $t \in I_\varrho(\sigma)$, $u \in \mathscr{K}_\varrho(\Phi)$ and $v, \bar{v} \in B_\varrho(\psi)$.

A set-valued function F is said to be *locally u.s.c. (l.s.c.) with respect to its last two variables* if for every $(\sigma, \Phi, \psi) \in D$ there is $\varrho > 0$ such that $I_\varrho(\sigma) \times \mathscr{K}_\varrho(\Phi) \times B_\varrho(\psi) \subset D$ and the restriction $F(t, \cdot, \cdot)$ to $\mathscr{K}_\varrho(\Phi) \times B_\varrho(\psi)$ is for each fixed $t \in I_\varrho(\sigma)$ u.s.c. (l.s.c.) on $\mathscr{K}_\varrho(\Phi) \times B_\varrho(\psi)$. If $F(t, \cdot, \cdot)$ is u.s.c. and l.s.c. on $\mathscr{K}_\varrho(\Phi) \times B_\varrho(\psi)$ we call it *locally continuous with respect to its last two variables.*

A set-valued function F is said to be *locally u.s.c.w. (l.s.c.w.) with respect to its last two variables* if for every $(\sigma, \Phi, \psi) \in D$ there exists $\varrho > 0$ such that $I_\varrho(\sigma) \times \mathscr{K}_\varrho(\Phi) \times B_\varrho(\psi) \subset D$ and so that for every $(u_0, v_0) \in \mathscr{K}_\varrho(\Phi) \times B_\varrho(\psi)$ and every sequence $\{(u_u, v_u)\}$ of $\mathscr{K}_\varrho(\Phi) \times B_\varrho(\psi)$ such that $|u_n - u_0|_0 \to 0$ and $v_n \rightharpoonup v_0$ in L_{0r} as $n \to \infty$ one has $\bar{h}(F(t, u_n, v_n), F(t, u_0, v_0)) \to 0$ $[\bar{h}(F(t, u_0, v_0), F(t, u_n, v_n)) \to 0]$ for a.e. $t \in I_\varrho(\sigma)$ as $n \to \infty$. If F is u.s.c.w. and l.s.c.w. on $\mathscr{K}_\varrho(\Phi) \times B_\varrho(\psi)$ it is called *locally continuous weakly with respect to its last two variables.*

Unfortunately, the continuities of set-valued functions defined above are very strong for NFDIs. They rule out even simple linear neutral differential-difference equations. It follows from the following example.

EXAMPLE 2.1. The function f: $\boldsymbol{R} \times H \to \boldsymbol{R}$ defined for $(t, \Phi, \psi) \in \boldsymbol{R} \times H$ by setting $f(t, \Phi, \psi) := a\Phi(0) + b\Phi(-1) + c\psi(-1)$, where $H = C([-1, 0], \boldsymbol{R}) \times L([-1, 0], \boldsymbol{R})$ and $a, b, c \in \boldsymbol{R}$ is not continuous with respect to its last two variables.

Indeed, let $\mathscr{S} \in C([-1, 0], \boldsymbol{R})$ be fixed and put for every $n \in \boldsymbol{N}$ and $j = 1, 2, \ldots$ $\ldots, n$, $F_{nj} = [2(j-1)/n, 2j/n]$. We have $\mu(F_{nj}) \leqslant 2/n$ for $n \in \boldsymbol{N}$ and $j = 1, 2, \ldots, n$. Denote now elements of the sequence $F_{11}, F_{21}, F_{22}, F_{31}, F_{32}, \ldots$ by $E_1, E_2, E_3, E_4, E_5, \ldots$ and put $u_n(t) = \chi_{E_n}(t)$ for $t \in [0, 2]$ and each $n \in \boldsymbol{N}$. Let $y^n(t) = \int_0^t u_n(\tau) d\tau$. For any $t \in [1, 2]$ and each $n \in \boldsymbol{N}$ we have $\dot{y}_t^n \in L([-1, 0], \boldsymbol{R})$ and

$$\int_{-1}^{0} |\dot{y}_t^n(s)| ds = \int_{-1}^{0} |\dot{y}^n(t+s)| ds = \int_{t-1}^{t} |\dot{y}^n(\tau)| d\tau \leqslant \int_0^2 \chi_{E_n}(t) dt = \mu(E_n).$$

Therefore, $\lim_{n \to \infty} |\dot{y}_t^n - \dot{y}^0|_0 = 0$, where $\dot{y}^0(s) \equiv 0$. On the other hand, for each fixed $t \in [0, 2]$ we have $\dot{y}^n(t) = 1$ for infinitely many values of n, so $\dot{y}^n(t-1) \to 0$ is always false.

But $|f(t, \mathscr{S}, \dot{y}_t^n) - f(t, \mathscr{S}, y^0)| = |c||\dot{y}^n(t-1) - 0|$ for each fixed $t \in [1, 2]$. Therefore,

$|\dot{y}_t^n - y^0|_0 \to 0$ as $n \to \infty$ do not imply $\lim_{n\to\infty} |f(t, \mathscr{S}, \dot{y}_t^n) - f(t, \mathscr{S}, \dot{y}^0)| = 0$. Thus $f(t, \mathscr{S}, \cdot)$ is not continuous at y^0. ∎

Therefore, we will consider also the following types of continuities of local integrable set-valued functions.

A locally integrable set-valued function $F \in \mathscr{M}(D, \mathrm{Comp}(\boldsymbol{R}^n))$ is said to be *locally $\mathscr{A}$-Lipschitz continuous with respect to its last two variables* if for every $(\sigma, \Phi, \psi) \in D$ there are a number $\varrho > 0$ and a continuous function K: $[0, \infty) \to \boldsymbol{R}^+$ with $K(0) = 0$ such that $I_\varrho(\sigma) \times \mathscr{K}_\varrho(\Phi) \times B_\varrho(\psi) \subset D$, $\int_{I_\varrho(\sigma)} \|F(t, u, v)\| dt < \infty$ for $(u, v) \in \mathscr{K}_\varrho(\Phi) \times B_\varrho(\psi)$ and so that $\int_\sigma^t h(F(\tau, u, v),\ F(\tau, \bar{u}, \bar{v}))d\tau \leqslant K(t-\sigma)\max(|u-\bar{u}|_0, |v-\bar{v}|_0)$ for $t \in [\sigma, \sigma+\varrho]$, $u, \bar{u} \in \mathscr{K}_\varrho(\Phi)$ and $v, \bar{v} \in B_\varrho(\psi)$.

In a similar way we define a *local $\mathscr{A}$-Lipschitz continuity of local integrable set-valued function $F \in \mathscr{M}(D, \mathrm{Comp}(\boldsymbol{R}^n))$ with respect to its last variable uniformly with respect to the second one.*

A locally integrable set-valued function $F \in \mathscr{M}(D, \mathrm{Comp}(\boldsymbol{R}^n))$ is said to be *locally hemi-s.-s.u.s.c. (hemi-s.-s.l.s.c.) with respect to its last two variables* if for every $(\sigma, \Phi) \in \boldsymbol{R} \times \mathrm{AC}_{0r}$ with $(\sigma, \Phi, \dot{\Phi}) \in D$ there are numbers $a > 0$ and $\lambda > 0$ such that $(t, (\Phi \oplus x)_t, (\dot{\Phi} \oplus z)_t) \in D$ for $t \in [\sigma, \sigma+a]$, $x \in \mathscr{K}_\lambda$ and $z \in B_\lambda$ and such that a local associated set-valued function G^F to F corresponding to (σ, Φ), $a > 0$ and $\lambda > 0$ is s.-s.u.s.c. (s.-s.l.s.c.) with respect to its last two variables. If F is locally hemi-s.-s.u.s.c. and hemi-s.-s.l.s.c. with respect to its last two variables we will call that F is *locally hemi-continuous with respect to its last two variables.*

A locally integrable set-valued function $F \in \mathscr{M}(D, \mathrm{Comp}(\boldsymbol{R}^n))$ is said to be *locally hemi-s.*-w.u.s.c. *(hemi-(s.-w.l.s.c.) with respect to its last two variables* if for every $(\sigma, \Phi) \in \boldsymbol{R} \times \mathrm{AC}_{0r}$ with $(\sigma, \Phi, \dot{\Phi}) \in D$ there are numbers $a > 0$ and $\lambda > 0$ such that $(t, (\Phi \oplus x)_t, (\dot{\Phi} \oplus z)_t) \in D$ for $t \in [\sigma, \sigma+a]$, $x \in \mathscr{K}_\lambda$ and $z \in B_\lambda$ and so that a local associated set-valued function G^F to F corresponding to (σ, Φ), $a > 0$ and $\lambda > 0$ is s.-w.u.s.c. (s.-w.l.s.c.) with respect to its last two variables.

In a similar way we also define *set-valued mappings locally hemi-w.-s.u.s.c. (hemi-w.-s.l.s.c.)*, and *locally hemi-w.-w.u.s.c. (hemi-w.-w.l.s.c.)*.

Remark 2.1. *The function f: $\boldsymbol{R} \times H \to \boldsymbol{R}$* defined in Example 2.1 is locally hemi-c. and locally hemi-w.-w.c. with respect to its last two variables. Indeed, $f \in \mathscr{M}(\boldsymbol{R} \times H, \boldsymbol{R})$ and it is locally integrable on $\boldsymbol{R}$. Furthermore, for every $(\sigma, \Phi, \dot{\Phi}) \in \boldsymbol{R} \times H$ we can take $\varrho > 0$ and $\lambda > 0$ arbitrarily. For every $(x, z) \in \mathscr{K}_\lambda \times B_\lambda$ we have $g^f(t, x, z) := a(\Phi \oplus x)_t(0) + b(\Phi \oplus x)_t(-1) + c(\dot{\Phi} \oplus z)_t(-1) = a(\Phi(0) + x(t)) + b\Phi(t-1-\sigma) + c\dot{\Phi}(t-1-\sigma)$ for $t \in [\sigma, \sigma+1]$ and $g^f(t, x, z) := a(\Phi(0)+x(t)) + b(\Phi(0)+x(t-1)) + c(\dot{\Phi}(0)+z(t-1))$ for $t \in [\sigma+1, \sigma+\varrho]$ if $\varrho > 1$. Then, for fixed $(\bar{x}, \bar{z}) \in \mathscr{K}_\lambda \times B_\lambda$ and any sequence $\{(x_n, z_n)\}$ such that $|x_n - \bar{x}|_\varrho \to 0$ and $|z_n - z|_\varrho \to 0$ or $z_n \rightharpoonup z$ as $n \to \infty$ we have $\int_\sigma^{\sigma+\varrho} |g^f(t, x_n, z_n) - g^f(t, \bar{x}, \bar{z})| dt \leqslant |a| x_n - \bar{x}|_\varrho$ if $0 < \varrho \leqslant 1$ and

$\int_\sigma^{\sigma+\varrho} |g^f(t, x_n, z_n) - g^f(t, \bar{x}, \bar{z})| \, dt \leqslant 2|a| |x_n - \bar{x}|_\varrho + |b| \int_\sigma^{\sigma+\varrho-1} |x_n(\tau) - \bar{x}(\tau)| \, d\tau + |c| \int_\sigma^{\sigma+\varrho-1} |z_n(\tau) - \bar{z}(\tau)| \, d\tau \leqslant (2|a| + |b|\varrho)|x_n - \bar{x}|_\varrho + |c|\varrho|z_n - z|_\varrho$ if $\varrho > 1$. Therefore, $\lim_{n\to\infty} \int_\sigma^{\sigma+\varrho} |g^f(t, x_n, z_n) - g^f(t, \bar{x}, \bar{z})| \, dt = 0$, whenever $\max(|x_n - \bar{x}|_\varrho, |z_n - z|_\varrho) \to 0$ as $n \to \infty$. Similarly, for every measurable set $E \subset [\sigma, \sigma+\varrho]$ one obtains $|\int_E [g^f(t, x_n, z_n) - g^f(t, \bar{x}, \bar{z})] \, dt| \leqslant |a| \, |x_n - \bar{x}|_\varrho$ if $0 < \varrho \leqslant 1$ and $|\int_F [g^f(t, x_n, z_n) - g^f(t, \bar{x}, \bar{z})] \, dt| \leqslant (2|a| + |b|\varrho)|x_n - \bar{x}|_\varrho + |c| \, |\int_{\tilde{E}} [z_n(\tau) - \bar{z}(\tau)] \, d\tau|$ if $\varrho > 1$, where $\tilde{E} = E \cap [\sigma, \sigma+\varrho-1]$. Hence, it follows that $\lim_{n\to\infty} |\int_E [g^f(t, x_n, z_n) - g^f(t, \bar{x}, \bar{z})] \, dt| = 0$ for every measurable set $E \subset [\sigma, \sigma+\varrho]$, whenever $|x_n - \bar{x}|_\varrho \to 0$ and $z_n \rightharpoonup \bar{z}$ as $n \to \infty$. ∎

Similarly as above assume D is a nonempty open subset of $\boldsymbol{R} \times C_{0r} \times L_{0r}$. We have the following results.

LEMMA 2.6. *If F: $D \to \mathrm{Comp}(\boldsymbol{R}^n)$ is locally Lipschitz continuous with respect to its last two variables (last variable uniformly with respect to the second one), then for every $(\sigma, \Phi) \in \boldsymbol{R} \times \mathrm{AC}_{0r}$ such that $(\sigma, \Phi, \dot{\Phi}) \in D$ there are numbers $a > 0$ and $\lambda > 0$ such that a local associated set-valued function G^F to F corresponding to (σ, Φ), $a > 0$ and $\lambda > 0$ is Lipschitz continuous with respect to its last two variables (last variable uniformly with respect to the second one).*

Proof. Let $\Omega = \{(\sigma, \Phi)\}$ and suppose $\varrho > 0$ and a L-integrable function k: $I_\varrho(\sigma) \to \boldsymbol{R}^+$ are such that $I_\varrho(\sigma) \times \mathscr{K}_\varrho(\Phi) \times B_\varrho(\dot{\Phi}) \subset D$ and $h(F(t, u, v), F(t, \bar{u}, \bar{v})) \leqslant k(t) \max(|u - \bar{u}|_0, |v - \bar{v}|_0)$ for a.e $t \in I_\varrho(\sigma)$, $u, \bar{u} \in \mathscr{K}_\varrho(\Phi)$ and $v, \bar{v} \in B_\varrho(\dot{\Phi})$. Similarly as in Lemma 2.5 we can select numbers $a > 0$ and $\lambda > 0$ such that $(t, (\Phi \oplus x)_t, (\Phi \oplus z)_t) \in I_\varrho(\sigma) \times \mathscr{K}_\varrho(\Phi) \times B_\varrho(\dot{\Phi}) \subset D$ for $t \in [\sigma, \sigma+a]$, $x \in \mathscr{K}_\lambda$ and $z \in B_\lambda$. Hence, for a.e. $t \in [\sigma, \sigma+a]$, $x, \bar{x} \in \mathscr{K}_\lambda$ and $z, \bar{z} \in B_\lambda$ one obtains

$$
\begin{aligned}
& h\big(G^F(t, x, z), G^F(t, \bar{x}, \bar{z})\big) \\
& := h\big(F\big(t, (\Phi \oplus x)_t, (\dot{\Phi} \oplus x)\big), F\big(t, (\Phi \oplus \bar{x})_t, (\Phi \oplus \bar{z})_t\big) \\
& \leqslant k(t) \max\big(|(\Phi \oplus x)_t - (\Phi \oplus \bar{x})_t|_0, |(\dot{\Phi} \oplus z)_t - (\dot{\Phi} \oplus \bar{z})_t|_0\big) \\
& = k(t) \max\Big(\sup_{t-r \leqslant \tau \leqslant t} |(\Phi \oplus x)(\tau) - (\Phi \oplus \bar{x})(\tau)|, \int_{t-r}^{t} |(\dot{\Phi} \oplus z)(\tau) - (\dot{\Phi} \oplus z)(\tau)| \, d\tau\Big) \\
& \leqslant k(t) \max\Big(\sup_{\sigma \leqslant \tau \leqslant t} |x(\tau) - \bar{x}(\tau)|, \int_\sigma^t |z(\tau) - \bar{z}(\tau)| \, d\tau\Big). \; ∎
\end{aligned}
$$

LEMMA 2.7. *If a local integrable set-valued function $F \in \mathscr{M}(D, \mathrm{Comp}(\boldsymbol{R}^n))$ is locally $\mathscr{A}$-Lipschitz continuous with respect to its last two variables (last variable uniformly with respect to the second one) then for every $(\sigma, \Phi) \in \boldsymbol{R} \times \mathrm{AC}_{0r}$ such that $(\sigma, \Phi, \dot{\Phi}) \in D$*

there are numbers $a > 0$ and $\lambda > 0$ such that a local associated set-valued function G^F to F corresponding to (σ, Φ), $a > 0$ and $\lambda > 0$ is $\mathscr{A}$-Lipschitz continuous with respect to its last two variables (last variable uniformly with respect to the second one) with Lipschitz constant $L < 1$.

Proof. Let $\Omega = \{(\sigma, \Phi)\}$ and suppose $\varrho > 0$ and a function K: $[0, \infty) \to \boldsymbol{R}^+$ are such that $I_\varrho(\sigma) \times \mathscr{K}_\varrho(\Phi) \times B_\varrho(\dot{\Phi}) \subset D$, $F(\cdot, u, v) \in \mathscr{A}(I_\varrho(\sigma), \boldsymbol{R}^n)$ and $\int_\sigma^t h(F(\tau, u, v), F(\tau, \bar{u}, \bar{v}))d\tau \leqslant K(t-\sigma)\max(|u-\bar{u}|_0, |v-\bar{v}|_0)$ for $u, \bar{u} \in \mathscr{K}_\varrho(\Phi)$, $v, \bar{v} \in B_\varrho(\dot{\Phi})$ and $t \in [\sigma, \sigma+\varrho]$. Similarly as in the proof of Lemma 2.5 we can find numbers $a > 0$ and $\lambda > 0$ such that $(t, (\Phi \oplus x)_t, (\dot{\Phi} \oplus z)_t) \in I_\varrho(\sigma) \times \mathscr{K}_\varrho(\Phi) \times B_\varrho(\dot{\Phi}) \subset D$ and such that $K(a) < 1$. Now, similarly as in the proof of Lemma 2.6 it follows

$$\int_\sigma^{\sigma+a} h\big(G^F(t, x, z), G^F(t, \bar{x}, \bar{z})\big)dt$$

$$\leqslant L\max\Big(\sup_{\sigma \leqslant \tau \leqslant t} |x(\tau)-\bar{x}(\tau)|, \int_\sigma^t |z(\tau)-\bar{z}(\tau)|\,d\tau\Big)$$

for $t \in [\sigma, \sigma+a]$, $x, \bar{x} \in \mathscr{K}_\lambda$ and $z, \bar{z} \in B_\lambda$, where $L = K(a)$. ■

Lemma 2.8. *Let F: $D \to \mathrm{Comp}(\boldsymbol{R}^n)$ be locally u.s.c. (l.s.c.) with respect to its last two variables. Then for every $(\sigma, \Phi) \in \boldsymbol{R} \times \mathrm{AC}_{0r}$ such that $(\sigma, \Phi, \dot{\Phi}) \in D$ there are numbers $a > 0$ and $\lambda > 0$ such that a local associated set-valued function G^F to F corresponding to (σ, Φ), $a > 0$ and $\lambda > 0$ is u.s.c. (l.s.c.) with respect to its last two variables.*

Proof. The proof is similar to the proof of Lemma 2.6. It is enough only to observe that if $\varrho > 0$ is such that $I_\varrho(\sigma) \times \mathscr{K}_\varrho(\Phi) \times B_\varrho(\dot{\Phi}) \subset D$ and such that $F(t, \cdot, \cdot)$ is u.s.c. (l.s.c.) on $\mathscr{K}_\varrho(\Phi) \times B_\varrho(\dot{\Phi})$ then by Lemma 2.5, for $\Omega = \{(\sigma, \Phi)\}$ there exist numbers $a > 0$ and $\varrho > 0$ such that $(t, (\Phi \oplus x)_t, (\Phi \oplus z)_t) \in D$ for $t \in [\sigma, \sigma+a]$ $x \in \mathscr{K}_\lambda$ and $z \in B_\lambda$. Hence, by continuity of mappings $\mathscr{R}(\Phi, \cdot)$ and $\mathscr{L}(\dot{\Phi}, \cdot)$ the results follow. ■

Since the mappings $\mathscr{R}(\Phi, \cdot)$ and $\mathscr{L}(\dot{\Phi}, \cdot)$ are also continuous weakly (see Theorem I.2.2) then similarly as above we also obtain the following lemma.

Lemma 2.9. *Let F: $D \to \mathrm{Comp}(\boldsymbol{R}^n)$ be locally u.s.c.w. (l.s.c.w.) with respect to its last two variables. Then for every $(\sigma, \Phi) \in \boldsymbol{R} \times \mathrm{AC}_{0r}$ such that $(\sigma, \Phi, \dot{\Phi}) \in D$ there are numbers $a > 0$ and $\lambda > 0$ such that a local associated set-valued function G^F to F corresponding to (σ, Φ), $a > 0$ and $\lambda > 0$ is u.s.c.w. (l.s.c.w.) with respect to its last two variables for fixed $t \in [\sigma, \sigma+a]$.* ■

Lemma 2.10. *Let $F \in \mathscr{M}(D, \mathrm{Comp}(\boldsymbol{R}^n))$ be locally uniformly integrable and locally* *u.s.c. (l.s.c.) with respect to its last two variables. Then for every* $(\sigma, \Phi) \in \boldsymbol{R} \times \mathrm{AC}_{0r}$ *such that* $(\sigma, \Phi, \dot{\Phi}) \in D$ *there are numbers* $a > 0$ *and* $\lambda > 0$ *such that a local associated set-valued function* G^F *to* F *corresponding to* (σ, Φ), $a > 0$ *and* $\lambda > 0$ *has the following properties*:

(i) $G^F \in \mathscr{H}_\lambda(I, \boldsymbol{R}^n)$ *with* $I := [\sigma, \sigma+a]$,

(ii) $\int_\sigma^{\sigma+a} \|G^F(t, x, z)\| dt \leqslant \lambda$ *for* $(x, z) \in \mathscr{K}_\lambda \times B_\lambda$,

(iii) G^F *is s-s.u.s.c. (s-s.l.s.c.) with respect to its last two varables.*

Proof. For the proof it is sufficient only to repeat the proofs of Lemmas 2.5 and 2.8 with $\Omega = \{(\sigma, \Phi)\}$. The results then follow from Lemma 2.8 and Vitali's theorem. ■

In a similar way we obtain

LEMMA 2.11. *Let* $F \in \mathscr{M}(D, \mathrm{Comp}(\boldsymbol{R}^n))$ *be locally uniformly integrable and locally u.s.c.w. (l.s.c.w.) with respect to its last two variables. Then for every* $(\sigma, \Phi) \in \boldsymbol{R} \times \mathrm{AC}_{0r}$ *such that* $(\sigma, \Phi, \dot{\Phi}) \in D$ *there are numbers* $a > 0$ *and* $\lambda > 0$ *such that a local associated set-valued function* G^F *to* F *corresponding to* (σ, Φ), $a > 0$ *and* $\lambda > 0$ *has the following properties*:

(i) $G^F \in \mathscr{H}_\lambda(I, \boldsymbol{R}^n)$ *with* $I := [\sigma, \sigma+a]$,

(ii) $\int_\sigma^{\sigma+a} \|G^F(t, x, z)\| dt \leqslant \lambda$ *for* $(x, z) \in \mathscr{K}_\lambda \times B_\lambda$,

(iii) G^F *is s.-w.u.s.c. (s.-w.l.s.c.) with respect to its last two variables.* ■

COROLLARY 2.3. *Each locally uniformly integrable set-valued function* $F \in \mathscr{M}(D, \mathrm{Comp}(\boldsymbol{R}^n))$ *locally u.s.c. (l.s.c.) with respect to its last two variables is locally hemi-s.-s.u.s.c. (s.-s.l.s.c.) with respect to its last two variables.* ■

COROLLARY 2.4. *Each locally uniformly integrable set-valued function* $F \in \mathscr{M}(D, \mathrm{Comp}(\boldsymbol{R}^n))$ *locally u.s.c.w. (l.s.c.w.) with respect to its last two variables is locally hemi-s.-w.u.s.c. (s.-w.l.s.c.) with respect to its last two variables.* ■

LEMMA 2.12. *Let* $F \in \mathscr{M}(D, \mathrm{Conv}(\boldsymbol{R}^n))$ *be locally uniformly integrable and locally l.s.c.w. (u.s.c.w.) with respect to its last two variables. Furthermore, suppose F satisfies anyone of the following local growth conditions*:

(γ_1) *For every* $(\sigma, \Phi, \psi) \in D$ *there exists* $\varrho > 0$ *such that* $I_\varrho(\sigma) \times \mathscr{K}_\varrho(\Phi) \times B_\varrho(\psi) \subset D$ *and such that for every* $p \in \boldsymbol{R}^n$ *there exists an L-integrable function* $\psi_p\colon I_\varrho(\sigma) \to \boldsymbol{R}$ *such that* $s(p, F(t, u, v)) \geqslant \psi_p(t)$ $[s(p, F(t, u, v)) \leqslant \psi_p(t)]$ *for a.e.* $t \in I_\varrho(\sigma)$, $(u, v) \in \mathscr{K}_\varrho(\Phi) \times B_\varrho(\psi)$ *and* $p \in \boldsymbol{R}^n$, *where* $s(\cdot, A)$ *is the support function of* $A \subset \boldsymbol{R}^n$.

(γ_2) *For every* $(\sigma, \Phi, \psi) \in D$ *there exists* $\varrho > 0$ *such that* $I_\varrho(\sigma) \times \mathscr{K}_\varrho(\Phi) \times B_\varrho(\psi) \subset D$ *and such that for every* $p \in \boldsymbol{R}^n$ *and* $\varepsilon > 0$ *there is an L-integrable function* $\psi_p^\varepsilon\colon I_\varrho(\sigma) \to \boldsymbol{R}$ *such that* $|v|_0 \leqslant \psi_p^\varepsilon(t) + \varepsilon s(p, F(t, u, v))$ $[|v|_0 \geqslant \psi\,(t) + \varepsilon s(p, F(t, u, v))]$ *for a.e.* $t \in I_\varrho(\sigma)$, $(u, v) \in \mathscr{K}_\varrho(\Phi) \times B_\varrho(\psi)$ *and* $p \in \boldsymbol{R}^n$.

Then for every $(\sigma, \Phi) \in \boldsymbol{R} \times \mathrm{AC}_{0r}$ *such that* $(\sigma, \Phi, \dot{\Phi}) \in D$ *there are numbers*

$a > 0$ and $\lambda > 0$ such that a local associated function G^F to F corresponding to (σ, Φ), $a > 0$ and $\lambda > 0$ has the following properties:

(i) $G^F \in \mathscr{H}_\lambda(I, \mathbf{R}^n)$ *with* $I := [\sigma, \sigma+a]$,

(ii) $\int_\sigma^{\sigma+a} \|G^F(t, x, z)\| dt \leqslant \lambda$ *for* $(x, z) \in \mathscr{K}_\lambda \times B_\lambda$,

(iii) *G^F has convex values and is w.-w.l.s.c. (w.-w.u.s.c.) with respect to its last two variables.*

Proof. Let $\Omega = \{(\sigma, \Phi)\}$ be fixed. Similarly as in the proof of Lemmas 2.8 and 2.10 we can find numbers $a > 0$ and $\lambda > 0$ such that G^F satisfies conditions (i) and (ii), G^F is l.s.c.w. with respect to its last two variables and so that G^F satisfies anyone of the following inequalities:

($\tilde{\gamma}_1$) $s(p, G^F(t, x, z)) \geqslant \psi_p(t)$ for a.e. $t \in [\sigma, \sigma+a]$, $(x, z) \in \mathscr{K}_\lambda \times B_\lambda$ and $p \in \mathbf{R}^n$,

($\tilde{\gamma}_2$) $|\Phi \oplus z)_t|_0 \leqslant \psi_p^\varepsilon(t) + \varepsilon s(p, G^F(t, x, z))$ for a.e. $t \in [\sigma, \sigma+a]$, $(x, z) \in \mathscr{K}_\lambda \times \times B_\lambda$ and $p \in \mathbf{R}^n$.

Hence, in particular it follows that for every $p \in \mathbf{R}^n$, every $(\bar{x}, \bar{z}) \in \mathscr{K}_\varrho \times B_\varrho$ and every sequence $\{(x_n, z_n)\}$ of $\mathscr{K}_\lambda \times B_\lambda$ such that $|x_n - \bar{x}|_a \to 0$ and $z_n \rightharpoonup \bar{z}$ as $n \to \infty$ one has

$$s\big(p, G^F(t, \bar{x}, \bar{z}\big) \leqslant \liminf_{n\to\infty} s\big(p, G^F(t, x_n, z_n)\big) \tag{2.6}$$

for a.e. $t \in [\sigma, \sigma+a]$. Furthermore, for every $p \in \mathbf{R}^n$ there is L-integrable function k_p: $[\sigma, \sigma+a] \to \mathbf{R}$ such that $s(p, G^F(t, x_n, z_n)) \geqslant k_p(t)$ for a.e. $t \in [\sigma, \sigma+a]$, $n = 1, 2, \ldots$ and $p \in \mathbf{R}^n$. Indeed, (2.6) follows immediately from Lemma II.1.3, because G^F is l.s.c.w. with respect to its last two variables. If (γ_1) is satisfied then taking $k_p = \psi_p|_{[\sigma,\sigma+a]}$, by ($\tilde{\gamma}_1$) one has $s(p, G^F(t, x_n, z_n)) \geqslant k_p(t)$ for a.e. $t \in [\sigma, \sigma+a]$, $n = 1, 2, \ldots$ and $p \in \mathbf{R}^n$. Under condition (γ_2) let us observe first that for every $t \in [\sigma, \sigma+a]$ and $n = 1, 2, \ldots$ we have $|(\dot{\Phi} \oplus z_n)_t|_0 \leqslant |\dot{\Phi}|_0 + \lambda$. Taking now in ($\gamma_2$) $\varepsilon = 1$, by ($\tilde{\gamma}_2$) one obtains $|(\dot{\Phi} \oplus z)_t|_0 \leqslant \psi_p^1(t) + s(p, G^F(t, x_n, z_n))$ for a.e. $t \in [\sigma, \sigma+a]$, $n = 1, 2, \ldots$ and $p \in \mathbf{R}^n$. Therefore, a function $k_p := \psi_p^1|_{[\sigma,\sigma+a]} - M$ with $M = |\dot{\Phi}|_0 + \lambda$ satisfies $k_p(t) \leqslant s(p, G^F(t, x_n, z_n))$ for a.e. $t \in [\sigma, \sigma+a]$, $n = 1, 2, \ldots$ and $p \in \mathbf{R}^n$.

Now, by virtue of Fatou's lemma, (2.6) implies

$$\int_E s\big(p, G^F(t, \bar{x}, \bar{z})\big)dt \leqslant \liminf_{n\to\infty} \int_E s\big(p, G^F(t, x_n, z_n)\big)dt \tag{2.7}$$

for $p \in \mathbf{R}^n$ and measurable set $E \subset [\sigma, \sigma+a]$. Hence, by Theorem II.3.21 for $p \in \mathbf{R}^n$ and a measurable set $E \subset [\sigma, \sigma+a]$ one has $s(p, \int_E G^F(t, \bar{x}, \bar{z})dt) \leqslant \liminf_{n\to\infty} s(p, \int_E G^F(t, x_n, z_n)dt)$. Hence, again by Lemma II.1.3 it follows $\bar{h}(\int_E G^F(t, \bar{x}, \bar{z})dt, \int_E G^F(t, x_n, z_n)dt) \to 0$ for each measurable set $E \subset [\sigma, \sigma+a]$ as $n \to \infty$. Thus G^F is w.-w.l.s.c. with respect to its last two variables. ∎

A set-valued function F: $D \to \mathrm{Comp}(\mathbf{R}^n)$ is said to be *locally convex (concave) with respect to its last variable* if for every $(\sigma, \Phi, \psi) \in D$ there is $\varrho > 0$ such that $I_\varrho(\sigma) \times \mathscr{K}_\varrho(\Phi) \times B_\varrho(\psi) \subset D$ and $F(t, u, \cdot)$ is convex (concave) on $B_\varrho(\psi)$ for every fixed $(t, u) \in I_\varrho(\sigma) \times \mathscr{K}_\varrho(\Phi)$, i.e. for every fixed $(t, u) \in I_\varrho(\sigma) \times \mathscr{K}_\varrho(\Phi)$, each $v_1, v_2 \in B_\varrho(\psi)$ and every $\alpha \in [0, 1]$ one has $\alpha F(t, u, v_1) + (1-\alpha) F(t, u, v_2) \subset F(t, u, \alpha v_1 + (1-\alpha) v_2)$ $[F(t, u, \alpha v_1 + (1-\alpha) v_2) \subset \alpha F(t, u, v_1) + (1-\alpha) F(t, u, v_2)]$.

LEMMA 2.13. *Let $F \in \mathscr{M}(D, \mathrm{Conv}(\boldsymbol{R}^n))$ be locally uniformly integrable, locally l.s.c. (u.s.c.) with respect to its last two variables and locally convex (concave) with respect to its last variable. Furthermore, suppose F satisfies anyone of the growth conditions* (γ_1) *or* (γ_2) *of Lemma* 2.12. *Then for every $(\sigma, \Phi) \in \boldsymbol{R} \times \mathrm{AC}_{0r}$ such that $(\sigma, \Phi, \dot{\Phi}) \in D$ there are numbers $a > 0$ and $\lambda > 0$ such that a local associated function G^F to F corresponding to (σ, Φ), $a > 0$ and $\lambda > 0$ satisfies conditions* (i)–(iii) *of Lemma* 2.12.

Proof. Similarly as in the proof of Lemma 2.12 we can find to $\Omega = \{(\sigma, \Phi)\}$ numbers $a > 0$, $\lambda > 0$ and for every $p \in \boldsymbol{R}^n$ an L-integrable function k_p: $[\sigma, \sigma+a] \to \boldsymbol{R}^n$ such that a local associated set-valued function G^F to F is l.s.c. with respect to its last two variables, satisfies conditions (i) and (ii) of Lemma 2.12 and is such that $s(p, G^F(t, x, z)) \geqslant k_p(t)$ for a.e. $t \in [\sigma, \sigma+a]$, $(x, z) \in \mathscr{K}_\lambda \times B_\lambda$ and $p \in \boldsymbol{R}^n$.

Let $(\bar{x}, \bar{z}) \in \mathscr{K}_\lambda \times B_\lambda$ be fixed and let $\{(x_n, z_n)\}$ be an arbitrary sequence of $\mathscr{K}_\lambda \times B_\lambda$ such that $|x_n - \bar{x}|_a \to 0$ and $z_n \rightharpoonup z$ as $n \to \infty$. Denote $H_p(t, x, z) := s(p, G^F(t, x, z))$ for $t \in [\sigma, \sigma+a]$, $(x, z) \in \mathscr{K}_\lambda \times B_\lambda$ and $p \in \boldsymbol{R}^n$. We shall show that for each measurable set $E \subset [\sigma, \sigma+a]$ and every $p \in \boldsymbol{R}^n$ one has

$$\int_E H_p(t, \bar{x}, \bar{z})\,dt \leqslant \liminf_{n\to\infty} \int_E H_p(t, x_n, z_n)\,dt. \tag{2.8}$$

Hence, similarly as in the proof of Lemma 2.12, condition (iii) of Lemma 2.12 for G^F will follow.

Let $p \in \boldsymbol{R}^n$ and a measurable set $E \subset [\sigma, \sigma+a]$ be fixed. Denote $i := \liminf_{n\to\infty} \int_E H_p(t, x_n, z_n)\,dt$ and $j_n := \int_E H_p(t, x_n, z_n)\,dt$ for $n = 1, 2, \ldots$ By taking a suitable subsequence, say (n_k) of (n) we may well assume that $j_{n_k} \to i$ as $|x_{n_k} - \bar{x}|_a \to 0$ and $z_{n_k} \rightharpoonup \bar{z}$ by $k \to \infty$. Hence, in particular it follows that $|(\Phi \oplus x_{n_k})_t - (\Phi \oplus \bar{x})_t|_0 \to 0$ and $(\dot{\Phi} \oplus z_{nk})_t \rightharpoonup (\dot{\Phi} \oplus \bar{z})_t$ in L_{0r} for $t \in [\sigma, \sigma+a]$ as $k \to \infty$.

Let $\delta_s := \max_{N \leqslant s+1} \max_{1 \leqslant k \leqslant N} |j^{(s)}_{(N,k)} - i|$ for $s = 1, 2, \ldots$, where $j^{(s)}_{(N,k)} := \int_E H_p(t, x_{n_{N+s}}, z_{n_{k+s}})\,dt$. We have $\delta_s \to 0$ as $s \to +\infty$. Furthermore, we have of course $|j^{(s)}_{(N,k)} - i| \leqslant \delta_s$ for $k = 1, 2, \ldots, N$ and $N, s = 1, 2, \ldots$

Now, for any $s = 1, 2, \ldots$ the sequence $(z_{n_{k+s}})$ converges weakly to $\bar{z}$ as $k \to \infty$. Then by virtue of Banach–Mazur's theorem there is a set of real numbers $C^s_{N_k} \geqslant 0$, $k = 1, 2, \ldots, N$, $N = 1, 2, \ldots$ with $\sum_{k=1}^{N} C^s_{N_k} = 1$ such that if $z^s_N(t) := \sum_{k=1}^{N} C^s_{N_k} (\dot{\Phi} \oplus z_{nk+s})(t)$, then $|z^{(s)}_N - z|_a \to 0$ as $N \to \infty$ for $s = 1, 2, \ldots$ Hence, in particular it follows that $(\dot{\Phi} \oplus z^{(s)}_N)_t \to (\dot{\Phi} \oplus z)_t$ in the norm topology of L_{0r} as $N \to \infty$ for $s = 1, 2, \ldots$ and $t \in [\sigma, \sigma+a]$.

Let us take $\psi = -k_p$, $\eta_N^{(s)}(t) := \sum_{k=1}^{N} C_{N_k}^{(s)} H_p(t, (\Phi \oplus x_{n_{N+s}})_t, (\Phi \oplus z_{n_{k+s}})_t)$ and note that $\eta_N^{(s)}(t) \geqslant -\psi(t)$ and $i - \delta_s \leqslant \int_E \eta_N^{(s)}(t) \leqslant i + \delta_s$ for all $s = 1, 2, \ldots, N = 1, 2, \ldots$ and a.e. $t \in [\sigma, \sigma + a]$, respectively. Then, for a.e. $t \in [\sigma, \sigma + a]$ and $s = 1, 2, \ldots$ we have $\liminf_{N\to\infty} \eta_N^{(s)}(t) \geqslant -\psi(t)$. Hence, by Fatou's lemma we obtain

$$\int_E \liminf_{N\to\infty} \eta_N^{(s)}(t)\,dt \leqslant \liminf_{N\to\infty} \int_E \eta_N^{(s)}(t)\,dt \leqslant i + \delta_s.$$

Thus $\eta^{(s)}(t) = \liminf_{N\to\infty} \eta_N^{(s)}(t)$ is for every $s = 1, 2, \ldots$ finite for a.e. $t \in [\sigma, \sigma + a]$. Taking $\eta(t) := \liminf_{s\to\infty} \eta^{(s)}(t)$ for a.e. $t \in [\sigma, \sigma + a]$ we obtain $\eta(t) \geqslant -\psi(t)$ for a.e. $t \in [\sigma, \sigma + a]$ and $\int_E \eta(t)\,dt \leqslant i$.

We shall show that for a.e. $t \in [\sigma, \sigma + a]$ we have also $H_p(t, \bar{x}, \bar{z}) \leqslant \eta(t)$. Indeed, by lower semicontinuity of $G^F(t, \cdot, \cdot)$ at $(\bar{x}, \bar{z})$, the function $H_p(t, \cdot, \cdot)$ is lower semicontinuous on $\mathscr{K}_\lambda \times B_\lambda$. Therefore, in particular we have

$$H_p(t, \bar{x}, \bar{z}) \leqslant \liminf_{N\to\infty} H_p\Big((t, x_{n_{N+s}}, \sum_{k=1}^{N} C_N^{(s)} z_{n_{k+s}}\Big)$$

for $s = 1, 2, \ldots$ Hence, for every $m = 1, 2, \ldots$ and $s = 1, 2, \ldots$ there is a positive integer $K_m^{(s)}$ such that for $N \geqslant K_m^{(s)}$ and a.e. $t \in [\sigma, \sigma + a]$ we have

$$H_p(t, \bar{x}, \bar{z}) - \frac{1}{m} < H_p\Big(t, x_{n_{N+s}}, \sum_{k=1}^{N} C_{N_k}^{(s)} z_{n_{k+s}}\Big).$$

Since $G^F(t, x, \cdot)$ is convex for each fixed $(t, x) \in [\sigma, \sigma + a] \times \mathscr{K}_\lambda$ then

$$H_p(t, \bar{x}, \bar{z}) - \frac{1}{m} < \sum_{k=1}^{N} C_{N_k}^{(s)} H_p(t, x_{n_{N+s}}, z_{n_{k+s}}) =: \eta_N^{(s)}(t)$$

for $s, m = 1, 2, \ldots, N > K_m^{(s)}$ and a.e. $t \in [\sigma, \sigma + 1]$. Therefore, for a.e. $t \in [\sigma, \sigma + a]$ and $m = 1, 2, \ldots$ one obtains

$$H_p(t, \bar{x}, \bar{z}) - \frac{1}{m} \leqslant \liminf_{s\to\infty} [\liminf_{N\to\infty} \eta_N^{(s)}(t)] =: \eta(t).$$

Therefore, for every measurable set $E \subset [\sigma, \sigma + a]$ and $p \in \boldsymbol{R}^n$ one has

$$\int_E H_p(t, \bar{x}, \bar{z})\,dt \leqslant \int_E \eta(t)\,dt \leqslant i.$$

Thus, (2.8) is satisfied. ■

2.3. Local existence theorems

Now as a consequence of Lemmas 2.6, 2.7 and Theorems III.3.4, III.3.5 we obtain the following local existence theorems for NFDI(D, F).

THEOREM 2.14. *Let $F \in \mathcal{M}(D, \mathrm{Comp}(\mathbf{R}^n))$ be locally uniformly integrable and locally Lipschitz continuous with respect to its last two variables. Then for every $(\sigma, \Phi) \in \mathbf{R} \times \times \mathrm{AC}_{0r}$ such that $(\sigma, \Phi, \dot{\Phi}) \in D$ there exists $a > 0$ such that* NFDI(D, F) *has on $[\sigma - r, \sigma + a]$ at least one solution through (σ, Φ).*

Proof. Similarly as in the proof of Lemma 2.10 we can show that for every $(\sigma, \Phi) \in \mathbf{R} \times \mathrm{AC}_{0r}$ such that $(\sigma, \Phi, \dot{\Phi}) \in D$ there are numbers $a > 0$ and $\lambda > 0$ such that a local associated set-valued function G^F to F corresponding to (σ, Φ), $a > 0$ and $\lambda > 0$ is Lipschitz continuous with respect to its last two variables and such that $G^F \in \mathcal{H}_\lambda(I, \mathbf{R}^n)$ with $I := [\sigma, \sigma+a]$ and $\int_\sigma^{\sigma+a} \|G^F(t, x, z)\| dt \leqslant \lambda$ for $(x, z) \in \mathcal{K}_\lambda \times B_\lambda$. Therefore, by Theorem III.3.4 there is $z \in B_\lambda$ such that $z(t) \in G^F(t, \mathcal{J}z, z)$ for a.e. $t \in I$. Taking $x := \mathcal{R}(\Phi, \mathcal{J}z)$ we can easily see that $x_\sigma = \Phi$ and $\dot{x}(t) \in F(t, x_t, \dot{x}_t)$ for a.e. $t \in I$. ∎

THEOREM 2.15. *Let $F \in \mathcal{M}(D, \mathrm{Comp}(\mathbf{R}^n))$ be locally uniformly integrable and locally $\mathcal{A}$-Lipschitz continuous with respect to its two last variables. Then for every $(\sigma, \Phi) \in \mathbf{R} \times \mathrm{AC}_{0r}$ such that $(\sigma, \Phi, \dot{\Phi}) \in D$ there exists $a > 0$ such that* NFDI(D, F) *has on $[\sigma - r, \sigma + a]$ at least one solution through (σ, Φ).*

Proof. Similarly as in the proof of Lemma 2.10 we can also show that for every $(\sigma, \Phi) \in \mathbf{R} \times \mathrm{AC}_{0r}$ such that $(\sigma, \Phi, \dot{\Phi}) \in D$ there are numbers $a > 0$ and $\lambda > 0$ such that a local associated set-valued function G^F to F corresponding to (σ, Φ), $a > 0$ and $\lambda > 0$ is $\mathcal{A}$-Lipschitz continuous with respect to its last two variables with Lipschitz constant $L < 1$ and such that $G^F \in \mathcal{H}_\lambda(I, \mathbf{R}^n)$ with $I := [\sigma, \sigma+a]$ and $\int_\sigma^{\sigma+a} \|G^F(t, x, z)\| dt \leqslant \lambda$ for $(x, z) \in \mathcal{K}_\lambda \times B_\lambda$. Now, the result follows from Theorem III.3.5 similarly as in the proof of Theorem 2.14. ∎

Imediately from Theorems III.3.6–III.3.10 the following local existence theorems follow.

THEOREM 2.16. *Let $F \in \mathcal{M}(D, \mathrm{Comp}(\mathbf{R}^n))$ be locally uniformly integrable and locally hemi-s.-w.l.s.c. with respect to its last two variables. Then for every $(\sigma, \Phi) \in \mathbf{R} \times \times \mathrm{AC}_{0r}$ such that $(\sigma, \Phi, \dot{\Phi}) \in D$ there exists $a > 0$ such that* NFDI(D, F) *has on $[\sigma - r, \sigma + a]$ at least one solution through (σ, Φ).* ∎

THEOREM 2.17. *Let $F \in \mathcal{M}(D, \mathrm{Conv}(\mathbf{R}^n))$ be locally uniformly integrable and locally hemi-w.-w.l.s.c. with respect to its last two variables. Then for every $(\sigma, \Phi) \in \mathbf{R} \times \mathrm{AC}_{0r}$ such that $(\sigma, \Phi, \dot{\Phi}) \in D$ there exists $a > 0$ such that* NFDI(D, F) *has on $[\sigma - r, \sigma + a]$ at least one solution theough (σ, Φ).* ∎

THEOREM 2.18. *Let $F \in \mathcal{M}(D, \mathrm{Conv}(\mathbf{R}^n))$ be locally uniformly integrable and locally hemi-w.-w.u.s.c. with respect to its last two variables. Then for every $(\sigma, \Phi) \in \mathbf{R} \times \mathrm{AC}_{0r}$*

there exists $a > 0$ *such that* NFDI(D, F) *has on* $[\sigma - r, \sigma + a]$ *at least one solution through* (σ, Φ). ■

THEOREM 2.19. *Let* $F \in \mathcal{M}(D, \mathrm{Conv}(\boldsymbol{R}^n))$ *be locally uniformly integrable, locally hemi-continuous with respect to second variable and locally* $\mathcal{A}$*-Lipschitz continuous with respect to its last variable uniformly with respect to the second one. Then for every* $(\sigma, \Phi) \in \boldsymbol{R} \times \mathrm{AC}_{0r}$ *such that* $(\sigma, \Phi, \dot{\Phi}) \in D$ *there exists* $a > 0$ *such that* NFDI(D, F) *has on* $[\sigma - r, \sigma + a]$ *at least one solution through* (σ, Φ). ■

REMARK 2.2. The function $f\colon \boldsymbol{R} \times H \to \boldsymbol{R}$ defined in Example 2.1 is locally uniformly integrable.

Indeed, let $(\sigma, \Phi, \psi) \in \boldsymbol{R} \times H$ be fixed and let $\varrho > 0$ be arbitrarily taken. For every measurable set $E \subset I_\varrho(\sigma)$ and every $(u, v) \in \mathcal{K}_\varrho(\Phi) \times B_\varrho(\psi)$ one has

$$\int_E |f(t, u, v)|\, dt \leqslant |a| \int_E |u(0)|\, dt + |b| \int_E |u(-1)|\, dt + |c| \int_E |v(-1)|\, dt$$

$$\leqslant (|a| + |b|)(|\Phi|_0 + \varrho)\mu(E) + |c| \int_E |w_v(-1)|\, dt.$$

where $w_v \in L_{01}$ is for every $v \in B_\varrho(\psi)$ such that $w_v(t) = v(t)$ for $t \in (-1, 0]$ and $w_v(-1) = 0$. Then, $\lim_{\mu(E) \to \infty} \int_E |f(t, u, v)|\, dt = 0$ uniformly with respect to $(u, v) \in \mathcal{K}_\varrho(\Phi) \times B_\varrho(\psi)$ with $E \subset I_\varrho(\sigma)$. ■

2.4. Continuation of solutions

Let D be a nonempty open subset of $\boldsymbol{R} \times C_{0r} \times L_{0r}$ and let $F \in \mathcal{M}(D, \mathrm{Comp}(\boldsymbol{R}^n))$ be such that for every $(\sigma, \Phi) \in \boldsymbol{R} \times \mathrm{AC}_{0r}$ with $(\sigma, \Phi, \dot{\Phi}) \in D$, NFDI$(D, F)$ has at least one local solution through (σ, Φ). We are interested now in the existence and some properties of noncontinuable solutions of NFDI(D, F) through given (σ, Φ). Recall, that a solution x of NFDI(D, F) on $[\sigma - r, b]$ is said to be *noncontinuable* if no continuation of x exists. The existence of noncontinuable solutions of NFDI(D, F) follows immediately from the Kuratowski–Zorn's lemma (see Section II.3.3). The proofs of such theorems are the same as those for ordinary differential equations, because they in fact are independent of the right-hand sides of equations. These theorems could also be proved for any function x which is absolutely continuous on the open interval $(\sigma - r, T)$ with $|x_t|_0$ defined for $t \in [\sigma, T)$.

We shall prove now the following theorems.

THEOREM 2.20. *Let* $F \in \mathcal{M}(D, \mathrm{Comp}(\boldsymbol{R}^n))$ *be such that for every* $(\sigma, \Phi) \in \boldsymbol{R} \times \mathrm{AC}_{0r}$ *satisfying* $(\sigma, \Phi, \dot{\Phi}) \in D$, NFDI$(D, F)$ *has at least one local solution through* (σ, Φ). *Then for every* $(\sigma, \Phi) \in \boldsymbol{R} \times \mathrm{AC}_{0r}$ *satisfying* $(\sigma, \Phi, \dot{\Phi}) \in D$ *there exists a noncontinuable solution of* NFDI(D, F) *through* (σ, Φ) *on the maximal interval* $[\sigma - r, T)$, *with* $T \in \boldsymbol{R}$ or $T = +\infty$.

Proof. Denote by $\mathscr{C}(\sigma, \Phi, F)$ the set of all functions x: $[\sigma-r, \lambda_x)$ such that each restriction x_μ of x to the interval $[\sigma-r, \mu]$ with $\mu \in (\sigma, \lambda_x)$ is a solution of NFDI(D, F) on $[\sigma-r, \mu]$ theough (σ, Φ). Let us introduce in $\mathscr{C}(\sigma, \Phi, f)$ an ordered relation $\precsim$ by setting $x \precsim y$ if and only if x is a restriction of y to any subinterval $[\sigma-r, \lambda_x)$ contained in the domain $[\sigma-r, \lambda_y)$ of y, for $x, y \in \mathscr{C}(\sigma, \Phi, F)$. It is clear that $(\mathscr{C}(\sigma, \Phi, F), \precsim)$ is a partially ordered system such that only $(X_x, \precsim)$ with X_x containing all restrictions x of $x_\mu \in \mathscr{C}(\sigma, \Phi, F)$ are totally ordered subsystem of $\mathscr{C}(\sigma, \Phi, F)$. Since for every $z \in X_x$ we have $z \precsim x$, then every totally ordered subsystem of $\mathscr{C}(\sigma, \Phi, F)$ has an upper bound in $\mathscr{C}(\sigma, \Phi, F)$. Thus, by virtue of Kuratowski–Zorn's lemma, there exists in $\mathscr{C}(\sigma, \Phi, F)$ a maximal element $x_{\max}$ defined on any interval $[\sigma-r, T)$ that is a noncontinuable solution of NFDI(D, F) through (σ, Φ). ∎

Let (D, A) be a conformable pair of nonempty open subsets of $\boldsymbol{R} \times C_{0r} \times L_{0r}$ and $\boldsymbol{R} \times \mathrm{AC}_{0r}$, respectively.

THEOREM 2.21. *Suppose that through every point (σ, Φ) of A there exists a local solution to* NFDI(D, F). *If x is a noncontinuable solution of* NFDI(D, A, F) *on $[\sigma, T)$ through (σ, Φ), then for any closed bounded set $\Omega \subset A$ there exists an $b \in [\sigma, T)$ such that $(t, x_t) \notin \Omega$ for $t \in [b, T)$.*

Proof. The case $T = +\infty$ is trivial because Ω is bounded. Suppose T is finite. Fix $t \in [\sigma, T)$ and let $\|x\|_{[\sigma-r, t]}$ be defined by

$$\|x\|_{[\sigma-r, t]} = |x(\sigma-r)| + \int_{\sigma-r}^{t} |\dot{x}(s)|\, ds,$$

The function $[\sigma, T) \in t \to \|x\|_{[\sigma-r, t]} \in \boldsymbol{R}^+$ is monotone, increasing and hence if it is bounded on $[\sigma, T)$ it must have a limit. If this is case then $\lim_{t \to T-0} x(t)$ exists. Putting $x(T) = \lim_{t \to T-0} x(t)$ we also get $\lim_{t \to T-0} x(t+s) = x(T+s)$ for every $s \in [-r, 0]$, i.e. $\lim_{t \to T-0} |x - x|_0 = 0$. If (t, x_t) did not leave and stay out Ω we would have $(T, x_T) \in \Omega \subset A$ which is a contradiction, because x is a noncontinuable solution of NFDI(D, A, F).

If the function $[\sigma, T) \in t \to \|x\|_{[\sigma-r, t]} \in \boldsymbol{R}$ is unbounded on $[\sigma, T)$ then its monotonicity is enough to conclude that (t, x_t) must leave and stay out of every closed bounded set Ω. ∎

3. PROPRTIES OF SETS OF SOLUTIONS OF NFDIs

Throughout this section (D, Ω) will be a conformable pair of subsets of $\boldsymbol{R} \times C_{0r} \times L_{0r}$ and $\boldsymbol{R} \times \mathrm{AC}_{0r}$, respectively such that D is open in $\boldsymbol{R} \times C_{0r} \times L_{0r}$ and Ω is closed in $\boldsymbol{R} \times C_{0r}$ and relatively weakly compact in $\boldsymbol{R} \times \mathrm{AC}_{0r}$.

Given $(\sigma, \Phi) \in \Omega$ and $F \in \mathscr{M}(D, \mathrm{Comp}(\boldsymbol{R}^n))$ by $\mathscr{C}(\sigma, \Phi, F)$ we denote the set of all local solutions of NFDI(D, F) on $[\sigma-r, \sigma+a]$ through (σ, Φ). We investigate here the properties of sets $\mathscr{C}(\sigma, \Phi, F)$ as well the properties of the mappings $\mathscr{C}(\cdot\,, \cdot, F)$

which associate with every $(\sigma, \Phi) \in \Omega$ the set $\mathscr{C}(\sigma, \Phi, F)$. As it happens in the case of ordinary differential equations solutions through different initial points or even different solutions through the same initial point may have unequal intervals of the existence. To avoid this difficulty we shall limit ourselves to local properties and consider the sets $\mathscr{C}(\sigma, \Phi, F)$.

3.1. Compactness and upper-semicontinuity

For given $(\sigma, \Phi) \in \boldsymbol{R} \times \mathrm{AC}_{0r}$ and $G \in \mathscr{H}_\lambda(I, \boldsymbol{R}^n)$ with $I = [\sigma, \sigma+a]$ by $S_\sigma(G)$ we shall denote the set of all fixed points of $\mathscr{F}(G \,\bar{\square}\, \mathscr{I})$. Similarly as above for given $F \in \mathscr{M}(D, \mathrm{Comp}(\boldsymbol{R}^n))$, $(\sigma, \Phi) \in \boldsymbol{R} \times \mathrm{AC}_{0r}$, $a > 0$ and $\lambda > 0$ by G^F we denote a local associated set-valued function to F defined by (2.6).

LEMMA 3.1. *Let (D, Ω) be a conformable pair of nonempty subsets of $\boldsymbol{R} \times C_{0r} \times L_{0r}$ and $\boldsymbol{R} \times \mathrm{AC}_{0r}$ with D open in $\boldsymbol{R} \times C_{0r} \times L_{0r}$. Let $F \in \mathscr{M}(D, \mathrm{Comp}(\boldsymbol{R}^n))$ be such that there exist $a > 0$ and $\lambda > 0$ such that for every $(\sigma, \Phi) \in \Omega$, $S_\sigma(G^F) \neq \emptyset$. Then for every $(\sigma, \Phi) \in \Omega$*

(i) $\mathscr{C}(\sigma, \Phi, F) = \mathscr{R}(\Phi, \mathscr{I}S_\sigma(G^F))$ *and*

(ii) $\mathscr{I}S_\sigma(G^F) = \mathscr{R}^-(\Phi, \mathscr{C}(\sigma, \Phi, F))$, *where $\mathscr{I}$, $\mathscr{R}$ and $\mathscr{R}^-$ are mappings defined by* (III.1.2), (2.2) *and* (2.3), *respectively.*

Proof. Let $(\sigma, \Phi) \in \Omega$ be fixed and suppose $y \in \mathscr{C}(\sigma, \Phi, F)$. Let $x := y|_{[\sigma,\sigma+a]} - \Phi(0)$. We have $x(\sigma) = y(\sigma) - \Phi(0) = 0$, because $y_\sigma = \Phi$ implies $y(\sigma) = \Phi(0)$. Furthermore, for every $t \in [\sigma, \sigma+a]$ we have $(t, y_t, \dot{y}_t) \in D$ and $\dot{y}(t) \in F(t, y_t, \dot{y}_t)$ for a.e. $t \in [\sigma, \sigma+a]$. Thus $\dot{x}(t) \in F(t, (\Phi \oplus x)_t, (\dot{\Phi} \oplus \dot{x})_t) := G^F(t, x, \dot{x})$ for a.e. $t \in [\sigma, \sigma+a]$. Then $x \in \mathscr{I}S_\sigma(G^F)$. Therefore, for every $y \in \mathscr{C}(\sigma, \Phi, F)$ there exists $x \in \mathscr{I}S_\sigma(G^F)$ such that $y = \Phi \oplus x = \mathscr{R}(\Phi, x)$, i.e., $y \in \mathscr{R}(\Phi, \mathscr{I}S_\sigma(G^F))$. Then $\mathscr{C}(\sigma, \Phi, F) \subset \mathscr{R}(\Phi, \mathscr{I}S_\sigma(G^F))$.

Suppose $x \in \mathscr{I}S_\sigma(G^F)$ and put $y = \Phi \oplus x$. Then $\dot{x}(t) \in G^F(t, \mathscr{I}\dot{x}, \dot{x}) = G^F(t, x, \dot{x}) := F(t, (\Phi \oplus x)_t, (\dot{\Phi} \oplus \dot{x})_t)$ and $\dot{x}(t) = \dot{y}(t)$ for a.e. $t \in [\sigma, \sigma+a]$. We have of course $y_\sigma = \Phi$, because for every $s \in [-r, 0]$, $\sigma + s \in [\sigma - r, \sigma]$ and therefore $y_\sigma(s) := y(\sigma + s) = (\Phi \oplus x)(\sigma+s) = \Phi(\sigma+s-\sigma) = \Phi(s)$. Then for every $x \in \mathscr{I}S_\sigma(G^F)$ there exists $y \in \mathscr{C}(\sigma, \Phi, F)$ such that $y = \mathscr{R}(\Phi, x)$. Thus $\mathscr{C}(\sigma, \Phi, F) \subset \mathscr{R}(\Phi, \mathscr{I}S_\sigma(G^F))$ and therefore (i) holds. In a similar way we also obtain $\mathscr{I}S_\sigma(G^F) = \mathscr{R}^-(\Phi, \mathscr{C}(\sigma, \Phi, F))$. ∎

Now, we obtain the following results.

THEOREM 3.2. *Let (D, Ω) be a conformable pair of nonempty subsets of $\boldsymbol{R} \times C_{0r} \times L_{0r}$ and $\boldsymbol{R} \times \mathrm{AC}_{0r}$, respectively such that D is open in $\boldsymbol{R} \times C_{0r} \times L_{0r}$ and Ω is closed in $\boldsymbol{R} \times C_{0r}$ and relatively weakly compact in $\boldsymbol{R} \times \mathrm{AC}_{0r}$. Let $F \in \mathscr{M}(D, \mathrm{Conv}(\boldsymbol{R}^n))$ be locally uniformly integrable and locally hemi-w.-w.u.s.c. with respect to its last two variables. Then there is $a > 0$ such for every $(\sigma, \Phi) \in \Omega$, $\mathscr{C}(\sigma, \Phi, F)$ is a nonempty compact subset of $C([\sigma - r, \sigma+a], \boldsymbol{R}^n)$ and weakly compact in $\mathrm{AC}([\sigma-r, \sigma+a], \boldsymbol{R}^n)$.*

Proof. Similarly as in Lemmas 2.5 and 2.10 we can easily verify that there are numbers $a > 0$ and $\lambda > 0$ such that a local associated set-valued function G^F to F corresponding to fixed $(\sigma, \Phi) \in \Omega$, $a > 0$ and $\lambda > 0$ belongs to $\mathscr{H}_\lambda^C(I, R^n)$ with $I = [\sigma, \sigma+a]$. Therefore, by Theorem III.3.13, $S_\sigma(G^F)$ is a nonempty weakly compact subset of P_λ. Now results follow immediately from Lemmas III.1.13 2.3, 3.1 and Proposition I. 2.1. ■

We shall consider now a set-valued mapping $\mathscr{C}(\cdot, \cdot, F)$: $\Omega \in (\sigma, \Phi) \to \mathscr{C}(\sigma, \Phi, F) \subset C_r(\Omega)$. Similarly as in Section III.3.3 (see Theorem III.3.18) we obtain the following lemma.

LEMMA 3.3. *Let $F \in \mathscr{M}(D, \mathrm{Conv}(R^n))$ be locally uniformly integrable and locally hemi-w.-w.u.s.c. with respect to its last two variables. Suppose (D, Ω) is a conformable pair such as in Theorem 3.2 and such that $\Pi_R(\Omega)$ is connected. Then there is $a > 0$ such that for every $(\sigma, \Phi) \in \Omega$ and every sequence $\{(\sigma_n, \Phi_n)\}$ of Ω converging to (σ, Φ) in the norm topology of $R \times C_{0r}$ as $n \to \infty$, every sequence (y_n) of $C_r(\Omega)$ such that $y_n \in \mathscr{C}(\sigma_n, \Phi_n, F)$ for $n = 1, 2, \ldots$ has a subsequence ϱ-converging to any $y \in \mathscr{C}(\sigma, \Phi, F)$.*

Proof. Suppose $a > 0$ and $\lambda > 0$ are such as in Lemma 2.5. Then for every $(\sigma_n, \Phi_n) \in \Omega$ a local associated set-valued function G_n^F to F corresponding to (σ_n, Φ_n), $a > 0$ and $\lambda > 0$ satisfies conditions (ii) of Lemma 2.5. Moreover $G^F \in \mathscr{H}_\lambda^C([\sigma_n, \sigma_n+ +a], R^n)$. Let $\{(\sigma_n, \Phi_n)\}$ and (y_n) be sequence of Ω and $C_r(\Omega)$, respectively such as above. By virtue of Lemma 3.1 there is a sequence of absolutely continuous functions x_n: $[\sigma_n, \sigma_n+a] \to R^n$ such that $x_n(\sigma_n) = 0$, $\dot{x}_n(t) \in G_n^F(t, x_n, \dot{x}_n)$ for a.e. $t \in [\sigma_n, \sigma_n+a]$ and so that $y_n = \Phi_n \oplus x_n$ for each $n = 1, 2, \ldots$ By virtue of (ii) of Lemma 2.5 for every $\varepsilon > 0$ there is a $\delta_1 > 0$ such that $\int_{E_n^1} |\dot{x}_n(t)|\,dt < \varepsilon/2$ for every measurable set $E_n^1 \subset [\sigma_n, \sigma_n+a]$; $n = 1, 2, \ldots$ with $\mu(E_n^1) < \delta_1$. By virtue of Lemma I.4.5 we have $\dot{\Phi}_n \rightharpoonup \dot{\Phi}$ as $n \to \infty$. Then to the given above $\varepsilon > 0$ there is $\delta_0 > 0$ such that $\int_{E_n^0} |\dot{\Phi}_n(s)|\,ds < \frac{1}{2}\varepsilon$ for every measurable set $E_n^0 \subset [-r, 0]$ with $\mu(E_n^0) < \delta_0$. Put $\delta = \min(\delta_0, \delta_1)$ and let $E_n \subset [\sigma_n - r, \sigma_n+a]$ be a measurable set such that $\mu(E) < \delta$. Since $\int_E |\dot{y}_n(t)|\,dt \leqslant \int_{E_n^0} |\dot{\Phi}_n(s)|\,ds + \int_{E_n^1} |\dot{x}_n(t)|\,dt$ with $E_n^0 := \{E \cap [\sigma_n - r, \sigma_n + a]\} - \sigma_n$ and $E_n^1 := E \cap [\sigma_n, \sigma_n+a]$ then $\int_E |\dot{y}_n(t)|\,dt < \varepsilon$ because $\mu(E_n^0) \leqslant \mu(E) < \delta_0$ and $\mu(E_n^1) \leqslant \mu(E) < \delta_1$. It is clear that (y_n) is a bounded sequence of $\mathscr{X}_r(\Omega)$. Therefore, by Theorem I.4.9, (y_n) is relatively d-weakly sequentially compact in $\mathscr{X}_r(\Omega)$. Hence it follows that it is also relatively ϱ-compact in $C_r(\Omega)$. Thus there is $(y;\ [\sigma - r, \sigma+ +a]) \in C_r(\Omega)$ and a subsequence of (y_n), say (y_k) such that $\varrho(y_k, y) \to 0$ as $k \to \infty$.

Let $\Pi_R(\Omega) = [m, M]$ and put $I := [m, M-a]$. Similarly as in Section III.3.3 we define $\mathscr{H}_I^C := \bigcup_{\sigma \in I} \mathscr{H}_\lambda^C(I_\sigma, R^n)$ with $I_\sigma = [\sigma, \sigma+a]$ for $\sigma \in I$. It is clear that $G_k^F \in \mathscr{H}_I^C$ for every $k = 1, 2, \ldots$

Let G^F denote an associated set-valued function to F corresponding to $(\sigma, \Phi) \in \Omega$ and above taken $a > 0$ and $\lambda > 0$. By the properties of F it is clear that a sequence

(G_k^F) of $\mathscr{H}_I^C$ is upper l-weakly converging to G^F. Therefore, by Remark III.3.3 a function x: $[\sigma, \sigma+a] \to \boldsymbol{R}^n$ defined by $x := y|_{[\sigma,\sigma+a]} - \Phi(0)$ is such that $x(\sigma) = 0$ and $\dot{x}(t) \in G^F(t, x, \dot{x})$ for a.e. $t \in [\sigma, \sigma+a]$. Since $y = \Phi \oplus x$, then $y \in \mathscr{C}(\sigma, \Phi, F)$. ■

Now we obtain the following continuous dependence theorem.

THEOREM 3.4. *Let $F \in \mathscr{M}(D, \mathrm{Conv}(\boldsymbol{R}^n))$ be locally uniformly integrable and locally hemi-w.-w.u.s.c. with respect to its last two variables. Assume (D, Ω) is a conformable pair of nonempty subsets of $\boldsymbol{R} \times C_{0r} \times L_{0r}$ and $\boldsymbol{R} \times \mathrm{AC}_{0r}$, respectively such that D is open and Ω is closed in $\boldsymbol{R} \times C_{0r}$, relatively weakly compact in $\boldsymbol{R} \times \mathrm{AC}_{0r}$ and such that $\Pi_{\boldsymbol{R}}(\Omega)$ is connected. Then there is $a > 0$ such that a set-valued mapping $\mathscr{C}(\cdot, \cdot, F)$: $\Omega \ni (\sigma, \Phi) \to \mathscr{C}(\sigma, \Phi, F) \subset C_r(\Omega)$ is u.s.c. on Ω.*

Proof. Select $a > 0$ and $\lambda > 0$ such as in Lemma 3.3. Let us observe that $\mathscr{C}(\cdot, \cdot, F)$ is a compact-valued mapping from $\Omega \subset \boldsymbol{R} \times C_{0r}$ into a space of nonempty subsets of a metric space $(C_r(\Omega), \varrho)$. Therefore, by Theorem II.2.2 it is enough only to verify that for every $(\sigma, \Phi) \in \Omega$ and every sequence $\{(\sigma_n, \Phi_n)\}$ of Ω, converging in the norm topology of $\boldsymbol{R} \times C_{0r}$ to (σ, Φ), every sequence (y_n) of $C(\Omega)$ satisfying $y_n \in \mathscr{C}(\sigma_n, \Phi_n, F)$ for $n = 1, 2, \ldots$ has a subsequence ϱ-converging to any $y \in \mathscr{C}(\sigma, \Phi, F)$. This follows immediately from Lemma 3.3. ■

COROLLARY 3.1. *If (D, Ω) and $F \in \mathscr{M}(D, \mathrm{Conv}(\boldsymbol{R}^n))$ are such as in Theorem 3.4 then there exists $a > 0$ such that $\mathscr{C}(\Omega, F) := \bigcup_{(\sigma,\Phi)\in\Omega} \mathscr{C}(\sigma, \Phi, F)$ is a compact subset of a metric space $(C_r(\Omega), \varrho)$.* ■

COROLLARY 3.2. *Assume (D, Ω) and $F \in \mathscr{M}(D, \mathrm{Conv}(\boldsymbol{R}^n))$ are such as in Theorem 3.4. Then there is a number $a > 0$ such that a mapping T: $\Omega \in (\sigma, \Phi) \to \{x_{\sigma+a}: x \in \mathscr{C}(\sigma, \Phi, F)\} \subset C_{0r}$ is u.s.c. on Ω. If, furthermore, Ω is convex and such that $T(\sigma, \Phi) \subset \Omega$ for every $(\sigma, \Phi) \in \Omega$ then there exists $(\sigma, \Phi) \in \Omega$ such that $(\sigma, \Phi) \in \overline{\mathrm{co}}\, T(\sigma, \Phi)$.* ■

3.2. Approximating and acyclic properties

A set-valued function F: $D \to \mathrm{Comp}(\boldsymbol{R}^n)$ is said to be *locally bounded* if for every $(\sigma, \Phi, \psi) \in D$ there are numbers $\varrho > 0$ and $M > 0$ such that $I_\varrho(\sigma) \times \mathscr{K}_\varrho(\Phi) \times B_\varrho(\psi) \subset D$ and $\|F(t, u, v)\| \leqslant M$ for a.e. $t \in I_\varrho(\sigma)$ and $(u, v) \in \mathscr{K}_\varrho(\Phi) \times B_\varrho(\psi)$.

As a corollary from Lemmas 2.5, 2.9 and Theorem III.4.5 we obtain the following lemma.

LEMMA 3.5. *Let $F \in \mathscr{M}(D, \mathrm{Conv}(\boldsymbol{R}^n))$ be locally bounded and locally u.s.c.w. with respect to its last two variables. Then for every $(\sigma, \Phi) \in \boldsymbol{R} \times \mathrm{AC}_{0r}$ such that $(\sigma, \Phi, \dot{\Phi}) \in D$ there are numbers $a > 0$ and $\lambda > 0$ such that a local associated set-valued function G^F to F corresponding to (σ, Φ), $a > 0$ and $\lambda > 0$ is such $\|G^F([\sigma, \sigma+a] \times \mathscr{K}_\lambda \times B_\lambda)\| \leqslant \lambda/a$ and a set-valued function $\mathscr{I}\mathscr{F}(G^F \,\square\, \mathscr{D})$ is σ-selectionable on $[\sigma, \sigma+a] \times \boldsymbol{K}_\lambda$, where $\boldsymbol{K}_\lambda = \mathscr{I}(\boldsymbol{P}_\lambda)$.* ■

Now, we prove the following theorems.

THEOREM 3.6. *Let $F \in \mathcal{M}(D, \mathrm{Conv}(\boldsymbol{R}^n))$ be locally bounded and locally u.s.c.w. with respect to its last two variables. Then for every $(\sigma, \Phi) \in \boldsymbol{R} \times \mathrm{AC}_{0r}$ such that $(\sigma, \Phi, \dot{\Phi}) \in D$ there are a number $a > 0$, a sequence (V^m) of compact convex subsets of any Banach space $(X, \|\cdot\|)$ and continuous functions s_n: $V^m \to C([\sigma-r, \sigma+a], \boldsymbol{R}^n)$ such that*

(i) $\mathscr{C}(\sigma, \Phi, F) \subset s_{m+1}(V^{m+1}) \subset s_m(V^m)$ *for* $m = 1, 2, \ldots,$

(ii) *for every $\varepsilon > 0$ there is a number $N_\varepsilon \geqslant 0$ such that $s_m(V^m) \subset \mathscr{C}(\sigma, \Phi, F) + \varepsilon\tilde{B}$ for $m \geqslant N_\varepsilon$, where $\tilde{B}$ denotes the closed unit ball of $C([\sigma-r, \sigma+a], \boldsymbol{R}^n)$.*

Furthermore, for every $x \in \mathscr{C}(\sigma, \Phi, F)$, the set $\{v \in V^m\colon s_m(v) = x\}$ is convex.

Proof. Let $(\sigma, \Phi) \in \boldsymbol{R} \times \mathrm{AC}_{0r}$ with $(\sigma, \Phi, \dot{\Phi}) \in D$ be fixed and let $a > 0$ and $\lambda > 0$ be such as in Lemma 3.5. By virtue of Theorem III.4.7 there exists a sequence (U^m) of compact convex subsets of any Banach space and a sequence (s_m) of continuous closed functions s_m: $U^m \to C([\sigma, \sigma+a], \boldsymbol{R}^n)$ such that conditions (i) and (ii) of Theorem III.4.7 are satisfied. Furthermore, for every $x \in \mathscr{I}S_\sigma(G^F)$ the set $\{u \in U^m\colon s_m(u) = x\}$ is convex.

Now, let $V^m = U^m$ and $c_m(u) = \mathscr{R}(\Phi, s_m(u))$ for $m = 0, 1, \ldots$ and $u \in V^m$. By Lemma 3.1 one has $\mathscr{C}(\sigma, \Phi, F) = \mathscr{R}(\Phi, \mathscr{I}S_\sigma(G^F))$ then, by (i) of Theorem III.4.7, condition (i) holds. Furthermore, by the continuity of $\mathscr{R}(\Phi, \cdot)$ it follows that a set-valued mapping $\mathrm{Comp}(C([\sigma, \sigma+a], \boldsymbol{R}^n)) \ni C \to \mathscr{R}(\Phi, C) \subset C([\sigma-r, \sigma+a], \boldsymbol{R}^n)$ is u.s.c. Then for every $\varepsilon > 0$ there is $\delta_\varepsilon > 0$ such that $\mathscr{R}(\Phi, C+\delta_\varepsilon B) \subset \mathscr{R}(\Phi, C) + \varepsilon\tilde{B}$, where B is a unit ball of $C([\sigma, \sigma+a], \boldsymbol{R}^n)$. Taking now, by (ii) of Theorem III.4.7, $N_\varepsilon \geqslant 0$ such that $s_m(U^m) \subset \mathscr{I}S_\sigma(G^F) + \delta_\varepsilon B$ for $m \geqslant N_\varepsilon$ we also obtain $c_m(V^m) = \mathscr{R}(\Phi, s_m(U^m)) \subset \mathscr{R}(\Phi, \mathscr{I}S_\sigma(G^F) + \delta_\varepsilon B) \subset \mathscr{R}(\Phi, \mathscr{I}S_\sigma(G^F)) + \varepsilon\tilde{B} = \mathscr{C}(\sigma, \Phi, F) + \varepsilon\tilde{B}$ for $m \geqslant N_\varepsilon$.

The convexity of the set $\{v \in V^m\colon c_m(v) = \tilde{x}\}$ for given $\tilde{x} \in \mathscr{C}(\sigma, \Phi, F)$ follows immediately from the convexity of the set $\{u \in U^m\colon s_m(u) = x\}$ for $x \in \mathscr{I}S_\sigma(G^F)$ such that $\tilde{x} = \mathscr{R}(\Phi, x)$, because $\mathscr{R}(\Phi, \cdot)$ is an affine mapping. ∎

THEOREM 3.7. *Let $F \in \mathcal{M}(D, \mathrm{Conv}(\boldsymbol{R}^n))$ be locally bounded and locally u.s.c.w. with respect to its last two variables. Then for every $(\sigma, \Phi) \in \boldsymbol{R} \times \mathrm{AC}_{0r}$ such that $(\sigma, \Phi, \dot{\Phi}) \in D$ there is a number $a > 0$ such that the set $\mathscr{C}(\sigma, \Phi, F)$ is acyclic.*

Proof. By Theorem 3.6 there is a number $a > 0$ such that $\mathscr{C}(\sigma, \Phi, F) = \bigcap_{m \geqslant 0} s_m(V^m)$, where s_m is a continuous surriective mapping of V^m onto $\mathscr{R}(\Phi, \mathscr{I}S_\sigma(H_m))$ with (H_m) defined in Theorem III.4.3 relative to G^F. Since V^m is convex, then by Vietoris–Begle theorem each $c_m(V_m)$ is acyclic. Thus, $\mathscr{C}(\sigma, \Phi, F)$ is also acyclic. ∎

3.3. Relaxation theorem

We give here some relaxation theorem for NFDI(D, F) with $F \in \mathcal{M}(D, \mathrm{Comp}(\boldsymbol{R}^n))$ satisfying the following *local Kamke conditions*:

(K) *For every* $(\sigma, \Phi, \psi) \in D$ *there are a number* $\varrho > 0$, *a continuous mapping* L: $[\sigma, \sigma+\varrho] \to \mathscr{L}(L_{0r}, \boldsymbol{R}^n)$ *and a Kamke function* ω: $[\sigma, \sigma+\varrho] \times \boldsymbol{R}^+ \to \boldsymbol{R}^+$ *such that* $I_\varrho(\sigma) \times K_\varrho(\Phi) \times S_\varrho(\psi) \subset D$ *and*

$$\sup_{\sigma \leqslant t \leqslant \sigma+\varrho} (\|L(t)\|_{\mathscr{L}}) h\big(F(t, u_1, \dot{u}_1), F(t, u_2, \dot{u}_2)\big)$$
$$\leqslant \omega\big(t, \max\{\|L(t)\|\, |u_1 - u_2|_0,\, |L(t)[\dot{u}_1 - \dot{u}_2]|\}$$

for $t \in [\sigma, \sigma+\varrho]$ *and* $u_1, u_2 \in \mathrm{AC}_{0r}$ *such that* $u_1, u_2 \in \mathscr{K}_\lambda(\Phi)$ *and* $\dot{u}_1, \dot{u}_2 \in B_\lambda(\psi)$.

THEOREM 3.8. *Let* $F \in \mathscr{M}(D, \mathrm{Comp}(\boldsymbol{R}^n))$ *be locally uniformly integrable and satisfies the local Kamke conditions* (K). *Then for every* $(\sigma, \Phi) \in \boldsymbol{R} \times \mathrm{AC}_{0r}$ *with* $(\sigma, \Phi, \dot{\Phi}) \in D$ *there is a number* $a > 0$ *such that* $\mathscr{C}(\sigma, \Phi, \mathrm{co}\, F) = \overline{[\mathscr{C}(\sigma, \Phi, F)]_C}$ *and* $\mathscr{C}(\sigma, \Phi, \mathrm{co}\, F) = \overline{[\mathscr{C}(\sigma, \Phi, F)]_{AC}^{w}}$.

Proof. Similarly as in the proof of Lemmas 2.5 and 2.6 to a given $(\sigma, \Phi) \in \boldsymbol{R} \times \times \mathrm{AC}_{0r}$ satisfying $(\sigma, \Phi, \dot{\Phi}) \in D$ we can select numbers $a > 0$ and $\lambda > 0$ such that a local associated set-valued function G^F to F corresponding to (σ, Φ), $a > 0$ and $\lambda > 0$ satisfies conditions of Theorem III.5.13. Therefore, one has $\mathscr{I}S_\sigma(\mathrm{co}\, G^F) = \overline{[\mathscr{I}S_\sigma(G^F)]_C} = \overline{[\mathscr{I}S_\sigma(G^F)]_{AC}^{w}}$. Hence, by Lemma 3.1 one obtains $\mathscr{C}(\sigma, \Phi, \mathrm{co}\, F) = \mathscr{R}(\Phi, \mathscr{I}S_\sigma(\mathrm{co}\, G^F)) = \mathscr{R}(\Phi, \overline{[\mathscr{I}S_\sigma(G^F)]_C}) = \overline{[\mathscr{R}(\Phi, \mathscr{I}S_\sigma(G^F))]_C} = \overline{[\mathscr{C}(\sigma, \Phi, F)]_C}$ because $\mathscr{R}(\Phi, \cdot)$ is a closed mapping. Similarly, we also obtain $\mathscr{C}(\sigma, \Phi, \mathrm{co}\, F) = \mathscr{R}(\Phi, \overline{[\mathscr{I}S_\sigma(G^F)]_{AC}^{w}}) = \overline{[\mathscr{R}(\Phi, \mathscr{I}S_\sigma(G^F))]_{AC}^{w}} = \overline{[\mathscr{C}(\sigma, \Phi, F)]_{AC}^{w}}$, because $\mathscr{R}(\Phi, \cdot)$ is an affine mapping. ■

4. Controllability theorems for NFDIs with convex-values functions

We give here some sufficient conditions for the controllability of systems described by NFDIs. The main results will follow from some viability theorems.

4.1. Notations and approximation theorem

Let $I := [\sigma, \sigma+\varrho]$ and K: $I \to \mathrm{Comp}(\boldsymbol{R}^n)$ be given. Similarly as in Section III.6.1 we define the *Bouligand's contingent cone* to $K(t)$ at $t \in I$ and $x(t) \in K(t)$ by setting

$$T_{K(t)}[x(t)] := \bigcap_{\varepsilon > 0} \bigcap_{\alpha > 0} \bigcup_{0 < h < \alpha} \left[\frac{1}{h}(K(t) - x(t)) + \varepsilon B\right], \tag{4.1}$$

where B is a closed unit ball of $\boldsymbol{R}^n$.

It is clear that for fixed $t \in I$ and $x(t) \in K(t)$, $v \in T_{K(t)}[x(t)]$ if and only if for every $\varepsilon > 0$ and every $\alpha > 0$ there are $u \in v + \varepsilon B$ and $h \in (0, \alpha)$ such that $x(t) + + hu \in K(t)$. Similarly as in Section III.6.1 we obtain the following lemma.

LEMMA 4.1. *Given* $t \in I$ *and* $x(t) \in K(t)$, $v \in T_{K(t)}[x(t)]$ *if and only if*

$$\liminf_{h \to 0^+} \frac{1}{h} d_{K(t)}\big(x(t) + hv\big) = 0. \;\blacksquare \tag{4.2}$$

Immediately from the definition of the distant function d_A it follows that for fixed $h > 0$, $\tau > 0$, $\tau \in I$ and $v \in \boldsymbol{R}^n$ a function $I \times \boldsymbol{C}(I, \boldsymbol{R}^n) \ni (t, x) \to d_{K(\tau)}(x(t) + hv) \in \boldsymbol{R}$ is continuous on $I \times \boldsymbol{C}(I, \boldsymbol{R}^n)$.

Given $\Phi \in \mathrm{AC}_{0r}$, $\alpha > 0$ and a nonempty set $H \subset \boldsymbol{R}^n$ such that $\Phi(0) \in H$ we define a subset $\boldsymbol{K}_\alpha(\Phi, H)$ of AC_{0r} by

$$K(\Phi, H) := \{\psi \in \mathrm{AC}_{0r}\colon \psi(0) \in H, \int_E |\dot{\psi}(s)|\,ds \leqslant \omega_\alpha(\Phi, \mu(E)) \quad \text{for every } E \in M_{0r}\} \tag{4.3}$$

with

$$\omega_\alpha(\Phi, \delta) := \alpha\delta + \sup_{E \in M_{0r}} \Big\{\int_E |\dot{\Phi}(s)|\,ds\colon \mu(E) \leqslant \delta\Big\}, \tag{4.4}$$

where M_{0r} denotes the family of all Lebesgue measurable subsets of $[-r, 0]$.

LEMMA 4.2. *For every $\Phi \in \mathrm{AC}_{0r}$, $\alpha > 0$ and a compact set $H \subset \boldsymbol{R}^n$ such that $\Phi(0) \in H$, $\boldsymbol{K}_\alpha(\Phi, H)$ is a nonempty compact subset of C_{0r} and weakly compact in* AC_{0r}.

Proof. We have $H \neq \emptyset$, because $\Phi(0) \in H$. Since for every measurable set $E \subset [-r, 0]$ we have $\int_E |\dot{\Phi}(s)|\,ds \leqslant \omega_\alpha(\Phi, \mu(E))$ then $\Phi \in \boldsymbol{K}_\alpha(\Phi, H)$ and $\boldsymbol{K}_\alpha(\Phi, H) \neq \emptyset$ It is clear that the set $\dot{\boldsymbol{K}}_\alpha(\Phi, H) := \{\dot{\psi}\colon \psi \in \boldsymbol{K}_\alpha(\Phi, H)\}$ is uniformly integrable on $[-r, 0]$. Hence and compactness of H it follows that it is bounded in L_{0r}. Thus, $\boldsymbol{K}_\alpha(\Phi, H)$ is a bounded and equiabsolutely continuous subset of AC_{0r}. Therefore, by Theorem I.4.4 and Eberlein–Šmulian's theorem, it is relatively weakly compact in AC_{0r}. Similarly as in the proof of Lemma 2.5 hence it follows that $\boldsymbol{K}_\alpha(\Phi, H)$ is relatively compact in $\boldsymbol{C}_{0r}$.

Suppose (ψ_n) is a sequence of $\boldsymbol{K}_\alpha(\Phi, H)$ converging in the norm topology of C_{0r} to $\psi \in \boldsymbol{C}_{0r}$. Since $\dot{\boldsymbol{K}}_\alpha(\Phi, H)$ is relatively weakly compact in L_{0r}, then by Lemma I.4.5, $\psi \in \mathbf{AC}_{0r}$ and $\dot{\psi}_n \rightharpoonup \dot{\psi}$ in L_{0r} as $n \to \infty$. Hence by Banach–Mazur's theorem there is a sequence $(\sum_{i=j}^{\infty} \lambda_i^j \dot{\psi}_{n_i})$ of convex combinations such that $|\sum_{i=j}^{\infty} \lambda_i^j \dot{\psi}_{n_i} - \dot{\psi}|_0 \to 0$ as $j \to \infty$. Now, for every measurable set $E \subset [-r, 0]$ and $j = 1, 2, \ldots$ we have $\int_E |\dot{\psi}(s)|\,ds \leqslant \int_E |\dot{\psi}(s) - \sum_{i=j}^{\infty} \lambda_i^j \dot{\psi}_{n_i}(s)|\,ds + \sum_{i=r}^{\infty} \lambda_i^j \int_E |\dot{\psi}_{n_i}(s)|\,ds \leqslant |\dot{\psi} - \sum_{i=j}^{\infty} \lambda_i^j \dot{\psi}_{n_i}|_0 + \omega_\alpha(\Phi, \mu(E))$. Therefore, $\int_E |\dot{\psi}(s)|\,ds \leqslant \omega_\alpha(\Phi, \mu(E))$ for every measurable set $E \subset [-r, 0]$. By the compactness of H, $\psi_n(0) \in H$ and $|\psi_n - \psi|_0 \to 0$ as $n \to \infty$ imply $\psi(0) \in H$. Therefore, $\psi \in \boldsymbol{K}_\alpha(\Phi, H)$. Similarly if (ψ_n) is a sequence of $\boldsymbol{K}_\alpha(\Phi, H)$ weakly convergent in AC_{0r} to $\psi \in \mathrm{AC}_{0r}$ we have at once $\psi(0) \in H$, because $\psi_n(0) \to \psi(0)$ as $n \to \infty$. Furthermore $(\dot{\psi}_n)$ is uniformly integrable. Thus there is a subsequence of $(\dot{\psi}_n)$, say again $(\dot{\psi}_n)$ weakly converging to any $u \in L_{0r}$. But $\int_{-r}^{s} \dot{\psi}(\tau)\,d\tau = \int_{-r}^{s} u(\tau)\,d\tau$ for every $s \in [-r, 0]$. Therefore $\dot{\psi} = u$. Then $\dot{\psi}_n \rightharpoonup \dot{\psi}$ as $n \rightharpoonup \infty$ which together with $\psi(0) \in H$ implies $\psi \in K_\alpha(\Phi, H)$. Thus $K_\alpha(\Phi, H)$ is sequentially weakly closed. Now the weak closedness of $K_\alpha(\Phi, H)$ follows from Šmulian's theorem. ■

REMARK 4.1. *For every* $\Phi \in AC_{0r}$, $H \subset R^n$, $\alpha > 0$, $\varrho > 0$ *and* $\lambda > 0$ *such that* $\Phi(0) \in H$ *and* $\lambda < \min(\varrho, r\alpha)$ *there is* $\psi_\lambda \in K_\alpha(\Phi, H)$ *such that* $\psi_\lambda \neq \Phi$, $\dot{\psi}_\lambda \neq \dot{\Phi}$, $\psi_\lambda \in [\mathscr{K}_\varrho(\Phi)]^0$ and $\dot{\psi}_\lambda \in [B_\varrho(\dot{\Phi})]^0$, *where* $[\mathscr{K}_\varrho(\Phi)]^0$ *and* $[B_\varrho(\Phi)]^0$ *denote the interiors of* $\mathscr{K}_\varrho(\Phi)$ *and* $B_\varrho(\dot{\Phi})$, *respectively.*

Indeed, for given above $\Phi \in AC_{0r}$ and $\varrho, \alpha, \lambda > 0$ let $\psi_\lambda\colon [-r, 0] \to R^n$ be defined by $\psi_\lambda(s) := \Phi(s) + \lambda s/r$. We have $\psi_\lambda(0) = \Phi(0) \in H$. Furthermore, $\int_E |\dot{\psi}(s)|\,ds \leqslant \int_E |\dot{\Phi}(s)|\,ds + \int_E (\lambda/r)\,ds \leqslant \sup_{F \in M_{0r}} \{\int_F |\dot{\Phi}(s)|\,ds\colon\ \mu(F) \leqslant \mu(E)\} + \alpha\mu(E) =: \omega_\alpha(\Phi, \mu(E))$ for every measurable set $E \subset [-r, 0]$. Therefore, $\psi_\lambda \in K_\alpha(\Phi, H)$. On the other hand, we have $|\Phi - \psi_\lambda|_0 = \sup_{-r \leqslant s \leqslant 0} |\Phi(s) - \Phi(s) - \lambda s/r| = \lambda$ and, $|\dot{\Phi} - \dot{\psi}_\lambda|_0 = \int_{r-}^{0} |\dot{\Phi}(s) - \dot{\Phi}(s) - \lambda/r|\,ds = \lambda$. Therefore, $\psi_\lambda \in [\mathscr{K}_\varrho(\Phi)]^0 \setminus \{\Phi\}$ and $\psi_\lambda \in [B_\varrho(\dot{\Phi})]^0 \setminus \{\dot{\Phi}\}$, because $\lambda \in (0, \varrho)$. ■

LEMMA 4.3. *Let D be an open subset of* $R \times C_{0r} \times L_{0r}$ *and suppose* $F\colon D \to \mathrm{Comp}(R^n)$ *is locally u.s.c.w. Assume* $(\sigma, \Phi) \in R \times AC_{0r}$ *and a compact set* $H \subset R^n$ *are such that* $\Phi(0) \in H$ *and* $(\sigma, \Phi, \dot{\Phi}) \in D$. *Let* $\alpha > 0$ *be given. There exists* $\varrho > 0$ *such that* $I_\varrho(\sigma) \times \mathscr{K}_\varrho(\Phi) \times B_\varrho(\dot{\Phi}) \subset D$ *and such that a local associated function* $G^F\colon [\sigma, \sigma+\varrho] \times B_\alpha(\varrho, \Phi, H) \to \mathrm{Comp}(R^n)$ *to F defined by setting*

$$G^F(t, \psi) := F(t, \psi, \dot{\psi}) \tag{4.5}$$

for $(t, \psi) \in [\sigma, \sigma+\varrho] \times B_\alpha(\varrho, \Phi, H)$, *with*

$$B_\alpha(\varrho, \Phi, H) := \{\psi \in AC_{0r}\colon\ \psi \in \mathscr{K}_\varrho(\Phi) \cap K_\alpha(\Phi, H),\ \dot{\psi} \in B_\varrho(\dot{\Phi}) \cap \dot{K}_\alpha(\Phi, H)\}$$

is u.s.c. on $[\sigma, \sigma+\varrho] \times B_\alpha(\varrho, \Phi, H)$, *as a mapping defined on a subset of* $R \times C_{0r}$.

Proof. Suppose $\varrho > 0$ is such that $I_\varrho(\sigma) \times \mathscr{K}_\varrho(\Phi) \times B_\varrho(\dot{\Phi}) \subset D$ and F is *u.s.c.w.* on $I_\varrho(\sigma) \times \mathscr{K}_\varrho(\Phi) \times B_\varrho(\dot{\Phi})$. By Remark 4.1 we have $B_\alpha(\varrho, \Phi, H) \neq \emptyset$. Then G^F can be defined. Fix $(t_0, \psi_0) \in [\sigma, \sigma+\varrho] \times B_\alpha(\varrho, \Phi, H)$ and let $\{(t_n, \psi_n)\}$ be any sequence of $[\sigma, \sigma+\varrho] \times B_\alpha(\varrho, \Phi, H)$ such that $\max(|t_n - t_0|, |\psi_n - \psi_0|_0) \to 0$ as $n \to \infty$. By virtue of Lemma I.4.5 we have $\dot{\psi}_n \to \dot{\psi}$ in L_{0r} as $n \to \infty$ because $\dot{K}_\alpha(\Phi, H)$ is relatively weakly compact in L_{0r} and $\psi_n \in K_\alpha(\Phi, H)$ for $n = 1, 2, \ldots$ Therefore, $\lim_{n\to\infty} \bar{h}(G^F(t_n, \psi_n),\ G^F(t_0, \psi_0)) = \lim_{n\to\infty} \bar{h}(F(t_n, \psi_n, \dot{\psi}_n),\ F(t_0, \psi_0, \dot{\psi}_0)) = 0$, i.e. G^F is u.s.c. at $(t_0, \psi_0) \in [\sigma, \sigma+\varrho] \times B_\alpha(\varrho, \Phi, H)$. ■

In a similar way we obtain

LEMMA 4.4. *Let D be an open subset of* $R \times C_{0r} \times L_{0r}$ *and suppose* $F \in \mathscr{M}(D, \mathrm{Comp}(R^n))$ *is locally continuous weakly with respect to its last two variables. Let* $(\sigma, \Phi) \in R \times AC_{0r}$ *and a compact set* $H \subset R^n$ *be such that* $\Phi(0) \in H$ *and* $(\sigma, \Phi, \dot{\Phi}) \in D$. *Then* *for every* $\alpha > 0$ *there exists* $\varrho > 0$ *such that* $I_\varrho(\sigma)\times\mathscr{K}_\varrho(\Phi)\times B_\varrho(\dot{\Phi}) \subset D$ *and such that a local associated function* G^F: $[\sigma, \sigma+\varrho]\times B_\alpha(\varrho, \Phi, H) \to \mathrm{Comp}(\boldsymbol{R}^n)$ *to* F *has the following properties*: $G^F(\cdot, \psi)$ *is measurable for fixed* $\psi \in B_\alpha(\varrho, \Phi, H)$ *and* $G^F(t, \cdot)$ *is continuous on* $B_\alpha(\varrho, \Phi, H)$ *for fixed* $t \in [\sigma, \sigma+\varrho]$ *as a mapping defined on a subset of* C_{0r}.

Proof. Suppose $\varrho > 0$ is such that $I_\varrho(\sigma)\times\mathscr{K}_\varrho(\Phi)\times B_\varrho(\dot{\Phi}) \subset D$ and $F(t, \cdot, \cdot)$ is continuous weakly on $\mathscr{K}_\varrho(\Phi)\times B_\varrho(\dot{\Phi})$ for any fixed $t \in [\sigma, \sigma+\varrho]$. By the definition of the measurability condition ($\mathscr{M}$) it follows that $F(\cdot, u, v)$ is measurable on $I_\varrho(\sigma)$ for fixed $(u, v) \in \mathscr{K}_\varrho(\Phi)\times B_\varrho(\dot{\Phi})$. Hence, in particular it follows that $G^F(\cdot, \psi)$ is measurable on $[\sigma, \sigma+\varrho]$ for fixed $\psi \in B_\alpha(\varrho, \Phi, H)$. Similarly as in the proof of Lemma 4.3, we can verify that for fixed $t \in [\sigma, \sigma+\varrho]$, any $\psi_0 \in B_\alpha(\varrho, \Phi, H)$ and an arbitrary sequence (ψ_n) of $B_\alpha(\varrho, \Phi, H)$ converging to ψ_0 in the norm topology of C_{0r} we have $\lim_{n\to\infty} h(G^F(t, \psi_n), G\ (t, \psi_0)) = 0$. Therefore, $G^F(t, \cdot)$ is continuous on $B_\alpha(\varrho, \Phi, H)$ for fixed $t \in [\sigma, \sigma+\varrho]$. ■

We shall prove now the following approximation theorem.

THEOREM 4.5. *Let* D *be an open subset of* $\boldsymbol{R}\times C_{0r}\times L_{0r}$ *and suppose* $F \in \mathscr{M}(D, \mathrm{Comp}(\boldsymbol{R}^n))$ *is locally continuous weakly with respect to its last two variables, and locally bounded. Let* $(\sigma, \Phi) \in \boldsymbol{R}\times \mathrm{AC}_{0r}$, $\alpha > 0$, $\varrho > 0$, *and a compact set* $H \subset \boldsymbol{R}^n$ *be such that* $\Phi(0) \in H$, $I_\varrho(\sigma)\times\mathscr{K}_\varrho(\Phi)\times B_\varrho(\dot{\Phi}) \subset D$, $F(I_\varrho(\sigma)\times\mathscr{K}_\varrho(\Phi)\times B_\varrho(\dot{\Phi})) \subset MB$ *and* $F(t, \cdot, \cdot)$ *is continuous weakly on* $\mathscr{K}_\varrho(\Phi)\times B_\varrho(\dot{\Phi})$ *for a fixed* $t \in [\sigma, \sigma+\varrho]$. *Moreover, suppose* $F(t, \psi, \dot{\psi})\cap T_H(\psi(0)) \neq \emptyset$ *for every* $t \in [\sigma, \sigma+\varrho]$ *and* $\psi \in B_\alpha(\varrho, \Phi, H)$. *Then there exists a sequence* (G_n) *of continuous set-valued functions* G_n: $[\sigma, \sigma+\varrho]\times B_\alpha(\varrho, \Phi, H) \to \mathrm{Comp}(\boldsymbol{R}^n)$ *such that*

(i) $\sup\{h(G_n(t, u), G^F(t, u)): u \in B_\alpha(\varrho, \Phi, H)\} \to 0$ *for a.e.* $t \in [\sigma, \sigma+\varrho]$ *as* $n \to \infty$ *with* G^F *defined by* (4.5),

(ii) $G_n([\sigma, \sigma+\varrho]\times B_\alpha(\varrho, \Phi, H)) \subset MB$ *for every* $n = 1, 2, \ldots$,

(iii) $G_n(t, \psi)\cap T_H(\psi(0)) \neq \emptyset$ *for* $t \in [\sigma, \sigma+\varrho]$, $n = 1, 2, \ldots$ *and* $\psi \in B_\alpha(\varrho, \Phi, H)$.

Proof. By virtue of Lemma 4.4, $G^F(\cdot, u)$ is measurable on $[\sigma, \sigma+\varrho]$ for fixed $u \in B_\alpha(\varrho, \Phi, H)$ and $G^F(t, \cdot)$ is continuous on $B(\varrho, \Phi, H)$ for fixed $t \in [\sigma, \sigma+\varrho]$. Let $I := [\sigma, \sigma+\varrho]$ and select, by Scorza-Dragoni–Castaing's theorem a sequence (E_n) of closed subsets $E_n \subset I$ with $\mu(I\setminus E_n) \leqslant (\frac{1}{2})^n$ so that restrictions of G^F to each $E_n\times B_\alpha(\varrho, \Phi, H)$ are continuous. Now, by Antosiewicz–Cellina's continuous extension theorem, there are continuous set-valued functions G_n: $I\times B_\alpha(\varrho, \Phi, H) \to \mathrm{Comp}(\boldsymbol{R}^n)$ such that $G_n = G^F$ on $E_n\times B_\alpha(\varrho, \Phi, H)$.

Let $E = \bigcap_{n=1}^{\infty} \bigcup_{k\geqslant n} (I\setminus E_k)$. We have $\mu(E) \leqslant \sum_{k=n}^{\infty} \mu(I\setminus E_k) \leqslant (\frac{1}{2})^{n-1}$ for every $n = 1, 2, \ldots$ Then $\mu(E) = 0$. Now, for every $t \in I\setminus E$ there exists n_t such that $t \in \bigcap_{k=n_t}^{\infty} E_k$. Therefore, $G_k(t, u) = G^F(t, u)$ for $k \geqslant n_t$ and $u \in B_\alpha(\varrho, \Phi, H)$. Then we have $h(G_k(t,u), G^F(t,u)) = 0$ for $k \geqslant n_t$ and $u \in B_\alpha(\varrho, \Phi, H)$, i.e. $h(G_k(t,u), G^F(t,u)) \to 0$ uniformly with respect to $u \in B_\alpha(\varrho, \Phi, H)$ and a.e. on I as $k \to \infty$.

By the proof of Antosiewicz–Cellina's extension theorem it follows that each set-valued function G_n is defined by $G_n(t,u) := G^F(t,u)$ for $t \in E_n$, $u \in B_\alpha(\varrho, \Phi, H)$ and

$$G_n(t,u) = \bigcup_{w \in L(t,u)} [G^F(s_w, u) + r_w(t,u)B] \cap S_n \tag{4.6}$$

for $t \in I \setminus E_n$, $u \in B_\alpha(\varrho, \Phi, H)$, where $L(t,u)$ is any finite set of indexes, $s_w \in E_n$, $\{r\ (t,u)\}_{w \in I \setminus E_n}$ is any equicontinuous family of real-valued functions on $(I \setminus E_n) \times \times B_\alpha(\varrho, \Phi, H)$ and $S_n = \mathrm{co} \bigcup \{G^F(s,u)\colon s \in E_n,\ u \in B_\alpha(\varrho, \Phi, H)\}$. Since for every $(t,u) \in A$ we have $G^F(t,u) \subset MB$ then, in paticular we also have $S_n \subset MB$ for every $n = 1, 2, \ldots$ But $G_n(t,u) \subset S_n$ for every $(t,u) \in I \times B_\alpha(\varrho, \Phi, H)$ and $n = 1, 2, \ldots$ Then, for every $n = 1, 2, \ldots$ and $(t,u) \in I \times B_\alpha(\varrho, \Phi, H)$, $G_n(t,u) \subset MB$, i.e. $G_n(I \times B_\alpha(\varrho, \Phi, H)) \subset MB$ for every $n = 1, 2, \ldots$

Finally, for every $n = 1, 2, \ldots$, $t \in E_n$ and $\psi \in B_\alpha(\varrho, \Phi, H)$ we have $G_n(t,\psi) \cap \cap T_H(\psi(0)) = G^F(t,\psi) \cap T_H(\psi(0)) \neq \emptyset$.

By the definition of G_n for every $t \in I \setminus E_n$ and $\psi \in B_\alpha(\varrho, \Phi, H)$ there are $s_w \in E_n$ such that (4.6) is satisfied. Since $G^F(s_w, \psi) \cap T_H(\psi(0)) \neq 0$ then also $\bigcup_{w \in L(t,\psi)} [G^F(s_w, \psi) + r_w(t,\psi)B] \cap T_H(\psi(0)) \neq 0$. Furthermore, for every $s \in E_n$, $n = 1, 2, \ldots$ and $\psi \in B_\alpha(\varrho, \Phi, H)$ we have $G^F(s,\psi) \cap T_H(\psi(0)) \neq 0$. Therefore, $S_n \cap T_H(\psi(0)) = [\mathrm{co} \bigcup_{s \in E_n} G^F(s,u)] \cap T_H(\psi(0)) \neq \emptyset$ for $n = 1, 2, \ldots$ and $\psi \in B_\alpha (\varrho, \Phi, H)$. Thus, for every $t \in I \setminus E_n$, $\psi \in B(\varrho, \Phi, H)$ and $n = 1, 2, \ldots$ we also have $G_n(t,\psi) \cap T_H(\psi(0)) \neq \emptyset$. ■

4.2. Viability theorems

We give here sufficient conditions for the existence of local solutions of NFDI(D, F) satisfying the following viable conditions: $(t, x_t) \in \Omega$ and $x(t) \in H$ for $t \in [\sigma, \sigma+\varrho]$, where Ω and H are given nonempty subsets of $\boldsymbol{R} \times \mathrm{AC}_{0r}$ and $\boldsymbol{R}^n$, respectively. Each such solution of NFDI(D, F) will be called by us a *(Ω, H)-viable trajectory of* NFDI(D, F) on $[\sigma-r, \sigma+\varrho]$. We begin with the following lemmas.

LEMMA 4.6. *Let $(\sigma, \Phi) \in \boldsymbol{R} \times \mathrm{AC}_{0r}$, $\varrho > 0$, $M > 0$ and a compact set $H \subset \boldsymbol{R}^n$ be given and let $\Phi(0) \in H$. Suppose $G\colon [\sigma, \sigma+\varrho] \times \boldsymbol{K}_\alpha(\Phi, H) \to \mathrm{Conv}(\boldsymbol{R}^n)$ with $\alpha = l+M$ is u.s.c. and such that $G([\sigma, \sigma+\varrho] \times \boldsymbol{K}_\alpha(\Phi, H)) \subset MB$. If $G(t,\psi) \cap T_H(\psi(0)) \neq \emptyset$ for $t \in [\sigma, \sigma+\varrho]$ and $\psi \in \boldsymbol{K}_\alpha(\Phi, H)$ then there exists an absolutely continuous function $x\colon [\sigma-r, \sigma+\varrho] \to \boldsymbol{R}^n$ such that*

$$\begin{aligned} &\dot{x}(t) \in G(t, x_t) && \text{for a.e. } t \in [\sigma, \sigma+\varrho], \\ &x_\sigma = \Phi, && \\ &x(t) \in H && \text{for } t \in [\sigma, \sigma+\varrho]. \end{aligned}$$

Proof. By virtue of Lemma 4.2, $\boldsymbol{K}_\alpha(\Phi, H) \subset \mathrm{AC}_{0r}$ is nonempty compact in C_{0r} and weakly compact in AC_{0r}. Let N_l be a positive integer such that $1/k \leqslant l$ for every $k \geqslant N_l$. Observe that by Lemma 4.1 the tangential condition $G(t, \psi) \cap T_H(\psi(0)) \neq \emptyset$ for $(t, \psi) \in I \times \boldsymbol{K}_\alpha(\Phi, H)$ implies, that for every $k \geqslant N_l$, every $\tau \in I$ and $\psi \in \boldsymbol{K}_\alpha(\Phi, H)$ there are $v_\psi^\tau \in G(\tau, \psi)$, $h_\psi^\tau \in (0, 1/k)$ such that $d_H(\psi(0) + h_\psi^\tau v_\psi^\tau) \leqslant h_\psi^\tau/3k$. We introduce the subsets $N(\tau, \psi) = \{x \in \boldsymbol{R} : d_H(x + h_\psi^\tau v_\psi^\tau) < h_\psi^\tau/2k\}$. Since a function $\boldsymbol{R}^n \ni x \to d_H(x + h_\psi^\tau v_\psi^\tau) \in \boldsymbol{R}^+$ is continuous, then $N(\tau, \psi)$ is for every $(\tau, \psi) \in I \times \boldsymbol{K}_\alpha(\Phi, H)$ an open subset of $\boldsymbol{R}^n$. Thus, for every $(\tau, \psi) \in I \times \boldsymbol{K}_\alpha(\Phi, H)$ there is an open ball $B(\tau, \psi)$ of $\boldsymbol{R}^n$ centred at $\psi(0)$ with a radius $\eta_\psi^\tau \in (0, 1/k)$ such that $B(\tau, \psi) \subset N(\tau, \psi)$. Let $I(\tau, \psi) = (\tau - \eta_\psi^\tau, \tau + \eta_\psi^\tau)$ and $S(\tau, \psi) = \{\Phi \in C_{0r} : |\Phi - \psi|_0 < \eta_\psi^\tau\}$. Let $g\colon I \times \boldsymbol{K}_\alpha(\Phi, H) \to \boldsymbol{R}^n$ be defined by $g(\tau, \psi) := \psi(0)$ for $(\tau, \psi) \in I \times \boldsymbol{K}_\alpha(\Phi, H)$. It is clear that Graph$(g)$ is compact in $\boldsymbol{R} \times C_{0r} \times \boldsymbol{R}^n$. Furthermore, for every $(\tau, \psi) \in I \times \boldsymbol{K}_\alpha(\Phi, H)$ we have $(\tau, \psi, g(\tau, \psi)) \in I(\tau, \psi) \times S(\tau, \psi) \times B(\tau, \psi)$. Then the family $\{I(\tau, \psi) \times S(\tau, \psi) \times B(\tau, \psi)\}_{(\tau, \psi) \in I \times K_\alpha(\Phi, H)}$ of subsets of $\boldsymbol{R} \times C_{0r} \times \boldsymbol{R}^n$ is an open covering of a compact set Graph$(g) \subset \boldsymbol{R} \times C_{0r} \times \boldsymbol{R}^n$. Therefore, there are $\tau_j \in I$ and $\psi_j \in \boldsymbol{K}_\alpha(\Phi, H)$ with $j = 1, 2, \ldots, q$ such that Graph$(g) \subset \bigcup_{j=1}^{q} I(\tau_j, \psi_j) \times S(\tau_j, \psi_j) \times B(\tau_j, \psi_j)$. For simplicity, we set $\eta_j := \eta_{\psi_j}^{\tau_j}$, $h_j := h_{\psi_j}^{\tau_j}$, $v_j := v_{\psi_j}^{\tau_j}$ and $h_0^k := \min_{1 \leqslant f \leqslant q} h_j$.

Let $(t, \psi) \in I \times \boldsymbol{K}_\alpha(\Phi, H)$ be fixed. It belongs to some $I(\tau_j, \psi_j) \times S(\tau_j, \psi_j)$ and $\psi(0) \in B(\tau_j, \psi_j) \subset N(\tau_j, \psi_j)$, because $(t, \psi, \psi(0)) \in$ Graph(g). Therefore, $|t - \tau_j| < \eta_j \leqslant 1/k$, $|\psi - \psi_j|_0 < \eta_j \leqslant 1/k$ and there is $x_j \in H$ such that $|v_j - (x_j - \psi(0))/h_j| \leqslant \frac{1}{h_j} d_H(\psi(0) + h_j v_j) + 1/2k \leqslant 1/k$. Let $u_j = (x_j - \psi(0))/h_j$. We have $u_j \in v_j + (1/k)B$ and $\psi(0) + h_j u_j \in H$. Finally, we see that for every $(t, \psi) \in I \times \boldsymbol{K}_\alpha(\Phi, H)$ there are $(\tau_j, \psi_j) \in I \times \boldsymbol{K}_\alpha(\Phi, H)$, $v_j \in G(\tau_j, \psi_j)$, $u_j \in v_j + (1/k)B$ and $h_j \in [h_0^k, 1/k]$ such that $|t - \tau_j| \leqslant 1/k$, $|\psi - \psi_j|_0 \leqslant 1/k$ and $\psi(0) + h_j u_j \in H$. Hence, in particular it follows that $(\tau_j, \psi_j, v_j) \in$ Graph G and $v_j \in G(I \times \boldsymbol{K}_\alpha(\Phi, H))$. Therefore for every $k \geqslant N_l$ and $(t, \psi) \in I \times \boldsymbol{K}_\alpha(\Phi, H)$ there are $h_j \in [h_0^k, 1/k]$ and $u_j \in G(I \times \boldsymbol{K}_\alpha(\Phi, H)) + lB$ such that $\psi(0) + h_j u_j \in H$ and $(t, \psi, u_j) \in$ Graph $G + (1/k)S$, where S denotes a closed unit ball of a normed space $(\boldsymbol{R} \times C_{0r} \times \boldsymbol{R}^n, \|\cdot\|)$ with $\|\cdot\|$ defined by $\|(t, \psi, u\| = \max(|t|, |\psi|_0, |u|)$ for $(t, \psi, u) \in \boldsymbol{R} \times C_{0r} \times \boldsymbol{R}^n$.

Let $k \geqslant N_l$ be fixed. Select for $(\sigma, \Phi) \in I \times \boldsymbol{K}_\alpha(\Phi, H)$, $h_0 \in [h_0^k, 1/k]$ and $u_0 \in G(I \times \boldsymbol{K}_\alpha(\Phi, H)) + lB$ such that $\Phi(0) + h_0 u_0 \in H$ and $(\sigma, \Phi, u_0) \in$ Graph $G + (1/k)S$. Put $t_1 = \sigma + h_0$ and define $l_0(t) := (t - \sigma)u_0$ for $t \in [\sigma, t_1]$. Let $z^1\colon [\sigma - r, t_1] \to \boldsymbol{R}^n$ be defined by

$$z^1(\tau) = \begin{cases} \Phi(\tau - \sigma) & \text{for } \tau \in [\sigma - r, \sigma), \\ \Phi(0) + l_0(\tau) & \text{for } \tau \in [\sigma, t_1]. \end{cases} \tag{4.7}$$

Then, let $\psi_1 = z_{t_1}^1$. We have $\psi_1 \in \mathrm{AC}_{0r}$ and $\psi_1(0) = z_{t_1}^1(0) = z^1(t_1) = \Phi(0) + u_0 h_0 \in H$. Furthermore, for every measurable set $E \subset [-r, 0]$ one has

$$\begin{aligned} \int_E |\dot\psi_1(s)|\, ds &= \int_E |\dot z_{t_1}^1(s)|\, ds = \int_{E \cap [-r, \sigma - t_1]} |\dot\Phi(t_1 + s - \sigma)|\, ds + \int_{E \cap]\sigma - t_1, 0]} (l + M)\, ds \\ &\leqslant \int_{E \cap [t_1 - \sigma - r, 0]} |\dot\Phi(\tau)|\, d\tau + \mu(E \cap [\sigma - t_1, 0])(l + M) \leqslant \omega_\alpha(\Phi, \mu(E)). \end{aligned}$$

Therefore, $\psi_1 \in \boldsymbol{K}_\alpha(\Phi, H)$.

Now, for $(t_1, \psi_1) \in I \times \boldsymbol{K}_\alpha(\Phi, H)$ we can again find $h_1 \in [h_0^k, 1/k]$ and $u_1 \in G(I \times \times \boldsymbol{K}_\alpha(\Phi, H)) + lB$ such that $\psi_1(0) + h_1 u_1 \in H$ and $(t_1, \psi_1, u_1) \in \operatorname{Graph} G + (1/k)S$. Put $t_2 = t_1 + h_1$ and define $l_1(t) = (t - t_1)u_1$ for $t \in [t_1, t_2]$. Let z^2: $[\sigma - r, t_2] \to \boldsymbol{R}^n$ be defined by

$$z^2(\tau) = \begin{cases} z^1(\tau) & \text{for } \tau \in [\sigma - r, t_1), \\ z^1(t^1) + l_1(\tau) & \text{for } \tau \in [t_1, t_2] \end{cases} \tag{4.8}$$

and put $\psi_2 = z^2_{t_2}$. Similarly as above we have $\psi_2 \in \mathrm{AC}_{0r}$ and $\psi_2(0) = z^2_{t_2}(0) = z^2(t_2) = z^1(t_1) + h_1 u_1 = z^1_{t_1}(0) + h_1 u_1 = \psi_1(0) + h_1 u_1 \in H$.
Furthermore, for every measurable set $E \subset [-r, 0]$ one has

$$\int_E |\dot{\psi}_2(s)|\, ds = \int_E |\dot{z}^2_{t_2}(s)|\, ds = \int_{E \cap [-r, \sigma - t_2]} |\dot{\Phi}(t_2 + s - \sigma)|\, ds + \int_{E \cap [\sigma - t_2, 0]} (l + M)\, ds$$

$$\leqslant \int_{E \cap [t_2 - \sigma - r, 0]} |\dot{\Phi}(\tau)|\, d\tau + \mu(E)(l + M) \leqslant \omega_\alpha(\Phi, \mu(E)).$$

Therefore, $\psi_2 \in \boldsymbol{K}_\alpha(\Phi, H)$.

Continuining this procedure we can find numbers $h_0, h_1, \ldots, h_{m_k+1} \in [h_0^k, 1/k]$ and points $u_0, u_1, \ldots, u_{m_{k+1}} \in G(I \times \boldsymbol{K}_\alpha(\Phi, H)) + lB$ such that $h_0 + h_1 + \ldots + h_{m_k} \leqslant \varrho < h_0 + h_1 + \ldots + h_{m_k} + h_{m_{k+1}}$. We can also define functions z : $[\sigma - r, t_i] \to \boldsymbol{R}^n$ with $t_i = t_{i-1} + h_{i-1}$, $t_0 = \sigma$, $i = 1, 2, \ldots, m_k$ such that $z^i(t) = z^{i-1}(t)$ for $t \in [\sigma - -r, t_{i-1})$, $z^i(t) = z^{i-1}(t_{i-1}) + (t - t_{i-1})u_{i-1}$ for $t \in [t_{i-1}, t_i]$, $(t_i, z^i_{t_k}) \in I \times \boldsymbol{K}_\alpha(\Phi, H)$, $z^{i-1}_{t_{k-1}}(0) + u_{i-1}h_{i-1} \in H$ and $(t_i, z^i_{t_i}, u) \in \operatorname{Graph} G + (1/k)S$ for $i = 1, \ldots, m_k$. Furthermore for $i = 0, 1, \ldots, m_k$ we have also $z^i_\sigma = \Phi$ and therefore $(t_0, z^0_{t_0}) \in I \times \boldsymbol{K}_\alpha(\Phi, H)$ and $(t_0, z^0_{t_0}, u_0) \in \operatorname{Graph} G + (1/k)S$.

If $\sum_{i=0}^{m_k} h_i = \varrho$, we define for every $k \geqslant N_l$ an absolutely continuous function x^k: $[\sigma - r, \sigma + \varrho] \to \boldsymbol{R}^n$ by taking $x^k = z^{m_k}$. If $\sum_{i=0}^{m_k} h_i < \varrho$ we define $z^{m_k+1}(t) = z^{m_k}(t)$ for $t \in [\sigma - r, t_{m_k})$ and $z^{m_k+1}(t) = z^{m_k}(t_{m_k}) + (t - t_{m_k})u_{m_k}$ for $t \in [t_{m_k}, \sigma + \varrho]$. Then we take $x^k = z^{m_k+1}$ for $k \geqslant N_l$. In the both cases we have of course $(t_i, x^k_{t_i}) \in I \times \times \boldsymbol{K}_\alpha(\Phi, H)$ and $(t_i, x^k_{t_i}, u_i) \in \operatorname{Graph} G + (1/k)S$ for $i = 1, \ldots, m_k$. Since $\dot{x}^1(t) = u_i$ for $t \in (t_{i-1}, t_i)$ and $i = 1, \ldots, m_k$; $\dot{x}^1(t) = u_{m_k}$ for $t \in (t_{m_k}, \sigma + \varrho)$ in the case $\sum_{i=0}^{m_k} h_i < \varrho$ and $u_j \in G(I \times \boldsymbol{K}_\alpha(\Phi, H)) + lB$ for $j = 0, 1, \ldots, m_k$, then in the both cases, we have $\dot{x}^k(t) \in G(I \times \boldsymbol{K}_\alpha(\Phi, H)) + lB$ for a.e. $t \in I$ and $k \geqslant N_l$. Hence, it follows that in the both cases we have $|\dot{x}^k(t)| \leqslant l + M$ for $k \geqslant N_l$ and a.e. $t \in I$. Therefore, there are an absolutely continuous function x: $[\sigma - r, \sigma + \varrho] \to \boldsymbol{R}^n$ and a subsequence, say again (x^k) of (x^k) such that $x_\sigma = \Phi$, $|x_k - x|_\varrho \to 0$ and $\dot{x}^k \rightharpoonup \dot{x}$ in $L(I, \boldsymbol{R}^n)$ as $k \to \infty$, where $|\cdot|_\varrho$ denotes the supremum norm of $C(I, \boldsymbol{R}^n)$.

By the definition of x^k we have $(t_i, x^k_{t_i}) \in I \times \boldsymbol{K}_\alpha(\Phi, H)$ and $(t_i, x^k_{t_i}, \dot{x}^k(t)) \in \in \operatorname{Graph} G + (1/k)S$ for $i = 1, 2, \ldots, m_k$ and $t \in (t_{i-1}, t_i)$. Since $\dot{x}^k \rightharpoonup \dot{x}$ as $k \to \infty$ then by Banach–Mazur's theorem there is a sequence $\{\sum_{i=j}^{\infty} \lambda^i_j \dot{x}^{k_i}\}$ of convex combinations converging a.e. in I to $\dot{x}$ as $j \to \infty$. Now, for every $\eta > 0$ there is $N_\eta \geqslant 1$ such that for $k \geqslant N_\eta$, $(t, x_t) \in [I \times \boldsymbol{K}_\alpha(\Phi, H)] + (1/k+\eta)Q$ for $t \in [t_{i-1}, t_i]$ and $(t, x_t, \dot{x}^k(t)) \in \operatorname{Graph} G + (2/k+\eta)S$ for $t \in (t_{i-1}, t_i)$ with $i = 1, \ldots, m_k$, where Q denotes a closed ball of $\boldsymbol{R} \times C_{0r}$. Indeed, for every $t \in [t_{i-1}, t_i]$ we have $t_i - t \leqslant h_i \leqslant 1/k$. Since $|x^k - x|_\varrho \to 0$ as $k \to \infty$ and $|x^k_{t_i} - x_{t_i}|_0 = \sup_{-r \leqslant s \leqslant 0} |x^k(t_i+s) - x(t_i+s)| = \sup_{t_i - r \leqslant \tau \leqslant t_i} |x^k(\tau) - x(\tau)| \leqslant |x^k - x|_\varrho$ then for every $\eta > 0$ there is $N_\eta \geqslant 1$ such that $|x^k_{t_i} - x_{t_i}|_0 \leqslant \frac{1}{2}\eta$ for $k \geqslant N_\eta$ and every $i = 1, \ldots, m_k$. Furthermore, we have $|x_{t_i} - x_t|_0 \leqslant \omega_x(t_i - t) \leqslant \omega_x(1/k)$, where ω_x is the modulus of continuity of x. Then for k sufficiently large, say for $k \geqslant N_\eta$ we have $\omega\ (1/k) \leqslant \frac{1}{2}\eta$. Therefore, for $k \geqslant N_\eta$ and $t \in [t_{i-1}, t_i]$ with $i = 1, \ldots, m_k$ we have $|x^k_{t_i} - x_t|_0 \leqslant |x^k_{t_i} - x_{t_i}|_0 + |x_{t_i} - x_t|_0 \leqslant \eta$. Thus $(t, x_t) \in (t_i, x^k_{t_i}) + (1/k+\eta)Q$ for $t \in [t_{i-1}, t_i]$ and $(t, x_t, \dot{x}^k(t)) \in (t_i, x^k_{t_i}, \dot{x}^k(t)) + (1/k+\eta)S$ for $t \in (t_{i-1}, t_i)$ with $i = 1, \ldots, m_k$ and $k \geqslant N_\eta$. Therefore, $(t, x_t) \in [I \times \boldsymbol{K}_\alpha(\Phi, H)] + (1/k+\eta)Q$ for $t \in [t_{i-1}, t_i]$ and $(t, x_t, \dot{x}^k(t)) \in \operatorname{Graph} G + (2/k+\eta)S$ for $t \in (t_{i-1}, t_i)$ with $i = 1, \ldots, m_k$ and $k \geqslant N_\eta$.

Now, if $\sum_{i=0}^{m_k} h_i = \varrho$, we have $(t, x_t) \in [I \times \boldsymbol{K}_\alpha(\Phi, H)] + (1/k+\eta)Q$ for $t \in [\sigma, \sigma+\varrho]$ and $(t, x_t, \dot{x}^k(t)) \in \operatorname{Graph} G + (2/k+\eta)S$ for a.e. $t \in [\sigma, \sigma+\varrho]$, $j = 1, 2, \ldots$ and $k \geqslant N^\eta$. Hence, similarly as in the proof of Theorem III.6.2 it follows that $(t, x_t) \in I \times \boldsymbol{K}_\alpha(\Phi, H)$ for $t \in I$ and $(t, x_t, \dot{x}(t)) \in \operatorname{Graph}(G)$ for a.e. $t \in I$. Therefore

$$\begin{aligned} &\dot{x}(t) \in G(t, x_t) \quad \text{for a.e. } t \in [\sigma, \sigma+\varrho], \\ &x_\sigma = \Phi, \\ &x_t(0) = x(t) \in H \quad \text{for } t \in [\sigma, \sigma+\varrho]. \end{aligned}$$

Suppose now $\sum_{i=0}^{m_k} h_i < \varrho$ and let $t \in [t_{m_k}, \sigma+\varrho]$. We have $|t - t_{m_k}| \leqslant 1/k$ and $|x_t - x^k_{t_{m_k}}| \leqslant \eta$ for $k \geqslant N_\eta$. Therefore $(t, x_t) \in (t_{m_k}, x^k_{t_{m_k}}) + (1/k+\eta)Q$ and $(t, x_t, \dot{x}^k(t)) \in (t_{m_k}, x^k_{t_{m_k}}, \dot{x}_k(t)) + (1/k+\eta)S$ if $t \in (t_{m_k}, \sigma+\varrho)$ for $k \geqslant N_\eta$.

Hence, similarly as in the case $\sum_{i=0}^{m_k} h_i = \varrho$, we obtain $(t, x_t) \in [I \times \boldsymbol{K}_\alpha(\Phi, H)] + (1/k+\eta)Q$ for $t \in [\sigma, \sigma+\varrho]$ and $(t, x_t, x^k(t)) \in \operatorname{Graph} G + (2/k+\eta)S$ for a.e. $t \in [\sigma, \sigma+\varrho]$ and $k \geqslant N_\eta$. Now, similarly as above, it also follows $(t, x_t) \in I \times \boldsymbol{K}_\alpha(\Phi, H)$ and $(t, x_t, \dot{x}(t)) \in \operatorname{Graph} G$ for a.e. $t \in [\sigma, \sigma+\varrho]$. ■

LEMMA 4.7. *Let $(\sigma, \Phi) \in \boldsymbol{R} \times \mathrm{AC}_{0r}$, $l > 0$, $\varrho > 0$, $M > 0$ and a compact set $H \subset \boldsymbol{R}^n$ be given and let $\Phi(0) \in H$. Suppose G: $[\sigma, \sigma+\varrho] \times B_\alpha(\varrho, \Phi, H) \to \mathrm{Conv}(\boldsymbol{R}^n)$ with $\alpha = l+M$ is u.s.c. and such that $G([\sigma, \sigma+\varrho] \times B_\alpha(\varrho, \Phi, H)) \subset MB$. If $G(t, \psi) \cap T_H(\psi(0)) \neq \emptyset$ for $t \in [\sigma, \sigma+\varrho]$ and $\psi \in B_\alpha(\varrho, \Phi, H)$ then there exists $a \in (0, \frac{1}{2}\varrho]$ and an absolutely contiouous function x: $[\sigma-r, \sigma+a] \to \boldsymbol{R}^n$ such that*

$$\begin{aligned} &\dot{x}(t) \in G(t, x_t) \quad \text{for a. e. } t \in [\sigma, \sigma+a], \\ &x_\sigma = \Phi, \\ &x(t) \in H \quad \text{for } t \in [\sigma, \sigma+a]. \end{aligned} \tag{4.9}$$

Proof. Let $C := [\sigma, \sigma+\frac{1}{2}\varrho]\times B_\alpha(\frac{1}{2}\varrho, \Phi, H)$ and $S := \overline{[\boldsymbol{R}\times C_{0r}\setminus[\sigma-\varrho, \sigma+ +\varrho]\times B_\alpha(\varrho, \Phi, H)]}_{\boldsymbol{R}\times C_{0r}}$. Let us observe that $B_\alpha(\lambda, \Phi, H)$ is a closed subset in C_{0r} for every $\lambda > 0$. Indeed, let (ψ_n) be a sequence of $B_\alpha(\lambda, \Phi, H)$ converging in the norm topology of C_{0r} to $\psi \in C_{0r}$. Hence by virtue of Lemma I.4.5 it follows $\dot{\psi}_n \rightharpoonup \dot{\psi}$ as $n \to \infty$. We have of course $\psi \in \mathscr{K}_\lambda(\Phi) \cap \boldsymbol{K}_\alpha(\Phi, H)$ and therefore $\dot{\psi} \in \dot{\boldsymbol{K}}_\alpha(\Phi, H)$. Furthermore we have $\dot{\psi} \in B_\lambda(\dot{\Phi})$ because $\dot{\psi}_n \in B_\lambda(\dot{\Phi})$ for $n = 1, 2, \ldots$ and $B_\lambda(\dot{\Phi})$ is weakly closed in L_{0r}. Therefore $\psi \in B_\alpha(\lambda, \Phi, H)$. Now, it is easily seen that sets C and S are closed subsets of a normal space $\boldsymbol{R}\times C_{0r}$ such that $C \cap S = \emptyset$. Then by Urysohn's theorem there exists a continuous function f: $\boldsymbol{R}\times \mathrm{AC}_{0r} \to [0, 1]$ such that $f(t, \psi) = 1$ for $(t, \psi) \in C$ and $f(t, \psi) = 0$ for $(t, \psi) \in S$. Let $\tilde{G}$: $\boldsymbol{R}\times C_{0r} \to \mathrm{Conv}(\boldsymbol{R}^n)$ be an upper semicontinuous extension of G on the whole space $\boldsymbol{R}\times C_{0r}$. Such extension exists by virtue of Theorem II.2.3. Let us define now a set-valued function G^f: $[\sigma, \sigma+\varrho]\times \boldsymbol{K}_\alpha(\Phi, H) \to \mathrm{Conv}(\boldsymbol{R}^n)$ be setting

$$G^f(t, \psi) = f(t, \psi)\tilde{G}(t, \psi) \tag{4.10}$$

for $(t, \psi) \in [\sigma, \sigma+\varrho]\times \boldsymbol{K}_\alpha(\Phi, H)$.

Obviously G^f is u.s.c. on $[\sigma, \sigma+\varrho]\to\boldsymbol{K}_\alpha(\Phi, H)$ and $G^f([\sigma, \sigma+\varrho]\times \boldsymbol{K}_\alpha(\Phi, H)) \subset MB$. Furthermore, for every $t \in [\sigma, \sigma+\varrho]$ and $\psi \in \boldsymbol{K}_\alpha(\Phi, H)\setminus B_\alpha(\varrho, \Phi, H)$ we have $G^f(t, \psi)\cap T_H(\psi(0)) \neq \emptyset$, because for $(t, \psi) \in D$, $G^f(t, \psi) = \{0\}$ and by the definition of $T_H(x)$ we have $0 \in T_H(x)$ for $x \in H$. For $t \in [\sigma, \sigma+\varrho]$ and $\psi \in B_\alpha(\varrho, \Phi, H)\setminus B_\alpha(\frac{1}{2}\varrho, \Phi, H)$ we have $G(t, \psi)\cap T_H(\psi(0)) \neq \emptyset$. Let $v \in G(t, \psi)\cap T_H(\psi(0))$ and let $v^f(t, \psi) = f(t, \psi)\cdot v$. We have $v^f(t, \psi) \in G\,(t, \psi)$ and $v^f(t, \psi) \in T_H(\psi(0))$ because $T_H(\psi(0))$ is a cone. Therefore, we have also $G^f(t, \psi)\cap T_H(\psi(0)) \neq \emptyset$. In a similar way we obtain $G^f(t, \psi)\cap T_H(\psi(0)) \neq \emptyset$ for $t \in (\sigma+\frac{1}{2}\varrho, \varrho]$ and $\psi \in B(\frac{1}{2}\varrho, \Phi, H)$. Finally, for $(t, \psi) \in C := [\sigma, \sigma+\frac{1}{2}\varrho]\times B_\alpha(\frac{1}{2}\varrho, \Phi, H)$ we have $G^f(t, \psi)\cap T_H(\psi(0)) = G(t, \psi)\cap T_H(\psi(0)) \neq \emptyset$.

Now, by virtue of Lemma 4.6, there exists an absolutely continuous function x^f: $[\sigma-r, \sigma+\varrho] \to \boldsymbol{R}^n$ such that

$$\begin{aligned}
&\dot{x}^f(t) \in f(t, x_t^f)\cdot \tilde{G}(t, x_t^f) \quad \text{for a.e. } t \in [\sigma, \sigma+\varrho],\\
&x_\sigma^f = \Phi,\\
&x^f(t) \in H \quad \text{for } t \in [\sigma, \sigma+\varrho].
\end{aligned}$$

Since $\Phi \in \mathscr{K}_{\varrho/2}(\Phi)$, $\dot{\Phi} \in B_{\varrho/2}(\dot{\Phi})$, $x_\sigma^f = \Phi$, $\dot{x}_\sigma^f = \dot{\Phi}$, $|x_t^f - x_\sigma^f|_0 \to 0$ and $|\dot{x}_t^f - \dot{x}_\sigma^f|_0 \to 0$ as $t \to \sigma+$ then there is $a \in (0, \frac{1}{2}\varrho]$ such that $|x_t^f - \Phi|_0 \leqslant \frac{1}{2}\varrho$ and $|\dot{x}_t^f - \dot{\Phi}|_0 \leqslant \frac{1}{2}\varrho$ for every $t \in [\sigma, \sigma+a]$, i.e., $x_t^f \in \mathscr{K}_\varrho(\Phi)$ and $\dot{x}_t^f \in B_\varrho(\dot{\Phi})$ for $t \in [\sigma, \sigma+a]$. By the proof of Lemma 4.6, we have $x_t^f \in \boldsymbol{K}_\alpha(\Phi, H)$ for every $t \in [\sigma, \sigma+\varrho]$. Then also $\dot{x}_t^f \in \dot{\boldsymbol{K}}_\alpha(\Phi, H)$ for $t \in [\sigma, \sigma+\varrho]$. Finally, hence it follows that for every $t \in [\sigma, \sigma+a]$ we have $(t, x_t^f) \in C = [\sigma, \sigma+\frac{1}{2}\varrho]\times B_\alpha(\frac{1}{2}\varrho, \Phi, H)$. Thus $f(t, x_t^f) = 1$ for all $t \in [\sigma, \sigma+a]$. Let x denote the restriction of x^f to the interval $[\sigma-r, \sigma+a]$. It is clear now that x satisfies conditions (49). ■

Now we prove the following viability theorems.

THEOREM 4.8. *Let* (D, Ω) *be a conformable pair of nonempty subsets of* $\boldsymbol{R}\times C_{0r}\times L_{0r}$ *and* $\boldsymbol{R}\times AC_{0r}$, *respectively such that there are* $(\sigma, \Phi)\in\Omega$, $\varrho > 0$, $l > 0$, $M > 0$ *and a compact set* $H \subset \boldsymbol{R}^n$ *with* $\Phi(0)\in H$ *and such that* $[\sigma, \sigma+\varrho]\times \boldsymbol{K}_\alpha(\Phi, H)\subset\Omega$ *with* $\alpha = l+M$. *If* $F\colon D\to \mathrm{Conv}(\boldsymbol{R})$ *is such that*

(i) F *is u.s.c.w. on* $[\sigma, \sigma+\varrho]\times \boldsymbol{K}_\alpha(\Phi, H)\times \dot{\boldsymbol{K}}_\alpha(\Phi, H)$,

(ii) $F([\sigma, \sigma+\varrho]\times \boldsymbol{K}_\alpha(\Phi, H)\times \dot{\boldsymbol{K}}_\alpha(\Phi, H))\subset MB$,

(iii) $F(t, \psi, \dot{\psi})\cap T_H(\psi(0))\neq \emptyset$ *for every* $t\in[\sigma, \sigma+\varrho]$ *and* $\psi\in \boldsymbol{K}_\alpha(\Phi, H)$,

then NFDI(D, F) *has on* $[\sigma-r, \sigma+\varrho]$, (Ω, H)*-viable trajectory through* (σ, Φ), *i.e., there is an absolutely continuous function* $x\colon [\sigma-r, \sigma+\varrho]\to \boldsymbol{R}^n$ *such that*

$$\begin{aligned}
&\dot{x}(t)\in F(t, x_t, \dot{x}_t) \quad \text{for a. e. } t\in[\sigma, \sigma+\varrho],\\
&x_\alpha = \Phi,\\
&(t, x_t)\in\Omega \quad \text{for } t\in[\sigma, \sigma+a],\\
&x(t)\in H \quad \text{for } t\in[\sigma, \sigma+a].
\end{aligned} \tag{4.11}$$

Proof. Let $G^F\colon [\sigma, \sigma+\varrho]\times \boldsymbol{K}_\alpha(\Phi, H)\to \mathrm{Conv}(\boldsymbol{R}_u)$ be defined by $G^F(t, \psi) := F(t, \psi, \dot{\psi})$ for $t\in[\sigma, \sigma+\varrho]$ and $\psi\in \boldsymbol{K}\,(\Phi, H)$. Similarly as in the proof of Lemma 4.3 it can be verified that G^F is u.s.c. on $[\sigma, \sigma+\varrho]\times \boldsymbol{K}_\alpha(\Phi, H)$. Immediately from the definition of G^F, (ii) and (iii) it follows that $G^F([\sigma, \sigma+\varrho]\times \boldsymbol{K}_\alpha(\Phi, H))\subset MB$ and $G^F(t, \psi)\cap T_H(\psi(0))\neq\emptyset$ for $t\in[\sigma, \sigma+\varrho]$ and $\psi\in \boldsymbol{K}_\alpha(\Phi, H)$. Then, by Lemma 4.6, there exists an absolutely continuous function $x\colon [\sigma-r, \sigma+\varrho]\to \boldsymbol{R}^n$ such that $\dot{x}(t)\in G^F(t, x_t)$ for a.e. $t\in[\sigma, \sigma+\varrho]$, $x_\sigma = \Phi$ and $x(t)\in H$ for $t\in[\sigma, \sigma+\varrho]$. Furthermore, by the proof of Lemma 4.6, it follows that $(t, x_t)\in[\sigma, \sigma+\varrho]\times \boldsymbol{K}_\alpha(\Phi, H)\subset\Omega$ for every $t\in[\sigma, \sigma+\varrho]$. Therefore, x satisfies conditions (4.11). ■

EXAMPLE 3.1. Let $D = \boldsymbol{R}\times C_{0r}\times L_{0r}$, $F\colon D\to \mathrm{Conv}(\boldsymbol{R}^n)$ be u.s.c.w. on D and $\Omega\subset \boldsymbol{R}\times AC_{0r}$ be such that there exists $(\sigma, \Phi)\in\Omega$ such that $(\sigma, \Phi_0)\in\Omega^0$, where $\Phi_0(s) = \Phi(0)$ for $s\in[-r, 0]$ and Ω^0 denotes the inferior of Ω in $\boldsymbol{R}\times C_{0r}$. Let $\lambda > 0$ be such that $B^0_\lambda(\sigma, \Phi_0)\subset\Omega$, where $B^0_\lambda(\sigma, \Phi_0)$ is an open ball of $\boldsymbol{R}\times C_{0r}$ with the centre at (σ, Φ_0) and a radius $\lambda > 0$. Assume furthermore, Φ is such that $|\dot{\Phi}|_0 < \frac{1}{2}\lambda$. Let $H\in \mathrm{Comp}(\boldsymbol{R}^n)$ be such that $\Phi(0)\in H$ and let $r > 0$, $l > 0$, $M > 0$ and F be such that $rl < \frac{1}{4}\lambda$, $rM < \frac{1}{4}\lambda$ and $F(\sigma, \Phi_0, 0)\subset \frac{1}{2}MB$. Since F is u.s.c.w. at $(\sigma, \Phi, 0)$ it is also u.s.c. there. Therefore, above $\lambda > 0$ can be so taken that we have $F(t, u, v)\subset F(\sigma, \Phi_0, 0)+\frac{1}{2}MB\subset MB$ for $t\in[\sigma, \sigma+\lambda]$, $(u, v)\in C_{0r}\times L_{0r}$ satisfying $|u-\Phi_0|_0 < \lambda$ and $|v|_0 < \lambda$. For every $\psi\in \boldsymbol{K}_\alpha(\Phi, H)$ with $\alpha = l+M$ we have $|\psi-\Phi_0|_0 = \sup_{-r\leqslant s\leqslant 0}|\psi(s)-\psi(0)|\leqslant \int_{-r}^{0}|\dot{\psi}(s)|\,ds\leqslant |\dot{\Phi}|_0+r(l+M) < \lambda$ and $|\dot{\psi}|_0 < \lambda$. Therefore, for every $(t, \psi)\in[\sigma, \sigma+\lambda]\times \boldsymbol{K}_\alpha(\Phi, H)$ we have $(t, \psi)\in B^0_\lambda(\sigma, \Phi_0)\subset\Omega$ and $F(t, \psi, \dot{\psi})\subset MB$, i.e., $[\sigma, \sigma+\lambda]\times \boldsymbol{K}_\alpha(\Phi, H)\subset\Omega$ and $F([\sigma, \sigma+\lambda]\times \boldsymbol{K}_\alpha(\Phi, H)\times \dot{\boldsymbol{K}}_\alpha(\Phi, H))\subset MB$. Finally, assume F is such that for every $(\tau, \tilde{u}, \tilde{v})\in D$ there is its neighbourhood $N(\tau, \tilde{u}, \tilde{v})$ such that $\bigcap\{F(t,u,v)\colon (t,u,v)\in N(\tau, \tilde{u}, \tilde{v})\}\neq\emptyset$. Suppose now, $\lambda > 0$ taken above and F are such that $0\in\bigcap\{F(t,u,v)\colon |t-\sigma| < \lambda,$

$|u - \Phi_0|_0 < \lambda$, $|v_0| < \lambda\}$. Hence it follows that $F(t, \psi, \dot{\psi}) \cap T_H(\psi(0)) \neq \emptyset$ for $t \in [\sigma, \sigma+\lambda]$, $\psi \in \boldsymbol{K}_\alpha(\Phi, H)$, because for every $\psi \in \boldsymbol{K}_\alpha(\Phi, H)$ we have $\psi(0) \in H$, $|\psi - \Phi_0| < \lambda$, $|\dot{\psi}|_0 < \lambda$ and $0 \in T_H(\psi(0))$. ■

THEOREM 4.9. *Let (D, Ω) be a conformable pair of nonempty subsets of $\boldsymbol{R} \times C_{0r} \times L_{0r}$ and $\boldsymbol{R} \times \mathrm{AC}_{0r}$, respectively and let D be open. Suppose F: $D \to \mathrm{Conv}(\boldsymbol{R}^n)$ is locally u.s.c.w. Assume there are $(\sigma, \Phi) \in \Omega$, $l > 0$, $\varrho > 0$, $M > 0$ and a compact set $H \subset \boldsymbol{R}^n$ such that $\Phi(0) \in H$, $I_\varrho(\sigma) \times \mathscr{K}_\varrho(\Phi) \times B_\varrho(\dot{\Phi}) \subset D$, $[\sigma, \sigma+\frac{1}{2}\varrho] \times B_\alpha(\frac{1}{2}\varrho, \Phi, H) \subset \Omega$ with $\alpha = l+M$, $F(I_\varrho(\sigma) \times \mathscr{K}_\varrho(\Phi) \times B_\varrho(\dot{\Phi})) \subset MB$, F is u.s.c.w. on $I_\varrho(\sigma) \times \mathscr{K}_\varrho(\Phi) \times B_\varrho(\dot{\Phi})$, and $F(t, \psi, \dot{\psi}) \cap T_H(\psi(0)) \neq 0$ for $t \in [\sigma, \sigma+\varrho]$ and $\psi \in B_\alpha(\varrho, \Phi, H)$. Then there exists $a \in (0, \frac{1}{2}\varrho]$ such that* NFDI(D, F) *has on $[\sigma - r, \sigma + a]$, (Ω, H)-viable trajectory through (σ, Φ).*

Proof. Let G^F: $[\sigma, \sigma+\varrho] \times B_\alpha(\varrho, \Phi, H) \to \mathrm{Conv}(\boldsymbol{R}^n)$ be defined by $G^F(t, \psi) = F(t, \psi, \dot{\psi})$ for $t \in [\sigma, \sigma+\varrho]$ and $\psi \in B_\alpha(\varrho, \Phi, H)$. By virtue of Lemma 4.3, G^F is n.s.c. on $[\sigma, \sigma+\varrho] \times B_\alpha(\varrho, \Phi, H)$. Furthermore, $G^F(t, \psi) \cap T_H(\psi(0)) \neq \emptyset$ and $G^F([\sigma, \sigma+\varrho] \times B_\alpha(\varrho, \Phi, H)) \subset MB$ for $t \in [\sigma, \sigma+\varrho]$ and $\psi \in B_\alpha(\varrho, \Phi, H)$. Then, by virtue of Lemma 4.7 there are $a \in (0, \frac{1}{2}\varrho]$ and an absolutely continuous function x: $[\sigma - r, \sigma + a] \to \boldsymbol{R}^n$ such that $\dot{x}(t) \in G^F(t, x_t)$ for a.e. $t \in [\sigma, \sigma+a]$, $x_\sigma = \Phi$ and $x(t) \in H$ for $t \in [\sigma, \sigma+a]$. Moreover, by the proof of Lemma 4.7 it follows that for every $t \in [\sigma, \sigma+a]$, $x_t \in B_\alpha(\frac{1}{2}\varrho, \Phi, H)$, i.e. $(t, x_t) \in [\sigma, \sigma+\frac{1}{2}\varrho] \times B_\alpha(\frac{1}{2}\varrho, \Phi, H) \subset \Omega$ for $t \in [\sigma, \sigma+a]$. Therefore x is a (Ω, H)-viable trajectory of NFDI(D, F) on $[\sigma - r, \sigma+a]$ through (σ, Φ). ■

THEOREM 4.10. *Let (D, Ω) be a conformable pair of nonempty subsets of $\boldsymbol{R} \times C_{0r} \times L_{0r}$* and $\boldsymbol{R} \times \mathrm{AC}_{0r}$, *respectively and let D be open. Suppose $F \in \mathscr{M}(D, \mathrm{Conv}(\boldsymbol{R}^n))$ is locally continuous weakly with respect to its last two variables. Assume there are $(\sigma, \Phi) \in \Omega$, $l > 0$, $\varrho > 0$, $M > 0$ and a compact set $H \subset \boldsymbol{R}^n$ such that $\Phi(0) \in H$, $I_\varrho(\sigma) \times \mathscr{K}_\varrho(\Phi) \times B_\varrho(\dot{\Phi}) \subset D$, $F(I_\varrho(\sigma) \times \mathscr{K}_\varrho(\Phi) \times B_\varrho(\dot{\Phi})) \subset MB$, $F(t, \cdot, \cdot)$ is continuous weakly on $\mathscr{K}_\varrho(\Phi) \times B_\varrho(\Phi)$ for fixed $t \in [\sigma, \sigma+\varrho]$, $[\sigma, \sigma+\frac{1}{2}\varrho] \times B_\alpha(\frac{1}{2}\varrho, \Phi, H) \subset \Omega$ with $\alpha = l + M$ and $F(t, \psi, \dot{\psi}) \cap T_H(\psi(0)) \neq \emptyset$ for every $t \in [\sigma, \sigma+\varrho]$ and $\psi \in B_\alpha(\varrho, \Phi, H)$. Then there exists $a \in (0, \frac{1}{2}\varrho]$ such that* NFDI(D, F) *has on $[\sigma - r, \sigma+a]$, (Ω, H)-viable trajectory through (σ, Φ).*

Proof. By virtue of Theorem 4.5 there exists a sequence (G_n) of continuous set-valued functions G_n: $[\sigma, \sigma+\varrho] \times B_\alpha(\varrho, \Phi, H) \to \mathrm{Comp}(\boldsymbol{R}^n)$ such that conditions (i)–(iii) of theorem are satisfied with G^F: $[\sigma, \sigma+\varrho] \times B_\alpha(\varrho, \Phi, H) \to \mathrm{Conv}(\boldsymbol{R}^n)$ defined by $G^F(t, \psi) := F(t, \psi, \dot{\psi})$ for $t \in [\sigma, \sigma+\varrho]$ and $\psi \in B_\alpha(\varrho, \Phi, H)$.

Let F_n: $[\sigma, \sigma+\varrho] \times B_\alpha(\varrho, \Phi, H) \to \mathrm{Conv}(\boldsymbol{R}^n)$ be defined by $F_n(t, \psi) = \mathrm{co}\, G_n(t, \psi)$ for $(t, \psi) \in [\sigma, \sigma+\varrho] \times B_\alpha(\varrho, \Phi, H)$ and $n = 1, 2, \ldots$ We have $\sup\{h(F_n(t, u), G^F(t, u)) \colon u \in B_\alpha(\varrho, \Phi, H)\} \to 0$ for a.e. $t \in [\sigma, \sigma+\varrho]$ as $n \to \infty$, because $h(\mathrm{co}\, G_n(t, u), G^F(t, u)) \leqslant h(G_n(t, u), G^F(t, u))$ for $(t, u) \in [\sigma, \sigma+\varrho] \times B_\alpha(\varrho, \Phi, H)$ and $n = 1, 2, \ldots$ Furthermore, we have $F_n(t, \psi) \subset MB$ and $F_n(t, \psi) \cap T_H(\psi(0)) \neq \emptyset$ for $(t, \psi) \in [\sigma, \sigma+\varrho] \times B_\alpha(\varrho, \Phi, H)$ and $n = 1, 2, \ldots$

Now, similarly as in the proof of Lemma 4.7 let $C := [\sigma, \sigma+\frac{1}{2}\varrho]\times B_\alpha(\frac{1}{2}\varrho, \Phi, H)$ and $S := \overline{[\boldsymbol{R}\times C_{0r}\setminus[\sigma-\varrho, \sigma+\varrho]\times B_\alpha(\varrho, \Phi, H)]}_{\boldsymbol{R}\times C_{0r}}$ and let $f\colon \boldsymbol{R}\times C_{0r}\to[0,1]$ be a continuous function such that $f(t,\psi)=1$ for $(t,\psi)\in C$ and $f(t,\psi)=0$ for $(t,\psi)\in S$. Put $F_n^f(t,\psi) := f(t,\psi)\cdot\tilde{F}_n(t,\psi)$ for $(t,\psi)\in[\sigma,\sigma+\varrho]\times \boldsymbol{K}_\alpha(\Phi,H)$ and $n = 1, 5, \ldots$, where $\tilde{F}_n$ denotes a continuous extension of F_n on $\boldsymbol{R}\times C_{0r}$. Similarly as in the proof of Lemma 4.7 we can verify that $F_n^f(t,\psi)\cap T_H(\psi(0))\neq\emptyset$ for $t\in[\sigma,\sigma+\varrho]$, $\psi\in\boldsymbol{K}_\alpha(\Phi,H)$ and $n=1,2,\ldots$ Obviously F_n^f is continuous on $[\sigma,\sigma+\varrho]\times\boldsymbol{K}_\alpha(\Phi,H)$, has compact convex values and $F_n^f([\sigma,\sigma+\varrho]\times\boldsymbol{K}_\alpha(\Phi,H))\subset MB$. Therefore, by virtue of Lemma 4.6, for every $n=1,2,\ldots$ there is an absolutely continuous function $x_n^f\colon[\sigma-r,\sigma+\varrho]\to\boldsymbol{R}^n$ such that $\dot{x}_n^f(t)\in F_n^f(t,(x_n^f)_t)$ for a.e. $t\in[\sigma,\sigma+\varrho]$, $(x_n^f)_\sigma=\Phi$ and $x_n^f(t)\in H$ for $t\in[\sigma,\sigma+\varrho]$. Furthermore, by the proof of Lemma 4.6, we also have $(x_n^f)_t\in\boldsymbol{K}_\alpha(\Phi,H)$ for $t\in[\sigma,\sigma+\varrho]$ and $n=1,2,\ldots$ Hence, in particular it follows that $|\dot{x}_n^f(t)|\leqslant M$ for $n=1,2,\ldots$ and a.e. $t\in[\sigma,\sigma+\varrho]$. Therefore, there is an absolutely continuous function $x^f\colon[\sigma-r,\sigma+\varrho]\to\boldsymbol{R}^n$ and a subsequence, say (x_k^f) of (x_n^f) such that $x_\sigma^f=\Phi$, $|x_k^f-x^f|_\varrho\to 0$ and $\dot{x}_k^f\rightharpoonup\dot{x}^f$ in $L([\sigma,\sigma+\varrho],\boldsymbol{R}^n)$ as $k\to\infty$. Similarly as in the proof of Lemma 4.7 we can prove that there exists $a\in(0,\frac{1}{2}\varrho]$ such that $(t,x_t^f)\in C := [\sigma,\sigma+\frac{1}{2}\varrho]\times B_\alpha(\frac{1}{2}\varrho,\Phi,H)$ for every $t\in[\sigma,\sigma+a]$. Therefore, for every measurable set $E\subset[\sigma,\sigma+a]$ and $\tau\in E$ we have $f(\tau,x_\tau^f)=1$.

Let us observe now that for every $t\in[\sigma,\sigma+\varrho]$ and $k=1,2,\ldots$ we have

$$
\begin{aligned}
&h(F_k^f(t,(x_k^f)_t), f(t,(x_k^f)_t)\tilde{G}^F(t,(x_k^f)_t))\\
&\leqslant f(t,(x_k^f)_t)\sup\{h(F_k(t,u),G^F(t,u))\colon u\in B_\alpha(\varrho,\Phi,H)\}\\
&\leqslant m\sup\{h(F_k(t,u),G^F(t,u))\colon u\in B_\alpha(\varrho,\Phi,H)\}\leqslant 2Mm,
\end{aligned}
$$

where $m=\sup\{f(t,u)\colon (t,u)\in[\sigma,\sigma+\varrho]\times B_\alpha(\varrho,\Phi,H)\}$ and $\tilde{G}^F(t,\psi) := F(t,\psi,\psi)$ for $t\in[\sigma,\sigma+\varrho]$ and $\psi\in\boldsymbol{K}_\alpha(\Phi,H)$. Indeed, it follows immediately from the properties of f, F_k and G^F. Hence it follows that $\int_E h(F_k^f(t,(x_k^f)_t, f(t,(x_k^f))\tilde{G}^F(t,(x_k^f)_t))dt\to 0$ as $k\to\infty$ for every measurable set $E\subset[\sigma,\sigma+a]$.

For every measurable set $E\subset[\sigma,\sigma+a]$ and every $k=1,2,\ldots$ we have

$$
\begin{aligned}
&\operatorname{dist}\Big(\int_E \dot{x}^f(t)\,dt, \int_E G^F(t,x_t^f)\,dt\Big)\\
&\leqslant \Big|\int_E \dot{x}^f(t)\,dt-\int_E \dot{x}_k^f(t)\,dt\Big| + \operatorname{dist}\Big(\int_E \dot{x}_k^f(t)\,dt, \int_E F_k^f(t,(x_k^f)_t)\,dt\Big)+\\
&\quad+\int_E h(F_k^f(t,(x_k^f)_t, f(t,(x_k^f)_t\cdot\tilde{G}^F(t,(x_k^f)_t))\,dt+\\
&\quad+\int_E h(f(t,(x_k^f)_t)\cdot\tilde{G}^F(t,(x_k^f)_t), f(t,(x_k^f)_t)\cdot\tilde{G}^F(t,x_t^f))\,dt+\\
&\quad+\int_E h(f(t,(x_k^f)_t)\cdot G^F(t,x_t^f), f(t,x_t^f)\cdot G^F(t,x_t^f))\,dt.
\end{aligned}
$$

Since,

$$\left|\int_E \dot{x}^f(t)dt - \int_E \dot{x}_k^f(t)dt\right| \to 0 \text{ as } k\to\infty, \quad \int_E \dot{x}_k^f(t)dt \in \int_E F_k^f(t,(x_k^f)_t)dt,$$

$$h\big(f(t,(x_k^t)_t)\cdot G^F(t,x_t^f), f(t,x_t^f)\cdot G^F(t,x_t^f)\big) \leqslant |f(t,(x_k^f)_t)-f(t,x_t^f)|M,$$
$$h\big(f(t,(x_k^f)_t)\cdot \tilde{G}^F(t,(x_k^f)_t), \quad f(t,(x_k^f)_t)\cdot \tilde{G}^F(t,x_t^f)\big) = 0 \quad \text{for}$$

$$(t,(x_k^f)_t)\in S \quad \text{and } h(f(t,(x_k^f)_t\cdot \tilde{G}^F(t,(x_k^f)_t)\cdot f(t,(x_k^f)_t)\cdot \tilde{G}^F(t,x_t^f)) \leqslant 2Mm,$$

then for every measurable set $E \subset [\sigma,\sigma+a]$ we have $\int_E \dot{x}(t)dt \in \int_E G^F(t,x_t^f)dt$, i.e. $\dot{x}^f(t)\in G^F(t,x_t^f)$ for a.e. $t\in[\sigma,\sigma+a]$.

Let x be the restriction of x^f to the interval $[\sigma-r,\sigma+a]$. We have proved that $\dot{x}(t)\in G^F(t,x_t)$ for a.e. $t\in[\sigma,\sigma+a]$, $x_\sigma=\Phi$, $(t,x_t^f)\in[\sigma,\sigma+\frac{1}{2}\varrho]\times B_\alpha(\frac{1}{2}\varrho, \Phi, H) \subset \Omega$ for $t\in[\sigma,\sigma+a]$. *It is clear that* $x(t)\in H$ *for* $t\in[\sigma,\sigma+a]$. Therefore, x is (Ω,H)-viable trajectory of NFDI(D,F) on $[\sigma-r,\sigma+a]$ through (σ,Φ). ■

Let (D,A) be a conformable pair of nonempty subsets of $\boldsymbol{R}\times C_{0r}\times L_{0r}$ and $\boldsymbol{R}\times \mathrm{AC}_{0r}$, respectively and $D_A=\{(t,x,\dot{y})\colon (t,x),(t,y)\in A\}$. Similarly as in Section III.6.2. for a given $\alpha>0$ we put $\Lambda_\alpha=\{u\in L(I,\boldsymbol{R}^n)\colon |u(t)|\leqslant\alpha$ for a.e. $t\in I\}$ and let $\mathscr{R}$ be the biaffine mapping defined by (2.5). For given $l>0$, $\Phi\in C_{0r}$ and a set-valued function $\mathscr{K}\colon [\sigma,\sigma+\varrho]\to \mathrm{Comp}(\boldsymbol{R}^n)$ by $K_\alpha(\Phi,\mathscr{K})$ we denote a set defined by $K_\alpha(\Phi,\mathscr{K}) := \{x\in\mathscr{I}(\Lambda_\alpha)\colon (t,\mathscr{R}(\Phi,x)_t(0))\in \mathrm{Graph}(\mathscr{K})+(\varrho+l)B'$ for $t\in(\sigma,\sigma+\varrho]\}$, where B' is a closed unit ball of $\boldsymbol{R}^{n+1}$.

THEOREM 4.11. *Let (D,A) be a conformable pair of nonempty subsets of $\boldsymbol{R}\times C_{0r}\times L_{0r}$ and $\boldsymbol{R}\times\mathrm{AC}_{0r}$, respectively. Suppose $F\colon D\to\mathrm{Conv}(\boldsymbol{R}^n)$ is u.s.c.w. on D_A. Let $\mathscr{K}\colon [\sigma,\sigma+\varrho]\to\mathrm{Comp}(\boldsymbol{R}^n)$ be u.s.c. on $[\sigma,\sigma+\varrho]$. Suppose there are $(\sigma,\Phi)\in A$ and numbers $\varrho>0$, $l>0$ and $M>0$ such that $K_\alpha(\Phi,\mathscr{K})\neq\emptyset$ with $\alpha=l+M$ and such that $(t,\mathscr{R}(\Phi,x)_t)\in A$ for $t\in I$ and $x\in K_\alpha(\Phi,\mathscr{K})$. If furthermore*

(i) *$F(t,(\Phi\oplus x)_t, (\dot{\Phi}\square\dot{x})_t)\subset MB$ for $t\in[\sigma,\sigma+\varrho]$ and $x\in K_\alpha(\Phi,\mathscr{K})$, where B is a closed unit ball of $\boldsymbol{R}^n$,*

(ii) *$F(t,(\Phi\oplus x)_t, (\dot{\Phi}\square x)_t)\cap T_{\mathscr{K}(t)}[(\Phi\oplus x)_t(0)]\neq\emptyset$ for $t\in[\sigma,\sigma+\varrho]$ and $x\in K_\alpha(\Phi,\mathscr{K})$,*

then there is $x\in\mathrm{AC}([\sigma-r,\sigma+\varrho],\boldsymbol{R}^n)$ such that

$$\dot{x}(t)\in F(t,x_t,\dot{x}_t) \quad \textit{for a. e. } t\in[\sigma,\sigma+\varrho],$$
$$(t,x_t)\in A \quad \textit{for } t\in[\sigma,\sigma+\varrho],$$
$$x_\sigma=\Phi,$$
$$x(t)\in\mathscr{K}(t) \quad \textit{for } t\in[\sigma,\sigma+\varrho].$$

Proof. Let $I:=[\sigma,\sigma+\varrho]$ and select $N_l\in\boldsymbol{N}$ such that $(3+l+m)/k\leqslant l$ for $k\geqslant N_l$. Define $G^F\colon I\times K_\alpha(\Phi,\mathscr{K})\to\mathrm{Conv}(\boldsymbol{R}^n)$ by setting $G^F(t,x):=F(t,(\Phi\oplus x)_t,(\Phi\oplus\dot{x})_t)$ for $(t,x)\in I\times K_\alpha(\Phi,\mathscr{K})$. It is clear that G^F is u.s.c. and bounded on $I\times K_\alpha(\Phi,\mathscr{K})$ and such that $G^F(I\times K_\alpha(\Phi,\mathscr{K}))\subset MB$. By Lemma 4.1, the tangential condition

(ii) implies that for every $k \geq N_l$ and $(\tau, y) \in I \times K_\alpha(\Phi, \mathscr{K})$ there are $v_y^\tau \in G^F(\tau, y))$ $h_y^\tau \in (0, 1/k)$ such that $d_{\mathscr{K}(\tau)}[(\Phi \oplus y)_\tau(0) + h_y^\tau v_y^\tau] \leq h_y^\tau/3k$. Similarly as in the proof of Theorem III.6.2 we introduce $N(\tau, y) = \{(t, x) \in I \times K_\alpha(\Phi, \mathscr{K}) \colon d_{\mathscr{K}(\tau)}(x(t) + h_y^\tau v_y^\tau) < h_y^\tau/2k\}$. In what follows we shall consider $I \times K_\alpha(\Phi, \mathscr{K})$ as a compact metric space with the relative topology generated by the metric of $\boldsymbol{R} \times C(I, \boldsymbol{R}^n)$. Since a mapping $I \times K_\alpha(\Phi, \mathscr{K}) \ni (t, x) \to d_{\mathscr{K}(\tau)}(x(t) + h_y^\tau v_y^\tau) \in \boldsymbol{R}^+$ is continuous then $N(\tau, y)$ is an open subset of $I \times K_\alpha(\Phi, \mathscr{K})$ for every $(\tau, y) \in I \times K_\alpha(\Phi, \mathscr{K})$. Then, for every $(\tau, y) \in I \times K_\alpha(\Phi, \mathscr{K})$ there is an open ball $S(\tau, y)$ of $I \times K_\alpha(\Phi, \mathscr{K})$ centred at (τ, y) and a radius $\eta_b^\tau \in (0, 1/k)$ such that $S(\tau, y) \subset N(\tau, y)$. Let $\mathscr{H} := \text{Graph } \mathscr{K} + (\varrho + l)B'$. It is a compact subset of $\boldsymbol{R}^{n+1}$ because $\mathscr{K}$ is u.s.c. on I. We shall also consider $\mathscr{H}$ as a compact metric space with the relative topology generated by the metric $\boldsymbol{R}^{n+1}$. By the definition of $K_\alpha(\Phi, \mathscr{K})$ we have $(\tau, (\Phi \oplus y)_\tau(0)) \in \mathscr{H}$ for every $(\tau, y) \in I \times K_\alpha(\Phi, \mathscr{K})$. Let $B(\tau, y)$ be for every $(\tau, y) \in I \times K_\alpha(\Phi, \mathscr{K})$ an open ball of $\mathscr{H}$ with the centred at $(\tau, (\Phi \oplus y)_\tau(0))$ and a radius $\eta_y^\tau \in (0, k/1)$. We have $I \times K_\alpha(\Phi, \mathscr{K}) \times \mathscr{H} = \bigcup \{S(\tau, y) \times B(\tau, y) \colon (\tau, y) \in I \times K_\alpha(\Phi, \mathscr{K})\}$. Then, by the compactness of $I \times K_\alpha(\Phi, \mathscr{K}) \times \mathscr{H}$ there are $(\tau_j, y_j) \in I \times K_\alpha(\Phi, \mathscr{K})$ with $j = 1, \ldots, q$ such that $I \times K_\alpha(\Phi, \mathscr{K}) \times \mathscr{H} = \bigcup_{j=1}^{q} S(\tau_j, y_j) \times B(\tau_j, y_j)$. Here, similar as above, we consider $I \times K_\alpha(\Phi, \mathscr{K}) \times \mathscr{H}$ as a metric space with a metric d defined by $d[(t_1, x_1, u_1), (t_2, x_2, u_2)] = \max(|t_1 - t_2|, |x_1 - x_2|_\varrho, |u_1 - u_2|)$ for (t_1, x_1, u_1), $(t_2, x_2, u_2) \in I \times K_\alpha(\Phi, \mathscr{K}) \times \mathscr{H}$. Then $S(\tau_j, y_j) \times B(\tau_j, y_j)$ is an open ball of $(I \times K_\alpha(\Phi, \mathscr{K}) \times \mathscr{H}, d)$ with the centre at $(\tau_j, y_j, \tau_j, (\Phi \oplus y_j)_{\tau_j}(0))$ and a radius $\eta_{y_j}^{\tau_j} \in (0, 1/k)$. For simplicity we set $\eta_j := \eta_{y_j}^{\tau_j}$, $h_j := h_{y_j}^{\tau_j}$, $v_j := v_{y_j}^{\tau_j}$ and $h_0^k = \min_{1 \leq j \leq q} h_j$.

Let $(t, x) \in I \times K_\alpha(\Phi, \mathscr{K})$ be fixed. We have $(t, x, t, (\Phi \oplus x)_t(0)) \in I \times K_\alpha(\Phi, \mathscr{K}) \times \mathscr{H}$. Then, there is $j \in \{1, \ldots, q\}$ such that $(t, x, t, (\Phi \oplus x)_t(0)) \in S(\tau_j, y_i) \times B(\tau_j, y_j) \subset S(\tau_j, y_j) \times N(\tau_j, y_j)$. Therefore, $|t - \tau_j| < \eta_j \leq 1/k$, $|x - y_j|_\varrho < \eta_j \leq 1/k$ and there is $z_j \in \mathscr{K}(\tau_j)$ such that $|v_j - (z_j - (\Phi \oplus x)_t(0))/h_j| \leq \frac{1}{h} d_{\mathscr{K}(\tau_j)}[(\Phi \oplus x)_t(0) + h_j v_j] + 1/2k \leq 1/k$. Let $u_j := [z_j - (\Phi \oplus x)_t(0)]/h_j$. We have $u_j \in v_j + (1/k)B$ and $(\Phi \oplus x)_t(0) + h_j u_j \in \mathscr{K}(\tau_j)$. Then for every $(t, x) \in I \times K_\alpha(\Phi, \mathscr{K})$ there are $(\tau_j, y_j) \in I \times K_\alpha(\Phi, \mathscr{K})$, $v_j \in G^F(\tau_j, y_j)$, $u_j \in v_j + (1/k)B$ and $h_l \in [h_0^k, 1/k]$ such that $|t - \tau_j| \leq 1/k$, $|x - y_j|_\varrho \leq 1/k$ and $(\Phi \oplus x)_t(0) + h_j u_j \in \mathscr{K}(\tau_j)$. Hence, in particular it follows that $(\tau_j, y_j, v_j) \in \text{Graph}\, G^F$ and $v_j \in G^F(I \times K_\alpha(\Phi, \mathscr{K}))$. Therefore, for every $k \geq N_l$ and $(t, x) \in I \times K\,(\Phi, \mathscr{K})$ there are $h_j \in [h_0^k, 1/k]$ and $u_j \in G^F(I \times K_\alpha(\Phi, \mathscr{K})) + lB$ such that $(t, (\Phi \oplus x)_t(0) + h_j u_j) \in \text{Graph}\, \mathscr{K} + (1/k)B'$ and $(t, x, u_j) \in \text{Graph}\, G^F + (1/k)S$ where S is a closed unit ball of $\boldsymbol{R} \times C(I, \boldsymbol{R}^n) \times \boldsymbol{R}^n$.

Similarly as in the proof of Theorem III.6.2, fix $k \geq N_l$ and $x^0 \in K_\alpha(\Phi, \mathscr{K})$ and then select for $(\sigma, x^0) \in I \times K_\alpha(\Phi, \mathscr{K})$, $h_0 \in [h_0^k, 1/k]$, $u_0 \in G^F(I \times K_\alpha(\Phi, \mathscr{K})) + lB$ such that $(\sigma, (\Phi \oplus x^0)_\sigma(0) + h_0 u_0) \in \text{Graph}\, \mathscr{K} + (1/k)B$ and $(\sigma, x^0, u_0) \in \text{Graph}\, G^F + (1/k)S$.

Let $t_1 = \sigma + h_0$ and define $l_0(t) = (t - \sigma)u_0$ for $t \in [\sigma, t_1]$. Let $z^1 \colon [\sigma - r, \sigma + \varrho] \to \boldsymbol{R}^n$ be defined by

$$z^1(\tau) = \begin{cases} \Phi(\tau-\sigma) & \text{for } \tau \in [\sigma-r, \sigma), \\ \Phi(0)+l_0(\tau) & \text{for } \tau \in [\sigma, t_1], \\ z^1(t_1) & \text{for } \tau \in (t_1, \sigma+\varrho] \end{cases}$$

and put $x^1 = \mathscr{I}(\dot{z}^1|_{[\sigma,\sigma+\varrho]})$,. We have $|\dot{x}^1(t)| = |u_0| \leqslant l+M$ for $t \in (\sigma, t_1)$ and $|\dot{x}^1(t)| = 0 \leqslant l+M$ for $t \in (t_1, \sigma+\varrho)$, i.e. $|\dot{x}^1(t)| \leqslant l+M$ for a.e. $t \in I$. Furthermore, $(t_1, (\Phi \oplus x^0)_{t_1}(0)) = (t_1, \Phi(0)+u_0 h_0) = (t_1, (\Phi \oplus x^0)_\sigma(0)+h_0 u_0) \in (\sigma, (\Phi \oplus x^0)_\sigma(0) + h_0 u_0) + \frac{1}{k} B' \subset \operatorname{Graph}\mathscr{K} + \frac{2}{k} B'$. Since, $|(t, (\Phi \oplus x^1)_t(0) - (t_1, (\Phi \oplus x')_{t_1})| \leqslant$ $\leqslant 1/k+(l+M)/k$ for $t \in [\sigma, t_1]$ then $(t, (\Phi \oplus x^1)_t(0) \in \operatorname{Graph}\mathscr{K} + \frac{1}{k}(3+l+M)B'$. Finally, for $t \in (t_1, \sigma+\varrho)$ we have $|(t, (\Phi \oplus x^1)_t(0) - (t_1, (\Phi \oplus x^1)_{t_1}|$ $\leqslant \varrho+(l+M)/k$. Then $(t, (\Phi \oplus x^1)_t(0)) \in \operatorname{Graph}\mathscr{K} + \varrho B' + \frac{1}{k}(2+l+M)B'$ for $t \in (t_1, \sigma+\varrho]$. Since $(3+l+M)/k \leqslant l$, then $(t, (\Phi \oplus x^1)_t(0)) \in \operatorname{Graph}\mathscr{K} + (\varrho + l)B'$ for $t \in [\sigma, \sigma+\varrho]$. Thus, $x^1 \in K_\alpha(\Phi, \mathscr{K})$. Now, for $(t_1, x^1) \in I \times K_\alpha(\Phi, \mathscr{K})$ we can find $h_1 \in [h_0^k, 1/k]$ and $u_1 \in G^F(I \times K_\alpha(\Phi, \mathscr{K})) + lB'$ such that $(t_1, (\Phi \oplus) \oplus x^1)_{t_1} + h_1 u_1) \in \operatorname{Graph}\mathscr{K} + \frac{1}{k} B$ and $(t_1, x^1, u_1) \in \operatorname{Graph} G^F + \frac{1}{k} S$. Let $t_2 = t_1 + h_1$ and define $l_1(t) = (t-t_1)u_1$ for $t \in [t_1, t_2]$. Let $z^2\colon [\sigma-r, \sigma+\varrho] \to R^n$ be define by

$$z^2(t) = \begin{cases} z^1(t) & \text{for } t \in [\sigma-r, t_1), \\ z^1(t_1)+l_1(t) & \text{for } t \in [t_1, t_2], \\ z^2(t_2) & \text{for } t \in (t_2, \sigma+\varrho]. \end{cases}$$

Put, $x^2 = \mathscr{I}(\dot{z}^2|_{[\sigma,\sigma+\varrho]})$. We have $|\dot{x}^2(t)| = |\dot{x}^1(t)| \leqslant l+M$ for $t \in (\sigma, t_1)$, $|\dot{x}^2(t)| = |u_1| \leqslant l+M$ for $t \in (t_1, t_2)$ and $|\dot{x}^2(t)| = 0 < l+M$ for $t \in (t_2, \sigma+\varrho)$. Thus $|\dot{x}^2(t)| \leqslant l+M$ for a.e. $t \in [\sigma, \sigma+\varrho]$. Since $x^2(t) = x^1(t)$ for $t \in [\sigma, t_1]$ then $(t, (\Phi \oplus x^2)_t(0)) = (t, (\Phi \oplus x^1)_t(0)) \in \operatorname{Graph}\mathscr{K} + (\varrho+l)B'$ for $t \in [\sigma, t_1]$. Furthermore, $(t_1, (\Phi \oplus x^2)_{t_2}(0)) = (t_1, \Phi(0)+x^2(t_2)) = (t_1, \Phi(0)+x^1(t_1)+h_1 u_1) = (t_1, (\Phi \oplus x^1)_{t_1}(0) + h_1 u_1) \in \operatorname{Graph}\mathscr{K} + \frac{1}{k} B$. Therefore $(t_2, (\Phi \oplus x^2)_{t_2}(0)) \in \operatorname{Graph}\mathscr{K} + \frac{2}{k} B'$. Hence, similarly as above, it follows that $(t, (\Phi \oplus x^2)_t(0)) \in \operatorname{Graph}\mathscr{K} + (\varrho+l)B'$ for $t \in (t_1, \sigma+\varrho]$. Therefore, $x^2 \in K_\alpha(\Phi, \mathscr{K})$.

Continuing this procedure we can find numbers $h_0, h_1, \ldots, h_{m_k} \in [h_0^k, 1/k]$ with $\sum_{i=0}^{m_k} h_i = \varrho$ and functions $x^1, \ldots, x^{m_k} \in K_\alpha(\Phi, \mathscr{K})$ such that $(t_i, (\Phi \oplus x^i)_{t_i}(0)) \in \operatorname{Graph}\mathscr{K} + \frac{2}{k} B'$ and $(t_i, x^i, x^i(t)) \in \operatorname{Graph} G^F + \frac{1}{k} S$ for $t \in (t_i, t_{i+1})$ and $x^{i+1}(t) = x^i(t)$ for $t \in [\sigma, t_i]$ with $i = 1, \ldots, m_{k-1}$. Taking for every $k \geqslant N_l$, $\xi^k = x^{m_k}$ similarly as in the proof of Theorem III.6.2. we can verify that $\xi^k \in K_\alpha(\Phi, \mathscr{K})$ $(t_i, \dot{\xi}^k, \xi^k(t)) \in \operatorname{Graph} G^F + \frac{1}{k} S$ for $t \in (t_{i-1}, t_i)$ with $i = 1, \ldots, m_k$ and $(t_i, (\Phi \oplus \xi^k)_{t_i}(0)) \in \operatorname{Graph}\mathscr{K} + \frac{2}{k} B'$ for $i = 1, \ldots, m_k$. Hence, similarly as in the proof of Theorem III.6.2 it follows the existence of $\xi \in K_\alpha(\Phi, \mathscr{K})$ such that $(t, \xi, \dot{\xi}(t)) \in \text{Graph}\, G^F$ for a. e. $t \in [\sigma, \sigma+\varrho]$ and $(t, (\Phi \oplus \zeta)_t\,(0) \in \text{Graph}\, \mathscr{K}$ for $t \in [\sigma, \sigma+\varrho]$, i.e. $\dot{\xi}(t) \in G^F(t, \xi) := F(t, (\Phi \oplus \xi)_t, (\Phi \Box \dot{\xi})_t)$ for a.e. $t \in I$ and $\Phi(0)+\xi(t) \in \mathscr{K}(t)$ for $t \in I$. Taking now $x = \Phi \oplus \xi$, we obtain $\dot{x}(t) \in F(t, x_t, \dot{x}_t)$ for a.e. $t \in I$ and $x(t) \in \mathscr{K}(t)$ for $t \in I$. Furthermore, we have $x_\sigma = \Phi$ and $(t, x_t) \in A$ for $t \in I$, because $\xi \in K_\alpha(\Phi, \mathscr{K})$ implies $(t, \mathscr{R}(\Phi, \xi)_t) \in A$ for each $t \in I$. ■

4.3. *η-approximation viability theorems*

It happens in practical problems of optimal control that instead of the tangential condition some approximation tangential condition is satisfied. It is natural in such cases to expect only the existence of approximation viable trajectories. Such results will follow from the following lemmas.

LEMMA 4.12. *Let* $(\sigma, \Phi) \in \boldsymbol{R} \times \text{AC}_{0r}$, $\varrho > 0$, $M > 0$, $\eta \in [0, 1)$ *and a compact set* $H \subset \boldsymbol{R}^n$ *be given and let* $\Phi(0) \in H$. *Suppose* $G\colon [\sigma, \sigma+\varrho] \times K_\alpha(\Phi, H) \to \text{Conv}(\boldsymbol{R}^n)$ *with* $\alpha = 1+M$ *is u.s.c. and such that* $G([\sigma, \sigma+\varrho] \times K_\alpha(\Phi, H)) \subset MB$. *If* $G(t, \psi) \cap \cap [T_H(\psi(0))]^\eta \neq \emptyset$ *for* $t \in [\sigma, \sigma+\varrho]$ *and* $\psi \in K_\alpha(\Phi, H)$, *where* $[T_H(\psi(0))]^\eta := T_H(\psi(0)) + +\eta B$, *then there exists an absolutely continuous function* $x^\eta\colon [\sigma-r, \sigma+\varrho] \to \boldsymbol{R}^n$ *such that*

$$\begin{aligned} &\dot{x}^\eta(t) \in G(t, x_t^\eta) + \eta B \quad \textit{for a. e. } t \in [\sigma, \sigma+\varrho], \\ &x_\sigma^\eta = \Phi, \\ &x^\eta(t) \in H. \end{aligned}$$

Proof. The proof follows immediately from Lemma 4.6. It is enough only to observe that for each $(t, \psi) \in [\sigma, \sigma+\varrho] \times K_\alpha(\Phi, H)$ an approximation tangential condition $G(t, \psi) \cap [T_H(\psi(0))]^\eta \neq \emptyset$ is equivalent to $G^\eta(t, \psi) \cap T_H\, \psi(0)) \neq \emptyset$ for $(t, \psi) \in [\sigma, \sigma+\varrho] \times K_\alpha(\Phi, H)$, where $G^\eta(t, \psi) := G(t, \psi) + \eta B$. Indeed, let $G(t, \psi) \cap \cap [T_H(\psi(0))]^\eta \neq \emptyset$. Then there is $v \in G(t, \psi)$ such that $v \in T_H(\psi(0)) + \eta B$. Thus there exists $v_\eta \in T_H(\psi(0))$ such that $|v - v_\eta| \leqslant \eta$. Hence it follows $v_\eta \in v + \eta B \subset G(t, \psi) + \eta B =: G^\eta(t, \psi)$. Therefore, $G(t, \psi) \cap [T_H(\psi(0))]^\eta \neq \emptyset$ implies $G^\eta(t, \psi) \cap \cap T_H(\psi(0)) \neq \emptyset$. Conversly, suppose $v_\eta \in G^\eta(t, \psi) \cap T_H(\psi(0))$ and let $v \in G(t, \psi)$ be such that $|v - v_\eta| \leqslant \eta$. We have $v \in G(t, \psi)$ and $v \in T_H(\psi(0)) + \eta B$, and therefore $G(t, \psi) \cap [T_H(\psi(0))]^\eta \neq \emptyset$.

Now let us observe that a set-valued mapping G^η satisfies the assumptions of Lemma 4.6 with $G^\eta([\sigma, \sigma+\varrho] \times K_\alpha(\Phi, H)) \subset (\eta+M)B$ instead of $G([\sigma, \sigma+\varrho] \times \times K_\alpha(\Phi, H)) \subset MB$. Moreover, for $k \geqslant 1/(1-\eta)$ we also have $G^\eta([\sigma, \sigma+\varrho] \times \times K_\alpha(\Phi, H)) + \dfrac{1}{k} B \subset (1+M)B$. Therefore the proof of Lemma 4.6 can be repeated for G^η. ■

LEMMA 4.13. *Let* $(\sigma, \Phi) \in \boldsymbol{R} \times \text{AC}_{0r}$, $\varrho > 0$, $M > 0$, $\eta \in [0, 1)$ *and a compact set* $H \subset \boldsymbol{R}^n$ *be given and let* $\Phi(0) \in H$. *Suppose* $G\colon [\sigma, \sigma+\varrho] \times B_\alpha(\varrho, \Phi, H) \to \text{Conv}(\boldsymbol{R}^n)$ *with* $\alpha = 1+M$ *is u.s.c. and such that* $G([\sigma, \sigma+\varrho] \times B_\alpha(\varrho, \Phi, H)) \subset MB$. *If* $G(t, \psi) \cap$

$\cap[T_H(\psi(0))]^\eta \neq \emptyset$ *for* $t \in [\sigma, \sigma+\varrho]$ *and* $\psi \in B_\alpha(\varrho, \Phi, H)$ *then there exist* $a \in (0, \frac{1}{2}\varrho]$ *and an absolutely continuous function* $x^\eta\colon [\sigma-r, \sigma+a] \to \boldsymbol{R}^n$ *such that*

$$\dot{x}^\eta(t) \in G(t, x_t^\eta)+\eta B \quad \text{for a. e. } t \in [\sigma, \sigma+a],$$
$$x_\sigma^\eta = \Phi,$$
$$x^\eta(t) \in H \quad \text{for } t \in [\sigma, \sigma+a]. \blacksquare$$

Now, we obtain the following λ-approximation viability theorems.

THEOREM 4.14. *Let* (D, Ω) *be a conformable pair of nonempty subsets of* $\boldsymbol{R}\times C_{0r}\times L_{0r}$ *and* $\boldsymbol{R}\times \mathrm{AC}_{0r}$, *respectively. Suppose* $(\sigma, \Phi) \in \Omega$, $\varrho > 0$, $M > 0$ *and a compact set* $H \subset \boldsymbol{R}^n$ *are such that* $\Phi(0) \in H$ *and* $[\sigma, \sigma+\varrho]\times \boldsymbol{K}_\alpha(\Phi, H) \subset \Omega$ *with* $\alpha = 1+M$. *If* $F\colon D \to \mathrm{Conv}(\boldsymbol{R}^n)$ *is such that*

(i) F *is u.s.c.w. on* $[\sigma, \sigma+\varrho]\times \boldsymbol{K}_\alpha(\Phi, H)\times \dot{\boldsymbol{K}}_\alpha(\Phi, H)$,

(ii) $F([\sigma, \sigma+\varrho]\times \boldsymbol{K}_\alpha(\Phi, H)\times \dot{\boldsymbol{K}}_\alpha(\Phi, H)) \subset MB$,

(iii) $F(t, \psi, \dot{\psi})\cap[T_H(\psi(0))]^\eta \neq \emptyset$ *for every* $t \in [\sigma, \sigma+\varrho]$, $\psi \in \boldsymbol{K}_\alpha(\Phi, H)$ *and any* $\eta \in [0, 1)$,

then NFDI(D, F) *has on* $[\sigma-r, \sigma+\varrho]$, *an* η-*approximation* (Ω, H)-*viable trajectory through* (σ, Φ), *i.e., there is an absolutely continuous function* $x^\eta\colon [\sigma-r, \sigma+\varrho] \to \boldsymbol{R}^n$ *such that*

$$\dot{x}^\eta(t) \in F(t, x_t^\eta, \dot{x}_t^\eta)+\eta B \quad \text{for a. e. } t \in [\sigma, \sigma+\varrho],$$
$$x_\sigma^\eta = \Phi,$$
$$(t, x_t^\eta) \in \Omega \quad \text{for } t \in [\sigma, \sigma+\varrho],$$
$$x^\eta(t) \in H \quad \text{for } t \in [\sigma, \sigma+\varrho]. \blacksquare$$

THEOREM 4.15. *Let* (D, Ω) *be a conformable pair of nonempty subsets of* $\boldsymbol{R}\times C_{0r}\times L_{0r}$ *and* $\boldsymbol{R}\times \mathrm{AC}_{0r}$, *respectively, and let* D *be open. Suppose* $F\colon D \to \mathrm{Conv}(\boldsymbol{R}^n)$ *is locally u.s.c.w. Assume* $(\sigma, \Phi) \in \Omega$, $\varrho > 0$, $M > 0$, *and a compact set* $H \subset \boldsymbol{R}^n$ *are such that* $\Phi(0) \in H$, $I_\varrho(\sigma)\times\mathscr{K}_\varrho(\Phi)\times B_\varrho(\dot{\Phi}) \subset D$, $[\sigma, \sigma+\frac{1}{2}\varrho]\times B_\alpha(\frac{1}{2}\varrho, \Phi, H) \subset \Omega$, *with* $\alpha = 1+M$, F *is u.s.c.w. on* $I_\varrho(\sigma)\times\mathscr{K}_\varrho(\Phi)\times B_\varrho(\dot{\Phi})$, $F(I_\varrho(\sigma)\times\mathscr{K}_\varrho(\Phi)\times B_\varrho(\dot{\Phi})) \subset MB$ *and* $F(t, \psi, \check{\psi})\cap[T_H(\psi(0))]^\eta \neq \emptyset$ *for* $t \in [\sigma, \sigma+\varrho]$, $\psi \in B_\alpha(\varrho, \Phi, H)$ *and any* $\eta \in [0, 1)$. *Then there exists* $a \in (0, \frac{1}{2}\varrho]$ *such that* NFDI(D, F) *has on* $[\sigma-r, \sigma+a]$, *an* η-*approximation* (Ω, H)-*viable trajectory through* (σ, Φ). $\blacksquare$

THEOREM 4.16. *Let* (D, Ω) *be a conformable pair of nonempty subsets of* $\boldsymbol{R}\times C_{0r}\times L_{0r}$ *and* $\boldsymbol{R}\times \mathrm{AC}_{0r}$, *respectively, and let* D *be open. Suppose* $F \in \mathscr{M}(D, \mathrm{Conv}(\boldsymbol{R}^n))$ *is locally continuous weakly with respect to its last two variables. Assume* $(\sigma, \Phi) \in \Omega$, $\varrho > 0$, $M > 0$ *and a compact set* $H \subset \boldsymbol{R}^n$ *are such that* $\Phi(0) \in H$, $I_\varrho(\sigma)\times\mathscr{K}_\varrho(\Phi)\times B_\varrho(\dot{\Phi}) \subset D$, $F(I_\varrho(\sigma)\times\mathscr{K}_\varrho(\Phi)\times B_\varrho(\dot{\Phi})) \subset MB$, $F(t, \cdot, \cdot)$ *is continuous weakly on* $\mathscr{K}_\varrho(\Phi)\times B_\varrho(\dot{\Phi})$ *for fixed* $t \in [\sigma, \sigma+\varrho]$, $[\sigma, \sigma+\frac{1}{2}\varrho]\times B_\alpha(\frac{1}{2}\varrho, \Phi, H) \subset \Omega$ *with* $\alpha = 1+M$ *and* $F(t, \psi, \dot{\psi})\cap[T_H(\psi(0))]^\eta \neq \emptyset$ *for every* $t \in [\sigma, \sigma+\varrho]$, $\psi \in B_\alpha(\varrho, \Phi, H)$ *and* $\eta \in [0, 1)$. *Then there exists* $a \in (0, \frac{1}{2}\varrho]$ *such that* NFDI(D, F) *has on* $[\sigma-r, \sigma+a]$, *an* η-*approximation* (Ω, H)-*viable trajectory through* (σ, Φ). $\blacksquare$

4.4. Controllability theorems

Given a conformable pair (D, Ω) of nonempty subsets of $\boldsymbol{R}\times C_{0r}\times L_{0r}$ and $\boldsymbol{R}\times \mathrm{AC}_{0r}$, respectively, $C \subset \boldsymbol{R}\times C_{0r}\times \boldsymbol{R}\times C_{0r}$ and a set-valued function $F\colon D \to \mathrm{Conv}(\boldsymbol{R}^n)$ we will investigate sufficient conditions for the existence of $\sigma \in \boldsymbol{R}$, $a > 0$ and an absolutely continuous function $x\colon [\sigma-r, \sigma+a] \to \boldsymbol{R}^n$ such that

$$\begin{aligned} &\dot{x}(t) \in F(t, x_t, \dot{x}_t) \quad \text{for a. e. } t \in [\sigma, \sigma+a],\\ &(t, x_t) \in \Omega \quad \text{for } t \in [\sigma, \sigma+a],\\ &(\sigma, x_\sigma, \sigma+a, x_{\sigma+a}) \in C. \end{aligned} \tag{4.12}$$

If such numbers $\sigma \in \boldsymbol{R}$, $a > 0$ and an absolutely continuous function $x\colon [\sigma-r, \sigma+a] \to \boldsymbol{R}^n$ exist, we will say that NFDI(D, F) is *(Ω, C)-controllable.*

We can also consider the controllability of NFDI(D, F) with a set $E \subset \boldsymbol{R}\times C_{0r}\times \boldsymbol{R}\times \boldsymbol{R}^n$ instead of C given above. In such case NFDI(D, F) is called (Ω, E)-Euclidean controllable instead of (Ω, E)-controllable. Precisely for given sets $\Omega \subset \boldsymbol{R}\times \mathrm{AC}_{0r}$, $E \subset \boldsymbol{R}\times C_{0r}\times \boldsymbol{R}\times \boldsymbol{R}^n$ and $F\colon D \to \mathrm{Conv}(\boldsymbol{R}^n)$ we will say that NFDI(D, F) is *(Ω, E)-Euclidean controllable* if there are numbers $\sigma \in \boldsymbol{R}$, $a > 0$ and an absolutely continuous function $x\colon [\sigma-r, \sigma+a] \to \boldsymbol{R}^n$ such that

$$\begin{aligned} &\dot{x}(t) \in F(t, x_t, \dot{x}_t) \quad \text{for } a. \text{ e. } t \in [\sigma, \sigma+a],\\ &(t, x_t) \in \Omega \quad \text{for } t \in [\sigma, \sigma+a],\\ &\big(\sigma, x_\sigma, \sigma+a, x(\sigma+a)\big) \in E. \end{aligned} \tag{4.13}$$

Similarly as in Section 1.2 we have the following graphic ilustrations (see Figs. 11 and 12) of trajectories of (Ω, C)-controllable and (Ω, E)-Euclidean controllable NFDIs, where for given $\psi \in \mathrm{AC}_{0r}$ and $T \in [\sigma, \sigma+a]$, by $\tilde{\psi}$ we have denoted a function defined on $[T-r, T]$ by setting $\tilde{\psi}(\tau) := \psi(\tau-T)$.

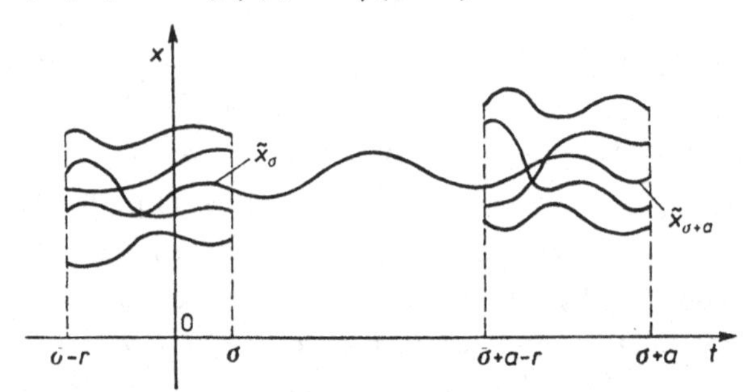

Fig. 11. (Ω, C)-controllable trajectory of NFDIs.

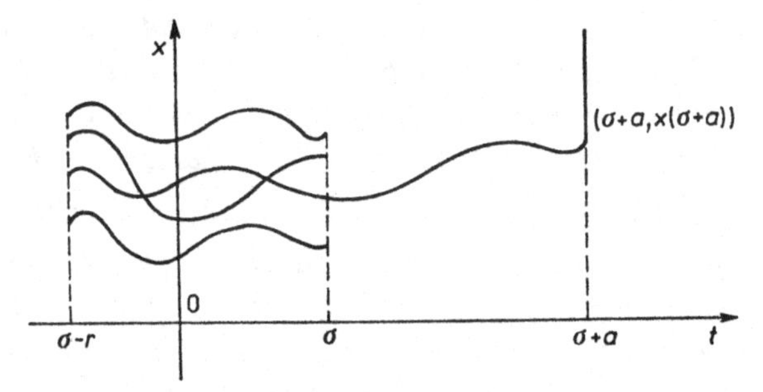

Fig. 12. (Ω, E)-Euclidean controllable trajectory of NFDIs.

Immediately from Theorem 4.8 the following controllability theorem follows.

THEOREM 4.17. *Let* (D, Ω) *be a conformable pair of nonempty subsets of* $\boldsymbol{R}\times C_{0r}\times L_{0r}$ *and* $\boldsymbol{R}\times AC_{0r}$, *respectively, and let* $C\subset \boldsymbol{R}\times C_{0r}\times\boldsymbol{R}\times C_{0r}$. *Suppose* Ω *and* C *are such that there are* $\Omega_0\subset\Omega$, $(\sigma, \Phi)\in\Omega_0$, $\varrho > r$, $l > 0$, $M > 0$ *and a compact set* $H\subset\boldsymbol{R}^n$ *such that* $\Phi(0)\in H$, $[\sigma, \sigma+\varrho]\times\boldsymbol{K}_\alpha(\Phi, H)\subset\Omega_0$ *with* $\alpha = l+M$ *and* $\Omega_0\times\times\Omega_0\subset C$. *If* $F\colon D\to \mathrm{Conv}(\boldsymbol{R}^n)$ *is such that*

(i) F *is u.s.c.w. on* $[\sigma, \sigma+\varrho]\times\boldsymbol{K}_\alpha(\Phi, H)\times\dot{\boldsymbol{K}}_\alpha(\Phi, H)$,

(ii) $F((\sigma, \sigma+\varrho]\times\boldsymbol{K}_\alpha(\Phi, H)\times\dot{\boldsymbol{K}}_\alpha(\Phi, H))\subset MB$,

(iii) $F(t, \psi, \dot{\psi})\cap T_H(\psi(0))\neq\emptyset$ *for every* $(t, \psi)\in[\sigma, \sigma+\varrho]\times\boldsymbol{K}_\alpha(\Phi, H)$ *then* NFDI(D, F) *is* (Ω, C)-*controllable.*

Proof. By virtue of Theorem 4.8 there is an absolutely continuous function $x\colon [\sigma-r, \sigma+\varrho]\to\boldsymbol{R}^n$ such that conditions (4.11) with Ω_0 instead of Ω are satisfied. Hence, in particular it follows that (σ, x_σ), $(\sigma+\varrho, x_{\sigma+\varrho})\in\Omega_0$. Therefore, $(\sigma, x_\sigma, \sigma+\varrho, x_{\sigma+\varrho})\in C$, because $\Omega_0\times\Omega_0\subset C$. Thus x satisfies conditions (4.12). ■

REMARK 4.2. Similar results for (Ω, C)-controllability from Theorems 4.9–4.11 can be obtained. ■

In a similar way we obtain

THEOREM 4.18. *Let* (D, Ω) *be a conformable pair of nonempty subsets of* $\boldsymbol{R}\times C_{0r}\times L_{0r}$ *and* $\boldsymbol{R}\times AC_{0r}$, *and let* $E\subset\boldsymbol{R}\times C_{0r}\times\boldsymbol{R}\times\boldsymbol{R}^n$. *Suppose* Ω *and* E *are such that there are* $\Omega_0\subset\Omega$, $(\sigma, \Phi)\in\Omega_0$, $\varrho > 0$, $l > 0$, $M > 0$ *and a compact set* $H\subset\boldsymbol{R}^n$ *such that* $\Phi(0)\in H$, $[\sigma, \sigma+\varrho]\times\boldsymbol{K}_\alpha(\Phi, H)\subset\Omega_0$ *with* $\alpha = l+M$ *and* $\Omega_0\times\{\sigma+\varrho\}\times H\subset E$.

If $F\colon D\to\mathrm{Conv}(\boldsymbol{R}^n)$ *is such that*

(i) F *is u.s.c.w. on* $(\sigma, \sigma+\varrho]\times\boldsymbol{K}_\alpha(\Phi, H)\times\dot{\boldsymbol{K}}_\alpha(\Phi, H)$,

(ii) $F((\sigma, \sigma+\varrho]\times\boldsymbol{K}_\alpha(\Phi, H)\times\dot{\boldsymbol{K}}_\alpha(\Phi, H_\alpha))\subset MB$,

(iii) $F(t, \psi, \dot{\psi})\cap T_H(\psi(0))$ *for* $(t, \psi)\in[\sigma, \sigma+\varrho]\times\boldsymbol{K}_\alpha(\Phi, H)$,

then NFDI(D, F) *is* (Ω, E)-*Euclidean controllable.*

Proof. Similarly as above, by virtue of Theorem 4.8 there is an absolutely continuous function $x\colon [\sigma-r, \sigma+\varrho]\to\boldsymbol{R}^n$ such that conditions (4.11) with Ω_0 instead of Ω, are satisfied. Hence, in particular it follows that $(\sigma, x_\sigma)\in\Omega_0$ and $x(\sigma+\varrho)\in H$. Therefore, x satisfies conditions (4.13). ■

REMARK 4.3. Similar results for (Ω, E)-Euclidean controllability from Theorems 4.9–4.11 can be obtained. ■

REMARK 4.4. Similarly as above, immediately from Theorems 4.14–4.16 we can obtain some sufficient conditions for the λ-approximation (Ω, C)-controllability or (Ω, E)-Euclidean controllability of NFDI(D, F). It is clear that such sufficient conditions can be obtained by changing in Theorems 4.17 and 4.18 of the tangential condition $F(t, \psi, \dot{\psi})\cap T_H(\Phi(0))\neq\emptyset$ by the λ-approximation tangential condition of the form $F(t, \psi, \dot{\psi})\cap[T_H(\psi(0))]^\eta\neq\emptyset$, where $[T_H(\Phi(0))]^\eta := T_H(\psi(0))+\eta B$. ■

4.5. Compactness of set of attainable trajectories

Let (D, Ω) be a conformable pair of nonempty subsets of $\boldsymbol{R}\times C_{0r}\times L_{0r}$ and $\boldsymbol{R}\times \times \mathrm{AC}_{0r}$, respectively, $C \subset \boldsymbol{R}\times C_{0r}\times \boldsymbol{R}\times C_{0r}$ and $E \subset \boldsymbol{R}\times C_{0r}\times \boldsymbol{R}\times \boldsymbol{R}^n$ be given. Suppose $F\colon D \to \mathrm{Conv}(\boldsymbol{R}^n)$, Ω, C and E are such that $\mathrm{NFDI}(D, F)$ is (Ω, C)-controllable or (Ω, E)-Euclidean controllable. Similarly as in Section 1.2, by $\mathscr{T}(D, F, \Omega, C)$ and $\mathscr{T}^e(D, F, \Omega, E)$ we shall denote the set of all absolutely continuous functions $x\colon (\sigma-r, \sigma+a] \to \boldsymbol{R}^n$ satisfying conditions (4.12) or (4.13), respectively. They will be called (Ω, C)-*attainable* and (Ω, E)-*attainable*, respectively sets of $\mathrm{NFDI}(D, F)$.

We shall consider $\mathscr{T}(D, F, \Omega, C)$ and $\mathscr{T}^e(D, F, \Omega, E)$ as subsets of the metric spaces $\mathscr{X}_r(\Omega)$ and $C_r(\Omega)$ defined in Section I.4. We shall assume here that Ω is bounded in $\boldsymbol{R}\times \mathrm{AC}_{0r}$ and such that $\Pi_{\boldsymbol{R}}(\Omega)$ is a compact interval of the real line, say $\Pi_{\boldsymbol{R}}(\Omega) = [\alpha, \beta]$. In the case of the (Ω, C)-controllability it is necessary also to assume that $\beta-\alpha \geqslant r > 0$.

Recall, given $r > 0$ and $\Omega \subset \boldsymbol{R}\times \mathrm{AC}_{0r}$ by $\mathscr{X}_r(\Omega)$ we denote the metric space $\bigcup_{\alpha\leqslant\sigma<\beta}\ \bigcup_{0<\varrho<\beta-\sigma} \mathrm{AC}([\sigma-r, \sigma+\varrho], \boldsymbol{R}^n)$ with the metric d defined by

$$d[(x_1;\ [\sigma_1-r, \sigma_1+\varrho_1]), (x_2;\ [\sigma_2-r, \sigma_2+\varrho_2])] := \max[|\sigma_1-\sigma_2|, |\varrho_1-\varrho_2|, \|x_1^\Omega-x_2^\Omega\|_{\alpha\beta}],$$

for $x_i \in \mathrm{AC}([\sigma_i-r, \sigma_i+\varrho_i], \boldsymbol{R}^n)$; $i = 1,2$, where $x_i^\Omega(t) = x_i(\sigma-r_i)$ for $t \in [\alpha-r, \sigma_i-r)$, $x_i^\Omega(t) = x_i(t)$ for $t \in (\sigma_i-r, \sigma_i+\varrho_i]$ and $x_i^\Omega(t) = x_i(\sigma_i+\varrho_i)$ for $t \in (\sigma_i+\varrho_i, \beta]$; $i = 1,5$. Here $\|\cdot\|_{\alpha\beta}$ denotes the norm of the Banach space $\mathrm{AC}([\alpha-r, \beta], \boldsymbol{R}^n)$.

Similarly we define the metric ϱ for $C_r(\Omega) := \bigcup_{\alpha\leqslant\sigma\leqslant\beta}\ \bigcup_{0<\varrho<\beta-\sigma} C([\sigma-r, \sigma+\varrho], \boldsymbol{R}^n)$. We have of course $\mathscr{X}_r(\Omega) \subset C_r(\Omega)$.

Let us also recall, we say that $K \subset \mathscr{X}_r(\Omega)$ is ϱ-*closed* (ϱ-*compact*) if it is closed (compact) in the ϱ-metric topology of $C_r(\Omega)$. Similarly, $K \subset \mathscr{X}_r(\Omega)$ is *relativelly d-sequentially weakly compact* if and only if it is bounded in $\mathscr{X}_r(\Omega)$ and for every $\varepsilon > 0$ there is a $\delta > 0$ such that for every $(x;\ (t_1^x-r, t_2^x]) \in K$ and every measurable set $E \subset (t_1^x-r, t_2^x]$ we have $\int_E |x(t)|\,dt \leqslant \varepsilon$ whenever $\mu(E) < \delta$.

We shall present here some sufficient conditions for the ϱ-compactness and the relative d-weak sequential compactness of attainable sets $\mathscr{T}(F, D, \Omega, C)$, $\mathscr{T}^e(F, D, \Omega, E) \subset \mathscr{X}_r(\Omega) \subset C_r(\Omega)$. We begin with the following lemma.

LEMMA 4.19. *Let $\Omega \subset \boldsymbol{R}\times \mathrm{AC}_{0r}$ be closed in $\boldsymbol{R}\times C_{0r}$, relatively weakly compact in $\boldsymbol{R}\times \mathrm{AC}_{0r}$ and such that $\Pi_{\boldsymbol{R}}(\Omega) = [\alpha, \beta]$ with $\beta-r \geqslant r > 0$. Suppose C is a closed subset of $\boldsymbol{R}\times C_{0r}\times \boldsymbol{R}\times C_{0r}$ and let $K(\Omega, C) \subset \mathscr{X}_r(\Omega)$ be defined by* $\mathrm{K}(\Omega, C) := \{(x;\ [\sigma-r, \sigma+\varrho]) \in \mathscr{X}_r(\Omega)\colon \varrho > 0,\ (t, x_t) \in \Omega$ *for* $t \in [\sigma, \sigma+\varrho]$ *and* $(\sigma, x_\sigma, \sigma+\varrho, x_{\sigma+\varrho}) \in C\}$. *The set* $\mathrm{K}(\Omega, C)$ *is ϱ-compact and relatively d-sequentially weakly compact.*

Proof. Suppose $\mathrm{K}(\Omega, C) \neq \emptyset$. Since Ω is bounded, then there is an $M > 0$ such that for every $(x;\ [\sigma-r, \sigma+\varrho]) \in \mathrm{K}(\Omega, C)$ and $t \in [\sigma, \sigma+\varrho]$ we have $|x(t-r)| + \int_{-r}^{0} |\dot{x}_t(s)|\,ds \leqslant M$. Hence, in particular it follows that $|x(\sigma-r)| \leqslant M$ and

$\int_{\sigma-r}^{\sigma+\varrho} |\dot{x}(\tau)|\,d\tau = \sup_{\sigma \leqslant t \leqslant \sigma+\varrho} \int_{t-r}^{t} |\dot{x}(\tau)|\,d\tau = \sup_{\sigma \leqslant t \leqslant \sigma+\varrho} \int_{-r}^{0} |\dot{x}_t(s)|\,ds \leqslant M$. Therefore, for every $(x;\ [\sigma-r, \sigma+\varrho]) \in \mathrm{K}(\Omega, C)$ we have $\|x\| := |x(\sigma-r)| + \int_{\sigma-r}^{\sigma+\varrho} |\dot{x}(\tau)|\,d\tau \leqslant 2M$. Then $\mathrm{K}(\Omega, C)$ is bounded in $\mathscr{X}_r(\Omega)$.

We shall show now that for every $\varepsilon > 0$ there is a $\delta > 0$ such that for every $(x;\ [\sigma-r, \sigma+\varrho]) \in \mathrm{K}(\Omega, C)$ and every measurable set $E \subset [\sigma-r, \sigma+\varrho]$ with $\mu(E) \leqslant \delta$ we have $\int_E |\dot{x}(\tau)|\,d\tau \leqslant \varepsilon$. Hence, by Theorem I.4.9, it will be follow the relative d-sequential weak compactness of $\mathrm{K}(\Omega, C)$ in $\mathscr{X}_r(\Omega)$.

Let $\gamma = (\beta-\gamma)/r$. By the definition for every $(x;\ [\sigma-r, \sigma+\varrho]) \in \mathrm{K}(\Omega, C)$ we have $\varrho \geqslant r$. Thus, there is a positive integer N_x such that $N_x - 1 < \varrho/r \leqslant N_x < \gamma+1$ and therefore $\sigma+(N_x-1)r < \sigma+\varrho \leqslant \sigma+N_x r$. Then $[\sigma-r, \sigma+\varrho] = \bigcup_{i=0}^{N_x-1} [\sigma+(i-1)r, \sigma+ir] \cup [\sigma+(N_x-1)r, \sigma+\varrho]$. By the relative sequential weak compactness of Ω there is a modulus of continuity, say ω such that for every $(\tau, \psi) \in \Omega$ and every measurable set $U \subset [-r, 0]$ we have $\int_E |\dot{\psi}(s)|\,ds \leqslant \omega(\mu(U))$. Let $\varepsilon > 0$ be an arbitrary number and select $\delta > 0$ such that $\omega(\delta) \leqslant \varepsilon/(\gamma+2)$. Now, for every $(x;\ [\sigma-r, \sigma+\varrho]) \in \mathrm{K}(\Omega, C)$ and every measurable set $E_i \subset [\sigma+(i-1)r,\ \sigma+ir]$ with $\mu(E_i) \leqslant \delta$ and $i = 0, 1, \ldots, N_x-1$ we have

$$\int_{E_i} |\dot{x}(\tau)|\,d\tau = \int_{E_i-(\sigma+ir)} |\dot{x}_{\sigma+ir}(s)|\,ds \leqslant \frac{\varepsilon}{\gamma+2},$$

because $(\sigma+ir, x_{\sigma+ir}) \in \Omega$, $E_i-(\sigma+ir) \subset [-r, 0]$ and $\mu(E_i-(\sigma+ir)) = \mu(E_i) \leqslant \delta$ for every $i = 0, 1, \ldots, N_x-1$.

Let us observe that for every set $E_\varrho \subset [\sigma+(N_x-1)r, \sigma+\varrho]$ we have $E_\varrho-(\sigma+\varrho) \subset [(N_x-1)r-\varrho, 0]$. Since $(N_x-1)r < \varrho \leqslant N_x r$ then $(N_x-1)r-\varrho < 0$ and $N_x r - \varrho \geqslant 0$. Thus $-r \leqslant -r+(N_x r-\varrho) = (N_x-1)r-\varrho$ and therefore $E_\varrho-(\sigma+\varrho) \subset [-r, 0]$. Now, similarly as above, for every measurable set $E_\varrho \subset [(N_x-1)r, \sigma+\varrho]$ with $\mu(E_\varrho) \leqslant \delta$ we have

$$\int_{E_\varrho} |\dot{x}(\tau)|\,d\tau = \int_{E_\varrho-(\sigma+\varrho)} |\dot{x}_{\sigma+\varrho}(s)|\,ds \leqslant \frac{\varepsilon}{\gamma+2}.$$

Hence, for every measurable set $E \subset [\sigma-r, \sigma+\varrho]$ with $\mu(E) \leqslant \delta$ we obtain

$$\int_E |\dot{x}(\tau)|\,d\tau = \sum_{i=0}^{N_x-1} \int_{E_i} |\dot{x}(\tau)|\,d\tau + \int_{E_\varrho} |\dot{x}(\tau)|\,d\tau,$$

where $E_i = E \cap [\sigma+(i-1)r, \sigma+ir]$ for $i = 0, 1, \ldots, N_x-1$ and $E_\varrho = E \cap [\sigma+(N_x-1)r, \sigma+\varrho]$. We have of course $\mu(E_i) \leqslant \delta$ for $i = 0, 1, \ldots, N_x-1$ and $\mu(E_\varrho) \leqslant \delta$. Then,

$$\int_E |\dot{x}(\tau)|\,d\tau \leqslant \sum_{i=0}^{N_x-1} \frac{\varepsilon}{\gamma+2} + \frac{\varepsilon}{\gamma+2} = \frac{N_x+1}{\gamma+2}\varepsilon \leqslant \varepsilon$$

and therefore, $K(\Omega, C)$ is relatively d-sequentially weakly compact in $\mathscr{X}_r(\Omega)$. Hence in particular it follows that $K(\Omega, C)$ is relatively ϱ-weakly compact in $C_r(\Omega)$. We will show now that it is ϱ-closed.

Indeed, suppose $\{(x^n; [\sigma_n - r, \sigma_n + \varrho_n])\}$ is a sequence of $K(\Omega, C)$ ϱ-converging to any $(x; [\sigma - r, \sigma + \varrho]) \in C_r(\Omega)$. Since, for every $t \in (\sigma, \sigma+\varrho)$ we have $t \in [\sigma_n, \sigma_n + \varrho_n]$ for n sufficiently large then we also have $|x_t^n - x_t|_0 \to 0$ as $n \to \infty$ for every $t \in (\sigma, \sigma+\varrho)$, because $|x_t^n - x_t|_0 = \sup_{t-r \leqslant \tau \leqslant t} |x^n(\tau) - x(\tau)| \leqslant \varrho(x^n, x)$ for n sufficiently large. Hence it follows that $(t, x_t) \in \Omega$ for every $t \in (\sigma, \sigma+\varrho)$, because $(t, x_t^n) \in \Omega$ for $t \in [\sigma_n, \sigma_n + \varrho_n]$ and Ω is closed in $\boldsymbol{R} \times C_{0r}$. Hence, by the continuity of the mapping $[\sigma, \sigma+\varrho] \ni t \to x_t \in C_{0r}$ (see Lemma III.5.1) it follows that $(\sigma, x_\sigma), (\sigma+\varrho, x_{\sigma+\varrho}) \in \Omega$. Therefore, we have $(t, x_t) \in \Omega$ for every $t \in [\sigma, \sigma+\varrho]$.

By the compactness of all extensions x^Ω of $(x; [\sigma-r, \sigma+\varrho]) \in K(\Omega, C)$ on $[\alpha - r, \beta]$ in $C([\alpha-r, \beta], \boldsymbol{R}^n)$ there exists a modulus of continuity, say ω_K such that $|x_{\sigma_k}^k - x_t^k|_0 \leqslant \omega_K(|\sigma_k - t|)$ and $|x_{\sigma_k+\varrho_k}^k - x_t^k|_0 \leqslant \omega_k(|\sigma_k + \varrho_k - t|)$ for $t \in (\sigma, \sigma+\varrho)$ and k sufficiently large. Since, $|x_{\sigma_k}^k - x_\sigma|_0 \leqslant |x_{\sigma_k}^k - x_t^k|_0 + |x_t^k|_0 + |x_t - x_\sigma|_0$ and $|x_{\sigma_k+\varrho_k}^k - x_{\sigma+\varrho}|_0 \leqslant |x_{\sigma_k+\varrho_k}^k + - x_t^k|_0 + |x_t^k - x_t|_0 + |x_t - x_{\sigma+\varrho}|_0$ for $t \in (\sigma, \sigma+\varrho)$ and $k = 1, 2, \ldots$ then by the results presented above we obtain $|x_{\sigma_k}^k - x_\sigma|_0 \to 0$ and $|x_{\sigma_k+\sigma_k}^k - x_{\sigma+\varrho}|_0 \to 0$ as $k \to \infty$. But $(\sigma_k, x_{\sigma_k}^k, \sigma_k + \varrho_k, x_{\sigma+\varrho_k}^k) \in C$ for $k = 1, 2, \ldots$ Then by the closedness of C it follows $(\sigma, x_\sigma, \sigma+\varrho, x_{\sigma+\varrho}) \in C$. It is clear that $\varrho \geqslant r$ and therefore $x \in K(\Omega, C)$. ■

In a similar way we obtain

LEMMA 4.20. *Let $\Omega \subset \boldsymbol{R} \times AC_{0r}$ be closed in $\boldsymbol{R} \times C_{0r}$ and relatively weakly compact in $\boldsymbol{R} \times AC_{0r}$ and such that $\Pi_R(\Omega) = [\alpha, \beta]$. Suppose E is a closed subset of $\boldsymbol{R} \times C_{0r} \times \boldsymbol{R} \times \boldsymbol{R}^n$. Then the set $K^e(\Omega, E) \subset \mathscr{X}_r(\Omega)$ defined by*

$$K^e(\Omega, E) := \{(x; [\sigma-r, \sigma+\varrho]) \in \mathscr{X}_r(\Omega):\ (t, x_t) \in \Omega \text{ for } t \in [\sigma, \sigma+\varrho] \text{ and } (\sigma, x_\sigma, \sigma+\varrho, x(\sigma+\varrho)) \in E\} \tag{4.12}$$

is ϱ-compact and relatively d-sequentially weakly compact.

Proof. Similarly as in the proof of Lemma 4.19 we can see that $K^e(\Omega, E)$ is bounded in $\mathscr{X}_r(\Omega)$ and that for every $\varepsilon > 0$ there is a $\delta > 0$ such that for every $(x; [\sigma - r, \sigma+\varrho] \in K^e(\Omega, E)$ with $\varrho > r > 0$ and every measurable set $E \subset [\sigma - r, \sigma+\varrho]$ with $\mu(E) \leqslant \delta$ we have $\int_E |\dot{x}(\tau)|\, d\tau \leqslant \varepsilon$. If $\varrho \leqslant r$, then for every measurable set $E \subset [\sigma - r, \sigma + \varrho]$ we get

$$\int_E |\dot{x}(\tau)|\, d\tau = \int_{E \cap [\sigma-r, \sigma]} |\dot{x}(\tau)|\, d\tau + \int_{E \cap [\sigma, \sigma+\varrho]} |\dot{x}(\tau)|\, d\tau = \int_{E_\sigma - \sigma} |\dot{x}_\sigma(s)|\, ds + \int_{E_\varrho - (\sigma+\varrho)} |\dot{x}_{\sigma+\varrho}(s)|\, ds,$$

where $E_\sigma = E \cap [\sigma - r, \sigma]$ and $E_\varrho = E \cap [\sigma, \sigma+\varrho]$. It is easy to see that $E_\sigma - \sigma \subset [-r, 0]$ and $E_\varrho - (\sigma+\varrho) \subset [-r, 0]$. Hence, similarly as in the proof of Lemma 4.19

we obtain that $\int_E |x(\tau)|\,d\tau \leqslant \varepsilon$ for every measurable set $E \subset [\sigma-r, \sigma+\varrho]$ with $\mu(E) \leqslant \delta$. Then $\mathrm{K}^e(\Omega, E)$ is relatively d-sequentially weakly compact in $\mathscr{X}_r(\Omega)$ and therefore also relatively compact in $C_r(\Omega)$. Similarly as in the proof of Lemma 4.19 we can verify that $\mathrm{K}^e(\Omega, E)$ is also ϱ-closed. ■

Now we can prove the following theorem.

THEOREM 4.21. *Let (D, Ω) be a conformable pair of nonempty subsets of $\boldsymbol{R}\times C_{0r}\times L_{0r}$ and $\boldsymbol{R}\times \mathrm{AC}_{0r}$, respectively, $C \subset \boldsymbol{R}\times C_{0r}\times \boldsymbol{R}\times C_{0r}$ and suppose $F\colon D \to \mathrm{Conv}(\boldsymbol{R}^n)$ is u.s.c.w. on D. Assume Ω is closed in $\boldsymbol{R}\times C_{0r}$, relatively weakly compact in $\boldsymbol{R}\times \mathrm{AC}_{0r}$ and such that $\Pi_R(\Omega) = [\alpha, \beta]$ with $\beta-\alpha \geqslant r > 0$. Let C be closed in $\boldsymbol{R}\times C_{0r}\times \boldsymbol{R}\times C_{0r}$. If furthermore F, Ω and C are such that there are $\Omega_0 \subset \Omega$, $(\sigma, \Phi) \in \Omega_0$, $\varrho \geqslant r$, $l, M > 0$ and a compact set $H \subset \boldsymbol{R}^n$ such that $\Phi(0) \in H$, $[\sigma, \sigma+\varrho]\times \boldsymbol{K}_\alpha(\Phi, H) \subset \Omega_0$ with $\alpha = l+M$ and $\Omega_0\times\Omega_0 \subset C$, $F(D) \subset MB$ and $F(t, \psi, \dot{\psi})\cap T_H(\psi(0)) \neq \emptyset$ for $t \in [\sigma, \sigma+\varrho]$ and $\psi \in \boldsymbol{K}_\alpha(\Phi, H)$ then $\mathscr{I}(D, F, \Omega, C)$ is a nonempty relatively d-sequentially weakly compact subset of $\mathscr{X}_r(\Omega)$ and ϱ-compact in $C_r(\Omega)$.*

Proof. Nonemptness of $\mathscr{T}(D, F, \Omega, C)$ follows immediately from Theorem 4.17. Furthermore, we have $\mathscr{T}(D, F, \Omega, C) \subset \mathrm{K}(\Omega, C)$. Therefore, by Lemma 4.19, $\mathscr{T}(D, F, \Omega, C)$ is relatively d-sequentially weakly compact in $\mathscr{X}_r(\Omega)$ and relatively ϱ-compact in $C_r(\Omega)$. Then it remains only to verify that $\mathscr{T}(D, F, \Omega, C)$ is ϱ-closed in $C_r(\Omega)$.

Suppose $\{(x^n\colon [\sigma_n-r, \sigma_n+\varrho_n]\}$ is a sequence of $\mathscr{T}(D, F, \Omega, C)$ ϱ-converging to $(x;\ [\sigma-r, \sigma+\varrho]) \in \mathrm{K}(\Omega, C)$. By the compactness in $C([\alpha-r, \beta], \boldsymbol{R}^n)$ of the set $\{(x^n)^\Omega;\ n = 1, 2, \ldots\}$ of all extensions of x^n on $[\alpha-r, \beta]$ and the uniform integrability of $\{(\dot{x}^m)\ \}$ in $L([\alpha-r, \beta], \boldsymbol{R}^n)$ there exists a subsequence, say $\{(x^k;\ [\sigma_k-r, \sigma_k+\varrho_k])\}$ of $\{(x^n;\ [\sigma_n-r, \sigma_n+\varrho_n])\}$ such that $|(x^k)^\Omega - x^\Omega|_{\alpha\beta} \to 0$ and $(\dot{x}^k)^\Omega \rightharpoonup x^\Omega$ as $k \to \infty$. Since $\sigma_k \to \sigma$ and $\varrho_k \to \varrho$ then for every $t \in (\sigma, \sigma+\varrho)$ and sufficiently large k we also have $t \in [\sigma_k, \sigma_k+\varrho_k]$. Therefore, $|x^k_t - x_t|_0 \to 0$ and $\dot{x}^k_t \rightharpoonup \dot{x}_t$ for every $t \in (\sigma, \sigma+\varrho)$ as $k \to \infty$. Hence, in particular it follows $\bar{h}(F(t, x^k_t, \dot{x}^k_t), F(t, x_t, \dot{x}_t)) \to 0$ for every $t \in (\sigma, \sigma+\varrho)$ as $k \to \infty$. Therefore, by Vitali's theorem, $\bar{h}(\int_E F(t, x^t_k, \dot{x}^t_k)\,dt, \int_E F(t, x_t, \dot{x}_t)\,dt) \to 0$ as $k \to \infty$ for every measurable set $E \subset (\sigma, \sigma+\varrho)$. But, for every measurable set $E \subset (\sigma, \sigma+\varrho)$ and k sufficiently large we have

$$\operatorname{dist}\Big(\int_E \dot{x}(t)\,dt, \int_E F(t, x^k_t, \dot{x}^k_t)\,dt\Big) \leqslant \Big|\int_E |\dot{x}(t)\,dt - \int_E \dot{x}^k(t)\,dt\Big| +$$

$$+\operatorname{dist}\Big(\int_E \dot{x}^k(t)\,dt, \int_E F(t, x^k_t, \dot{x}^k_t)\,dt\Big) + \bar{h}\Big(\int_E F(t, x^k_t, \dot{x}^k_t)\,dt, \int_E F(t, x_t, \dot{x}_t)\,dt\Big).$$

Thus, $\dot{x}(t) \in F(t, x_t, \dot{x}_t)$ for a.e. $t \in [\sigma, \sigma+\varrho]$ and therefore $(x;\ [\sigma-r, \sigma+\varrho]) \in \mathscr{T}(D, F, \Omega, C)$. ■

In a similar way we obtain

THEOREM 4.22. *Let (D, Ω) be a conformable pair of nonempty subsets of $\boldsymbol{R}\times C_{0r}\times L_{0r}$ and $\boldsymbol{R}\times \mathrm{AC}_{0r}$, respectively, $E\subset \boldsymbol{R}\times C_{0r}\times \boldsymbol{R}\times \boldsymbol{R}^n$ and suppose $F\colon D\to \mathrm{Conv}(\boldsymbol{R}^n)$ is u.s.c.w. on D. Assume, Ω is closed in $\boldsymbol{R}\times C_{0r}$ and relatively weakly compact in $\boldsymbol{R}\times \mathrm{AC}_{0r}$ and such that $\Pi_{\boldsymbol{R}}(\Omega) = [\alpha, \beta]$. Let E be closed in $\boldsymbol{R}\times C_{0r}\times \boldsymbol{R}\times \boldsymbol{R}^n$. If furthermore, F, Ω and E are such that there are $\Omega_0\subset\Omega$, $(\sigma, \Phi)\in\Omega_0$, $l, M>0$, $\varrho>0$ and a compact set $H\subset \boldsymbol{R}^n$ such that $\Phi(0)\in H$, $[\sigma, \sigma+\frac{1}{2}\varrho]\times B_\alpha(\frac{1}{2}\varrho, \Phi, H)\subset\Omega_0$ with $\alpha = l+M$, $\Omega_0\times[\sigma,\sigma+\frac{1}{2}\varrho]\times H\subset E$, $F(t,\psi,\dot\psi)\subset MB$ and $F(t,\psi,\dot\psi)\cap T_H(\psi(0))\neq\emptyset$ for $t\in[\sigma,\sigma+\varrho]$ and $\psi\in B_\alpha(\varrho,\Phi,H)$ then $\mathscr{T}^e(D,F,\Omega,E)$ is a nonempty relatively d-sequentially weakly compact subset of $\mathscr{X}_r(\Omega)$ and ϱ-compact in $C_r(\Omega)$.* ■

REMARK 4.5. Given $\eta\in[0,1)$ we can obtain similarly as above sufficient conditions for d-sequential weak compactness and ϱ-compactness of the sets $\mathscr{T}_\eta(D,F,\Omega,C)$ and $\mathscr{T}^e_\eta(D,F,\Omega,E)$ of all (Ω, C)-attainable and (Ω, E)-attainable, respectively η-approximation trajectories of NFDI(D,F). It is enough only to change in Theorems 4.21 and 4.22, the tangential condition $F(t,\psi,\dot\psi)\cap T_H(\psi(0))\neq\emptyset$ by its η-approximation form, i.e., by $F(t,\psi,\dot\psi)\cap[T_H(\psi(0))]^\eta\neq\emptyset$, where $[T_H(\psi(0))]^\eta = T_H(\psi(0)) +\eta B$. ■

5. Relaxation theorem for attainable sets

Theorems 4.21 and 4.22 are not in general true for set-valued mappings F having compact but not convex values. It follows from the following example.

EXAMPLE 5.1. Let $n=1$, $\Omega=[0,1]\times[-1,1]$, $F(t,x)=\{z;\ -1\leqslant z\leqslant 1\}$ and $(x^k;\ [0,1])$, $k=1,2,\dots$ be defined by $x^k(t)=t-ik^{-1}$ if $ik^{-1}\leqslant t\leqslant ik^{-1}+(2k)^{-1}$, $x^k(t)=(i+1)k^{-1}-t$ if $ik^{-1}+(2k)^{-1}\leqslant t\leqslant(i+1)k^{-1}$, $i=0,1,\dots,k-1$. Then $x_k(t)\to x_0(t)=0$ uniformly in $[0,1]$. On the other hand, for the two sets of intervals above, then $\dot x_0(t)=0$, $\dot x^k(t)\in F(t,x^k(t))$ and $\dot x_0(t)\in F(t,x_0(t))$ for a.e. $t\in[0,1]$. Here $F(t,x)$ is a compact and convex set. If we had taken $F(t,x)=\{z\colon z=-1$ and $z=1\}$ then obviously $\dot x^k(t)\in F(t,x^k(t))$ for a.e. $t\in[0,1]$ and $\dot x_0(t)\notin F(t,x_0(t))$. Here $F(t,x)$ is compact but not convex. ■

It is natural to consider in the non-convex case the attainable sets of the relaxed control systems, i.e., sets $\mathscr{T}(D,\mathrm{co}F,\Omega,C)$ and $\mathscr{T}^e(D,\mathrm{co}F,\Omega,E)$ instead of $\mathscr{T}(D,F,\Omega,C)$ and $\mathscr{T}^e(D,F,\Omega,E)$, respectively. From the practical point of view it is very important to know when the sets $\mathscr{T}(D,F,\Omega,C)$ and $\mathscr{T}^e(D,F,\Omega,E)$ with F having compact but not necessarily convex values, are nonempty and dense in $\mathscr{T}(D,\mathrm{co}F,\Omega,C)$ and $\mathscr{T}^e(D,\mathrm{co}F,\Omega,E)$, respectively. We give here some results concerning this problem.

5.1. Controllability theorems for NFDIs *with non-convex-valued functions*

We shall consider here NFDI(D,F) with F having compact but not necessarily convex values. Similarly as in the convex case, controllability problems can be reduced to the viability ones.

Similarly as in Section III.6.3 we obtain the following result.

LEMMA 5.1. *Let* $(\sigma, \Phi) \in \boldsymbol{R} \times \mathrm{AC}_{0r}$, $\varrho > 0$, $l > 0$, $M > 0$ *and a compact set* $H \subset \boldsymbol{R}^n$ *be given such that* $\Phi(0) \in H$. *Suppose* $G\colon [\sigma, \sigma+\varrho] \times \boldsymbol{K}_\alpha(\Phi, H) \to \mathrm{Comp}(\boldsymbol{R}^n)$ *with* $\alpha = l+M$ *is such that* $G(\cdot, u)$ *is measurable on* $I := [\sigma, \sigma+\varrho]$ *and* $G(t, \cdot)$ *is l.s.c. on* $\boldsymbol{K}_\alpha(\Phi, H)$. *If furthermore* $G(I \times \boldsymbol{K}_\alpha(\Phi, H)) \subset MB$ *and* $G(t, \psi) \subset T_H(\psi(0))$ *for* $t \in I$ *and* $\psi \in \boldsymbol{K}_\alpha(\Phi, H)$, *then there is an absolutely continuous function* $x\colon [\sigma-r, \sigma+$ $+\varrho] \to \boldsymbol{R}^n$ *such that*

$$\dot{x}(t) \in G(t, x_t) \quad \text{for a. e. } t \in I,$$
$$x_\sigma = \Phi,$$
$$x(t) \in H \quad \text{for } t \in I.$$

Proof. By virtue of Theorem III.2.17 there is a continuous function $g\colon \boldsymbol{K}_\alpha(\Phi, H) \to L(I, \boldsymbol{R}^n)$ such that $g(u)(t) \in G(t, u)$ for $u \in \boldsymbol{K}_\alpha(\Phi, H)$ and a.e. $t \in I$. Furthermore $g(\psi)(t) \in T_H(\psi(0))$ for $\psi \in \boldsymbol{K}_\alpha(\Phi, H)$ and a.e. $t \in I$. It is also easy to see that for every set $K \subset C([\sigma-r, \sigma+\varrho], \boldsymbol{R}^n)$ such that $(t, x_t) \in \boldsymbol{K}_\alpha(\Phi, H)$ for each $x \in K$ and $t \in I$ a function $f\colon I \times K \to \boldsymbol{R}^n$ defined by $f(t, x) = g(x_t)(t)$ for $t \in I$ and $x \in K$ is a Carathéodory selector having Volterra's property with respect to its last variable of $\tilde{G}$ defined by $\tilde{G}(t, x) := G(t, x_t)$ for $(t, x) \in I \times K$. Now, from Theorem III.6.5 the existence of a viable solution of $\dot{x}(t) \in G(t, x_t)$ follows. ∎

LEMMA 5.2. *Let* $(\sigma, \Phi) \in \boldsymbol{R} \times \mathrm{AC}_{0r}$, $M > 0$, $l > 0$, $\varrho > 0$ *and a compact set* $H \subset \boldsymbol{R}^n$ *be such that* $\Phi(0) \in H$. *Suppose* $G\colon [\sigma, \sigma+\varrho] \times B_\alpha(\varrho, \Phi, H) \to \mathrm{Comp}(\boldsymbol{R}^n)$ *with* $\alpha = l+M$ *is such that* $G(\cdot, u)$ *is measurable on* $I = [\sigma, \sigma+\varrho]$ *and* $G(t, \cdot)$ *is l.s.c. on* $B_\alpha(\varrho, \Phi, H)$. *If furthermore* $G([\sigma, \sigma+\varrho] \times B_\alpha(\varrho, \Phi, H)) \subset MB$ *and* $G(t, \psi)$ $\subset T_H(\psi(0))$ *for* $t \in [\sigma, \sigma+\varrho]$ *and* $\psi \in B_\alpha(\varrho, \Phi, H)$, *then there exist* $a \in (0, \frac{1}{2}\varrho)$ *and an absolutely continuous function* $x\colon [\sigma-r, \sigma+a] \to \boldsymbol{R}^n$ *such that*

$$\dot{x}(t) \in G(t, x_t) \quad \text{for a. e. } t \in [\sigma, \sigma+a],$$
$$x_\sigma = \Phi,$$
$$x(t) \in H \quad \text{for } t \in [\sigma, \sigma+a]. \ \blacksquare$$

Similarly we also obtain the following lemma.

LEMMA 5.3. *Let* $(\sigma, \Phi) \in \boldsymbol{R} \times \mathrm{AC}_{0r}$, $\varrho > 0$, $l > 0$, $M > 0$ *and a compact set* $H \subset \boldsymbol{R}^n$ *be such that* $\Phi(0) \in H$. *Suppose* $G\colon [\sigma, \sigma+\varrho] \times \boldsymbol{K}_\beta(\Phi, H) \to \mathrm{Comp}(\boldsymbol{R}^n)$ *with* $\beta = l+$ $+2M$ *is such that* $G(\cdot, u)$ *is measurable on* I *and* $G(t, \cdot)$ *is l.s.c. on* $\boldsymbol{K}_\beta(\Phi, H)$. *If furthermore* $G([\sigma, \sigma+\varrho] \times \boldsymbol{K}_\beta(\Phi, H)) \subset MB$ *and* $G(t, \psi) \subset T_H(\psi(0))$ *for* $t \in [\sigma, \sigma+$ $+\varrho]$ *and* $\psi \in \boldsymbol{K}_\beta(\Phi, H)$ *then for every* $(\sigma_1, \Phi_1) \in [\sigma, \sigma+\varrho) \times \boldsymbol{K}_\alpha(\Phi, H)$ *with* $\alpha = l+M$ *there is* $x \in \mathrm{AC}([\sigma_1-r, \sigma+\varrho], \boldsymbol{R}^n)$ *such that*

$$\begin{aligned} &\dot{x}(t) \in G(t, x_t) \quad \text{for a. e. } t \in [\sigma_1, \sigma+\varrho],\\ &x_{\sigma_1} = \Phi_1, \\ &x(t) \in H \quad \text{for } t \in [\sigma_1, \sigma+\varrho]. \end{aligned} \tag{5.1}$$

Proof. Let $(\sigma_1, \Phi_1) \in [\sigma, \sigma+\varrho] \times \boldsymbol{K}_\alpha(\Phi, H)$ be given. We can repeat the first part of the proof of Lemma 5.1 with $\boldsymbol{K}_\beta(\Phi, H)$ instead of $\boldsymbol{K}_\alpha(\Phi, H)$ (see also the proof of Theorem III.6.5).

Then, for $(\sigma_1, \Phi_1) \in I \times \boldsymbol{K}_\alpha(\Phi, H) \subset I \times \boldsymbol{K}_\beta(\Phi, H)$, $k \geqslant N_l$ we select $h_0 \in (h_0^k, 1/k]$ and $u_0 \in G(I \times \boldsymbol{K}_\beta(\Phi, H)) + lB$ such that $\Phi_1(0) + h_0 u_0 \in H$ and $|u_0 - \frac{1}{\eta}\int_{\sigma_1}^{\sigma_1+\eta} g(\Phi_1)(s)ds|$ $\leqslant 1/k + \frac{1}{3}\varepsilon$ for $\eta \in (0, \eta_0)$. Then we define z^1: $[\sigma_1 - r, t_1] \to \boldsymbol{R}^n$ by

$$z^1(\tau) = \begin{cases} \Phi_1(\tau - \sigma_1) & \text{for } \tau \in [\sigma_1 - r, \sigma_1), \\ \Phi_1(0) + l_0(\tau) & \tau \in [\sigma_1, t_1], \end{cases}$$

with $l_0(t) := (t - \sigma_1)u_0$ for $t \in [\sigma_1, t_1]$ and $t_1 = \sigma_1 + h_0$. Hence it follows that $z^1_{t_1}(0) = \Phi_1(0) + h_0 u_0 \in H$. Furthermore, for every measurable set $E \subset [-r, 0]$ we obtain

$$\int_E |\dot{z}^1_{t_1}(s)|\, ds \leqslant \int_{E \cap (-r, \sigma_1 + t_1) + (t_1 - \sigma_1)} |\dot{\Phi}_1(\tau)|\, d\tau + \mu(E)(l + M)$$
$$\leqslant \omega_\alpha(\Phi, \mu(E)) + \mu(E)(l + M) = \omega_\beta(\Phi, \mu(E)).$$

Then, $(t_1, z^1_{t_1}) \in \boldsymbol{K}_\beta(\Phi, H)$ and therefore we can define a function z^2: $[\sigma_1 - r, t_1] \to \boldsymbol{R}^n$. Continuing this procedure finally we find a function $x \in \mathrm{AC}([\sigma_1 - r, \sigma + \varrho], \boldsymbol{R}^n)$ such that conditions (5.1) are satisfied. ■

Let $\Omega \subset \boldsymbol{R} \times \mathrm{AC}_{0r}$ be nonempty and such that $\Pi_{\boldsymbol{R}}(\Omega)$ is connected. The set Ω is said to be *decomposable* if for every interval $[\sigma, \sigma+\varrho] \subset \Pi_{\boldsymbol{R}}(\Omega)$, every $\sigma_1 \in [\sigma, \sigma + \varrho)$, $\Phi \in \mathrm{AC}_{0r}$ and ξ: $[\sigma_1 - r, \sigma+\varrho] \to \boldsymbol{R}^n$ such that $(\sigma_1, \Phi) \in \Omega$ and $(t, \xi_t) \in \Omega$ for $t \in [\sigma_1, \sigma+\varrho]$ we also have $(t, (\Phi \oplus \xi^1)_t) \in \Omega$ for $t \in [\sigma_1, \sigma+\varrho]$, where ξ^1 denotes the restriction of ξ to the interval $[\sigma_1, \sigma+\varrho]$, $(\Phi \oplus \xi^1)(t) = \Phi(t - \sigma_1)$ for $t \in [G_1 - r_1, G_1]$ and $(\Phi \oplus \xi^2)(t) = \Phi(0) - \xi(\sigma) + \xi^1(t)$ for $t \in [\sigma_1, \sigma+\varrho]$.

Let $\mathscr{K}(\Omega) = \{x \in \mathrm{AC}([\sigma_1 - r, \sigma+\varrho], \boldsymbol{R}^n)$: $(t, x_t) \in \Omega$ for $t \in [\sigma_1, \sigma+\varrho]\}$ and $K(\Omega) = \{u_x$: $x \in \mathscr{K}(\Omega)\}$, where for a given $x \in \mathscr{K}(\Omega)$, u_x denote vector-valued function from $[\sigma_1, \sigma+\varrho]$ into C_{0r} defined by $u_x(t) = x_t$ for $t \in [\sigma_1, \sigma+\varrho]$. For a given set $H \subset \boldsymbol{R}^n$ and $\beta > 0$ by $P(H)$ we shall denote a set of all constant functions from $[\sigma_1, \sigma+\varrho]$ into $\boldsymbol{K}_\beta(\Phi, H)$, i.e. $u \in P(H)$ if and only if there is $\psi \in \boldsymbol{K}_\beta(\Phi, H)$ such that $u(t) = \psi$ for $t \in [\sigma_1, \sigma+\varrho]$.

LEMMA 5.4. *Let $X = K(\Omega) \cup P(H)$, where $K(\Omega)$ and $P(H)$ are such as above. If $H \in \mathrm{Comp}(\boldsymbol{R}^n)$, $\Omega \subset \boldsymbol{R} \times \mathrm{AC}_{0r}$ is closed in $\boldsymbol{R} \times C_{0r}$ and relatively weakly compact in $\boldsymbol{R} \times \mathrm{AC}_{0r}$ and $(\sigma, \Phi) \in \Omega$ is such that $\Phi(0) \in H$ then X is a nonempty compact subset of the Banach space $C([\sigma_1, \sigma+\varrho], C_{0r})$.*

Proof. Let us observe that it is enough only to verify that $K(\Omega)$ is compact in $C([\sigma_1, \sigma+\varrho], C_{0r})$ because by Lemma 4.2 $\boldsymbol{K}_\beta(\Phi, H)$ is compact in C_{0r}. It is clear that $\mathscr{K}(\Omega) \neq \emptyset$. Similarly as in the proof of Lemma 4.19 we can also verify that $\mathscr{K}(\Omega)$ is compact in $C([\sigma_1 - r, \sigma+\varrho], \boldsymbol{R}^n)$.

By the compactness of $\mathscr{K}(\Omega)$ there are a number $\alpha > 0$ and a modulus of continuity, say w such that $\|x\|_\varrho \leqslant \alpha$ and $|x(\tau_1) - x(\tau_2)| \leqslant w(|\tau_1 - \tau_2|)$ for every $x \in \mathscr{K}(\Omega)$

and $\tau_1, \tau_2 \in [\sigma_1, \sigma+\varrho]$, where $||x||_\varrho = \sup_{\sigma_1 - r \leqslant \tau \leqslant \sigma+\varrho} |x(\tau)|$. Therefore, for every $u_x \in K(\Omega)$ and $t_1, t_2 \in [\sigma_1, \sigma+\varrho]$ we have $||u_x|| = \sup_{\sigma_1 \leqslant t \leqslant \sigma+\varrho} |u_x(t)|_0 = \sup_{\sigma_1 \leqslant t \leqslant \sigma+\varrho} \sup_{-r \leqslant s \leqslant 0} |x(t+s)| = \sup_{\sigma_1 \leqslant t \leqslant \sigma+\varrho} \sup_{t-r \leqslant \tau \leqslant t} |x(\tau)| \leqslant ||x||_\varrho \leqslant \alpha$ and $|u_x(t_1) - u_x(t_2)|_0 = \sup_{-r \leqslant s \leqslant 0} |x(t_1+s) + -x(t_2+s)| \leqslant w(|t_1 - t_2|)$. Therefore, $K(\Omega)$ is a bounded and equicontinuous subset of $C([\sigma_1, \sigma+\varrho], C_{0r})$. It is also clear that for every $t \in [\sigma_1, \sigma+ +\varrho]$, a set $K(\Omega)(t) := \{u(t): u \in K(\Omega)\}$ is relatively compact in C_{0r}. Therefore, by the Ascoli–Arzela theorem, $K(\Omega)$ is relatively compact in $C([\sigma_1, \sigma+\varrho], C_{0r})$.

We shall show now that $K(\Omega)$ is also closed in $C([\sigma_1, \sigma+\varrho], C_{0r})$. Indeed, let a sequence (u_n) of $K(\Omega)$ and $u \in C([\sigma_1, \sigma+\varrho], C_{0r})$ be such that $\sup_{\sigma_1 \leqslant t \leqslant \sigma+\varrho} |u_n(t) - -u(t)|_0 \to 0$ as $n \to \infty$. By the definition of $K(\Omega)$ there is a sequence (x^n) of $\mathscr{K}(\Omega)$ such that $u_n(t) = x^n_t$ for every $t \in [\sigma_1, \sigma+\varrho]$. By the compactness of $\mathscr{K}(\Omega)$ there is a $x \in \mathscr{K}(\Omega)$ and a subsequence, say (x^k) of (x^n) such that $||x^k - x||_\varrho \to 0$ as $k \to \infty$. Hence, in particular it follows $\sup_{1 \leqslant t \leqslant \sigma+\varrho} |x^k_t - x_t|_0 \to 0$ as $k \to \infty$. Then $u(t) = x_t$ for every $t \in [\sigma_1, \sigma+\varrho]$. Thus $u \in K(\Omega)$. ■

LEMMA 5.5. *Let $(\sigma, \Phi) \in \boldsymbol{R} \times \mathrm{AC}_{0r}$, $\varrho > 0$, $l > 0$, $M > 0$, $\Omega \subset \boldsymbol{R} \times \mathrm{AC}_{0r}$ and $H \subset \boldsymbol{R}^n$ be such that $(\sigma, \Phi) \in \Omega$, $\Phi(0) \in H$, $[\sigma, \sigma+\varrho] \times \boldsymbol{K}_\beta(\Phi, H) \subset \Omega$ with $\beta = l+2M$. Suppose H is compact in $\boldsymbol{R}^n$ and Ω is decomposable, closed in $\boldsymbol{R} \times C_{0r}$ and relatively weakly compact in $\boldsymbol{R} \times \mathrm{AC}_{0r}$. Assume G: $\Omega \to \mathrm{Comp}(\boldsymbol{R}^n)$ is such that*

(i) *$G(t, \psi) \subset MB$ for $(t, \psi) \in [\sigma, \sigma+\varrho] \times \boldsymbol{K}_\beta(\Phi, H)$,*

(ii) *$G(\cdot, \psi)$ is measurable on $[\sigma, \sigma+\varrho]$ for fixed $\psi \in \boldsymbol{K}_\beta(\Phi, H)$,*

(iii) *there are a continuous mapping L: $[\sigma, \sigma+\varrho] \to \mathscr{L}(L_{0r}, \boldsymbol{R}^n)$ and a Kamke function ω: $[\sigma, \sigma+\varrho] \times \boldsymbol{R}^+ \to \boldsymbol{R}^+$ such that*

$$\sup_{\sigma \leqslant t \leqslant \sigma+\varrho} (||L(t)||_{\mathscr{L}}) h(G(t, \psi_1), G(t, \psi_2))$$
$$\leqslant \omega(t, \max\{||L(t)||_{\mathscr{L}} |\psi_1 - \psi_2|_0, |L(t)[\dot\psi_1 - \dot\psi_2]|\})$$

for every (t, ψ_1), $(t, \psi_2) \in \Omega$ with $t \in [\sigma, \sigma+\varrho]$,

(iv) *$G(t, \psi) \subset T_H(\psi(0))$ for $t \in [\sigma, \sigma+\varrho]$ and $\psi \in \boldsymbol{K}_\beta(\Phi, H)$.*

Then, for every $\varepsilon > 0$ and $\xi \in \mathrm{AC}([\sigma-r, \sigma+\varrho], \boldsymbol{R}^n)$ such that $(t, \xi_t) \in \Omega$ for $t \in [\sigma, \sigma+ +\varrho]$ and $\dot\xi(t) \in \mathrm{co}\, G(t, \xi_t)$ for a.e. $t \in [\sigma, \sigma+\varrho]$ there is $\eta > 0$ such that for every $\sigma_1 \in [\sigma, \sigma+\varrho]$ and $\Phi_1 \in \boldsymbol{K}_\alpha(\Phi, H)$ with $\alpha = l+M$ such that $|\Phi_1 - \xi_{\sigma_1}|_0 < \frac{1}{2}\eta$ there exists $x \in \mathrm{AC}([\sigma_1 - r, \sigma+\varrho], \boldsymbol{R}^n)$ satisfying conditions (5.1) and such that $L \cdot \sup_{\sigma_1 - r \leqslant t \leqslant \sigma+\varrho} |x(t) - \xi(t)| \leqslant \varepsilon$, where $L := \sup_{\sigma_1 \leqslant t \leqslant \sigma+\varrho} ||L(t)||_{\mathscr{L}}$.

Proof. Suppose $L > 0$. Let ξ and $\sigma_1 \in [\sigma, \sigma+\varrho)$ be such as above. Define a set-valued function $\mathscr{G}$: $[\sigma_1, \sigma+\varrho] \times X \to \mathrm{Comp}(\boldsymbol{R}^n)$ by taking $\mathscr{G}(t, u) := G(t, u(t))$ for $t \in [\sigma_1, \sigma+\varrho]$ and $u \in X$, where X is such as in Lemma 5.4. Observe that for every $u \in X$ we have $(t, u(t)) \in \Omega$. Indeed, for every $u \in K(\Omega) \cup P(H)$ there is $x \in \mathscr{K}(\Omega)$ or $\psi \in \boldsymbol{K}_\beta(\Phi, H)$ such that $u(t) = x_t$ or $u(t) = \psi$ for every $t \in [\sigma_1, \sigma+\varrho]$. It is clear that $\mathscr{G}$ is bounded and such that $\mathscr{G}(\cdot, u)$ is measurable for fixed $u \in X$. Furthermore, by (iii) we have

$$\begin{aligned}
&\sup_{\sigma_1\leqslant t\leqslant\sigma+\varrho} (\|L(t)\|_{\mathscr{L}})h(\mathscr{G}(t,u),\mathscr{G}(t,v))\\
&=\sup_{\sigma_1\leqslant t\leqslant\sigma+\varrho} (\|L(t)\|_{\mathscr{L}})h\big(G(t,u(t)),G(t,v(t))\big)\\
&\leqslant \omega\big(t,\max\{\|L(t)\|_{\mathscr{L}}|u(t)-v(t)|_0,\ |L(t)[\dot u(t)-\dot v(t)]|\}\big)\\
&\leqslant \omega\big(t,\max\{\|L(t)\|_{\mathscr{L}}|u(t)-v(t)|_0,\ \sup_{\sigma_1\leqslant\tau\leqslant t}|L(\tau)[\dot u(t)-\dot v(t)]|\}\big)\\
&=\omega\big(t,\boldsymbol{R}(t,u,v)\big)
\end{aligned}$$

for a.e. $t\in[\sigma_1,\sigma+\varrho]$ and $u,v\in X$, where $\boldsymbol{R}(t,u,v)$ is such as in Theorem III.5.10.

Let us observe now, that a set-valued mapping $\mathscr{F}(\mathrm{co}\,\mathscr{G})$: $X\ni u\to\mathscr{F}(\mathrm{co}\,\mathscr{G})(u)$ is l.s.c. on X (see Lemma III.2.9). Therefore, for given above $\xi\in\mathscr{K}(\Omega)$ and an $\varepsilon>0$ there is $\eta\in(0,\frac{1}{2}\varepsilon L^{-1})$ such that $\mathscr{F}(\mathrm{co}\,\mathscr{G})(u)\cap S^0(\psi_\xi,\frac{1}{4}\varepsilon L^{-1})\neq\emptyset$ for every $u\in X$ satisfying $\|u-v_\xi\|<\eta$, where $\psi_\xi\in L([\sigma_1,\sigma+\varrho],\boldsymbol{R}^n)$, $v_\xi\in X$ are defined by $\psi_\xi(t)=\dot\xi(t)$ for a.e. $t\in[\sigma_1,\sigma+\varrho]$ and $v_\xi(t)=\xi_t$ for $t\in[\sigma_1,\sigma+\varrho]$. Moreover $S^0(\psi\ ,\frac{1}{4}\varepsilon L^{-1})$ denotes an open ball of $L([\sigma_1,\sigma+\varrho],\boldsymbol{R}^n)$ with the centre at ψ_ξ and a radius $\frac{1}{4}\varepsilon L^{-1}>0$. Let $\Phi_1\in\boldsymbol{K}_\alpha(\Phi,H)$ be such that $|\Phi_1-\xi_{\sigma_1}|_0<\frac{1}{2}\eta$ and define a set $\mathscr{K}_1(\Omega)$ by taking $\mathscr{K}_1(\Omega)=\{x\in\mathscr{K}(\Omega)\colon x_{\sigma_1}=\Phi_1\}$. It is clear by Lemma 5.3 that $\mathscr{K}_1(\Omega)\neq\emptyset$. Similarly as above let $K_1(\Omega)=\{u_x\colon x\in\mathscr{K}_1(\Omega)\}$ and then put $X_1:=K_1(\Omega)\cup P(H)$. We have $X_1\subset X$. For given above ξ, $\sigma_1\in[\sigma,\sigma+\varrho]$ and Φ_1 let $\xi^1:=(\Phi_1\oplus\xi|_{[\sigma_1,\sigma+\varrho]})$ and then put $v_\xi^1(t)=\xi_t^1$ for every $t\in[\sigma_1,\sigma+\varrho)$. Since $(\sigma_1,\Phi_1)\in[\sigma,\sigma+\varrho]\times\boldsymbol{K}_\beta(\Phi,H)\subset\Omega$ and $(t,\xi_t)\in\Omega$ for $t\in[\sigma_1,\sigma+\varrho]$ then for every $t\in[\sigma_1,\sigma+\varrho]$ we also have $(t,\xi_t^1)\in\Omega$, because Ω is decomposable. Then $v_\xi^1\in X$. It is clear that $\xi_{\sigma_1}^1=\Phi_1$. Therefore, $v_\xi^1\in X_1$. We also have $\|v_\xi^1-v_\xi\|<\eta$, because

$$\begin{aligned}
\|v_\xi^1-v_\xi\| &= \sup_{\sigma_1\leqslant t\leqslant\sigma+\varrho}|\xi_t^1-\xi_t|_0=\sup_{\sigma_1\leqslant t\leqslant\sigma+\varrho}\{\sup_{-r\leqslant s\leqslant 0}|\xi^1(t+s)-\xi(t+s)|\}\\
&\leqslant 2|\Phi_1-\xi_{\sigma_1}|_0<\eta.
\end{aligned}$$

Therefore, $\mathscr{F}(\mathrm{co}\,\mathscr{G})(v_\xi^1)\cap S^0(\psi_\xi,\frac{1}{4}\varepsilon L^{-1})\neq\emptyset$. Then there is $\Psi\in\mathscr{F}(\mathrm{co}\,\mathscr{G})(v_\xi^1)$ such that $\int_{\sigma_1}^{\sigma+\varrho}|\Psi(t)-\dot\xi(t)|\,dt<\frac{1}{4}\varepsilon L^{-1}$.

Let us observe that $\mathscr{K}_1(\Omega)$ is a closed subset of $\mathscr{K}(\Omega)$. Therefore X_1 is a compact subset of $C([\sigma_1,\sigma+\varrho],C_{0r})$. Now, by virtue of Theorem III.5.10 for given above $\varepsilon>0$, $\Phi_1\in L_{0r}$, $v_\xi^1\in X_1$ and $\Psi\in\mathscr{F}(\mathrm{co}\,\mathscr{G})(v_\xi^1)$ we can find a continuous function $g\colon X_1\to L([\sigma_1,\sigma+\varrho],\boldsymbol{R}^n)$ such that $g(u)\in\mathscr{F}(\mathscr{G})(u)$ for $u\in X_1$ and

$$\begin{aligned}
&\max\Big\{\|L(t)\|_{\mathscr{L}}\sup_{\sigma_1\leqslant s\leqslant t}\Big|\int_{\sigma_1}^{s}[g(u)(\tau)-\Psi(\tau)]\,d\tau\Big|,\\
&|L(t)[\dot\Phi_1\oplus g(u)_t-\dot\xi_t^1]|\Big\}\leqslant\tfrac{1}{4}\varepsilon+\int_{\sigma_1}^{t}\omega\big(s,\boldsymbol{R}(s,u,v_\xi^1)\big)\,ds
\end{aligned}\tag{5.2}$$

for $t\in[\sigma_1,\sigma+\varrho]$, where $\boldsymbol{R}(s,u,v_\xi^1)$ is such as in Theorem III.5.10.

Let g_1 denote the restriction of g to the set $P(H)$. For every $u\in P(H)$ there is $\psi\in\boldsymbol{K}_\beta(\Phi,H)$ such that $u\equiv\psi$. Therefore, we have defined a continuous mapping

$\tilde{g}$: $K_\beta(\Phi, H) \to L([\sigma_1, \sigma+\varrho], \boldsymbol{R}^n)$ such that $\tilde{g}(\psi)(t) := g_1(u_\psi)(t) \in \mathscr{G}(t, u_\psi) := G(t, \psi)$ for a.e. $t \in [\sigma_1, \sigma+\varrho]$ and $\psi \in K_\beta(\Phi, \Phi, H)$, where $u_\psi \in P(H)$ is such that $u_\psi \equiv \psi$. Now, by virtue of Lemma 5.3, there is $x \in \mathrm{AC}([\sigma_1, \sigma+\varrho], \boldsymbol{R}^n)$ such that $\dot{x}(t) = \dot{g}(x_t) := g(u_x)(t) \in \mathscr{G}(t, u_x) := G(t, x_t)$ for a.e. $t \in [\sigma_1, \sigma+\varrho]$, $x_{\sigma_1} = \Phi_1$ and $x(t) \in H$ for $t \in [\sigma_1, \sigma+\varrho]$. Furthermore, by the proof of Lemma 5.1 we also have $(t, x_t) \in [\sigma_1, \sigma+\varrho] \times K_\beta(\Phi, H) \subset \Omega$ for $t \in [\sigma_1, \sigma+\varrho]$. Then, $u_x \in X_1$ and therefore, by (5.2) for $t \in [\sigma_1, \sigma+\varrho]$ one has

$$\|L(t)\|_{\mathscr{L}} \sup_{\sigma_1 \leqslant s \leqslant t} \left| \int_{\sigma_1}^{s} [\dot{x}(\tau) - \Psi(\tau)] d\tau \right| \leqslant \tfrac{1}{4}\varepsilon + \int_{\sigma_1}^{t} \omega\big(s, R(s, u_x, v_\xi^1)\big) ds$$

and

$$|L(t)[\dot{x}_t - \dot{\xi}_t^1]| \leqslant \tfrac{1}{4}\varepsilon + \int_{\sigma_1}^{t} \omega\big(s, R(s, u_x, v_\xi^1)\big) ds,$$

where

$$R(s, u_x, v_\xi^1) := \max\{\|L(t)\|_{\mathscr{L}} |x_t - \xi_t^1|_0, \sup_{\sigma_1 \leqslant \tau \leqslant t} |L(\tau)[\dot{x}_\tau - \dot{\xi}_\tau^1]|\}.$$

Hence for $t \in [\sigma_1, \sigma+\varrho]$ we obtain

$$\|L(t)\|_{\mathscr{L}} \sup_{\sigma_1 \leqslant s \leqslant t} \left| \int_{\sigma_1}^{s} [\dot{x}(\tau) - \dot{\xi}(\tau)] d\tau \right| \leqslant \|L(t)\|_{\mathscr{L}} \sup_{\sigma_1 \leqslant s \leqslant t} \left| \int_{\sigma_1}^{t} [\dot{x}(\tau) - \Psi(\tau)] d\tau \right| +$$

$$+ \|L(t)\|_{\mathscr{L}} \int_{\sigma_1}^{\sigma+\varrho} |\Psi(\tau) - \dot{\xi}(\tau)| d\tau \leqslant \tfrac{1}{2}\varepsilon + \int_{\sigma_1}^{t} \omega\big(s, R(s, u_x, v_\xi^1)\big) ds,$$

$$|x_t - \xi_\xi^1|_0 := \sup_{-r \leqslant \tau \leqslant 0} |x(t+\tau) - \xi^1(t+\tau)| = \sup_{\sigma_1 \leqslant s \leqslant t} |x(s) - \xi^1(s)|$$

$$= \sup_{\sigma_1 \leqslant s \leqslant t} |[x(s) - \xi(s)] - [x(\sigma_1) - \xi(\sigma_1)]| = \sup_{\sigma_1 \leqslant s \leqslant t} \left| \int_{\sigma_1}^{t} [\dot{x}(\tau) - \dot{\xi}(\tau)] d\tau \right|$$

and

$$\sup_{\varrho_1 \leqslant \tau \leqslant t} |L(\tau)[\dot{x}_\tau - \dot{\xi}]_\tau^1| \leqslant \tfrac{1}{2}\varepsilon + \int_{\sigma_1}^{t} \omega\big(s, R(s, u_x, v_\xi^1)\big) ds.$$

Therefore, for every $t \in [\sigma_1, \sigma+\varrho]$ we have

$$R(t, u_x, v_\xi^1) \leqslant \tfrac{1}{2}\varepsilon + \int_{\sigma_1}^{t} \omega\big(s, R(s, u_x, v_\xi^1)\big) ds.$$

Hence, by the properties of the Kamke function ω it follows that $R(t, u_x, v_\xi^1) \leqslant \varepsilon$ for $t \in [\sigma_1, \sigma+\varrho]$. Therefore, in particular, we get

$$\|L(t)\|_{\mathscr{L}} \sup_{\sigma_1 \leqslant s \leqslant t} \left| \int_{\sigma_1}^{s} [\dot{x}(\tau) - \dot{\xi}(\tau)] d\tau \right| \leqslant \tfrac{1}{2}\varepsilon$$

for $t \in [\sigma_1, \sigma+\varrho]$. But

$$\sup_{\sigma_1 \leqslant s \leqslant t} |x(s)-\xi(s)|-|x(\sigma_1)-\xi(\sigma_1)| \leqslant \sup_{\sigma_1 \leqslant s \leqslant t} |[x(s)-\xi(s)]-[x(\sigma_1)-\xi(\sigma_1)]|$$

$$= \sup_{\sigma_1 \leqslant s \leqslant t} \left| \int_{\sigma_1}^{s} [\dot{x}(\tau)-\dot{\xi}(\tau)]d\tau \right|.$$

Therefore, for every $t \in [\sigma_1, \sigma+\varrho]$ one has

$$||L(t)||_{\mathscr{L}} \sup_{\sigma_1 \leqslant t \leqslant \sigma+\varrho} |x(t)-\xi(t)| \leqslant L|\Phi_1(0)-\xi_{\sigma_1}(0)|+\tfrac{1}{2}\varepsilon \leqslant \tfrac{1}{2}L\eta+\tfrac{1}{2}\varepsilon,$$

i.e.,

$$\sup_{\sigma_1 \leqslant t \leqslant \sigma+\varrho} |x(t)-\xi(t)| \leqslant \tfrac{1}{2}\eta+\tfrac{1}{2}\varepsilon L^{-1}$$

where $L = \sup_{\iota_1 \leqslant t \leqslant \sigma+\varrho} ||L(t)||_{\mathscr{L}}$. Hence, it follows $||x-\xi||_\varrho := \sup_{\sigma_1-r \leqslant t \leqslant \sigma+\varrho} |x(t)+$ $+-\xi(t)| \leqslant |x_{\sigma_1}-\xi_{\sigma_1}|_0+\tfrac{1}{2}\eta+\tfrac{1}{2}\varepsilon L^{-1} \leqslant \eta+\tfrac{1}{2}\varepsilon L^{-1} \leqslant \varepsilon/L$ because $\eta \in (0, \tfrac{1}{2}\varepsilon L^{-1})$. ■

Now, similarly as in Section 4.3 we obtain the following viability theorems.

THEOREM 5.6. *Let (D, Ω) be a conformable pair of nonempty subsets of $\boldsymbol{R} \times C_{0r} \times L_{0r}$ and $\boldsymbol{R} \times \mathrm{AC}_{0r}$, respectively. Suppose there are $(\sigma, \Phi) \in \Omega$, $\varrho > 0$, $l > 0$, $M > 0$ and a compact set $H \subset \boldsymbol{R}^n$ such that $\Phi(0) \in H$ and $[\sigma, \sigma+\varrho] \times \boldsymbol{K}_\alpha(\Phi, H) \subset \Omega$ with $\alpha = l+M$. If F: $D \to \mathrm{Comp}(\boldsymbol{R}^n)$ is such that*

(i) *$F(\cdot, u, v)$ is measurable on $[\sigma, \sigma+\varrho]$ for fixed $(u, v) \in \boldsymbol{K}_\alpha(\Phi, H) \times \dot{\boldsymbol{K}}_\alpha(\Phi, H)$,*

(ii) *$F(t, \cdot, \cdot)$ is l.s.c.w. on $\boldsymbol{K}_\alpha(\Phi, H) \times \dot{\boldsymbol{K}}_\alpha(\Phi, H)$ for fixed $t \in [\sigma, \sigma+\varrho]$,*

(iii) *$F([\sigma, \sigma+\varrho] \times \boldsymbol{K}_\alpha(\Phi, H) \times \dot{\boldsymbol{K}}_\alpha(\Phi, H)) \subset MB$,*

(iv) *$F(t, \psi, \dot{\psi}) \subset T_H(\psi(0))$ for every $t \in [\sigma, \sigma+\varrho]$ and $\psi \in \boldsymbol{K}_\alpha(\Phi, H)$,*

then NFDI(D, F) *has on $[\sigma-r, \sigma+\varrho]$, (Ω, H)-viable trajectory through (σ, Φ).*

Proof. Let G^F: $[\sigma, \sigma+\varrho] \times \boldsymbol{K}_\alpha(\Phi, H) \to \mathrm{Comp}(\boldsymbol{R}^n)$ *be defined by* $G^F(t, \psi) := F(t, \psi, \dot{\psi})$ for $t \in [\sigma, \sigma+\varrho]$ and $\psi \in \boldsymbol{K}_\alpha(\Phi, H)$. It is easily seen that G^F satisfies the assumptions of Lemma 5.1 and therefore the result follows immediately from this lemma ■

Similarly from Lemma 5.2 we obtain

THEOREM 5.7. *Let (D, Ω) be a conformable pair of nonempty subsets of $\boldsymbol{R} \times C_{0r} \times L_{0r}$ and $\boldsymbol{R} \times \mathrm{AC}_{0r}$ respectively and let D be open. Assume there are $(\sigma, \Phi) \in \Omega$, $l > 0$, $M > 0$, $\varrho > 0$ and a compact set $H \subset \boldsymbol{R}^n$ such that $\Phi(0) \in H$, $I_\varrho(\sigma) \times \mathscr{K}_\varrho(\Phi) \times B_\varrho(\dot{\Phi}) \subset D$, $[\sigma, \sigma+\tfrac{1}{2}\varrho] \times B_\alpha(\tfrac{1}{2}\varrho, \Phi, H) \subset \Omega$ with $\sigma = l+M$, $F(\cdot, u, v)$ is measurable on $[\sigma, \sigma+\varrho]$ for fixed $(u, v) \in \mathscr{K}_\varrho(\Phi) \times B_\varrho(\dot{\Phi})$, $F(t, \cdot, \cdot)$ is l.s.c.w. on $\mathscr{K}_\varrho(\Phi) \times B_\varrho(\dot{\Phi})$ for fixed $t \in [\sigma, \sigma+\varrho]$, $F(I_\varrho(\sigma) \times \mathscr{K}_\varrho(\Phi) \times B_\varrho(\dot{\Phi})) \subset MB$ and $F(t, \psi, \dot{\psi}) \subset T_H(\psi(0))$ for $t \in [\sigma, \sigma+\varrho]$ and $\psi \in B_\alpha(\varrho, \Phi, H)$. Then there exists $a \in (0, \tfrac{1}{2}\varrho]$ such that* NFDI(D, F) *has on $[\sigma-r, \sigma+a]$, (Ω, H)-viable trajectory through (σ, Φ).* ■

REMARK 5.1. Similarly as in Section 4.4 we can obtain some sufficient conditions for the existence of η-approximation (Ω, H)-viable trajectories for NFDI(D, F) with F: $D \to \operatorname{Comp}(\boldsymbol{R}^n)$. It is enough only to change in Theorems 5.6 and 5.7 the tangential condition $F(t, \psi, \dot{\psi}) \subset T_H(\psi(0))$ by its following η-approximation from $F(t, \psi, \dot{\psi}) \subset [T_H(\psi(0))]^\eta$, where $[T_H(\psi(0))]^\eta = T_H(\psi(0)) + \eta B$ and $\eta \in [0, 1)$. ∎

Now as corollaries from Theorems 5.6 and 5.7 we obtain the following controllability theorems.

THEOREM 5.8. *Let (D, Ω) be a conformable pair of nonempty subsets of $\boldsymbol{R} \times C_{0r} \times L_{0r}$ and $\boldsymbol{R} \times \mathrm{AC}_{0r}$, respectively and let $C \subset \boldsymbol{R} \times C_{0r} \times \boldsymbol{R} \times C_{0r}$. Suppose Ω, C and F: $D \to \operatorname{Comp}(\boldsymbol{R}^n)$ are such that there are $\Omega_0 \subset \Omega$, $(\sigma, \Phi) \in \Omega_0$, $\varrho \geqslant r$, $l > 0$, $M > 0$ and a compact set $H \subset \boldsymbol{R}^n$ such that $\Phi(0) \in H$, $[\sigma, \sigma+\varrho] \times \boldsymbol{K}_\alpha(\Phi, H) \subset \Omega_0$ with $\alpha = l+M$, $\Omega_0 \times \Omega_0 \subset C$ and*

(i) *$F(\cdot, u, v)$ is measurable on $[\sigma, \sigma+\varrho]$ for fixed $(u, v) \in \boldsymbol{K}_\alpha(\Phi, H) \times \dot{\boldsymbol{K}}_\alpha(\Phi, H)$,*

(ii) *$F(t, \cdot, \cdot)$ is l.s.c.w. on $\boldsymbol{K}_\alpha(\Phi, H) \times \dot{\boldsymbol{K}}_\alpha(\Phi, H)$ for fixed $t \in [\sigma, \sigma+\varrho]$*

(iii) *$F([\sigma, \sigma+\varrho] \times \boldsymbol{K}_\alpha(\Phi, H) \times \dot{\boldsymbol{K}}_\alpha(\Phi, H)) \subset MB$,*

(iv) *$F(t, \psi, \dot{\psi}) \subset T_H(\psi(0))$ for $t \in [\sigma, \sigma+\varrho]$ and $\psi \in \boldsymbol{K}_\alpha(\Phi, H)$.*

Then NFDI(D, F) *is (Ω, C)-controllable.* ∎

THEOREM 5.9. *Let (D, Ω) be a conformable pair of nonempty subsets of $\boldsymbol{R} \times C_{0r} \times L_{0r}$ and $\boldsymbol{R} \times \mathrm{AC}_{0r}$, respectively, let D be open and $E \subset \boldsymbol{R} \times C_{0r} \times \boldsymbol{R} \times \boldsymbol{R}^n$. Suppose Ω, E and F: $D \to \operatorname{Comp}(\boldsymbol{R}^n)$ are such that there are $\Omega_0 \subset \Omega$, $(\sigma, \Phi) \in \Omega_0$, $M > 0$, $l > 0$, $\varrho > 0$ and a compact set $H \subset \boldsymbol{R}^n$ such that $\Phi(0) \in H$, $I_\varrho(\sigma) \times \mathscr{K}_\varrho(\Phi) \times B_\varrho(\dot{\Phi}) \subset D$, $[\sigma, \sigma+\frac{1}{2}\varrho] \times B_\alpha(\frac{1}{2}\varrho, \Phi, H) \subset \Omega_0$ with $\alpha = l+M$, $\Omega_0 \times [\sigma, \sigma+\frac{1}{2}\varrho] \times H \subset E$ and*

(i) *$F(\cdot, u, v)$ is measurable on $[\sigma, \sigma+\varrho]$ for fixed $(u, v) \in \mathscr{K}_\alpha(\Phi) \times B_\varrho(\dot{\Phi})$,*

(ii) *$F(t, \cdot, \cdot)$ is l.s.c.w. on $\mathscr{K}_\varrho(\Phi) \times B_\varrho(\dot{\Phi})$ for fixed $t \in [\sigma, \sigma+\varrho]$*

(iii) *$F(I_\varrho(\sigma) \times \mathscr{K}_\varrho(\Phi) \times B_\varrho(\dot{\Phi})) \subset MB$,*

(iv) *$F(t, \psi, \dot{\psi}) \subset T_H(\psi(0))$ for $t \in [\sigma, \sigma+\varrho]$ and $\psi \in B_\alpha(\varrho, \Phi, H)$.*

Then NFDI(D, F) *is (Ω, E)-Euclidean controllable.* ∎

REMARK 5.2. Similarly as above we can obtain some sufficient conditions for the η-approximating (Ω, E)-Euclidean controllability of NFDI(D, F). ∎

5.2. Relaxation theorem

We shall consider here NFDI(D, F) with F taking compact and not necessarily convex values. Given sets $\Omega \subset \boldsymbol{R} \times \mathrm{AC}_{0r}$, $C \subset \boldsymbol{R} \times C_{0r} \times \boldsymbol{R} \times C_{0r}$ and $E \subset \boldsymbol{R} \times C_{0r} \times \boldsymbol{R} \times \boldsymbol{R}^n$ we shall consider sets $\mathscr{T}(D, F, \Omega, C)$, $\mathscr{T}(D, \operatorname{co} F, \Omega, C)$, $\mathscr{T}^e(D, F, \Omega, E)$ and $\mathscr{T}^e(D, \operatorname{co} F, \Omega, E)$ as subsets of the metric spaces $\mathscr{X}_r(\Omega)$ and $C_r(\Omega)$.

LEMMA 5.10. *Let (D, Ω) be a conformable pair of nonempty subsets of $\boldsymbol{R} \times C_{0r} \times L_{0r}$ and $\boldsymbol{R} \times \mathrm{AC}_{0r}$, respectively and let $C \subset \boldsymbol{R} \times C_{0r} \times \boldsymbol{R} \times C_{0r}$. Suppose Ω, C and F:*

$D \to \mathrm{Comp}(\boldsymbol{R}^n)$ are such that there are $(\sigma, \Phi) \in \Omega$, $\varrho > 0, l > 0, M > 0$ and a compact set $H \subset \boldsymbol{R}^n$ such that $\Phi(0) \in H$, $[\sigma, \sigma+\varrho] \times \boldsymbol{K}_\beta(\Phi, H) \subset \Omega$ with $\beta = l+2M$, $\Omega \times \times \Omega \subset C$ and

(i) *$F(t, \psi, \dot{\psi}) \subset MB$ for $(t, \psi) \in [\sigma, \sigma+\varrho] \times \boldsymbol{K}_\beta(\Phi, H)$,*

(ii) *$F(\cdot, \psi, \dot{\psi})$ is measurable on $[\sigma, \sigma+\varrho]$ for fixed $\psi \in \boldsymbol{K}_\beta(\Phi, H)$,*

(iii) *there are a continuous mapping L: $[\sigma, \sigma+\varrho] \to \mathscr{L}(L_{0r}, \boldsymbol{R}^n)$ and a Kamke function ω: $[\sigma, \sigma+\varrho] \times \boldsymbol{R}^+ \to \boldsymbol{R}^+$ such that*

$$\sup_{\sigma \leqslant t \leqslant \sigma+\varrho} (||L(t)||_{\mathscr{L}}) h(F(t, \psi_1, \dot{\psi}_1), F(t, \psi_2, \dot{\psi}_2))$$
$$\leqslant \omega(t, \max\{||L(t)||_{\mathscr{L}} |\psi_1 - \psi_2|_0, |L(t)[\dot{\psi}_1 - \dot{\psi}_2]|\}$$

for every (t, ψ_1), $(t, \psi_2) \in \Omega$ with $t \in [\sigma, \sigma+\varrho]$,

(iv) *$F(t, \psi, \dot{\psi}) \subset T_H(\psi(0))$ for $t \in [\sigma, \sigma+\varrho]$ and $\psi \in \boldsymbol{K}_\beta(\Phi, H)$.*

If furthermore Ω is decomposable, closed in $\boldsymbol{R} \times C_{0r}$ and relatively weakly compact in $\boldsymbol{R} \times \mathrm{AC}_{0r}$ and such that $\Pi_{\boldsymbol{R}}(\Omega)$ is connected, then for every $\varepsilon > 0$ and $\xi \in \mathrm{AC}([\sigma- -r, \sigma+\varrho], \boldsymbol{R}^n)$ such that $(t, \xi_t) \in \Omega$ for $t \in [\sigma, \sigma+\varrho]$ and $\dot{\xi}(t) \in \mathrm{co}\, F(t, \xi_t, \dot{\xi}_t)$ for a.e. $t \in [\sigma, \sigma+\varrho]$ there is $\eta > 0$ such that for every $\sigma_1 \in [\sigma, \sigma+\varrho]$ and $\Phi_1 \in \boldsymbol{K}_\alpha(\Phi, H)$ with $\alpha = l+M$ satisfying $|\Phi_1 - \xi_{\sigma_1}|_0 < \frac{1}{2}\eta$ there exists $x \in \mathrm{AC}(\sigma_1 - r, \sigma+\varrho], \boldsymbol{R}^n)$ such that

$$\begin{aligned} &\dot{x}(t) \in F(t, x_t, \dot{x}_t) \quad \text{for a. e. } t \in [\sigma_1, \sigma+\varrho], \\ &x_{\sigma_1} = \Phi_1, \\ &(t, x_t) \in \Omega \quad \text{for } t \in [\sigma_1, \sigma+\varrho], \\ &(\sigma_1, x_{\sigma_1}, \sigma+\varrho, x_{\sigma+\varrho}) \in C, \\ &||x - \xi_\varrho|| \leqslant \varepsilon/L, \end{aligned} \tag{5.3}$$

where $L = \sup_{\sigma_1 \leqslant t \leqslant \sigma+\varrho} ||L(t)||_{\mathscr{L}}$ and $||\cdot||_\varrho$ is the supremum norm of $C([\sigma_1 - r, \sigma+\varrho], \boldsymbol{R}^n)$.

Proof. The proof follows immediately from Lemma 5.5, with G defined by $G(t, \psi) := F(t, \psi, \dot{\psi})$ for $(t, \psi) \in \Omega$. ■

Now we can prove the following relaxation theorem.

THEOREM 5.11. *Let (D, Ω) be a conformable pair of nonempty subsets of $\boldsymbol{R} \times C_{0r} \times L_{0r}$ and $\boldsymbol{R} \times \mathrm{AC}_{0r}$, respectively and let C be a closed subset of $\boldsymbol{R} \times C_{0r} \times \boldsymbol{R} \times C_{0r}$ such that $\Omega \times \Omega \subset C$. Suppose Ω is decomposable, closed in $\boldsymbol{R} \times C_{0r}$, relatively weakly compact in $\boldsymbol{R} \times \mathrm{AC}_{0r}$ and such that $\Pi_{\boldsymbol{R}}(\Omega) = [a, b]$ with $b - a \geqslant r > 0$. Assume F: $D \to \mathrm{Comp}(\boldsymbol{R}^n)$ and Ω are such that there is $M > 0$ such that $F(t, \psi, \dot{\psi}) \subset MB$ for every $(t, \psi) \in \Omega$. Moreover suppose that for every $(\sigma, \Phi) \in \Omega$ there are numbers $l > 0, \varrho > 0$, a compact set $H \subset \boldsymbol{R}^n$, a continuous mapping L: $[\sigma, \sigma+\varrho] \to \mathscr{L}(L_{0r}, \boldsymbol{R}^n)$ and a Kamke function ω: $[\sigma, \sigma+\varrho] \to \boldsymbol{R}^+$ such that:*

(i) *$B^\circ(\Phi(0), \varrho) \subset H$, where $B^\circ(\Phi(0), \varrho)$ is an open ball of $\boldsymbol{R}^n$ with the centre at $\Phi(0)$ and a radius ϱ,*

(ii) *$[\sigma, \sigma+\varrho] \times \boldsymbol{K}_\beta(\Phi, H) \subset \Omega$ with $\beta = l+2M$,*

(iii) *$F(t, \psi, \dot{\psi}) \subset T_H(\psi(0))$ for $(t, \psi) \in [\sigma, \sigma+\varrho] \times \boldsymbol{K}_\beta(\Phi, H)$,*

(iv) *$F(\cdot, \psi, \dot{\psi})$ is measurable on $[\sigma, \sigma+\varrho]$ for fixed $\psi \in \boldsymbol{K}_\beta(\Phi, H)$,*

(v) $\sup_{\sigma \leqslant t \leqslant \sigma+\varrho} (\|L(t)\|_{\mathscr{L}}) h(F(t, \psi_1, \dot{\psi}_1), F(t, \psi_2, \dot{\psi}_2))$
$\leqslant \omega(t, \max\{\|L(t)_{\mathscr{L}}|\psi_1-\psi_2|_0, |L(t)[\dot{\psi}_1-\dot{\psi}_2]|\}$

for $t \in [\sigma, \sigma+\varrho]$ and $\psi_1, \psi_2 \in \boldsymbol{K}_\beta(\Phi, H)$.

Then

$$\mathscr{T}(D, \operatorname{co}F, \Omega, C) = \overline{[\mathscr{I}(D, F, \Omega, C)]}_{C_r(\Omega)}. \tag{5.4}$$

Proof. It is clear by Theorem 5.8 that $\mathscr{T}(D, F, \Omega, C) \neq \emptyset$. Therefore, also $\mathscr{T}(D, \operatorname{co}F, \Omega, C) \neq \emptyset$ because $\mathscr{T}(D, F, \Omega, C) \subset \mathscr{T}(D, \operatorname{co}F, \Omega, C)$. By Theorem 4.21, $\mathscr{T}(D, \operatorname{co}F, \Omega, C)$ is ϱ-compact in $C_r(\Omega)$. Therefore, $\overline{[\mathscr{I}(D, F, \Omega, C)]}_{C_r(\Omega)} \subset \mathscr{T}(D, \operatorname{co}F, \Omega, C)$.

Suppose, $(\xi, [\sigma-r, \sigma+\varrho]) \in \mathscr{T}(D, \operatorname{co}F, \Omega, C)$. Then $\xi \in \mathrm{AC}([\sigma-r, \sigma+\varrho], \boldsymbol{R}^n)$ with $\varrho \geqslant r$ is such that $(t, \xi_t) \in \Omega$ for $t \in [\sigma, \sigma+\varrho]$ and $\dot{\xi}(t) \in \operatorname{co}F(t, \xi_t, \dot{\xi}_t)$ for a.e. $t \in [\sigma, \sigma+\varrho]$. Hence, in particular it follows that $\int_E |\dot{\xi}(\tau)|\,d\tau \leqslant \tilde{\omega}(\mu(E))$ for every measurable set $E \subset [\sigma-r, \sigma+\varrho]$, where $\tilde{\omega}(\delta) := \sup_{E \in \mathscr{L}(J)} \{\int_E |\dot{\xi}(\tau)|\,d\tau\colon \mu(E) \leqslant \delta\}$ with $J := [\sigma-r, \sigma+\varrho]$, where $\mathscr{L}(J)$ denotes the Lebesque σ-algebra of subsets of J.

Let $g(t) = \xi_t$ for each $t \in [\sigma, \sigma+\varrho]$. It is clear that such defined function g is continuous on $[\sigma, \sigma+\varrho]$ and therefore has a closed graph. The set Graph(g) is also compact in $\boldsymbol{R} \times C_{0r}$ because Graph(g) $\subset \Omega$ and Ω is compact in $\boldsymbol{R} \times C_{0r}$.

Fix $(t, \psi_t) \in$ Graph(g) and let $l_t > 0$, $\varrho_t > 0$ and a compact set $H_t \subset \boldsymbol{R}^n$ be such that conditions (i)–(iv) are satisfied. Denote by $S^\circ(t)$ an open ball of $\boldsymbol{R} \times C_{0r}$ with the centre at (t, ψ_t) and a radius $\frac{1}{2}\varrho_t$. We have Graph(g) $\subset \bigcup_{t \in [\sigma, \sigma+\varrho]} S^\circ(t)$. Hence, by the compactness of Graph(g) there is a set $\{t_1, t_2, \ldots, t_p\} \subset [\sigma, \sigma+\varrho]$ such that Graph(g) $\subset \bigcup_{i=1}^{p} S^\circ(t_i)$. Put, $l_i := l_{t_i}$, $\varrho_i := \varrho_{t_i}$, $\xi_i := \xi_{t_i}$, $H_i := H_{t_i}$, $S_i^\circ := S^\circ(t_i)$, $\boldsymbol{K}_i := K_{\alpha_i}(\xi_i, H_i)$ with $\alpha_i = l_i + M$ and $\varrho^0 := \min_{1 \leqslant i \leqslant p} \varrho_i$.

Suppose $t_1 < t_2 < \ldots < t_p$. Let us observe that for $\psi \in \mathrm{AC}_{0r}$ and fixed $i = 1, \ldots, p$, $\psi \in \boldsymbol{K}_i$ if and only if $\psi(0) \in H$ and $\int_E |\dot{\psi}(s)|\,ds \leqslant (l_i+M)\mu(E) + \tilde{\omega}(\mu(E))$ for each measurable set $E \subset [-r, 0]$, where $\tilde{\omega}$ is a modulus of continuity defined above.

Furthermore, observe that by virtue of Lemma 5.10 for every $\varepsilon > 0$ and given above $(\xi; [\sigma-r, \sigma+\varrho])$ there is $\eta \in (0, \frac{1}{2}\varepsilon L^{-1}]$ such that for every $\sigma_1 \in [t_i, t_i+\varrho_i]$ and $\Phi_1 \in \boldsymbol{K}_i$ satisfying $|\Phi_1 - \xi_{\sigma_1}| < \frac{1}{2}\eta$ there exists $x \in \mathrm{AC}([\sigma_1-r, t_i \times \varrho_i], \boldsymbol{R}^n)$ such that conditions (5.3) of Lemma 5.10 are satisfied with $\sigma = t_i$, $\varrho = \varrho_i$ and $\Phi = \xi_{t_i}$. Take $\varepsilon \in (0, \frac{1}{2}L\varrho^0)$ and denote by $\eta(\varepsilon)$ a number $\eta \in (0, \frac{1}{2}\varepsilon L^{-1}]$ defined by Lemma 5.10 corresponding to so taken $\varepsilon > 0$. Let n be a positive integer such that $n > \varrho/\varrho^0$ and let $\varepsilon_n \in (0, \eta(\varepsilon)]$. Select now $\varepsilon_{n-1} \in (0, \frac{1}{2}L\eta(\varepsilon_n))$, where $L = \sup_{\sigma \leqslant t \leqslant \sigma+\varrho} \|L(t)\|_{\mathscr{L}}$. We have $\varepsilon_{n-1} < \frac{1}{2}L\eta(\varepsilon_n) \leqslant \frac{1}{2}L \cdot \frac{1}{2}\varepsilon_n/L = \frac{1}{4}\varepsilon_n$. Obviously, $\varepsilon_n \leqslant \frac{1}{2}\varepsilon L^{-1} < \frac{1}{4}\varrho^0$. It is clear that we can defined a set $\{\varepsilon_1, \varepsilon_2, \ldots, \varepsilon_n\}$ of positive numbers such that $\varepsilon_i < \frac{1}{2}L\eta(\varepsilon_{i+1})$ and $\varepsilon_i < \frac{1}{4}\varepsilon_{i+1}$ for $i = 1, 2, \ldots, n-1$. Hence, in particular it follows that

$$\varepsilon_i < \frac{1}{4^{n-1}} \frac{\varepsilon}{2} < \frac{L}{4^{n+1-i}} \cdot \frac{\varrho^0}{2}.$$

Put now $\Phi_0 = \xi_\sigma$. Wihout any loss of generality we can assume that $\sigma = t_1$. If it is not true we can consider a finite open covering $\{S_0^o, S_1^o, \ldots, S_p^o\}$ of Graph(g) instead of $(S_1^o, \ldots, S_p^o\}$, where S_0^o is an open ball of $\boldsymbol{R} \times C_{0r}$ with the centre at (σ, Φ_0) and a radius $\varrho^0 > 0$. We have of course $\Phi_0 \in K_{\alpha_1}(\Phi_0, H_1)$ with $\alpha_1 = l_1 + M$. Therefore, by virtue of Lemma 5.10 to given above $\varepsilon_1 > 0$ we can find an absolutely continuous function x^1: $[\sigma - r, t_1 + \varrho_1] \to \boldsymbol{R}^n$ such that conditions (5.3) of Lemma 5.10 are satisfied with $\varepsilon = \varepsilon_1$, $\sigma_1 = \sigma$, $\sigma = t_1$, $\varrho = \varrho_1$ and $\Phi_1 = \Phi_0$. Put now $\Phi_1 = x^1_{t_1+\varrho_1}$ and suppose $(t_1 + \varrho_1, \Phi_1) \in S_2$. By Lemma 5.10 we have $|\Phi_1 - \xi_{t_1+\varrho_1}|_0 \leqslant \varepsilon_1/L$ and $|\xi_{t_1+\varrho_1} - \xi_2|_0 < \frac{1}{2}\varrho^0$ because $(t_1+\varrho_1, \xi_{t_1+\varrho_1}) \in S_2$. Therefore, in particular $|\Phi_1(0) - \xi_2(0)| < \varrho^0$. Thus $\Phi_1(0) \in B^o(\xi_2(0), \varrho_2) \subset H_2$. Furthermore, for every measurable set $E \subset [-r, 0]$ one has

$$\int_E |\dot{\Phi}_1(s)| ds \leqslant \int_{(E+t_1+\varrho_1)\cap[\sigma-r,0]} |\dot{\xi}(\tau)| d\tau + \int_{(E+t_1+\varrho_1)\cap[\sigma,\sigma+\varrho]} |\dot{x}^1(\tau)| d\tau$$
$$\leqslant M\mu(E) + \tilde{\omega}(\mu E)) < (l_2 + M)\mu(E) + \tilde{\omega}(\mu(E)).$$

Therefore, $\Phi_1 \in K_2$. Now, similarly as above we can find an absolutely continuous function x^2: $[t_1 + \varrho_1 - r, t_2 + \varrho_2] \to \boldsymbol{R}^n$ such that $x^2_{t_1+\varrho_1} = \Phi_1$, $(t, x_t^2) \in \Omega$ for $t \in [t_1 + \varrho_1, t_2 + \varrho_2]$, $\dot{x}^2(t) \in F(t, x_t^2, \dot{x}_t^2)$ for a.e. $t \in [t_1 + \varrho_1, t_2 + \varrho_2]$, $(t_1 + \varrho_1, x^2_{t_1+\varrho_1}, t_2 + \varrho_2, x^2_{t_2+\varrho_2}) \in C$ and $\sup_{t_1+\varrho_1 \leqslant t \leqslant t_2+\varrho_2} |x^2(t) - \xi(t)| \leqslant \varepsilon_2/L$. Continuing the above procedure, finally we define an absolutely continuous function $x : [\sigma - r, \sigma + \varrho] \to \boldsymbol{R}^n$ such that $x^\varepsilon \in \mathcal{T}(D, F, \Omega, C)$ and $\|x^\varepsilon - \xi\|_\varrho \leqslant \varepsilon/L$, where $\|\cdot\|_\varrho$ is the supremum norm of the space $C([\sigma - r, \sigma + \varrho], \boldsymbol{R})$. Taking now $\varepsilon = 1/k$ with $k \in \boldsymbol{N}$ sufficiently large, we obtain a sequence (x_k) of $\mathcal{T}(D, F, \Omega. C)$ converging in the metric topology of $C_r(\Omega)$ to ξ. Therefore, $\xi \in [\overline{\mathcal{T}(D, F, \Omega, C)}]_{C_r(\Omega)}$. ∎

REMARK 5.3. In a similar way we can obtain the relaxation theorem for $\mathcal{T}^e(D, F, \Omega, E)$ with a closed $E \subset \boldsymbol{R} \times C_{0r} \times \boldsymbol{R} \times \boldsymbol{R}^n$. ∎

6. NOTES AND REMARKS

In recent years much work has been done with respect to functional-differential equations. The texts of Bellman and Cooke [1], Elsgolc [1], Myshkis [1] and Hale [1], among other bear testimony to it. These equations were classified into three main categories, i.e., equations with delay, of neutral type and advanced arguments. There is also a large constantly growing literature on the subject of functional-differential inclusions. Some remarks concerning such type differential inclusions are given in Aubin and Cellina [1]. Extensive investigations of the viability problems for functional-differential inclusions $\dot{x}(t) \in F(t, x_t)$ are contained in Haddad [1], whereas the properties of their fixed points in the σ-selectinable case are contained in Haddad [2] and Haddad and Lasry [1]. The first results concerning neutral-functional differential inclusions $\dot{x}(t) \in F(t, x_t, \dot{x}_t)$ have arisen in Kisielewicz [1] and [2]. Later on, these first results have been extended in Kisielewicz [5] on the case of functional-differential inclusions of the form $\dot{x}(t) \in F(t, x, \dot{x})$. Existence theorems for random neutral functional-differential inclusions of the form $\dot{x}(\omega, t) \in F(\omega, t, x(\omega, \cdot), \dot{x}(\omega, \cdot))$ and $\dot{x}(\omega, t) \in F(\omega, t, x_t(\omega, \cdot), \dot{x}_t(\omega, \cdot))$ are contained in Papageorgiou [2]. Some optimal control problems for functional-differential equations have been, among others, extensively investigated by F. Colonius (see Colonius [1]). Comprehensive investigations dealing with the controllability of linear neutral systems are contained in Jacobs and Langenhop [1]. The question of controllability is in Jacobs and Langenhop [1] examined only for neutral differential-difference equations of the form $\dot{x}(t) = A_{-1}\dot{x}(t-h)+A_0x(t)+A_1x(t-h)+Bu(t)$, where h is a positive constant, the A_i, $i = \pm 1, 0$, are $n \times n$ constant real matrixes and B is an $n \times m$ constant real matrix. Some results are also true for quite general linear neutral systems $\frac{d}{dt}\mathscr{D}(x_t) = L(x_t)+Bu(t)$, where $\mathscr{D}$ is Hale-type difference operator (see Hale [1]). Necessary conditions for optimal control of a systems constrained by nonlinear differential-difference inclusions are given in Clarke and Watkins [1]. Examples 1.1–1.6 arise from Aubin and Cellina [1], Leapes [1] and Bellman and Cooke [1], respectively, whereas Examples 2.1–2.4 are taken from Cesari [1]. The idea of the proofs of all viability theorems given in this chapter has been taken from Haddad [1] and Aubin and Cellina [1]. Others methods investigations of controllability of systems described by differential inclusions can be find in Aubin, Frankowska and Olech [1] and Frankowska [1]. Definitions and properties some others then Bouligand's cone are given in Frankowska [2].

Chapter V

Optimal control problems for systems described by NFDIs

The existence theorems for optimal control problems for systems described by neutral functional-differential inclusions can be obtained as a consequence of the compactness of the set of attainable trajectories and lower or upper semicontinuities of given functionals.

In this chapter we will present some applications of the properties of NFDIs in the existence theory of some optimal control problems. In particular, existence theorems for Mayer and Lagrange optimal control problems for systems described by NFDIs are given. Moreover we shall consider some applications of relaxed theorems for approximation of such optimal control problems without the convexity of given set-valued functions. Finally, we present some results concerning the existence of solutions of NFDIs having minimal norms.

1. Existence theorems

We consider here the Mayer and Lagrange optimal control problems for systems described by neutral functional-differential inclusions (see Section IV.1). The main results concern the existence of the optimal trajectories of such problems.

1.1. Weak sequential lower semicontinuity of integrals

Let (D, Ω) be a conformable pair of nonempty subsets of $\boldsymbol{R}\times C_{0r}\times L_{0r}$ and $\boldsymbol{R}\times \mathrm{AC}_{0r}$, respectively. Suppose Ω is bounded and such that $\Pi_{\boldsymbol{R}}(\Omega) = [\alpha, \beta]$ with $\beta-\alpha \geqslant r > 0$. Let $C \subset \boldsymbol{R}\times C_{0r}\times \boldsymbol{R}\times C_{0r}$ and $E \subset \boldsymbol{R}\times C_{0r}\times \boldsymbol{R}\times \boldsymbol{R}^n$ be given. We shall consider here sets $\mathrm{K}(\Omega, C)$ and $\mathrm{K}^e(\Omega, E)$ defined in Section IV.4.5. If Ω is closed in $\boldsymbol{R}\times C_{0r}$ and relatively weakly compact in $\boldsymbol{R}\times \mathrm{AC}_{0r}$, C and E are closed, then by Lemmas IV.4.19 and IV.4.20, they are relatively d-sequentially weakly compact and ϱ-compact.

We will consider here a functional $I(f)\colon \mathrm{K}(\Omega, C) \to \boldsymbol{R}$ defined by

$$I(f)[x] = \int_{\sigma}^{\sigma+a} f(t, x_t, \dot{x}_t)\,dt \tag{1.1}$$

for $(x;\ [\sigma-r, \sigma+a]) \in \mathrm{K}(\Omega, C)$, where $f\colon D \to \boldsymbol{R}$ is given.

We shall assume that for $f\colon D \to \boldsymbol{R}$ the following conditions are satisfied:

(i) for every $\varepsilon > 0$ there is a compact set $E_\varepsilon \subset [\alpha, \beta]$ with $\mu([\alpha, \beta]\setminus E_\varepsilon) < \varepsilon$ and such that the restriction $\tilde{f}_\varepsilon$ of a function $\tilde{f}_\varepsilon\colon \boldsymbol{R}\times C_{0r}\times L_{0r} \to \boldsymbol{R}$ defined by

$$\tilde{f}(t, x, z) = \begin{cases} f(t, x, z) & \text{for } (t, x, z) \in D, \\ +\infty & \text{for } (t, x, z) \in \boldsymbol{R} \times C_{0r} \times L_{0r} \setminus D \end{cases}$$

to the set $E_\varepsilon \times C_{0r} \times L_{0r}$ is Borel measurable,

(ii) for almost all fixed $t \in [\alpha, \beta]$, the function $\tilde{f}(t, \cdot, \cdot)$: $C_{0r} \times L_{0r} \to \boldsymbol{R}$ has values finite or $+\infty$, and is lower semicontinuous on $C_{0r} \times L_{0r}$,

(iii) $\tilde{f}(t, x, \cdot)$: $L_{0r} \to \boldsymbol{R}$ is convex for each fixed $(t, x) \in [\alpha, \beta] \times C_{0r}$,

(iv) f satisfies alternative one from the following conditions (L_i) $(i = 1, 2, 3)$:

(L_1) there is a function $\psi \in L([\alpha, \beta], \boldsymbol{R}^+)$ such that $f(t, x, z) \geqslant -\psi(t)$ for $(t, x, z) \in D$ with $t \in [\alpha, \beta]$,

(L_2) there is a function $\psi \in L([\alpha, \beta], \boldsymbol{R}^+)$ and a number $C > 0$ such that $f(t, x, z) \geqslant -\psi(t) - C|z|_0$ for $(t, x, z) \in D$ with $t \in [\alpha, \beta]$,

(L_3) there are functions $\psi \in L([\alpha, \beta], \boldsymbol{R})$ and $\Phi \in L^*_{0r}([-r, 0], \boldsymbol{R}^n)$ such that $f(t, x, z) \geqslant -\psi(t) - [\Phi, z]$ for $(t, x, z) \in \mathrm{D}$ with $t \in [\alpha, \beta]$, where $[\Phi, z] = \int_{-r}^{0} \Phi(s) \cdot z(s) ds$ and "$\cdot$" denotes the inner product in $\boldsymbol{R}^n$.

We shall prove here that for such functions f: $D \to \boldsymbol{R}$ the functional $I(f)$ is d-sequentially weakly lower semicontinuous on $\mathrm{K}(\Omega, C)$, i.e. such that for every $(x; [\sigma-r, \sigma+a]) \in \mathrm{K}(\Omega, C)$ and every sequence $\{(x^n; [\sigma_n - r, \sigma_n + a_n])\}$ of $\mathrm{K}(\Omega, C)$ d-weakly converging to $(x; [\sigma-r, \sigma+a])$ we have $I(f)[x] \leqslant \liminf_{n\to\infty} I(f)[x^n]$.

We begin with the following lemmas.

LEMMA 1.1. *Let (D, Ω) be a conformable pair of nonempty subsets of $\boldsymbol{R} \times C_{0r} \times L_{0r}$ and $\boldsymbol{R} \times \mathrm{AC}_0$, respectively and let Ω be closed in $\boldsymbol{R} \times C_{0r}$, relatively compact in $\boldsymbol{R} \times \mathrm{AC}_{0r}$ and such that $\Pi_{\boldsymbol{R}}(\Omega) = [\alpha, \beta]$. Let C be a closed subset of $\boldsymbol{R} \times C_{0r} \times \boldsymbol{R} \times C_{0r}$. Suppose f: $D \to \boldsymbol{R}$ is such that conditions* (i)–(iv) *are satisfied. Then for every sequence $\{(x^n; [\sigma-r, \sigma+a])\}$ of $\mathrm{K}(\Omega, C)$ weakly converging in $\mathrm{AC}([\sigma-r, \sigma+a], \boldsymbol{R}^n)$ to $(x; [\sigma-r, \sigma+a]) \in \mathrm{K}(\Omega, C)$ there are a subsequence $\{\{x^{n_k}; [\sigma-r, \sigma+a])\}$ of $\{(x^n; [\sigma-r, \sigma+a])\}$ and a set of real numbers $C^{(s)}_{N_k} \geqslant 0$, $k = 1, \ldots, N$, $N = 1, 2, \ldots$ and $s = 1, 2, \ldots$ with $\sum_{k=1}^{N} C^{(s)}_{N_k} = 1$ for each $s = 1, 2, \ldots$ and such that*

$$f(t, x_t, \dot{x}_t) \leqslant \liminf_{s\to+\infty} [\liminf_{N\to+\infty} \sum_{k=1}^{N} C^{(s(}_{N_k} f(t, x_t^{n_{N+s}}, \dot{x}_t^{n_{k+s}})] \tag{1.2}$$

for a.e. $t \in [\sigma, \sigma+a]$.

Proof. By virtue of Lemma IV.4.19, $\mathrm{K}(\Omega, C)$ is bounded in $\mathscr{X}_r(\Omega)$. Let $\{(x^n; [\sigma-r, \sigma+a])\}$ be any sequence of $\mathrm{K}(\Omega, C)$ weakly converging in $\mathrm{AC}([\sigma-r, \sigma+a], \boldsymbol{R}^n)$ to $(x; [\sigma-r, \sigma+a])$. Then, by virtue of Lemma I.4.4 it is equiabsolutely continuous and therefore by Theorem I.4.3 sequence $\{(\dot{x}^n; [\sigma-r, \sigma+a])\}$ is uniformly integrable. Then, there is a subsequence, say $\{(x^{n_k}; [\sigma-r, \sigma+a])\}$ of $\{(x^n; [\sigma-r, \sigma+a])\}$ such that $\sup_{\sigma-r\leqslant t\leqslant\sigma+a} |x^{n_k}(t) - x(t)| \to 0$ and $\dot{x}^{n_k} \rightharpoonup \dot{x}$ in $L([\sigma-r, \sigma+a], \boldsymbol{R}^n)$ as $k \to +\infty$.

Now, by the Banach–Saks–Mazur theorem, there is a set of real numbers $C_{N_k}^{(s)} \geq 0$, $k = 1, \ldots, N$, $N, s = 1, 2, \ldots$ with $\sum_{k=1}^{N} C_{N_k}^{(s)} = 1$ such that $\sum_{k=0}^{N} C_{N_k}^{(s)} \dot{x}^{n_{k+s}} \to \dot{x}$ in $L([\sigma-r, \sigma+a], R^n)$ as $N \to +\infty$ for each fixed $s = 1, 2, \ldots$ Furthermore, for fixed $s = 1, 2, \ldots$ we have $\sup_{\sigma - r \leq t \leq +a} |x^{n_{N+s}}(t) - x(t)| \to 0$ as $N \to +\infty$. Hence, in particular it follows $|x_t^{n_{N+s}} - x_t|_0 \to 0$ and $\left|\sum_{k=1}^{N} C_{N_k}^{(s)} \dot{x}_t^{n_{k+s}} - \dot{x}_t\right|_0 \to 0$ for $t \in [\sigma, \sigma+a]$ and $s = 1, 2, \ldots$ as $N \to +\infty$. Then, by the lower semicontinuity of $\tilde{f}(t, \cdot, \cdot)$ hence it follows $f(t, x_t, \dot{x}_t) \leq \liminf_{N \to +\infty} f(t, x_t^{n_{N+s}}, \sum_{k=1}^{N} C_{N_k}^{(s)} \dot{x}_t^{n_{k+s}})$ for $t \in [\sigma, \sigma+a]$ and $s = 1, 2, \ldots$ Thus, for every $t \in [\sigma, \sigma+a]$, $s = 1, 2, \ldots$ and $m \geq 1$ there is $K \geq 1$ such that for $N \geq K$ we have $f(t, x_t, \dot{x}_t) - 1/m < f(t, x_t^{n_{N+s}}, \sum_{k=1}^{N} C_{N_k}^{(s)} \dot{x}_t^{n_{k+s}})$. Since $f(t, x, \cdot)$ is convex then $f(t, x_t, \dot{x}_t) + -1/m < \sum_{k=1}^{N} C_{N_k}^{(s)} f(t, x_t^{n_{N+s}}, \dot{x}_t^{n_{k+s}})$ for fixed $m \geq 1$, $t \in [\sigma, \sigma+a]$, $s = 1, 2, \ldots$ and $N \geq K$. Therefore, $f(t, x_t, \dot{x}_t) - 1/m \leq \liminf_{N \to +\infty} \sum_{k=1}^{N} C_{N_k}^{(s)} f(t, x_t^{n_{N+s}}, \dot{x}_t^{n_{k+s}})]$ for $s = 1, 2, \ldots$ and $t \in [\sigma, \sigma+a]$. ■

LEMMA 1.2. *Suppose the assumptions of Lemma* 1.1 *are satisfied. Then for every sequence* $\{(x^n; [\sigma-r, \sigma+a])\}$ *of* $\mathrm{K}(\Omega, C)$ *weakly converging in* $\mathrm{AC}([\sigma-r, \sigma+a], R^n)$ *to* $(x; [\sigma-r, \sigma+a]) \in \mathrm{K}(\Omega, C)$ *there exists a function* $\eta \in L([\sigma-r, \sigma+a], R^n)$ *such that*

$$f(t, x_t, \dot{x}_t) \leq \eta(t) \quad \text{for a. e. } t \in [\sigma, \sigma+a] \tag{1.3}$$

and

$$\int_{\sigma}^{\sigma+a} \eta(t)dt \leq \liminf_{n \to +\infty} \int_{\sigma}^{\sigma+a} f(t, x_t^n, \dot{x}_t^n)dt. \tag{1.4}$$

Proof. Let $\{x^n; [\sigma-r, \sigma+a]\}$ be any sequence of $\mathrm{K}(\Omega, C)$ weakly converging in $\mathrm{AC}([\sigma-r, \sigma+a], R^n)$ to $(x; [\sigma-r, \sigma+a]) \in \mathrm{K}(\Omega, C)$. Observe that by the properties of the function f and the continuity of mappings $[\sigma, \sigma+a] \in t \to x_t^n \in C_{0r}$ and $[\sigma, \sigma+a] \in t \to \dot{x}_t^n \in L_{0r}$ each integral $\int_{\sigma}^{\sigma+a} f(t, x_t^n, \dot{x}_t^n)dt$ exists and is finite or $+\infty$ for each $n = 1, 2, \ldots$ Let $i = \liminf_{n \to +\infty} \int_{\sigma}^{\sigma+a} f(t, x_t^n, \dot{x}_t^n)dt$. If $i = +\infty$, then for $\eta(t) = f(t, x_t, \dot{x}_t)$ we have $\int_{\sigma}^{\sigma+a} \eta(t)dt \leq i$.

Suppose $i < +\infty$. We shall prove that we have also $i > -\infty$. Let us observe first that there exists a number $M > 0$ such that for each $n = 1, 2, \ldots$ and $t \in [\sigma, \sigma+ +a]$ we have $|\dot{x}_t^n|_0 \leq M$. Indeed, by the properties of the set $\mathrm{K}(\Omega, C)$ the sequence $(\dot{x}^n)$ is bounded in $L([\sigma-r, \sigma+a], R^n)$. Hence it follows that $|\dot{x}_t^n|_0 = \int_{-r}^{0} |\dot{x}_t^n(s)| ds$

$= \int_{t-r}^{t} |x(\tau)|\,d\tau \leqslant \int_{\sigma-r}^{\sigma+a} |\dot{x}^n(\tau)|\,d\tau \leqslant M$ for every $t \in [\sigma, \sigma+a]$, $n = 1, 2, \ldots$ and any $M > 0$.

Suppose now f satisfies the condition (L_1) of (iv). Then for every $n = 1, 2, \ldots$ we have $-\infty < -\int_{\sigma}^{\sigma+a} \psi(t)\,dt \leqslant \int_{\sigma}^{\sigma+a} f(t, x_t^n, \dot{x}_t^n)\,dt$. Thus, $-\infty < -\int_{\sigma}^{\sigma+a} \psi(t)\,dt \leqslant \liminf_{n\to\infty} \int_{\sigma}^{\sigma+a} f(t, x_t^n, \dot{x}_t^n)\,dt = i$. If f satisfies the condition (L_2), for $n = 1, 2, \ldots$ we obtain $-MCa - \int_{\sigma}^{\sigma+a} \psi(t)\,dt \leqslant -C \int_{\sigma}^{\sigma+a} |\dot{x}_t^n|_0\,dt - \int_{\sigma}^{\sigma+a} \psi(t)\,dt \leqslant \int_{\sigma}^{\sigma+a} f(t, x_t^n, \dot{x}_t^n)\,dt$. Then also $i > -\infty$. Finally, if f satisfies (L_3) of (iv), then for $n = 1, 2, \ldots$ we have $-ML - \int_{\sigma}^{\sigma+a} \psi(t)\,dt \leqslant \int_{\sigma}^{\sigma+a} f(t, x_t^n, \dot{x}_t^n)\,dt$, where $L = \operatorname*{ess\,sup}_{-r\leqslant s\leqslant 0} |\Phi(s)|$. Therefore, $i > -\infty$.

Let us observe now that in any case, mentioned above there is an increasing subsequence (n_k) of (n) and a sequence (λ_{n_k}) of Lebesgue integrable functions $\lambda_{n_k}\colon [\sigma, \sigma+ +a] \to \boldsymbol{R}^1$ weakly converging to any $\lambda \in L([\sigma, \sigma+a], \boldsymbol{R}^1)$ and such that $f(t, y, \dot{x}_t^{n_k}) \geqslant \lambda_{n_k}(t)$ for $(t, y, \dot{x}_t^{n_k}) \in D$ with $k = 1, 2, \ldots$ and $t \in [\sigma, \sigma+a]$. Indeed, if f satisfies the condition (L_1), we can take $\lambda_n(t) = \lambda(t) = -\psi(t)$ for $t \in [\sigma, \sigma+a]$ and $n = 1, 2, \ldots$ Suppose (L_2) is satisfied and let $\lambda_n(t) = -\psi(t) - C|\dot{x}_t^n|_0$ for $n = 1, 2, \ldots$ and $t \in [\sigma, \sigma+a]$. Since $(\dot{x}^n)$ is uniformly integrable on $[\sigma-r, \sigma+a]$, there is a modulus of continuity ω such that for every measurable set $E \subset [\sigma-r, \sigma+a]$ we have $\int_E |\dot{x}^n(\tau)|\,d\tau \leqslant \omega(\mu(E))$ for $n = 1, 2, \ldots$ Now, for every measurable set $F \subset [\sigma, \sigma+a]$ and $n = 1, 2, \ldots$ we have $\int_F |\dot{x}_t^n|_0\,dt = \int_F \left\{\int_{-r}^{t} |\dot{x}^n(t+s)|\,ds\right\} dt = \int_F \left\{\int_{t-r}^{t} |\dot{x}^n(\tau)|\,d\tau\right\} dt \leqslant \int_F \omega(r)\,dt = \mu(F)\omega(r)$. Then a family of continuous functions $v_n\colon [\sigma, \sigma+a] \ni t \to |\dot{x}_t^n|_0 \in \boldsymbol{R}^+$ is uniformly integrable on $[\sigma, \sigma+a]$ and hence also relatively weakly compact in $L([\sigma, \sigma+a], \boldsymbol{R}^1)$. Therefore there is an increasing subsequence (n_k) of (n) and a function $v \in L([\sigma, \sigma+a], \boldsymbol{R}^1)$ such that $v_{n_k} \rightharpoonup v$ in $L([\sigma, \sigma+a], \boldsymbol{R}^1)$ as $k \to \infty$. Taking $\lambda_{n_k}(t) = -\psi(t) - C|\dot{x}_t^{n_k}|_0$ and $\lambda(t) = -\psi(t) - Cv(t)$ we get $\lambda_{n_k} \rightharpoonup \lambda$ in $L([\sigma, \sigma+a], \boldsymbol{R}^1)$ as $k \to \infty$ and $f(t, y, \dot{x}_t^{n_k}) \geqslant \lambda_{n_k}(t)$ for $(t, y, x_t^{n_k}) \in D$ with $k = 1, 2, \ldots$ and $t \in [\sigma, \sigma+a]$. Similarly, if f satisfies (L_3) we can take $\lambda_{n_k}(t) = - -\psi(t) - L|x_t^{n_k}|_0$ and $\lambda(t) = -\psi(t) - Lv(t)$ for $k = 1, 2, \ldots$ and $t \in [\sigma, \sigma+a]$, where v is such as above and $L = \operatorname*{ess\,sup}_{-r\leqslant s\leqslant 0} |\Phi(s)|$.

Let $j_n = \int_{\sigma}^{\sigma+a} f(t, x_t^n, \dot{x}_t^n)\,dt$ for $n = 1, 2, \ldots$ By taking a suitable subsequence, say again (n_k) of (n) we may well assume that $j_{n_k} \to i$, $|x_t^{n_k} - x_t|_0 \to 0$, $\dot{x}_t^{n_k} \rightharpoonup \dot{x}_t$ in L_{0r} for $t \in [\sigma, \sigma+a]$ and $\lambda_{n_k} \rightharpoonup \lambda$ in $L([\sigma, \sigma+a], \boldsymbol{R}^1)$ as $k \to \infty$. Here $i \in (-\infty, +\infty)$ so that if $\delta_s := \max_{N\geqslant s+1} \max_{1\leqslant k\leqslant N} |j_{(N,k)}^{(s)} - i|$, for $s = 1, 2, \ldots$ with $j_{(N,k)}^{(s)} = \int_{\sigma}^{\sigma+a} f(t, x_t^{n_{N+s}}, \dot{x}_t^{n_{k+s}})\,dt$, we have $\delta_s \to 0$ as $s \to +\infty$. We have of course $|j_{(N,k)}^{(s)} - i| \leqslant \delta_s$ for $k = 1, \ldots, N$ and $N, s = 1, 2, \ldots$ Now, for any $s = 1, 2, \ldots$, the sequence $(\lambda_{n_{k+s}}\}$

$\dot{x}^{n_{k+s}})\}$ of pair functions defined on $[\sigma, \sigma+a]$ converges weakly to $(\lambda, \dot{x})$ as $k \to +\infty$. Then, by the Banach–Saks–Mazur theorem there is a set of real numbers $C_{N_k}^{(s)} \geqslant 0$, $k = 1, \ldots, N$, $N = 1, 2, \ldots$ with $\sum_{k=1}^{N} C_{N_k}^{(s)} = 1$ such that if $\lambda_N^{(s)}(t) = \sum_{k=1}^{N} C_{N_k}^{(s)} \lambda_{n_{k+s}}(t)$ and $\xi_N^{(s)}(t) = \sum_{k=1}^{N} C_{N_k}^{(s)} \dot{x}^{n_{k+s}}(t)$ then $(\lambda_N^{(s)}, \xi_N^{(s)}) \to (\lambda, \dot{x})$ in $L([\sigma-r, \sigma+a], \boldsymbol{R}^{n+1})$ as $N \to +\infty$. Let us take $\eta_N^{(s)}(t) = \sum_{k=1}^{N} C_{N_k}^{(s)} f(t, x_t^{n_{N+k}}, \dot{x}^{u_{k+s}})$ and note that $\eta_N^{(s)}(t) \geqslant \lambda_N^{(s)}$ for $t \in [\sigma, \sigma+a]$ and $N, s = 1, 2, \ldots$ Indeed, $x^{n_{N+k}}, x^{n_{k+s}} \in \mathrm{K}(\Omega, C)$ imply $(t, x_t^{n_{N+k}})$, $(t, x_t^{n_{k+s}}) \in \Omega$ for $t \in [\sigma, \sigma+a]$ and $k = 1, \ldots, N$, $N, s = 1, 2, \ldots$ Therefore, for $t \in [\sigma, \sigma+a]$, $k = 1, \ldots, N$ and $N, s = 1, 2, \ldots$ we have $(t, x_t^{n_{N+k}}, \dot{x}_t^{n_{k+s}}) \in D$. Hence by the definition of the sequence $(\lambda_{n_{k+s}})$ we get $f(t, x_t^{n_{N+l}}, \dot{x}_t^{n_{k+s}}) \geqslant \lambda_{n_{k+s}}(t)$ for $t \in [\sigma, \sigma+a]$, $k = 1, \ldots, N$ and $N, s = 1, 2, \ldots$ Thus $\eta_N^{(s)}(t) \geqslant \lambda_N^{(t)}(t)$ for $N, s = 1, 2, \ldots$ and $t \in [\sigma, \sigma+a]$. Since $\lambda_N^{(s)} \to \lambda$ in $L([\sigma, \sigma+a], \boldsymbol{R}^1)$ as $N \to +\infty$ for each $s = 1, 2, \ldots$, then there is a subsequence, say again $(\lambda_N^{(s)})$ of $(\lambda_N^{(s)})$ such that $\lambda_N^{(s)}(t) \to \lambda(t)$ for a.e. $t \in [\sigma, \sigma+a]$ and $s = 1, 2, \ldots$ as $N \to +\infty$. Thus for $s = 1, 2, \ldots$ and a.e. $t \in [\sigma, \sigma+a]$ we have $\liminf_{N\to+\infty} \eta_N^{(s)}(t) \leqslant \lambda(t)$. Therefore, by Fatou's lemma we obtain

$$\int_{\sigma}^{\sigma+a} \liminf_{N\to+\infty} \eta_N^{(s)}(t)\,dt \leqslant \liminf_{N\to+\infty} \int_{\sigma}^{\sigma+a} \eta_N^{(s)}(t)\,dt$$

for $s = 1, 2, \ldots$ Furthermore, for $k = 1, \ldots, N$, $N \geqslant s+1$ and $s = 1, 2, \ldots$ we have $i-\delta_s \leqslant \int_{\sigma}^{\sigma+a} f(t, x_t^{n_{k+s}}, \dot{x}_t^{n_{k+s}})\,dt \leqslant i+\delta_s$. Then, we also have $i-\delta_s \leqslant \int_{\sigma}^{\sigma+a} \eta_N^{(s)}(t)$ $\leqslant i+\delta_s$ for $N \geqslant s+1$ and $s = 1, 2, \ldots$ Therefore

$$\int_{\sigma}^{\sigma+a} \liminf_{N\to+\infty} \eta_N^{(s)}(t)\,dt \leqslant \liminf_{N\to+\infty} \int_{\sigma}^{\sigma+a} \eta_N^{(s)}(t)\,dt \leqslant i+\delta_s$$

for every $s = 1, 2, \ldots$ Thus $\eta^{(s)}(t) := \liminf_{N\to+\infty} \eta_N^{(s)}(t)$ is for every $s = 1, 2, \ldots$ finite a.e. in $[\sigma, \sigma+a]$. Finally, taking $\eta(t) := \liminf_{s\to+\infty} \eta^{(s)}(t)$ for a.e. $t \in [\sigma, \sigma+a]$ we obtain $\eta(t) \geqslant \lambda(t)$ for a.e. $t \in [\sigma, \sigma+a]$ and $\int_{\sigma}^{\sigma+a} \eta(t)\,dt \leqslant i$. Hence by Lemma 1.1 it follows that (1.3) and (1.4) are satisfied. ∎

Now, we can prove that the assumptions of Lemma 1.1 imply the d-sequential lower semicontinuity of the functional $I(f)\colon \mathrm{K}(\Omega, C) \to \boldsymbol{R}$ defined by (1.1).

THEOREM 1.3. *Let (D, Ω) be a conformable pair of nonempty subsets of $\boldsymbol{R} \times C_{0r} \times L_{0r}$ and $\boldsymbol{R} \times \mathrm{AC}_{0r}$, respectively. Assume Ω is closed in $\boldsymbol{R} \times C_{0r}$ and relatively weakly compact in $\boldsymbol{R} \times \mathrm{AC}_{0r}$ and such that $\Pi_{\boldsymbol{R}}(\Omega) = [\alpha, \beta]$ with $\beta - \alpha \geqslant r > 0$. Let C be closed in $\boldsymbol{R} \times C_{0r} \times \boldsymbol{R} \times C_{0r}$ and suppose Ω and C are such that $\mathrm{K}(\Omega, C) \neq \emptyset$. Let*

$f: D \to \boldsymbol{R}$ be such that conditions (i)–(iv) *are satisfied. Then the functional $I(f)$: $\mathrm{K}(\Omega, C) \to \boldsymbol{R}$ defined by* (1.1) *is d-sequentially weakly lower semicontinuous on* $\mathrm{K}(\Omega, C)$, *i.e., for every $(x;\ [\sigma-r, \sigma+a]) \in \mathrm{K}(\Omega, C)$ and every sequence $\{(x^n;\ [\sigma_n - r, \sigma_n+a_n])\}$ of $\mathrm{K}(\Omega, C)$ d-weakly converging to $(x;\ [\sigma-r, \sigma+a])$ we have $I[x] \leqslant \liminf_{n\to\infty} I[x^n]$.*

Proof. Let $(x;\ [\sigma-r, \sigma+a])$ be an arbitrary point of $\mathrm{K}(\Omega, C)$ and let $\{(x^n;\ [\sigma_n - r, \sigma_n+a_n])\}$ be any sequence d-weakly converging to $(x;\ [\sigma-r, \sigma+a])$. By the definition of the d-weak convergence, in particular we have $\sigma_n \to \sigma$ and $a_n \to a$ as $n \to \infty$. Then for any $\delta \in (0, \frac{1}{2} a)$ the interval $[\sigma+\delta, \sigma+a-\delta]$ is contained in all intervals $[\sigma_n, \sigma_n+a_n]$ with n sufficiently large, say with $n \geqslant N(\delta)$. Let (n_k) be an increasing subsequence of (n) with $n_1 \geqslant N(\delta)$ and denote by (x^k) the restrictions of all $(x^{n_k};\ [\sigma_{n_k} - r, \sigma_{n_k}+a_{n_k}])$ to $[\sigma+\delta-r, \sigma+a-\delta]$. It is clear that (x^k) is weakly convergent in $\mathrm{AC}([\sigma+\delta-r, \sigma+a-\delta], \boldsymbol{R})$ to the restriction of $(x;\ [\sigma-r, \sigma+a])$ to the interval $[\sigma+\delta-r, \sigma+a-\delta]$. Then, by virtue of Lemma 1.2 we have $f(t, x_t, \dot{x}_t) \leqslant \eta(t)$ for a.e. $t \in [\sigma+\delta-r, \sigma+a-\delta]$ and $\int_{\sigma+\delta}^{\sigma+a-\delta} \eta(t)\,dt \leqslant \liminf_{k\to+\infty} \int_{\sigma+\delta}^{\sigma+\delta-\delta} f(t, x_t^k, \dot{x}_t^k)\,dt$, where $\eta \in L([\sigma+\delta, \sigma+a-\delta], \boldsymbol{R})$ is defined in Lemma 1.2 corresponding to the sequence (x^k). Hence it follows $\int_{\sigma+\delta}^{\sigma+a-\delta} f(t, x_t, \dot{x}_t)\,dt \leqslant \liminf_{k\to+\infty} \int_{\sigma+\delta}^{\sigma+a-\delta} f(t, x_t^k, \dot{x}_t^k)\,dt$. Now, let us observe that under any of the conditions (L_i) $(i = 1, 2, 3)$ of (iv) there is any constant C such that $f_k(t) := f(t\, x_t^k, \dot{x}_t^k) \geqslant -\psi(t) - C|\dot{x}_t^k|_0$. Since a sequence (v_k) of real-valued functions v_k: $[\sigma+\delta,\ \sigma+a-\delta] \ni t \to |\dot{x}_t^k|_0 \in \boldsymbol{R}$ is uniformly integrable then to a given $\varepsilon > 0$ we can find a $\delta > 0$, sufficiently small so that $I_k^{(1)} := \int_{\sigma_{n_k}}^{\sigma+\delta} f_k(t)\,dt \geqslant -\varepsilon$, $I_k^{(2)} := \int_{\sigma+a-\delta}^{\sigma_{n_k}+a_{n_k}} f_k(t)\,dt \geqslant -\varepsilon$ for all k sufficiently large. For $I_k^\sigma := \int_{\sigma+\delta}^{\sigma+a-\delta} f(t)\,dt$ we have now

$$
\begin{aligned}
-\varepsilon + \liminf_{k\to+\infty} I_k^\sigma - \varepsilon &\leqslant \liminf_{k\to+\infty} I_k^{(1)} + \liminf_{k\to+\infty} I_k^\sigma + \liminf I_k^{(2)} \\
&\leqslant \liminf_{k\to+\infty} (I_k^{(1)} + I_k^\sigma + I_k^{(2)}) = i,
\end{aligned}
$$

where $i = \liminf_{i\to\infty} \int_{\sigma_n}^{\sigma_n+a_n} f(t, x_t^n, \dot{x}_t^n)\,dt$. Hence it follows that $\liminf_{k\to+\infty} I_k^\sigma \leqslant i + 2\varepsilon$. From the above we have now

$$
\int_{\sigma+\delta}^{\sigma+a-\delta} f(t, x_t, \dot{x}_t)\,dt \leqslant \liminf_{k\to+\infty} \int_{\sigma+\delta}^{\sigma+-\delta} f(t, x_t^k, \dot{x}_t^k)\,dt \leqslant i + 2\varepsilon,
$$

where $\delta > 0$ can be taken as small as we want. Again, $f(t, x_t, \dot{x}_t) \geqslant -\psi(t) - C|\dot{x}_t|_0$ and is L-integrable function, so the limit of the first integral as $\delta \to 0^+$ exists, is finite, and equal to the Lebesgue integral $\int_\sigma^{\sigma+a} f(t, x_t, \dot{x}_t)dt$. Since $\varepsilon > 0$ is arbitrary, we have $\int_\sigma^{\sigma+a} f(t, x_t, \dot{x}_t)dt \leqslant i$, i.e. $I(f)[x] \leqslant \liminf_{n\to\infty} I(f)[x^n]$. ■

1.2. Existence theorems for Mayer problem of optimal control

Let $D \subset \boldsymbol{R}\times C_{0r}\times L_{0r}$, $\Omega \subset \boldsymbol{R}\times \mathrm{AC}_{0r}$ and $C \subset \boldsymbol{R}\times C_{0r}\times \boldsymbol{R}\times C_{0r}$ be given. We present here some sufficient conditions for the existence of optimal trajectory of the Mayer problem $\mathscr{M}(D, F, \Omega, C, g)$ of (Ω, C)-controllable NFDI(D, F). Here g: $C \to \boldsymbol{R}$ is given and a functional $J(g)$ is defined on $\mathrm{K}(\Omega, C)$ by setting

$$J(g)[x] := g(\sigma, x_\sigma, \sigma+a, x_{\sigma+a}) \quad \text{for } (x;\ [\sigma-r, \sigma+a]) \in \mathrm{K}(\Omega, C). \tag{1.5}$$

In a similar way we can consider the Euclidean form $\mathscr{M}^e(D, F, \Omega, E, g)$ of the Mayer problem, with $E \subset \boldsymbol{R}\times C_{0r}\times \boldsymbol{R}\times \boldsymbol{R}^n$. Let us recall (see Section IV.1.2) a trajectory $(\tilde{x};\ [\tilde{\sigma}-r, \tilde{\sigma}+\tilde{a}] \in \mathscr{T}(D, F, \Omega, C)$ is said to be the *optimal trajectory of the Mayer problem* $\mathscr{M}(D, F, \Omega, C, g)$ if $J(g)[x] = \inf\{J(g)[x]\colon x \in \mathscr{T}(D, F, \Omega, C)\}$, where $\mathscr{T}(D, F, \Omega, C)$ denotes the set of all (Ω, C)-attainable trajectories of NFDI(D, F). We shall prove here the following existence theorems.

THEOREM 1.4. *Let (D, Ω) be conformable pair of nonempty subsets of $\boldsymbol{R}\times C_{0r}\times L_{0r}$ and $\boldsymbol{R}\times \mathrm{AC}_{0r}$, respectively and let $C \subset \boldsymbol{R}\times C_{0r}\times \boldsymbol{R}\times C_{0r}$. Suppose F: $D \to \mathrm{Conv}(\boldsymbol{R}^n)$ is u.s.c.w. on D. Assume Ω is closed in $\boldsymbol{R}\times C_{0r}$ and relatively weakly compact in $\boldsymbol{R}\times\times \mathrm{AC}_{0r}$ and such that $\Pi_{\boldsymbol{R}}(\Omega) = [\alpha, \beta]$ with $\beta-\alpha \geqslant r > 0$. Let C be closed in $\boldsymbol{R}\times\times C_{0r}\times \boldsymbol{R}\times C_{0r}$. If furthermore F, Ω and C are such that there are $\Omega_0 \subset \Omega$, $(\sigma, \Phi) \in \Omega_0$, $\varrho \geqslant r$, $l > 0$, $M > 0$ and a compact set $H \subset \boldsymbol{R}^n$ so that $\Phi(0) \in H$, $[\sigma, \sigma++\varrho]\times \boldsymbol{K}_\alpha(\Phi, H) \subset \Omega_0$ with $\alpha = l+M$, $\Omega_0\times\Omega_0 \subset C$, $F(t, \psi, \dot{\psi}) \subset MB$ and $F(t, \psi, \dot{\psi})\cap T_H(\psi(0)) \neq \emptyset$ for $t \in [\sigma, \sigma+\varrho]$ and $\psi \in \boldsymbol{K}_\alpha(\Phi, H)$, then for every lower semicontinuous function g: $C \to \boldsymbol{R}$ there is an optimal trajectory of the Mayer optimal control problem $\mathscr{M}(D, F, \Omega, C, g)$.*

Proof. Let us observe that Ω is compact in $\boldsymbol{R}\times C_{0r}$. By virtue of Theorem IV.4.21, $\mathscr{T}(D, F, \Omega, C)$ is nonempty, relatively d-sequentially weakly compact in $\mathscr{X}(\Omega)$ and ϱ-compact in $C(\Omega)$. Then, $\mathrm{K}(\Omega, C) \neq \emptyset$ and $(\Omega\times\Omega)\cap C \neq \emptyset$. Thus $(\Omega\times\times\Omega)\cap C$ is a nonempty compact subset of $\boldsymbol{R}\times C_{0r}\times \boldsymbol{R}\times C_{0r}$. Since g is lower semicontinuous then by Weierstrass theorem there is a $\gamma > 0$ such that $\inf\{g(z)\colon z \in (\Omega\times\Omega)\cap C\} \geqslant -\gamma$. Hence it follows that for every $x \in \mathscr{T}(D, F, \Omega, C)$ we have $J(g)[x] \geqslant -\gamma$, where $J(g)$ is defined by (1.5). Let $i := \inf\{J(g)[x]\colon x \in \mathscr{T}(D, F, \Omega, C)\}$. We have $i \geqslant -\gamma > -\infty$. Since $\mathscr{T}(D, F, \Omega, C) \neq \emptyset$, then we also have $i < +\infty$. Thus $i \in (-\infty, +\infty)$ and therefore there exists a minimizing sequence, say $\{(x^n;\ [\sigma^n-r, \sigma_n+a_n])\}$ for i, i.e., the sequence of $\mathscr{T}(D, F, \Omega, C)$ such that $i = \lim_{n\to\infty} J[x^n]$. Since $\mathscr{T}(D, F, \Omega, C)$ is relatively d-sequentially weakly compact in $\mathscr{X}_r(\Omega)$ then there is a subsequence, say $\{(x^k;\ [\sigma_k-r, \sigma_k+a_k\})\}$ of $\{(x^n;\ [\sigma_n--r, \sigma_n+a_n])\}$, d-weakly convergent to any $(x;\ [\sigma-r, \sigma+a]) \in \mathscr{T}(D, F, \Omega, C)$. By the ϱ-compactness of $\mathscr{T}(D, F, \Omega, C)$ in $C_r(\Omega)$ we can well assume that $(\sigma_k, x^k_{\sigma_k}, \sigma_k+$

$+a_k, x^k_{\sigma_k+a_k}) \to (\sigma, x_\sigma, \sigma+a, x_{\sigma+a})$ in $\mathbf{R}\times C_{0r}\times\mathbf{R}\times C_{0r}$ as $k\to\infty$. By the lower semicontinuity of g it implies $J(g)[x] := g(\sigma, x_\sigma, \sigma+a, x_{\sigma+a}) \leqslant \liminf_{k\to\infty} g(\sigma_k, x^k_{\sigma_k}, \sigma_k+$ $+a_k, x^k_{\sigma_k+a_k}) = i$. Since $(x; [\sigma-r, \sigma+a]) \in \mathscr{T}(D, F, \Omega, C)$, we have also $J(g)[x] \geqslant i$. Therefore, $J(g)[x] = i$. ■

THEOREM 1.5. *Let (D, Ω) be a conformable pair of nonempty subsets of $\mathbf{R}\times C_{0r}\times L_{0r}$ and $\mathbf{R}\times \mathrm{AC}_{0r}$, respectively, $E\subset \mathbf{R}\times C_{0r}\times\mathbf{R}\times\mathbf{R}^n$ and suppose $F\colon D\to \mathrm{Conv}(\mathbf{R}^n)$ is u.s.c.w. on D. Assume Ω is closed in $\mathbf{R}\times C_{0r}$ and relatively weakly compact in $\mathbf{R}\times$ $\times\mathrm{AC}_{0r}$ and such that $\Pi_{\mathbf{R}}(\Omega)$ is connected. Let E be compact in $\mathbf{R}\times C_{0r}\times\mathbf{R}\times\mathbf{R}^n$. If furthermore, F, Ω and E are such that there are $\Omega_0\subset\Omega$, $(\sigma,\Phi)\in\Omega_0$, $M>0$, $l>0$, $\varrho>0$ and a compact set $H\subset\mathbf{R}^n$ such that $\Phi(0)\in H$, $[\sigma,\sigma+\frac{1}{2}\varrho]\times B_\alpha(\frac{1}{2}\varrho,\Phi,H)\subset$ $\subset\Omega_0$, $\Omega_0\times[\sigma,\sigma+\frac{1}{2}\varrho]\times H\subset E$, $F(t,\psi,\dot\psi)\subset MB$ and $F(t,\psi,\dot\psi)\cap T_H(\psi(0))\neq\emptyset$ for $t\in[\sigma,\sigma+\varrho]$ and $\psi\in B_\alpha(\varrho,\Phi,H)$, then for every lower semicontinuous function $g\colon E\to\mathbf{R}$ there is an optimal trajectory of the Euclidean form of the Mayer optimal control problem $\mathscr{M}^e(D,F,\Omega,E,g)$ with the functional $J(g)[x]\colon \mathrm{K}^e(\Omega,E)\to\mathbf{R}$ defined by*

$$J(g)[x] := g\big(\sigma, x_\sigma, \sigma+a, x(\sigma+a)\big) \quad \textit{for } (x;\ [\sigma-r,\sigma+a]\in \mathrm{K}^e(\Omega,E). \tag{1.6}$$

Proof. Similarly as in the proof of Theorem 1.4 we can easily see that $(\Omega\times\mathbf{R}\times$ $\times\mathbf{R}^n)\cap E$ is a nonempty compact subset of $\mathbf{R}\times C_{0r}\times\mathbf{R}\times\mathbf{R}^n$. Furthermore, by Theorem IV.4.21, $\mathscr{T}^e(D,F,\Omega,E)$ is nonempty, relatively d-sequentially weakly compact in $\mathscr{X}_r(\Omega)$ and ϱ-compact in $C_r(\Omega)$. Therefore, similarly as in the proof of Theorem 1.4, we can find $(\tilde x;\ [\sigma-r,\sigma+a])\in\mathscr{T}(D,F,\Omega,E)$ such that $J(g)[\tilde x] = \inf\{J(g)[x];$ $x\in\mathscr{T}^e(D,F,\Omega,E)\}$. ■

1.3. Existence theorems for Lagrange problem of optimal control

We shall consider here the Lagrange optimal control problem $\mathscr{L}(D,F,\Omega,C,g,f)$ and its Euclidean form $\mathscr{L}^e(D,F,\Omega,E,g,f)$, where D, Ω, F, C, E and g are such as above and $f\colon D\to\mathbf{R}$ satisfies conditions (i)–(iv). We shall consider now a functional I defined on $\mathrm{K}(\Omega,C)$ or $\mathrm{K}^e(\Omega,E)$ by

$$I[x] = g(\sigma, x_\sigma, \sigma+a, x_{\sigma+a}) + \int_\sigma^{\sigma+a} f(t, x_t, \dot x_t)\,dt \tag{1.7}$$

or

$$I[x] := g\big(\sigma, x_\sigma, \sigma+a, x(\sigma+a)\big) + \int_\sigma^{\sigma+a} f(t, x_t, \dot x_t)\,dt \tag{1.8}$$

for $(x;\ [\sigma-r,\sigma+a])$ from $\mathrm{K}(\Omega,C)$ or $\mathrm{K}^e(\Omega,E)$, respectively.

It will be more convenient to consider $I[x]$ as the sum of the functionals $J(g)[x]$ defined by (1.5) or (1.6), respectively and a functional $I(f)$ defined on $\mathrm{K}(\Omega,C)\cup$ $\cup\mathrm{K}^e(\Omega,E)$ by setting

$$I(f)[x] := \int_\sigma^{\sigma+a} f(t, x_t, x_t)\,dt \tag{1.9}$$

for $(x; [\sigma-r, \sigma+a]) \in K(\Omega, C) \cup K^e(\Omega, E)$. Similarly as in Section V.1.2 we obtain the following existence theorem.

THEOREM 1.6. *Let D, Ω, C, F: $D \to \operatorname{Conv}(\boldsymbol{R}^n)$ and g: $C \to \boldsymbol{R}$ be such as in Theorem* 1.4 *and suppose f: $D \to \boldsymbol{R}$ satisfies conditions* (i)–(iv). *Then there is an optimal trajectory of the Lagrange optimal control problem $\mathscr{L}(D, F, \Omega, C, g, f)$.*

Proof. By virtue of Theorem 1.3, a functional $I(f)$ is d-sequentially weakly lower semicontinuous on $K(\Omega, C)$. By Theorem IV.4.21, $\mathscr{T}(D, F, \Omega, C)$ is nonempty, relatively d-sequentially weakly compact in $\mathscr{X}_r(\Omega)$ and ϱ-compact in $C_r(\Omega)$. Then, by the Weierstrass theorem, $I(f)$ is bounded below and has an absolutely minimum in $\mathscr{T}(D, F, \Omega, C)$. Then there is a number $N > 0$ such that $I(f)[x] \geqslant -N$ for $x \in \mathscr{T}(D, F, \Omega, C)$. We can also find a number $\gamma > 0$ such that $J(g)[x] \geqslant -\gamma$ for $x \in \mathscr{T}(D, F, \Omega, C)$. Then $I[x] = J(g)[x] + I(f)[x] \geqslant -(\gamma+N)$ for each $x \in \mathscr{T}(D, F, \Omega, C)$. Let $i = \inf\{I[x]: x \in \mathscr{T}(D, F, \Omega, C)\}$ where $I[x]$ is defined by (1.7). We have $i \in (-\infty, +\infty)$ and therefore there exists a minimizing sequence, say $\{(x^n; [\sigma_n - r, \sigma_n + a_n])\}$ for i. We can assume $i \leqslant I[x^n] = J(g)[x_n] + I(f)[x_n] \leqslant i+1$ for $n = 1, 2, \ldots$ By a suitable selection we can find a subsequence $\{(x^k; [\sigma_k - r, \sigma_k + a_k])\}$ of $\{(x^n; [\sigma_n - r, \sigma_n + a_n])\}$ d-weakly convergent to any $(x; [\sigma - r, \sigma + a]) \in \mathscr{T}(D, F, \Omega, C)$ and such that $J(g)[x^k] \to i_1$, $I(f)[x^k] \to i_2$ as $k \to +\infty$; $i_1 + i_2 = i$, where i_1, i_2 are suitable numbers (neither of which needs be the infimum of $J(g)$ or $I(f)$). Now, by the properties of $J(g)$ and $I(f)$ we obtain $J(g)[x] \leqslant \liminf_{k\to+\infty} J(g)[x^k] = i_1$ and $I(f)[x] \leqslant \liminf_{k\to+\infty} I(f)[x\ ] = i_2$. Therefore, $I[x] = J(g)[x] + I(f)[x] \leqslant i_1 + i_2 = i$. On the other hand $i \leqslant I[x]$, because $x \in \mathscr{T}(D, F, \Omega, C)$. Therefore, $I[x] = i$. ■

In a similar way we obtain

THEOREM 1.7. *Let D, Ω, E, F: $D \to \operatorname{Conv}(\boldsymbol{R}^n)$ and g: $E \to \boldsymbol{R}$ be such as in Theorem* 1.5 *and suppose f: $D \to \boldsymbol{R}$ satisfies conditions* (i)–(iv). *Then there is an optimal trajectory of the Euclidean form of the Lagrange optimal control problem $\mathscr{L}^e(D, F, \Omega, E, g, f)$.* ■

1.4. Relaxed optimal control problems for systems described by NFDIs

Existence of optimal trajectories for Mayer and Lagrange optimal control problems $\mathscr{M}(D, F, \Omega, C, g)$ and $\mathscr{L}(D, F, \Omega, C, g, f)$ has been proved by the important convexity assumption on the set-valued function F: $D \to \operatorname{Comp}(\boldsymbol{R}^n)$. The importance convexity hypothesis in Theorems 1.4–1.7 can be illustrated by the following example where the optimal control does not exist.

EXAMPLE 1.1. Consider the control process in the plane $\dot{x} = -y^2 + u^2$, $\dot{y} = u$ with restraint $|u(t)| \leqslant 1$. We wish to steer $x(0) = y(0) = 0$ to the segment $X_1 = \{x = 1, |y| \leqslant 1\}$ in the minimum time $t^* > 0$.

There is a uniform bound $|x(t)| + |y(t)| \leqslant 12$ on $0 \leqslant t \leqslant 2$ for all measurable controllers u satisfying the restraint. Since $\dot{x}(t) \leqslant 1$, we have a lower bound of $t^* \geqslant 1$. That is for each controller $u(t)$ on $0 \leqslant t \leqslant t_1$ we compute $x(t_1) = \int_0^{t_1} [(u(t))^2 - -(y(t))^2]dt = 1$ only when $t_1 > 1$. To construct a minimizing sequence of controls, devide the interval $[0, 2]$ onto segments of length $1/k$ and let $u\ (t)$ be $+1$ or -1 on alternate segments. Then the corresponding response satisfies

$$|y_k(t)| \leqslant \frac{1}{k} \quad \text{and} \quad \dot{x}_k(t) \geqslant 1 - \frac{1}{k^2} \quad \text{for } k = 1, 2, \ldots$$

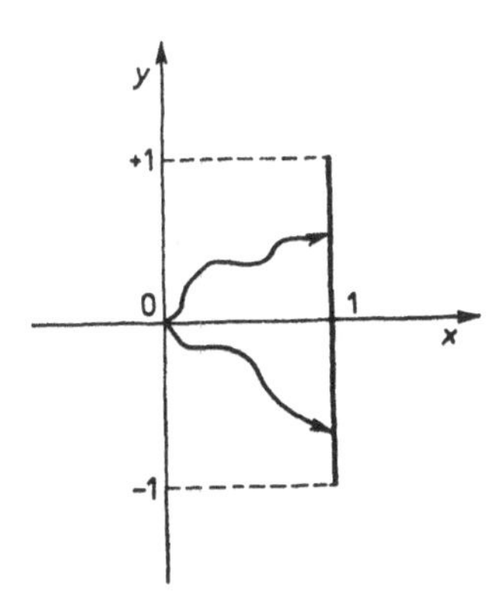

Fig. 13.

The response reaches the target X_1 at a time t_1^k on $1 < t_1^k \leqslant k^2/(k^2 - 1)$ and $\lim t_1^k = 1$. Thus the minimal optimal time $t^* = 1$ is never achieved by any admissible controller. We note that the velocity set $F = \{1, u, -y^2 + u^2\}$ is not convex in $\boldsymbol{R}^3$ and so the basic existence theorems for optimal controllers is not applicable. ∎

Although the above given problem has no optimal solution, the mathematical problem and the corresponding set of solutions can be modified in such a way that an optimal solution exists, and yet neither the system of trajectories nor the corresponding values of the cost functional are essentially modified. The modified problem, as usual is called a *relaxed* one. This new problem and its solutions are of interest in themselves, and often have relevant physical interpretations. Moreover, they have a theoretical relevance.

In the connection with the example presented above, intuitively, an "almost optimal" controller should hop rapidly back and forth between $u = +1$ and $u = -1$ so that $y(t) = \int_0^t u(s)ds$ is nearly zero and $x(t) = \int_0^t [u^2(s) - y^2(s)]ds$ is nearly t. In each time interval $u(t)$ should spend half the time at $u = +1$ and half at $u = -1$; that is $u(t) = +1$ with probability $\frac{1}{2}$ and $u(t) = -1$ with probability $\frac{1}{2}$ at each instant time.

As it was noted in Section IV.1.2 the Mayer relaxed optimal control problem $\mathcal{M}(D, \overline{\text{co}}\, F, \Omega, C, g)$ can be expressed as a problem of finding maxima or minima in $\mathscr{X}_r(\Omega)$ of the functional $J(g)[x]$ defined by (1.5) with the constraints

$$\begin{aligned} &\dot{x}(t) \in \overline{\text{co}}\, F(t, x_t, \dot{x}_t) \quad \text{for a.e. } t \in [\sigma, \sigma+a], \\ &(t, x_t) \in \Omega \quad \text{for } t \in [\sigma, \sigma+a], \\ &(\sigma, x_\sigma, \sigma+a, x_{\sigma+a}) \in C. \end{aligned} \tag{1.10}$$

The Euclidean form of the relaxed Mayer optimal control problem $\mathscr{M}^e(D, \overline{\text{co}}\, F, \Omega, E, g)$ with $E \subset \boldsymbol{R} \times C_{0r} \times \boldsymbol{R} \times \boldsymbol{R}^n$ is defined similarly.

It is clear that the existence of optimal trajectories of $\mathscr{M}(D, \overline{\text{co}}\, F, \Omega, C, g)$ and $\mathscr{M}^e(D, \overline{\text{co}}\, F, \Omega, E, g)$ follows immediately from Theorems 1.4 and 1.5. We will call them the *relaxed trajectories* of the Mayer and the Euclidean form of the Mayer optimal control problems. Hence and the relaxed Theorem IV.5.11 we obtain the following approximation theorem.

THEOREM 1.8. *Let (D, Ω) be conformable pair of nonempty subsets of $\boldsymbol{R} \times C_{0r} \times L_{0r}$ and $\boldsymbol{R} \times \text{AC}_{0r}$, respectively and let C be a closed subset of $\boldsymbol{R} \times C_{0r} \times \boldsymbol{R} \times C_{0r}$ such that $\Omega \times \Omega \subset C$. Assume Ω is decomposable, closed in $\boldsymbol{R} \times C_{0r}$, relatively weakly compact in $\boldsymbol{R} \times \text{AC}_{0r}$ and such that $\Pi_{\boldsymbol{R}}(\Omega) = [a, b]$ with $b-a \geqslant r > 0$. Suppose there is $M > 0$ such that $F(t, \psi, \dot{\psi}) \subset MB$ for every $(t, \psi) \in \Omega$. Assume F: $D \to \text{Comp}(\boldsymbol{R}^n)$ and Ω are such that for every $(\sigma, \Phi) \in \Omega$ there are numbers $l > 0$, $\varrho > 0$, a compact set $H \subset \boldsymbol{R}^n$, a continuous mapping L: $[\sigma, \sigma+\varrho] \to \mathscr{L}(L_{0r}, \boldsymbol{R}^n)$ and a Kamke function ω: $[\sigma, \sigma+\varrho] \to \boldsymbol{R}^+$ such that:*

(i) *$B^o(\Phi(0), \varrho) \subset H$, where $B^o(\Phi(0), \varrho)$ is an open ball of $\boldsymbol{R}^n$ with the centre at $\Phi(0)$ and a radius $\varrho > 0$,*

(ii) *$[\sigma, \sigma+\varrho] \times \boldsymbol{K}_\beta(\Phi, H) \subset \Omega$ with $\beta = l+2M$,*

(iii) *$F(t, \psi, \dot{\psi}) \subset T_H(\psi(0))$ for $(t, \psi) \in [\sigma, \sigma+\varrho] \times \boldsymbol{K}_\beta(\Phi, H)$,*

(iv) *$F(\cdot, \psi, \dot{\psi})$ is measurable on $[\sigma, \sigma+\varrho]$ for fixed $\psi \in \boldsymbol{K}_\beta(\Phi, H)$,*

(v) $\sup_{\sigma \leqslant t \leqslant \sigma+\varrho} (\|L(t)\|_{\mathscr{L}}) h(F(t, \psi_1, \dot{\psi}_1), F(t, \psi_2, \dot{\psi}_2)) \leqslant \omega(t, \max\{\|L(t)\|_{\mathscr{L}} |\psi_1 - \psi_2|_0, |L(t)[\dot{\psi}_1 - \dot{\psi}_2]|\})$ *for $t \in [\sigma, \sigma+\varrho]$ and $\psi_1, \psi_2 \in \boldsymbol{K}_\beta(\Phi, H)$,*

(vi) *g: $C \to \boldsymbol{R}$ is lower semicontinuous on C.*

Then there is a relaxed optimal trajectory of the Mayer optimal control problem $\mathscr{M}(D, F, \Omega, C, g)$. Furthermore, for every relaxed optimal trajectory $(x;\ [\sigma-r, \sigma+a])$ of $\mathscr{M}(D, \text{co}\, F, \Omega, C, g)$ there is a sequence $\{(x^n;\ [\sigma_n - r, \sigma_n + a_n])\}$ of $\mathscr{T}(D, F, \Omega, C)$ such that $\varrho(x_n, x) \to 0$ as $n \to \infty$.

Proof. It is clear that a set-valued function $\text{co}\, F$ satisfies conditions of Theorem 1.4. Therefore, the relaxed optimal trajectory of $\mathscr{M}(D, F, \Omega, C, g)$ exists. But, by virtue of Theorem IV.5.11 we $\mathscr{T}$have $(D, \text{co}\, F, \Omega, C) = [\mathscr{T}(D, F, \Omega, C)]_{C_r(\Omega)}$. Therefore, for every relaxed optimal trajectory $(x;\ [\sigma-r, \sigma+a]) \in \mathscr{T}(D, \text{co}\, F, \Omega, C)$ there is a sequence $\{(x^n;\ [\sigma_n - r, \sigma_n + a_n])\}$ of $\mathscr{T}(D, F, \Omega, C)$ such that $\varrho(x^n, x) \to 0$ as $n \to \infty$. ■

REMARK 1.1. In a similar way we can obtain on analogously approximation theorem for relaxed optimal trajectories of the Lagrange optimal control problem. ■

2. Attainable trajectories with minimal selection property

We shall consider here an (Ω, C)-controllable NFDI(D, F) and give some sufficient conditions for the existence of $(x; [\sigma-r, \sigma+a]) \in \mathscr{T}(D, F, \Omega, C)$ such that

$$|\dot{x}|_a = \inf\{|u_a|: u \in \mathscr{F}(F)(x, \dot{x})\}, \tag{2.1}$$

where $|v_a| := \int_\sigma^{\sigma+a} |v(t)|\, dt$ for $v \in L([\sigma, \sigma+a], \boldsymbol{R}^n)$ and $\mathscr{F}(F)(x, \dot{x})$ denotes the subtrajectory integrals of $F(\cdot, x_., \dot{x}_.)$. Every solution $(x; [\sigma, \sigma+a]) \in \mathscr{T}(D, F, \Omega, C)$ satisfying (2.1) is said to be a *(Ω, C)-attainable trajectory of* NFDI(D, F) with minimal selection property.

2.1. Dissipative set-valued functions

Let X be real Banach space, $\|\cdot\|$ its norm and X^* the dual space and K be any subset of X. Let $\mathscr{P}(X)$ be a family of all nonempty subsets of X.

A set-valued function $\mathscr{F}: K \to \mathscr{P}(X)$ is said to be *dissipative* if for every $x, y \in K$, $\lambda > 0$, $u \in \mathscr{F}(x)$ and $v \in \mathscr{F}(y)$ the following inequality is satisfied:

$$\|x-y\| \leqslant \|x-y-\lambda(u-v)\|. \tag{2.2}$$

A dissipative set-valued function $\mathscr{F}: K \to \mathscr{P}(X)$ is called *maximal dissipative* if there is no any dissipative set-valued function $\mathscr{G}: K \to \mathscr{P}(X)$ such that Graph$(\mathscr{F})$ $\subset$ Graph$(\mathscr{G})$.

A dissipative set-valued function $\mathscr{F}: K \to \mathscr{P}(X)$ is called *m-dissipative* if $\mathscr{R}(I-\mathscr{F}) = X$, where I is the identity mapping in X and $\mathscr{R}(I-\mathscr{F})$ denotes the range of the set-valued function $I-\mathscr{F}$.

We have the following lemmas.

LEMMA 2.1. *A dissipative set-valued function $\mathscr{F}: K \to \mathscr{P}(X)$ is m-dissipative, if and only if for every $\lambda > 0$, $\mathscr{R}(I-\lambda\mathscr{F}) = X$.*

Proof. Suppose $\mathscr{F}$ is m-dissipative. Let $y \in X$ and $\lambda > 0$. The inclusion $y \in (I-\lambda\mathscr{F})(x)$ may be written as $x = (I-\mathscr{F})^{-1}(y/\lambda+(1-1/\lambda)x)$, because $\mathscr{R}(I-\mathscr{F}) = X$. Moreover, the operator $(I-\mathscr{F})^{-1}$ is nonexpansive on X. Thus the operator defined by the right-hand side of the last equation is Lipschitzean on X with Lipschitz constant $|1-1/\lambda|$. By applying the standard fixed-point contraction theorem we conclude that the last equation has a unique solution $x \in X$. Since y is arbitrary we may conclude that $\mathscr{R}(I-\lambda\mathscr{F}) = X$ for every $\lambda > 0$.

Conversely, suppose that $\mathscr{F}$ is dissipative and that for some $\lambda > 0$, $\mathscr{R}(I-\mathscr{F}) = X$. Let y be arbitrarily given in X and let $\mu > 0$. The inclusion $y \in \mu x-\mathscr{F}(x)$ can be written in the following equivalent form: $x = (I-\lambda\mathscr{F})^{-1}((1-\lambda\mu)x+\lambda y)$. Since $(I-\lambda\mathscr{F})^{-1}$ is nonexpansive on X, we see that the above equation has a unique solution x whenever $0 < \mu < 2/\lambda$. In particular, $(I-\frac{3}{4}\lambda\mathscr{F})^{-1}$ is nonexpansive on X.

Now taking $(I-\frac{3}{4}\lambda\mathscr{F})^{-1}$ instead of $(I-\lambda\mathscr{F})^{-1}$ we find that $\mathscr{R}(\mu I-\mathscr{F}) = X$ for $0 < \mu < \frac{8}{3}\lambda$. Continuing in this way we finally obtain $\mathscr{R}(I-\mathscr{F}) = X$. ∎

LEMMA 2.2. *Any m-dissipative set-valued function $\mathscr{F}: K \to \mathscr{P}(X)$ is maximal dissipative.*

Proof. Suppose $\mathscr{F}$ is m-dissipative. We must prove that $\mathscr{F}$ is maximal dissipative. Suppose that this is not the case. Then there exists $(x_0, y_0) \in K \times X$ such that $y_0 \notin \mathscr{F}(x_0)$ and $||x_0 - x|| \leqslant ||x_0 - x - \lambda(y_0 - y)||$ for every $(x, y) \in \text{Graph}(\mathscr{F})$ and $\lambda > 0$. Since $\mathscr{R}(I - \mathscr{F}) = X$, we may choose $(x_1, y_1) \in \text{Graph}(\mathscr{F})$ such that $x_1 - y_1 = x_0 - y_0$. Hence, in particular for $\lambda = 1$ we obtain $||x_0 - x_1|| \leqslant ||x_0 - x_1 - (y_0 - y_1)|| = 0$. Thus $x_0 = x_1$, $y_0 = y_1$ and therefore $y_0 \in \mathscr{F}(x_0)$. A contradiction. ■

Given a Banach space $(X, ||\cdot||)$ we define on X a dual set-valued function F: $X \to \mathscr{P}(X^*)$ by setting $F(x) := \{x^* \in X^*: x^*(x) = ||x||^2 = ||x^*||^2\}$.

EXAMPLE 2.1. Let $X = L([\alpha, \beta], \boldsymbol{R}^n)$ and let F: $X \to \mathscr{P}(X^*)$ be defined by $F(x) = \{F_x\} \in L^\infty([\alpha, \beta], \boldsymbol{R}^n)$ with $F_x(y) = |x| \int_\beta^\alpha \frac{x(t) \cdot y(t)}{|x(t)|} dt$ for $y \in X$ if $x(s) \neq 0$ for all $s \in [\alpha, \beta]$. For every $x \in X \setminus \{0\}$ and $F_x \in F(x)$ one has

$$F_x(t) = \begin{cases} |x| |x(t)|^{-1} x(t) & \text{for } t \in \operatorname{supp} x, \\ a & \text{for } r \in [\alpha, \beta] \setminus \operatorname{supp} x, \end{cases}$$

where $a \in \boldsymbol{R}$ is such that $|a| \leqslant |x|$. Here $|\cdot|$ is the norm of X and "$\cdot$" denotes the inner product in $\boldsymbol{R}^n$. ■

We have the following equivalence lemma.

LEMMA 2.3. *A set-valued function $\mathscr{F}: K \to \mathscr{P}(X)$ is dissipative if and only if for every $x, y \in K$, every $u \in \mathscr{F}(x)$ and $v \in \mathscr{F}(y)$ there exists $F_{x-y} \in F(x-y)$ such that $F_{x-y}(u-v) \leqslant 0$, where F: $X \to \mathscr{P}(X^*)$ is a dual set-valued function.*

Proof. Let $x, y \in K$ be arbitrarily fixed and suppose for every $u \in \mathscr{F}(x)$ and $v \in \mathscr{F}(y)$ thre exists $F_{x-y} \in F(x-y)$ such that $F_{x-y}(u-v) \leqslant 0$. By the definition of F one obtains $||x-y||^2 = F_{x-y}(x-y) \leqslant F_{x-y}[(x-y) - \lambda(u-v)] \leqslant ||(x-y) - \lambda(u-v)|| \, ||F_{x-y}|| = ||(x-y) - \lambda(u-v)|| \, ||x-y||$ for every $\lambda > 0$. Hence (2.2) follows for every $\lambda > 0$.

Suppose now for every $x, y \in K$, $\lambda > 0$ and $u \in \mathscr{F}(x)$ and $v \in \mathscr{F}(y)$, (2.2) is satisfied. Let f_λ be an arbitrary element in $F((x-y) - \lambda(u-v))$ for every $\lambda > 0$. Without any loss of generality we may assume that $x \neq y$. Then $f_\lambda \neq 0$ for all $\lambda > 0$. Let $= g_\lambda f_\lambda / ||f_\lambda||$. Since, by Aloaglu theorem the unit sphere of the dual space X^* is compact in the weak-* topology of X^*, we may assume that $\lim_{\lambda \to 0} g_\lambda(z) = g(z)$ for each $z \in X$, where $g \in X^*$. Next, by (2.2) one obtains $||x-y|| \leqslant ||(x-y) - \lambda(u-v)|| = g_\lambda((x-y) - \lambda(u-v)) \leqslant ||x-y|| - \lambda g_\lambda(u-v)$ for each $\lambda > 0$. Hence, $g_\lambda(u-v) \leqslant 0$ for all $\lambda > 0$ and therefore also $g(u-v) \leqslant 0$. But, we also have $||x-y|| \leqslant g_\lambda(x-y) - \lambda g_\lambda(u-v)$. Hence it follows $||x-y|| \leqslant g(x-y)$. Since $||g_\lambda|| = 1$ for every $\lambda > 0$ then $||g|| \leqslant 1$. Therefore $||x-y|| \leqslant g(x-y) \leqslant ||g|| \, ||x-y|| \leqslant ||x-y||$. Thus $g(x-y) = ||x-y||$. Taking $F_{x-y} := ||x-y|| g$ we obtain $F_{x-y} \in F(x-y)$ and $F_{x-y}(u-v) \leqslant 0$. ■

EXAMPLE 2.2. Let $\mathscr{F}: (-\infty, 1] \to \mathscr{P}(\boldsymbol{R})$ and $\mathscr{G}: -(\infty, 1] \to \mathscr{P}(\boldsymbol{R})$ be set-valued mappings defined by

$$\mathscr{F}(x) = \begin{cases} -2^x+2 & \text{for } x < 0, \\ \{-1, 1\} & \text{for } x = 0, \\ -\frac{1}{2}x-1 & \text{for } 0 < x \leqslant 1 \end{cases}$$

and

$$\mathscr{G}(x) = \begin{cases} -2^x+2 & \text{for } x < 0, \\ [-1, 1] & \text{for } x = 0, \\ -\frac{1}{2}x-1 & \text{for } 0 < x < 1, \\ [-\infty, -\frac{3}{2}) & \text{for } x = 1. \end{cases}$$

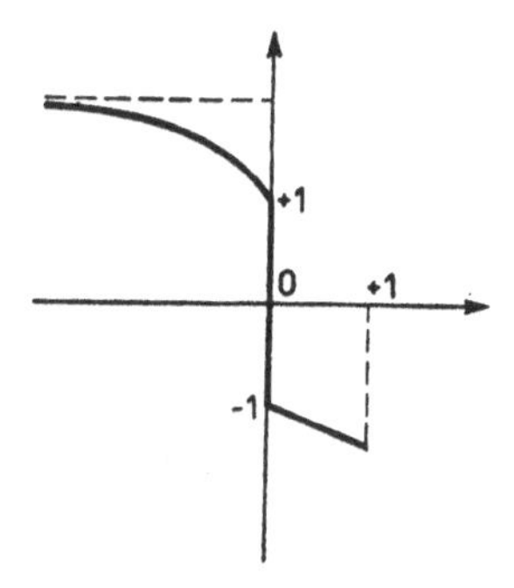

Fig. 14. Graph($\mathscr{F}$).

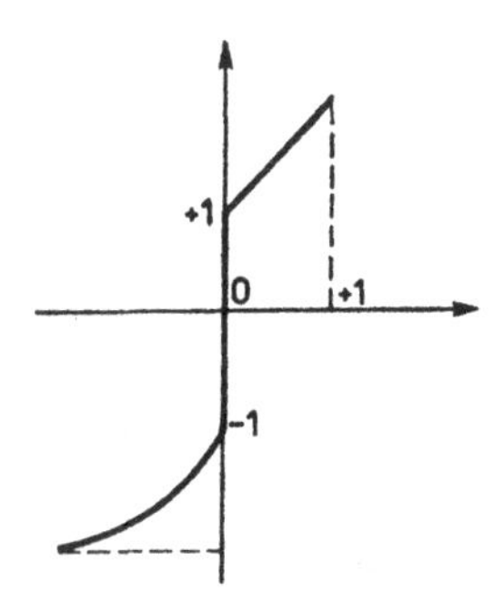

Fig. 15. Graph(I–$\mathscr{F}$).

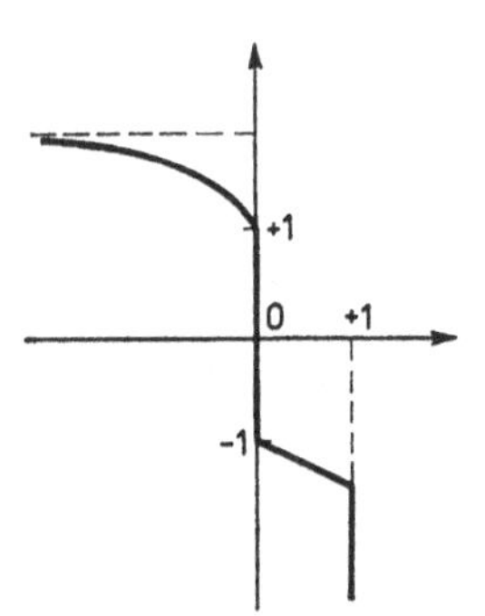

Fig. 16. Graph($\mathscr{G}$).

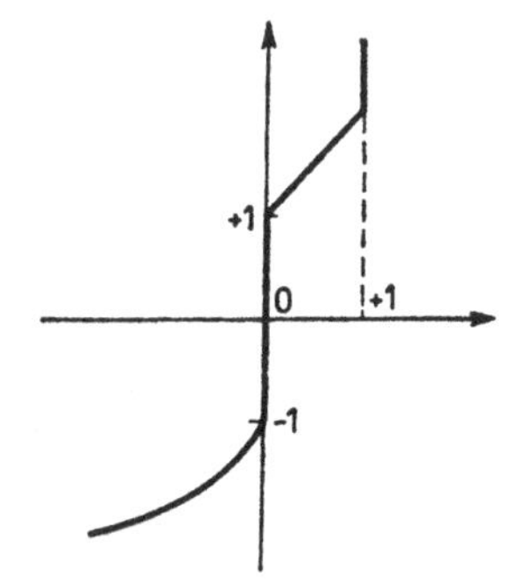

Fig. 17. Graph(I–$\mathscr{G}$).

A set-valued function $\mathscr{F}$ is dissipative but it is not m-dissipative because $\mathscr{R}(I-\mathscr{F}) = (-\infty, -1] \cup [1, \frac{3}{2}] \neq \boldsymbol{R}$. Hence it follows that $\mathscr{F}$ is also not maximal dissipative. It is easy to see that $\mathscr{G}$ is an m-dissipative extension of $\mathscr{F}$. ■

REMARK 2.1. Let $\mathscr{F}: K \to \mathscr{P}(l_2^p)$, with $K \subset l_2^p$, $p \geqslant 1$. A set-valued function $\mathscr{F}$ is dissipative if for every $(x_1, y_1), (x_2, y_2) \in \text{Graph}(\mathscr{F})$ we have $y_1 - y_2 \in H$, where H denotes the subspace defined by the support function to the unit ball at $x_1 - x_2$ (see Fig. 18). ■

EXAMPLE 2.3. Let X be a Banch space and $f\colon X \to (-\infty, +\infty]$ be convex and not equal $+\infty$ for each $x \in X$. Let $\partial f(x)$ be defined for fixed $x \in X$ by setting $\partial f(x) := \{x^* \in X^*\colon f(x)-f(y) \leqslant x^*(x-y)\}$. Each $x^* \in \partial f(x)$ is called the *subgradient of f at $x \in X$* and $\partial f(x)$ denotes its *subdifferential at x*. It can be verified that if X is a Hilbert space and f is convex and lower semicontinuous on X then $-\partial f$ is a maximal dissipative set-valued function on X. ■

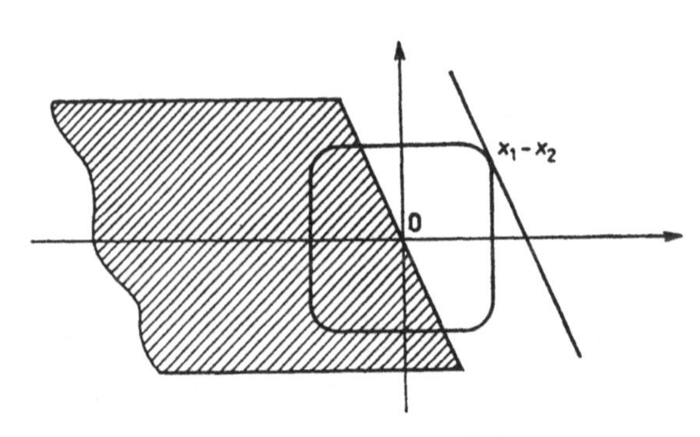

Fig. 18.

Let $\mathscr{F}\colon K \to \mathscr{P}(X)$ be m-dissipative. Since for every $\lambda > 0$, $x, y \in K$, $u \in \mathscr{F}(x)$ and $v \in \mathscr{F}(y)$ we have $||x-y|| \leqslant ||(x-\lambda u)-(y-\lambda v)||$ then for every $x \neq y$ and $\lambda > 0$ we get $(x-\lambda\mathscr{F}(x))\cap(y-\lambda\mathscr{F}(y)) = \Phi$. Thus we can define a mapping $(I-\lambda\mathscr{F})^{-1}\colon X \to K$. It is nonexpansive, because $||(I-\lambda\mathscr{F})^{-1}(x)+(I-\lambda\mathscr{F})^{-1}(y)|| \leqslant ||x-y||$ for $x, y \in X$.

In what follows we will need the following operators J_λ and $\mathscr{F}_\lambda$ defined by

$$J_\lambda(x) := (I-\lambda\mathscr{F})^{-1}(x) \quad \text{for } x \in X \tag{2.3}$$

and

$$\mathscr{F}_\lambda(x) := \frac{1}{\lambda}(J_\lambda - I)(x) \quad \text{for } x \in X. \tag{2.4}$$

The mappings J_λ and $\mathscr{F}_\lambda$ defined above are called the *resolvent* and *Yosida approximation*, respectively of $\mathscr{F}$. We have the following lemma.

LEMMA 2.4. *Let $\mathscr{F}\colon K \to \mathscr{P}(X)$ be an m-dissipative set-valued function and let $\lambda > 0$. Then*

(i) $||J_\lambda(x)-J_\lambda(y)|| \leqslant ||x-y||$ *for* $x, y \in X$,
(ii) $\mathscr{F}_\lambda$ *is dissipative and Lipschitz continuous with Lipschitz constant* $2/\lambda$,
(iii) $\mathscr{F}_\lambda(x) \in \mathscr{F}(J_\lambda(x))$ *for* $x \in X$,
(iv) $||\mathscr{F}_\lambda(x)|| \leqslant \inf\{||u||\colon u \in \mathscr{F}(x)\}$ *for* $x \in K$,
(v) $||J_\lambda x - x|| \leqslant \lambda \inf\{||u||\colon u \in \mathscr{F}(x)\}$ *for* $x \in K$.

Proof. The properties (i), (ii) are immediately clear.

(iii) Let $x \in X$. By the definition of $J_\lambda(x)$ and $\mathscr{F}_\lambda(x)$ we obtain $\mathscr{F}_\lambda(x) \in \lambda^{-1}[J_\lambda(x)-(I-\lambda\mathscr{F})(J_\lambda(x))] = \mathscr{F}(J_\lambda(x))$, because $x \in (I-\lambda\mathscr{F})(J_\lambda(x))$.

(iv) Let $x \in K$. Then $\mathscr{F}_\lambda(x) = \lambda^{-1}[J_\lambda(x)-J_\lambda((I-\lambda\mathscr{F})(x))]$. Since J_λ is nonexpansive, this implies that $\|\mathscr{F}_\lambda(x)\| \leqslant \|y\|$ for any $y \in \mathscr{F}_\lambda(x)$, i.e. $\|\mathscr{F}_\lambda(x)\| \leqslant \inf\{\|u\|: u \in \mathscr{F}(x)\}$.

(v) For every $x \in K$ we have $\|J_\lambda(x)-x\| = \lambda\|\mathscr{F}_\lambda(x)\| \leqslant \lambda\inf\{\|u\|: u \in \mathscr{F}(x)\}$. ■

COROLLARY 2.1. *Suppose $\mathscr{F}: K \to \mathscr{P}(X)$ is m-dissipative and such that* dist$(0, \mathscr{F})$ *is bounded, then $\lim_{\lambda\to 0^+} \|J_\lambda(x)-x\| = 0$ uniformly with respect to $x \in K$.*

Indeed, by (v) of Lemma 2.4 we have $\|J_\lambda(x)-x\| \leqslant \lambda M$, where $M > 0$ is such that $\sup\{\|u\|: u \in \mathscr{F}(x)\} \leqslant M$ for each $x \in K$. Hence $\sup_{x\in K} \|J_\lambda(x)-x\| \leqslant \lambda M$ and therefore $\sup_{x\in K} \|J_\lambda(x)-x\| \to 0$ as $\lambda \to 0^+$. ■

2.2. Integral-dissipative set-valued functions

We will consider here the Banach space $X = L(I, \boldsymbol{R}^n)$, with the usual norm $|\cdot|_a$ defined by $|u|_a = \int_I |u(t)|\,dt$, where I is a compact interval of $\boldsymbol{R}$, say $I = [\sigma, \sigma+a]$ with $a > 0$. Put $I^0 = (\sigma, \sigma+a)$, $s \in I^0$ and denote by O_s a neighbourhood of s contained in I. We will assume that $s \notin O_s$. Furthermore, for fixed $s \in I^0$ and $\delta > 0$ we define $O_s^\delta := \{t \in I: 0 < |t-s| < \delta\}$.

Given $x: I \to \boldsymbol{R}^n$ and $\delta > 0$ by $E(x)$, $\tilde{E}(x, \delta)$ and $E(x, \delta)$ we denote the sets of the form:

$$E(x) = \{s \in I: \dot{x}(s) \text{ exists and } \dot{x}(s) \neq 0\},$$

$$\tilde{E}(x, \sigma) = \{s \in E(x): x(t) \neq x(s) \text{ for every } t \in 0_s^\delta\},$$

$$E(x, \delta) = \tilde{E}(x, \delta) \cap [\sigma, \sigma+a-\delta].$$

LEMMA 2.5. *For every $x: I \to \boldsymbol{R}^n$ a family $\{E(x, \delta)\}_{\delta>0}$ of subsets of I is nonincreasing, i.e. for every $\delta_1 \geqslant \delta_2 > 0$, $E(x, \delta_1) \subset E(x, \delta_2)$. Furthermore, for every decreasing sequence $\{\delta_n\}$ of positive numbers we have $E(x) = \bigcup_{n=1}^{\infty} E(x, \delta_n) = \lim_{n\to\infty} E(x, \delta_n)$.*

Proof. The first assertion follows immediately from the definition of the set $E(x, \delta)$. Now we prove the second one. If $E(x) = \emptyset$ the proof is trivial. Suppose $E(x) \neq \emptyset$ and let $s \in E(x)$ be fixed. There exists a neighbourhood, say O_s of s such that for every $t \in O_s$ we have $x(t) \neq x(s)$. Indeed, if it is not true, it will be possible to find a sequence (t_n) of I such that $t_n \to s$ as $n \to \infty$ and $x(t_n) = x(s)$ for each $n = 1, 2, \ldots$ which implies that $\dot{x}(s) = 0$ and therefore that $s \notin E(x)$. Select now $\delta_1 > 0$ such that $s \in [\sigma, \sigma+a-\delta_1]$ and let δ_2 denote the radius of O_s. Taking $\delta_0 = \min(\delta_1, \delta_2)$ we have $s \in E(x, \delta_0)$. Thus, by the property of the family $\{E(x, \delta)\}_{\delta>0}$ we get $s \in E(x, \delta)$ for every $\delta \leqslant \delta_0$. Then $E(x) \subset \lim_{n\to\infty} E(x, \delta_n)$. On the other hand we have $E(x, \delta) \subset E(x)$ for every $\delta > 0$. Therefore $E(x) = \lim_{n\to\infty} E(x, \delta_n)$. ■

Let $\mathscr{F}: K \to \mathscr{P}(X)$ be given, where $K \subset X$, with $X = L(I, \boldsymbol{R}^n)$. A set-valued function $\mathscr{F}$ is said to be *integral-dissipative* if for every $x \in K$ there exists $\delta_0 > 0$ such that for every $\delta \in (0, \delta_0)$, $h \in (0, \delta)$, $f, g \in \mathscr{F}(x)$ and $\lambda > 0$ the following inequality is satisfied

$$\int_{\sigma}^{\sigma+a-\delta} \left(\int_{s}^{s+h} |x(t)-x(s)|\,dt \right) ds \leqslant \int_{\sigma}^{a+\sigma-\delta} \left(\int_{s}^{s+h} |x(t)-x(s)-\lambda(f(t)-g(s))|\,dt \right) ds.$$

Given $x: I \to \boldsymbol{R}^n$ and any subinterval $[s, s+h] \subset I$ let $x_{s,h}$ denote the restriction of x to $[s, s+h]$, i.e. $x_{s,h} := x|_{[s,s+h]}$. Furthermore, put $|x|_{s,h} = \int_{s}^{s+h} |x(t)|\,dt$ and $G_{s,h}(x_{s,h}) := |x|_{s,h}^{-1} F(x_{s,h})$, where F is a dual set-valued function defined on $L([s, s+h], \boldsymbol{R}^n)$. In what follows we shall denote elements of $G_{s,h}(z)$ by $g^{s,h}$. We prove now the following lemma.

Lemma 2.6. *Let $x, y \in L(I, \boldsymbol{R}^n)$ and suppose $\delta, h \in \boldsymbol{R}$ are such that $0 < \delta < a$ and $0 \leqslant h \leqslant \delta$. Then for every measurable set $A \subset [\sigma, \sigma+a-\delta]$ the following conditions are equivalent:*

(i) $\int_A |x|\,ds_{s,h} \leqslant \int_A |x-\lambda y|\,ds_{s,h}$ *for every* $\lambda > 0$,

(ii) *for every $s \in [\sigma, \sigma+a-\delta]$ there exist functionals $g^{s,h} \in G_{s,h}(x_{s,h})$ such that* $\int_A g^{s,h}(y_{s,h})\,ds \leqslant 0$.

Proof. Let x, y, δ, h and $A \subset [\sigma, \sigma+a-\delta]$ be such as above and let (i) be satisfied. For every $s \in [\sigma, \sigma+a-\delta]$ select $g_\lambda^{s,h} \in G_{s,h}(x_{s,h} + \lambda y_{s,h})$. Then $\|g_\lambda^{s,h}\| = 1$ and $g_\lambda^{s,h}(x_{s,h} - \lambda y_{s,h}) = |x-\lambda y|_{s,h}$. Since the unit ball of $L([s, s+h], \boldsymbol{R}^n)$ is weakly-* compact then we can assume (taking eventually appropriate subsequences) that $g_\lambda^{s,h}(z) \to g^{s,h}(z)$ for every $z \in L([s, s+h], \boldsymbol{R}^n)$ as $\lambda \to 0^+$. We will show that $g^{s,h} \in G_{s,h}(x_{s,h})$. Indeed, we have

$$\begin{aligned} |x|_{s,h} &= \lim_{\lambda \to 0^+} |x-\lambda y|_{s,h} = \lim_{\lambda \to 0^+} g_\lambda^{s,h}(x_{s,h} - \lambda y_{s,h}) \\ &= \lim_{\lambda \to 0^+} g_\lambda^{s,h}(x_{s,h}) - \lim_{\lambda \to 0^+} \lambda g_\lambda^{s,h}(y_{s,h}) = g^{s,h}(x_{s,h}) \end{aligned}$$

because $g_\lambda^{s,h}(y_{s,h}) \leqslant |y|_{s,h} \leqslant |y|_a$ for every $\lambda > 0$. Since $\|g_\lambda^{s,h}\| = 1$ then $\|g^{s,h}\| \leqslant 1$. Therefore, $g^{s,h} \in G_{s,h}(x_{s,h})$. Furthermore, by (i) we have

$$\begin{aligned} \int_A |x|_{s,h}\,ds &\leqslant \int_A |x-\lambda y|_{s,h}\,ds = \int_A g_\lambda^{s,h}(x_{s,h} - \lambda y_{s,h})\,ds \\ &\leqslant \int_A |x|_{s,h}\,ds - \lambda \int_A g_\lambda^{s,h}(y_{s,h})\,ds. \end{aligned}$$

Hence $\int_A g_\lambda^{s,h}(y_{s,h})\,ds \leqslant 0$ for each $\lambda > 0$. For every $s \in [\sigma, \sigma+a-\delta]$ we have $g_\lambda^{s,h}(y_{s,h}) \leqslant |y_a|$ and $g_\lambda^{s,h}(y_{s,h}) \to g^{s,h}(y_{s,h})$ as $\lambda \to 0^+$. Then by virtue of the Lebesgue dominanted convergence theorem, we get

$$\int_A g^{s,h}(y_{s,h})\,ds = \lim_{\lambda \to^+} \int_A g_\lambda^{s,h}(y_{s,h})\,ds \leqslant 0.$$

Suppose (ii) is satisfied and let $\lambda > 0$ be arbitrarily fixed. By the definition of $g^{s,h}$ we obtain

$$\int_A |x|_{s,h} ds = \int_A g^{s,h}(x_{s,h}) ds \leq \int_A [g^{s,h}(x_{s,h}) - \lambda g^{s,h}(y_{s,h})] ds$$
$$\leq \int_A |x_{s,h} - \lambda y_{s,h}| ds = \int_A |x - \lambda y|_{s,h} ds.$$

Thus (i) is also satisfied. ■

COROLLARY 2.2. *A set-valued function $\mathscr{F}\colon K \to \mathscr{P}(L(I, R^n))$ with $K \subset L(I, R^n)$ is integral dissipative if and only if for every $x \in K$ there exists a $\delta_0 > 0$ such that for every $f, g \in \mathscr{F}(x)$, $\delta \in (0, \delta_0)$, $h \in [0, \delta]$ and $s \in [\sigma, \sigma + a - \delta]$ there exist functionals $g^{s,h} \in G_{s,h}(x_{t,h} - \overline{x(s)})$ such that $\int_\sigma^{\sigma+a-\delta} g^{s,h}(f_{s,h} - \overline{g(s)}) ds \leq 0$, where $\overline{u(s)}$ denotes the constant function with the value equal to $u(s)$ at each point of the interval $[s, s+ +h]$.* ■

We will need the following lemma.

LEMMA 2.7. *Let $x \in \mathrm{AC}([\sigma, \sigma + a], R^n)$ and $\delta \in (0, a)$ be given. For every $s \in E(x, \delta)$, $h \in [0, \delta]$ and $g_{s,h} \in G_{s,h}(x_{s,h} - \overline{x(s)})$ we have $g_{s,h}(\dot{x}_{s,h}) = |x(s+h) - x(s)|$.*

Proof. By virtue of Example 2.1, for every $s \in E(x, \delta)$ and $g_{s,h} \in G_{s,h}(x_{s,h} - \overline{x(s)})$ we have

$$g_{s,h}(\dot{x}_{s,h}) = \int_s^{s+h} \frac{[x(t) - x(s)] \cdot \frac{d}{dt} x(t)}{|x(t) - x(s)|} dt$$
$$= \int_s^{s+h} \frac{[x(t) - x(s)] \cdot \frac{d}{dt} [x(t) - x(s)]}{|x(t) - x(s)|} dt,$$

where " $\cdot$ " denotes the inner product in R^n. Really, because we have $|x| \in \mathrm{AC}(I, R)$ if $x \in \mathrm{AC}(I, R^n)$ and if $t \in I$ is such that $x(t)$ and $d|x|(t)/dt$ exists, then

$$\frac{d|x|(t)}{dt} = \frac{x(t) \cdot \dot{x}(t)}{|x(t)|}.$$

Therefore, $g_{s,h}(\dot{x}_{s,h}) = \int_s^{s+h} \left(\frac{d}{dt} |x(t) - x(s)| \right) dt = |x(s+h) - x(s)|$. ■

EXAMPLE 2.4. Let B be a dissipative set-valued function from R^n into $\mathscr{P}(R^n)$ such that for each $x \in L(I, R^n)$, $B(x(\cdot))$ is Aumann integrable on I. Then a set-valued function $\mathscr{F}(B(x(\cdot))\colon L(I, R^n) \to \mathscr{P}(L(I, R^n))$, where as usual, $\mathscr{F}(B(x(\cdot)))$ denotes subtrajectory integrals of $B(x(\cdot))$ is integral dissipative. Indeed, for every $t, s \in I$, $\lambda > 0$,

$f(t) \in B(x(t))$ and $g(s) \in B(x(s))$ we have $|x(t)-x(s)| \leqslant |x(t)-x(s)-\lambda(f(t)-g(s))|$. Hence, it follows

$$\int_{\sigma}^{\sigma+a-\delta} \left(\int_{s}^{s+h} |x(t)-x(s)|\, dt \right) ds \leqslant \int_{\sigma}^{\sigma+a-\delta} \left(\int_{s}^{s+h} |x(t)-x(s)-\lambda\big(f(t)-g(s)\big)|\, dt \right) ds$$

for $f, g \in \mathscr{F}(B(x(\cdot)))$ and every $\delta \in (0, a)$, $\lambda > 0$, $h \in (0, \delta)$ and $x \in L(I, \boldsymbol{R}^n)$. ∎

EXAMPLE 2.5. Let $f(t) = t$ for $t \in [a, b]$ and let $\mathscr{F}$: $L([a, b], \boldsymbol{R}) \to \mathscr{P}(\boldsymbol{R})$ be defined by $\mathscr{F}(x) = f$ for every $x \in L([a, b], \boldsymbol{R})$. It is clear that $\mathscr{F}$ is dissipative in $L([a, b], \boldsymbol{R})$ because $||x-y|| \leqslant ||x-y-\lambda(f-f)||$.

We will show that $\mathscr{F}$ is not integral dissipative. Indeed, by the definition for $x = f \in L([a, b], \boldsymbol{R}^n)$ and $\lambda = 1$ we obtain

$$\int_{a}^{b-\delta} \left[\int_{s}^{s+h} |f(t)-f(s)|\, dt \right] ds \leqslant \int_{a}^{b-\delta} \left[\int_{s}^{s+h} |f(t)-f(s)-[f(t)-f(s)]|\, dt \right] ds,$$

It is false because the right-hand side of the last inequality is equal is equal to zero, while

$$\int_{a}^{b-\delta} \left[\int_{s}^{s+h} |t-s|\, dt \right] ds = \tfrac{1}{2}h^2(b-a-\delta) > 0. \ ∎$$

EXAMPLE 2.6. Let $K = \{x \in L([a, b], \boldsymbol{R})$: x is increasing almost everywhere in $[a, b]\}$. For each $x \in K$ let $\mathscr{F}(x)$ be equal to arbitrary almost everywhere decreasing function in $L([a, b], \boldsymbol{R})$. Then we have $|x(t)-x(s)| \leqslant |x(t)-x(s)-\lambda[\mathscr{F}(x)(t)- -\mathscr{F}(x)(s)]|$ for every $\lambda > 0$ and almost all $t, s \in [a, b]$. Then $\mathscr{F}$ is integral dissipative.

To see that $\mathscr{F}$ need not to be dissipative in $L([a, b], \boldsymbol{R})$ it is enough to take any x belonging to K, $y = x+1$ and put

$$\mathscr{F}(x) = -x, \quad \mathscr{F}(x) = -x+1.$$

Taking in the definition of dissipativity $\lambda = 1$ we obtain $||x-y|| = b-a > 0$ while $||x-y-[\mathscr{F}(x)-\mathscr{F}(y)]|| = 0$. ∎

2.3. *Controllability of* NFDIs *with m-dissipative right-hand sides*

Let D, Ω and C be nonempty subsets of $\boldsymbol{R} \times C_{0r} \times L_{0r}$, $\boldsymbol{R} \times \mathrm{AC}_{0r}$ and $\boldsymbol{R} \times C_{0r} \times \boldsymbol{R} \times \times C_{0r}$, respectively. We will say (D, Ω, C) has a *nonempty intersection property* if the following conditions are satisfied:

(i) for every $(t, x), (t, y) \in \Omega$ we have $(t, x, \dot{y}) \in D$,

(ii) $(\Omega \times \Omega) \cap C \neq \emptyset$,

(iii) there are $(\sigma, \Phi) \in \Omega$ and a number $\varrho > 0$ such that $K_{\Phi}(\Omega, C) := \{x \in \mathrm{AC}^0([\sigma, \sigma+\varrho], \boldsymbol{R}^n)$: $(t, (\Phi \oplus x)_t) \in \Omega$ for $t \in [\sigma, \sigma+\varrho]$, $(\sigma, x_\sigma, \sigma+\varrho, x_{\sigma+\rho}) \in C\}$ is nonempty.

Given $\Omega \subset \boldsymbol{R} \times \mathrm{AC}_{0r}$ and $C \subset \boldsymbol{R} \times C_{0r} \times \boldsymbol{R} \times C_{0r}$ a set-valued function $F \in \mathscr{M}(D, \mathrm{Comp}(\boldsymbol{R}^n))$ is said to be *m-dissipative with respect to* (Ω, C) if (D, Ω, C)

has a nonempty intersection property and a set-valued mapping $\mathscr{F}(G^F)\colon K_\Phi(\Omega, C) \to \mathscr{P}(L([\sigma, \sigma+\varrho], \boldsymbol{R}^n))$ is m-dissipative on $K_\Phi(\Omega, C)$ where $G^F(t, x) := F(t, (\Phi \oplus x)_t, (\dot{\Phi} \oplus \dot{x})_t)$ for $t \in [\sigma, \sigma+\varrho]$ and $x \in K_\Phi(\Omega, C)$.

Let $F \in \mathscr{M}(D, \mathrm{Comp}(\boldsymbol{R}^n))$ be m-dissipative with respect to (Ω, C). We say that $\mathscr{F}(G^F)\colon K_\Phi(\Omega, C) \to \mathscr{P}(L([\sigma, \sigma+\varrho], \boldsymbol{R}^n))$ has a strong-weak closed graph if for every sequence (x_n) of $K_\Phi(\Omega, C)$ and every sequence (u_n) of $L([\sigma, \sigma+\varrho], \boldsymbol{R}^n)$ such that $u_n \in \mathscr{F}(G^F)(x_n)$ for $n = 1, 2, \ldots$, $|x_n - x|_\varrho \to 0$ and $u_n \rightharpoonup u$ in $L([\sigma, \sigma+\varrho], \boldsymbol{R}^n)$ as $n \to \infty$ we also have $u \in \mathscr{F}(G^F)(x)$.

We have the following lemma.

LEMMA 2.8. *Let $F \in \mathscr{M}(D, \mathrm{Comp}(\boldsymbol{R}^n))$ be bounded, m-dissipative with respect to (Ω, C) and such that $\mathscr{F}(G^F)$ has a strong-weak closed graph. Then for every $\eta > 0$, $\mathscr{F}(G^F_\eta)\colon K_\Phi(\Omega, C) \to \mathscr{P}(L([\sigma, \sigma+\varrho], \boldsymbol{R}^n))$ with $G^F_\eta(t, x) := G^F(t, x) + \eta C$ for $(t, x) \in [\sigma, \sigma+\varrho] \times K_\Phi(\Omega, C)$ has also a strong-weak closed graph.*

Proof. Suppose (x_n) and (v_n) are arbitrary sequences of $K_\Phi(\Omega, C)$ and $L(I, \boldsymbol{R}^n)$ with $I := [\sigma, \sigma+\varrho]$ such that $v_n \in \mathscr{F}(G^F_\eta)(x_n)$ for $n = 1, 2, \ldots$, $|x_n - x|_\varrho \to 0$ and $v_n \rightharpoonup v$ in $L(I, \boldsymbol{R}^n)$ as $n \to \infty$. Since $v_n \in \mathscr{F}(G^F_\eta)(x)$ then $\mathrm{dist}(v_n(t), G^F(t, x_n)) \leqslant \eta$ for a.e. $t \in I$ and $n = 1, 2, \ldots$ By Teorem II.3.13 for every $n = 1, 2, \ldots$ there is $u_n \in \mathscr{F}(G^F)(x_n)$ such that $|v_n(t) - u_n(t)| = \mathrm{dist}(v_n(t), G^F(t, x_n)) \leqslant \eta$ for a.e. $t \in I$. By the boundedness of F, a family $\{G^F(\cdot, x)\}$, $x \in K_\Phi(\Omega, C)$ is uniformly integrable on I. Then, by Corollary III.1.1, the sequence (u_n) is relatively weakly compact in $L(I, \boldsymbol{R}^n)$. Thus there exists $u \in L(I, \boldsymbol{R}^n)$ and a subsequence, say (u_k) of (u_n) such that $u_k \rightharpoonup u$ as $k \to \infty$. We have of course $|v_k(t) - u_k(t)| \leqslant \eta$ for $k = 1, 2, \ldots$ and a.e. $t \in I$. Since $v - u_k \rightharpoonup v - u$ as $k \to \infty$, then by the Banach–Mazur theorem there exists a set of real numbers $C_{N_k} \geqslant 0$, $k = 1, 2, \ldots, N$, $N = 1, 2, \ldots$ with $\sum_{k=1}^{N} C_{N_k} = 1$ and an increasing sequence (n_k) of positive integers such that $\sum_{k=1}^{N} C_{N_k}[v_{n_k}(t) - u_{n_k}(t)] \to v(t) - u(t)$ for a.e. $t \in I$ as $N \to \infty$. But for every $N = 1, 2, \ldots$ and a.e. $t \in I$ we have

$$\Big|\sum_{k=1}^{N} C_{N_k}[v_{n_k}(t) - u_{n_k}(t)]\Big| \leqslant \sum_{k=1}^{N} C_{N_k}|v_{n_k}(t) - u_{n_k}(t)| \leqslant \sum_{k=1}^{N} C_{N_k}\eta = \eta.$$

Then, for a.e. $t \in I$ we also have $|v(t) - u(t)| \leqslant \eta$.

Now, by the property of the graph of $\mathscr{F}(G^F)$ we have $u \in \mathscr{F}(G^F)(x)$, i.e. $u(t) \in G^F(t, x)$ for a.e. $t \in I$. Therefore, for a.e. $t \in I$ we have $v(t) \in G^F(t, x) + \eta B$, i.e. $v \in \mathscr{F}(G^F)(x)$. ■

Now, we shall prove the following controllability theorem.

THEOREM 2.9. *Let $F \in \mathscr{M}(D, \mathrm{Comp}(\boldsymbol{R}^n))$ be bounded, m-dissipative with respect to (Ω, C) with $\Omega \subset \boldsymbol{R} \times \mathrm{AC}_{0r}$, $C \subset \boldsymbol{R} \times C_{0r} \times \boldsymbol{R} \times C_{0r}$ and such that $\mathscr{F}(G^F)$ has a strong-weak closed graph. Moreover, suppose Ω, C and F are such that there are numbers $l > 0$, $M > 0$ and a compact set $H \subset \boldsymbol{R}^n$ such that $(\mathscr{I}u)(t) \in H - \Phi(0)$ for every $t \in I$ and $u \in \Lambda_\alpha := \{u \in L(I, \boldsymbol{R}^n)\colon |u(t)| \leqslant \alpha$ for a.e. $t \in I\}$ with $\alpha = l + M$,*

$G^F(I\times K_\Phi(\Omega, C)) \subset MC$, $K_\alpha \subset K_\Phi(\Omega, C)$ *where* $K_\alpha = \mathscr{J}\Lambda_\lambda$, *and so that for every* $\eta > 0$ *there is* $\delta_\eta > 0$ *such that for every* $x \in K_\alpha$ *and* $y \in K_\Phi(\Omega, C)$ *satisfying* $|x-y|_\varrho \leqslant \delta_\eta$ *we have* $F(t, (\Phi\oplus y)_t, (\dot\Phi\oplus \dot y)_t) \subset T_H[(\Phi\oplus x)_t(0)]+\eta B$ *for a.e.* $t\in I$. *If furthermore for every* $x, y \in K_\Phi(\Omega, C)$, $\lambda > 0$, $u \in \mathscr{F}(G^F)(x)$ *and* $v \in \mathscr{F}(G^F)(y)$ *we have*

$$\sup_{\sigma\leqslant\tau\leqslant t} |x(\tau)-y(\tau)| \leqslant \sup_{\sigma\leqslant\tau\leqslant t} |x(\tau)-y(\tau)-\lambda\big(u(\tau)-v(\tau)\big)| \tag{2.5}$$

for $t\in I$, *then* NFDI(D, F) *is* (Ω, C)-*controllable.*

Proof. Let us observe that by the m-dissipativity of F with respect to (Ω, C) the resolvente J_λ and the Yosida approximation $\mathscr{F}_\lambda$ of $\mathscr{F}(G^F)$ can be defined on $L(I, \boldsymbol{R}^n)$. They satisfy conditions (i)–(v) of Lemma 2.4. For every fixed $\lambda > 0$ we define f_λ: $I\times K_\Phi(\Omega, C) \to \boldsymbol{R}^n$ by setting $f_\lambda(t, x) := \mathscr{F}_\lambda(x)(t)$ for $t \in I$ and $x \in K_\Phi(\Omega, C)$. We have $f_\lambda(\cdot, x) = \mathscr{F}_\lambda(x) \in L(I, \boldsymbol{R}^n)$ for every $x\in K_\Phi(\Omega, C)$.

Now let $z_1 = J_\lambda(x)$ and $z_2 = J_\lambda(y)$ for fixed $x, y\in K_\Phi(\Omega, C)$. By the definition of J_λ we obtain $z_1, z_2 \in K_\Phi(\Omega, C)$ and $x\in z_1-\lambda\mathscr{F}(G^F)(z_1)$, $y\in z_2-\lambda\mathscr{F}(G^F)(z_2)$. Select now $u_1 \in \mathscr{F}(G^F)(z_1)$ and $u_2\in\mathscr{F}(G^F)(z_2)$ such that $x = z_1-\lambda u_1$ and $y = z_2-\lambda u_{\_}$. By (2.5) for fixed $t\in I$ we obtain

$$\sup_{\sigma\leqslant\tau\leqslant t} |z_1(\tau)-z_2(\tau)| \leqslant \sup_{\sigma\leqslant\tau\leqslant t} |z_1(\tau)-z_2(\tau)-\lambda\big(u_1(\tau)-u_2(\tau)\big)|$$

i.e.

$$\sup_{\sigma\leqslant\tau\leqslant t} |J_\lambda(x)(\tau)-J_\lambda(y)(\tau)| \leqslant \sup_{\sigma\leqslant\tau\leqslant t} |x(\tau)-y(\tau)|.$$

Hence, by the definition of $\mathscr{F}_\lambda$ it follows

$$\begin{aligned}\sup_{\sigma\leqslant\tau\leqslant t} |\mathscr{F}_\lambda(x)(\tau)-\mathscr{F}_\lambda(y)(\tau)| &= \frac{1}{\lambda}\sup_{\sigma\leqslant\tau\leqslant t} |[J_\lambda(x)(\tau)-J_\lambda(y)(\tau)]-[x(\tau)-y(\tau)]| \\ &\leqslant \frac{2}{\lambda}\sup_{\sigma\leqslant\tau\leqslant t}|x(\tau)-y(\tau)|\end{aligned}$$

for $t\in I$. Therefore,

$$|f_\lambda(t, x)-f_\lambda(t, y)| \leqslant \frac{2}{\lambda}|x-y|_t \tag{2.6}$$

for $t \in I$, $\lambda > 0$ and $x, y\in K_\Phi(\Omega, C)$, where $|x-y|_t := \sup_{\sigma\leqslant\tau\leqslant t}|x(\tau)-y(\tau)|$. Hence, in particular it follows that f_λ is for every fixed $\lambda > 0$ continuous and has the Volterra's property with respect to its second variable. By virtue of (iii) of Lemma 2.4, we have $f_\lambda(\cdot, x) = \mathscr{F}_\lambda(x) \in \mathscr{F}(G^F)(J_\lambda(x))$ for $x\in K_\Phi(\Omega, C)$. Hence, in particular, it follows that $|f_\lambda(t, x)| \leqslant M$ for $x\in K_\Phi(\Omega, C)$ and a.e. $t\in I$. Since $\mathscr{F}_\lambda(G^F)$ is bounded in $L(I, \boldsymbol{R}^n)$, then by Corollary 2.1 we have $\lim_{\lambda\to 0^+}\sup\{|J_\lambda(x)-x|_\varrho\colon x\in K_\Phi(\Omega, C)\} = 0$. Therefore, for every $\delta_0 > 0$ there is $\lambda_0 > 0$ such that $|J_\lambda(x)-x|_\varrho \leqslant \delta_0$ for $x\in K_\Phi(\Omega, C)$ and $\lambda\in(0, \lambda_0)$.

Now, let $\varepsilon > 0$ be given and suppose $\delta_0 > 0$ is so taken that $F(t, (\Phi\oplus y)_t, (\dot\Phi\oplus\dot y))\subset T_H[(\Phi\oplus x)_t(0)]+\varepsilon B$ for a.e. $t\in I$, every $x\in K_\alpha$ and $y\in K_\Phi(\Omega, B)$ satisfying $|x-y|_\varrho \leqslant \delta_0$. Then, for every $x\in K_\alpha$, $\lambda\in(0, \lambda_0)$ and a.e. $t\in I$ we have $G^F(t, J_\lambda(x)) \subset T_H[(\Phi\oplus x)_t(0)]+\varepsilon B$. Therefore, we also have $f_\lambda(t, x) = \mathscr{F}_\lambda(x)(t)$

$\in G^F(t, J_\lambda(x)) \subset T_H[(\Phi \oplus x)_t(0))]+\varepsilon B$ for every $x \in K_\alpha$, $\lambda \in (0, \lambda_0)$ and a.e. $t \in I$. Now, by Remark III.6.2, for every $\varepsilon > 0$ and every $\lambda \in (0, \lambda_0)$ there is $x_\lambda \in K_\alpha$ such that $\dot{x}_\lambda(t) \in G^F(t, x_\lambda)+\varepsilon B$ for a.e. $t \in I$. Let $\lambda = 1/k$ and select N such that $1/k < \lambda_0$ for $k \geqslant N$ and put $x^k := x_{1/k}$ for $k \geqslant N$. Since K_α is compact and $|\dot{x}^k(t)| \leqslant l+M$ for a.e. $t \in I$ and $k \geqslant N$ then there is $x_\varepsilon \in K_\alpha$ and a subsequence, say again (x^k) of (x^k) such that $|x^k - x_\varepsilon|_\varrho \to 0$ and $\dot{x}^k \rightharpoonup \dot{x}_\varepsilon$ as $k \to \infty$. We have of course $\dot{x}^k \in \mathscr{F}(G_\varepsilon^F)(x_k)$ for $k = 1, 2, \ldots$, where $G_\varepsilon^F(t, x) := G^F(t, x)+\varepsilon B$ for $(t, x) \in I \times K_\alpha$. By virtue of Lemma 2.8, hence it follows $\dot{x}_\varepsilon \in \mathscr{F}(G_\varepsilon^F)(x_\varepsilon)$ for every $\varepsilon > 0$. Taking $\varepsilon = 1/k$ and $x^k := x_{1/k}$ we can again find $x \in K_\alpha$ and a subsequence of (x^k), say again (x^k) such that $|x^k - x|_\varrho \to 0$ and $\dot{x}^k \rightharpoonup \dot{x}$ as $k \to \infty$. Since $\dot{x}^k \in \mathscr{F}(G_{1/k}^F)(x^k)$ for every $k = 1, 2, \ldots$ then $\dot{x} \in \mathscr{F}(G_{1/k}^F)(x)$ for every $k = 1, 2, \ldots$ Therefore, $\dot{x} \in \mathscr{F}(G^F)(x)$. Put now, $y = \Phi \oplus x$. We have $\dot{y}(t) \in F(t, y_t, \dot{y}_t)$ for a.e. $t \in I$, $(t, y_t) \in \Omega$ and $(\sigma, y_\sigma, \sigma+\varrho, y_{\sigma+\varrho}) \in C$, because $x \in K_\alpha \subset K_\Phi(\Omega, C)$. Then NFDI$(D, F)$ is (Ω, C)-controllable. ■

2.4. Existence of attainable trajectories with minimal selection property

We shall consider here (Ω, C)-controllable NFDI(D, F) with D, Ω, C and F such as in Theorem 2.9. Furthermore we will assume that F is such that $\mathscr{F}(G\ )$ is integral dissipative. For simplicity we will say F is integral dissipative with respect to $K_\Phi(\Omega, C)$.

Now, we prove the following existence theorem.

THEOREM 2.10. *Suppose $F \in \mathscr{M}(D, \mathrm{Comp}(\boldsymbol{R}^n))$, Ω and C are such as in Theorem 2.9. If furthermore F is integral dissipative with respect to $K_\Phi(\Omega, C)$, then there is an absolutely continuous function z: $[\sigma - r, \sigma+\varrho] \to \boldsymbol{R}_n$ such that*

$$\begin{aligned} &\dot{z}(t) \in F(t, z_t, \dot{z}_t) \quad \text{for a. e. } t \in [\sigma, \sigma+\varrho], \\ &(t, z_t) \in \Omega \quad \text{for } t \in [\sigma, \sigma+\varrho], \\ &(\sigma, z, \sigma_\sigma+\varrho, z_{\sigma+\varrho}) \in C, \\ &|\dot{z}|_\varrho = \inf\{|u|_a\colon u \in \mathscr{F}(F)(z, \dot{z})\}, \end{aligned} \tag{2.7}$$

where $\mathscr{F}(F)(z, \dot{z})$ denotes the subtrajectory integrals of a set-valued function $[\sigma, \sigma+ +\varrho] \ni t \to F(t, z_t, \dot{z}_t) \subset \boldsymbol{R}^n$.

Proof. By virtue of Theorem 2.9, there exists $x \in K_\Phi(\Omega, C)$ such that a function $z = \Phi \oplus x$ satisfies the first three conditions of (2.7). Let $\eta > 0$ be arbitrarily fixed and let $\varepsilon > 0$ be such that

$$\int_{E(x)} |\dot{x}(s)|\, ds - \eta \leqslant \int_{E(x, \delta)} |\dot{x}(s)|\, ds$$

whenever $\mu(E(x) \setminus E(x, \delta)) \leqslant \varepsilon$. It is possible select such $\varepsilon > 0$, because $\dot{x} \in L([\sigma, \sigma+ +a], \boldsymbol{R}^n)$ and $E(x, \delta) \to E(x)$ as $\delta \to 0$. By virtue of Lemma 2.5 for every $\varepsilon > 0$ there is a $\delta_0 > 0$ such that for $\delta \in (0, \delta_0)$ we have $\mu(E(x) \setminus E(x, \delta)) \leqslant \varepsilon$. Then for all such $\delta \in (0, \delta_0)$ we have

$$|x|_a - \eta = \int_{E(x)} |\dot{x}(s)|\, ds - \eta \leqslant \int_{E(x, \delta)} |\dot{x}(s)|\, ds. \tag{2.8}$$

Let us observe that for every $h \in (0, \delta)$ we can define on $E(x, \delta)$ functionals f_h by setting

$$f_h(s) = \left| \frac{x(s+h)-x(s)}{h} \right|.$$

It is clear that a family $(f_h)_{h\in(0,\delta)}$ satisfies conditions of Fatou's lemma. Therefore

$$\int_{E(x,\delta)} |\dot{x}(s)|\,ds = \int_{E(x,\delta)} \liminf_{h\to 0^+} \left| \frac{x(s+h)-x(s)}{h} \right| ds$$
$$\leqslant \liminf_{h\to 0^+} \int_{E(x,\delta)} \left| \frac{x(s+h)-x(s)}{h} \right| ds. \tag{2.9}$$

Let $y \in \mathscr{F}(G^F)(x)$ be arbitrarily taken. Since $\dot{x} \in \mathscr{F}(G^F)(x)$ and F is integral dissipative with respect to $K_\Phi(\Omega, C)$ then there is a $\delta_1 \in (0, \delta_0)$ such that for every $\delta \in (0, \delta_1)$, $h \in (0, \delta)$ and $\lambda > 0$ we have

$$\int_{\sigma}^{\sigma+a-\delta} |x-\overline{x(s)}|_{s,h}\,ds \leqslant \int_{\sigma}^{\sigma+a-\sigma} |x-\overline{x(s)}-\lambda(\dot{x}-\overline{y(s)})|_{s,h}\,ds.$$

By virtue of Corollary 2.2, there exist functionals $g^{s,h} \in G_{s,h}(x_{s,h}-x(s))$ such that $\int_{\sigma}^{\sigma+a-\delta} g^{s,h}(\dot{x}_{s,h}-\overline{y(s)})\,ds \leqslant 0$. Then

$$\int_{E(x,\delta)} g^{s,h}(\dot{x}_{s,h})\,ds \leqslant \int_{\sigma}^{\sigma+a-\delta} g^{s,h}(\overline{y(s)})\,ds - \int_{E_1(x,\delta)} g^{s,h}(\dot{x}_{s,h})\,ds, \tag{2.10}$$

where $E_1(x, \delta) = [\sigma, \sigma+a-\delta] \quad E(x, \delta)$.

By Lemma 2.7 we have

$$g^{s,h}(x_{sh}) = |x(s+h)-x(s)|.$$

Now, from (2.8)–(2.10) we obtain

$$|\dot{x}|_a - \eta \leqslant \lim_{h\to 0^+} \frac{1}{h} \int_{E(x,\delta)} |x(s+h)-x(s)|\,ds$$
$$\leqslant \lim_{h\to 0^+} \frac{1}{h} \int_{\sigma}^{\sigma+a-\delta} g^{s,h}(\overline{y(s)})\,ds - \lim_{h\to 0^+} \int_{E_1(x,\delta)} g^{s,h}(\dot{x}_{s,h})\,ds$$
$$\leqslant \lim_{h\to 0^+} \frac{1}{h} \int_{\sigma}^{\sigma+a-\delta} |\overline{y(s)}|\,ds - \lim_{h\to 0^+} \frac{1}{h} \int_{E_1(x,\delta)} g^{s,h}(\dot{x}_{s,h})\,ds$$
$$= \lim_{h\to 0^+} \frac{1}{h} \int_{\sigma}^{\sigma+a-\delta} h|y(s)|\,ds - \lim_{h\to 0^+} \frac{1}{h} \int_{E_1(x,\delta)} g^{s,h}(\dot{x}_{s,h})\,ds$$
$$\leqslant |y|_a + \int_{E_1(x,\delta)} |\dot{x}(s)|\,ds.$$

Hence it follows $|\dot{x}|_a \leqslant |y|_a$ for each $y \in \mathscr{F}(G^F)(x)$, because x is absolutely continuous and we have $E_1(x, \delta) \to \{s \in I: \dot{x}(s) = 0$ or $\dot{x}(s)$ does not exist$\}$ as $\delta \to 0$. On the other hand, we have $|x|_a \geqslant |y_0|_a$ for any $y_0 \in \mathscr{F}(G^F)(x)$, because $\dot{x} \in \mathscr{F}(G^F)(x)$. Therefore, finally we have $|\dot{x}|_a = \inf\{|u| : u \in \mathscr{F}(G^F)(x)\}$. Since $\mathscr{F}(G^F)(x) = \mathscr{F}(F)(z, \dot{z})$ for $z = \Phi \oplus x$, then $|\dot{z}|_a = \inf\{|u|_a : u \in \mathscr{F}(F)(z, \dot{z})\}$.

6. Notes and remarks

The results presented in this chapter extend some results of Cesari [1] on the case of the Mayer and Lagrange optimal control problems for the systems described by neutral functional-differential inclusions $\dot{x}(t) \in F(t, x_t, \dot{x}_t)$. Some extensions of such results on the case of the neutral functional-differential inclusions of the form $\frac{d}{dt}\mathscr{D}(x_t) \in F(t, x_t)$, where $\mathscr{D}$ is the Hale difference operator, are given in Angell [1]. In particular, the idea of the proofs of Theorems 1.4–1.6 arise from Cesari [1], whereas Example 1.1 is taken from Lee and Markus [1]. The definition and basic properties of dissipative set-valued functions are taken from Barbu [1], whereas Example 2.1 is taken from Deimling [1]. The definition of integral-dissipative set-valued functions and the proofs of Lemmas 2.5 and 2.6 arise from Motyl [1]. Some necessary and sufficient conditions for strong-weak lower semicontinuity of integral functional are given in Balder [1].

Bibliography

Alexiewicz, A.

[1] *Functional Analysis*, Monografie Matematyczne, **49**, Polish Scientific Publishers, Warszawa 1969.

Angell, T. S.

[1] 'On controllability for nonlinear hereditary systems: a fixed point approach', *Nonlinar Anal. Theory, Meth. Appl.* **4** (3) (1980), 93–107.

Antosiewicz, H. A., Cellina, A.

[1] 'Continuous selection and differential relations', *J. Diff. Equ.* **19** (1975), 386–398.

[2] 'Continuous extensions of multifunctions', *Ann. Polon. Math.* **34** (1977), 107–111.

Arino, O., Gautier, S., Penot, J. P.

[1] 'A fixed point theorem for sequentially continuous mappings with application to ordinary differential equations', *Func. Ekv.* **27** (1984), 273–279.

Artestein, Z.

[1] 'Set-valued measures', *Trans. Amer. Math. Soc.* **165** (1972), 103–124.

[2] 'Weak convergence of set-valued functions and control', *SIAM J. Control* **13** (4) (1975), 865–878.

Artstein, Z., Prikry, K.

[1] 'Carathédory selections and the Scorza–Dragoni property', *J. Math. Anal. Appl.* **127** (1987), 540–547.

Aubin, J. P., Cellina, A.

[1] *Differential Inclusions*, Springer-Verlag, Berlin–Heidelberg–New York–Tokio 1984.

Aubin, J. P., Frankowska, H., Olech, C.

[1] 'Controllability of convex processes', *SIAM J. Control and Optim.* **24** (1986), 1192–12111

Aumann, R. J.

[1] 'Integrals of set-valued functions', *Journal Math. Anal. Appl.* **12** (1965), 1–12.

Balder, E. J.

[1] 'Necessary and sufficient conditions for L_1-strong-weak lower semicontinuity of integral functional', *Nonlinear Anal. Theory Meth. Appl.* **11** (1987), 1399–1404.

Barbu, V.

[1] *Nonlinear Semigropus and Differential Equations in Banach Spaces*, Noordhoff, International Publishing, Leyden 1976.

Bellman, R., Cook, K. L.

[1] *Differential-Difference Equations*, Acad. Press, New York 1963.

Bourgin, M.

[1] *Modern Algebraic Topology*, Mac Millan, New York 1963.

Bridgland, Jr., T. F.

[1] 'Trajectory integrals of set-valued functions', *Pac. Journal Math.* **33**(1) (1970), 43–67.

Castaing, M. C.

[1] 'Une nouvelle extension du thèorème de Dragoni–Scorza', *C. R. Acad. Sc. Paris* **271** (1970), 396–398.

Castaing, M. C., Valadier, M.

[1] *Convex Analysis and Measurable Multifunctions*, Lecture Notes in Math. **580** (1977).

Cesari, L.

[1] *Optimization Theory and Application*, Appl. of Math. **17**, Springer-Verlag 1983.

Clarke, F. H., Watkins, G. G.

[1] 'Necessary conditions, controllability and value function for differential-difference inclusions', *Nonlinear Anal. Theory Meth. Appl.* **10** (1986), 1155–1179.

Covitz, H., Nadler, J ., S. B.

[1] 'Multivalued contraction mappings in generalized metric spaces', *Israel Journal Math.* **8** (1970), 5–11.

Colonius, F.

[1] 'Optimality for periodic control of functional differential systems', *Journal Math. Anal. Appl.* **120** (1986), 119–149.

Deimling, K.

[1] *Ordinary Differential Equations in Banach Spaces*, Lecture Notes in Math. **596** (1977).

Diestel, J., Uhl, Jr., J. J.

[1] *Vector Measures*, Amer. Math. Soc., Providence, Rhode Island 1977.

Dunford, N., Schwartz, J. T.

[1] *Linear Operators I*, Int. Publ., INC., New York 1967.

Edwards, R. E.

[1] *Functional Analysis Theory and Applications*, Holt, Rinehart and Winston, New York 1965.

Elsgolts, L. E.

[1] *Qualitative Methods in Mathematical Analysis*, Moscow 1955 (in Russian).

Fillipov, A. F.

[1] *Differential Equations with Discontinuous Right Hand Sides*, Moscow 1985 (in Russian).

[2] 'On certain questins in the theory of optimal control', *SIAM J. Control* **1** (1962), 76–84.

Frankowska, H.

[1] 'Local controllability and infinitesimal generators of semigroups of set-valued maps', *SIAM J. Control and Optim.* **25** (1987), 412–432.

[2] 'The maximum principle for optimal solution to a differential inclusion with end points constraints', *SIAM J. Control and Optim.* **25** (1987), 145–157.

Fryszkowski, A.

[1] 'Continuous selections for a nonconvex multivalued maps', *Studia Math.* **76** (1983), 163–174.

[2] 'Existence of solutions of functional-differential inclusions in nonconvex case', *Ann. Polon. Math.* **45** (1985), 121–124.

Haddad, G.

[1] 'Momotone viable trajectories for functional-differential inclusions', *Journal Diff. Equations* **42** (1981), 1–24.

[2] 'Topological properties of the sets of solutions for functional-differential inclusions', *Nonlinear Anal. Theory Meth. Appl.* **5**(12) (1981), 1349–1366.

Haddad, G., Lasry, J. M.

[1] 'Periodic solutions of functional differential inclusions and fixed points of σ-selectionable correspondences', *Journal Math. Anal. Appl.* **96** (1983), 295–312.

Hale, J.

[1] *Theory of Functional Differential Equations*, Springer-Verlag, New York 1977.

Hermes, H.

[1] 'Calculus of set valued functions and control', *Journal of Math. and Mech.* **18** (1968), 47–59,

[2] 'On the structure of attainable sets for generalized differential equations and control systems'. *J. Diff. Equ.* **9** (1971), 141–154.

Hermes, H., La Salle, J. P.

[1] *Functional Analysis and Time Optimal Control*, Acad. Press, New York 1969.

Hiai, F. and Unegaki, H.

[1] 'Integrals, conditional expections and martingals of multivalued functions', *J. Mult. Anal.* **7** (1977), 149–182.

Hildenbrand, W.

[1] *Core and Equilibria of a Large Economy*, Princ. Studies in Mathematical Economics, Princ. Univ. Press, New Jersey 1974.

Himmelberg, C. J.

[1] 'Measurable relations', *Fund. Math.* **87** (1975), 59–71.

Himmelberg, C. J., Van Vleck, F. S.

[1] 'Lipschitzean generalized differential equations', *Rend. Sem. Mat. Univ. Padova* **48** (1973), 159–169.

Jacobs, M. Q.

[1] 'Measurable multivalued mappings and Lusin's theorem', *Trans. Amer. Math. Soc.* **134** (1968), 471–481.

Jacobs, M. Q., Langenhop, C. E.

[1] 'Criteria for function space controllability of linear neutral systems', *SIAM J. Control and Optim.* **14**, (1976) 1009–1048.

Kisielewicz, M.

[1] 'Existence theorem for generalized functional-differential equations of neutral type', *J. Math. Anal. Appl.* **78** (1980), 173–182.

[2] 'On the trajectories of generalized functional differential systems of neutral type', *J. Optimization Theory Appl.* **33** (1981), 255–266.

[3] 'Generalized functional-differential equations of neutral type', *Ann. Polon. Math.* **42** (1983), 140–148.

[4] 'Subtrajectory integrals of set-valued functions and neutral functional-differential inclusions', *Funk. Ekv.* **32** (2) (1989), 163–189.

[5] 'Viability theorem for neutral functional-differential inclusions', *Funk. Ekv.* (to appear).

[6] *Nonlinear Functional Differential Equations of Neutral Type*, PWN—Polish Scientific Publishers, Poznań–Warszawa (1984).

Kisielewicz, M., Rybiński, L.

[1] 'Generalized fixed point theorem', *Demonstration Math.* **16** (1983), 1037–1041.

Kuratowski, K., Ryll-Nardzewski, C.

[1] 'A general theorem on selectors', *Bull. Polon. Acad. Sci.* **13** (1965), 397–403.

Lee, E. B., Markus, L.

[1] *Foundations of Optimal Control Theory*, John Wiley and Sons, Inc. New York 1968.

Leopes, O.

[1] 'Forced oscilations in nonlinear neutral differential equations', *SIAM J. Appl. Math.* **29** (1975), 196–207.

Lindenstrauss, J.

[1] 'A short proof of Liapunoff's convexity theorem', *Journal of Math. and Mech.* **15** (1966), 971–972.

Melvin, W. R.

[1] 'A class of neutral functional differential equations', *Journal Diff. Equ.* **12** (1972), 524–534.

[2] 'Some extensions of the Krasnosielskii fixed point theorem', *Journal Diff. Equ.* **11** (1972), 335–348.

Michael, E.

[1] 'Continuous selections I', *Annales Math.* **63** (1956), 361–382.

Motyl, J.

[1] 'Integral dissipative set-valued functions', *Disc. Math.* **9** (to appear).

Myshkis, A. D.

[1] 'General theory of differential equations with a retarded argument', *Usp. Mat. Nauk*, **4**, **5**(33) (1949), 99–141 (in Russian).

Olech, C.

[1] 'A contribution to the time optimal control problem', *Abhandlungen Deutsch. Akad. Wiss. Berlin, Kl. Phys. Tech.* **2** (1965), 438–446.

[2] 'Extermal solution of a control system', *J. Diff. Equ.* **2** (1966), 74–101.

[3] '*Decomposability as a substitute for convexity*', Lect. Notes in Math. **1091** (1983), 193–205.

Papageorgiou, N. S.

[1] 'On the theory of Banach space valued multifunctions Part 1: Integrations and conditional expection', *J. Multiv. Anal.* **17** (1985), 185–207.

[2] 'On measurable multifunctions with applications to random multivalued equations', *Math. Japonica* **3** (1987), 437–464.

[3] 'A relaxation theorem for differential inclusions in Banach spaces', *Tohoku Math. J.* **39** (1987), 505–517.

Parthasarathy, T.

[1] *Selection Theorem and Their Applications*, Springer-Verlag, Berlin–Heilderberg–New York 1972.

Pianigiani, G.

[1] 'On the fundamental theory of multivalued differential equations', *Journal Diff. Equ.* **25** (1977) 30–38.

Pliś, A.

[1] 'Remark on measurable set valued functions', *Bull. Acad. Polon. Sci.* **13** (1965), 565–569.

Rybiński, L.

[1] 'Multivalued contraction with parameter', *Ann. Polon. Math.* **45** (1985), 275–282.

[2] 'Multivalued contraction mappings with parameters and random fixed point theorems', *Disc. Math.* **8** (1986), 101–108.

Rzepecki, B.

[1] 'A fixed point theorem of Krasnosielski type for multivalued mappings', *Demonstratio Math.* **17** (1984), 767–776.

Schwartz, L.

[1] *Analyse Mathématique I*, Hermann, Paris 1967.

Taylor, A. E.

[1] *General Theory of Functions and Integration*, Blaisdell Publ. Comp., New York 1965.

Tolstonogov, A. A.

[1] *Differential Inclusions in Banach Spaces*, Izd. Nauka, Novosibirsk 1986 (in Russian).

Tsoy-Wo Ma

[1] 'Topological degrees of set-valued compact fields in locally convex space', *Diss. Math.* **92** 1972.

Wagner, D. H.

[1] 'Survey of measurable selection theorems', *SIAM J. Control and Optim.* **15** (1977), 859–903.

Index

Absolutely continuous function, 15
Acyclic sets, 9
Admissible control of, 149
pair for, 149
trajectory of, 149
Affine mapping, 9
Alaoglu theorem, 7
Alexiewicz, A., 22
Angel, T. S., 233
Antosiewicz, H. A., 64, 139
Approximatelly continuous set-valued function 54
Arino, O., 64
Artstein, Z., 65, 139, 140
Ascoli theorem, 15
Attainable trajectory with minimal selection property, 221
Aubin, J. R., 64, 65, 140, 208, 209
Aumann, R. J., 65
integral, 49
theorem, 53
$\mathscr{A}$-continuous mapping, 82
$\mathscr{A}$-Lipschitz continuous mapping, 83

Balder, E. J., 233
Banach fixed point theorem, 21
Banach–Mazur theorem, 8
Banach space, 6
Barbu, V., 233
Base topology, 1
Bellman, R., 209
Bochner integrable function, 11
Bolza problem of optimal control, 153
Bouligand's contingent cone, 129, 175
Bourgin, M., 22
Bridgland, Jr., T. F., 64, 139

Cantor theorem, 5
Carathéodory theorem, 8
Castaing, M. C., 64, 65
Cauchy sequence, 5
Cellina, A., 64, 65, 140, 208, 209
Cesari, L., 22, 209, 233
Clarke, F. H., 209
Closed convex hull of, 8
set-valued function, 30
Compact set, 3
Compactification, 3
Compatible system of, 150
Complete metric space, 5
Completely continuous mapping, 76
Conformable pair of sets, 155
Connected set, 3
Continuous function, 2
Controllable system of, 150
Convex hull of, 8
set, 7
Cook, K. L., 208
Covitz, H., 65
Covitz–Nadler theorem, 61
Colonius, F., 209
μ-continuous vector measure, 13

Decomposable sets, 50, 89
set-valued function, 92
Deimling, K., 233
Dense set, 3
Diameter of set, 5
Diestel, J., 22
Dissipative set-valued function, 221
Distant function, 5
Dunford, N., 22
Dunford–Schwartz theorem, 9
Dunford theorem, 12

Eberlein–Šmulian theorem, 7
Edwards, R. E., 22
Egoroff theorem, 10
Elsgolc, L. E., 208
Equiabsolutely continuous set of, 16
Equicontinuous set of, 15
Equivalent topologies, 2
Extreme point of, 8

Fatou lemma, 22
Fillipov, A. F., 65
Frankowska, H.. 209
Fryszkowski, A., 139
Function of bounded variation, 14

Gautier, S., 64
Generated topology, 4
Gronwall lemma, 22
Growth conditions, 164, 166

Haddad, G., 208
Hale, J., 208
Hausdorff distance, 23
topology, 2, 24
Hermes, H., 22, 65
Hiai, F., 65, 139
Hildenbrand, W., 65
Himmelberg, C. J., 64
H-lower semicontinuous set-valued function, 35
H-upper semicontinuous set-valued function, 28

Integrably bounded set-valued function, 49
Integral dissipative set-valued function, 226

Jacobs, M. Q., 65, 209

Kakutani–Ky Fan theorem, 62
Kakutani theorem, 61
Kamke function, 119
Kisielewicz, M., 65, 66, 139, 209
Krein–Milman theorem, 9
Krein–Šmulian theorem, 8
Kuratowski, K., 65, 46, 47
Kuratowski–Ryll-Nardzewski theorem, 47
Kuratowski–Zorn's lemma, 46

Lagrange problem of optimal control, 152
Langenhop, C. E., 209
Lasry, J. M., 209
LaSalle, J. P., 22, 65
Leapes, O., 209
Lebesgue dominated convergence theorem, 12
point of, 54
Lee, E. B., 233
Lindenstrauss, J., 22
Linear operator, 6
Lipschitz continuity of, 81
L-measurable function, 39
Local associated set-valued function to, 157
hemi-upper (lower) semicontinuous set-valued functions, 161
Lipschitz continuous set-valued function, 160
measurability condition ($\mathscr{M}$) of, 141
measurability condition (M) of, 154
solution of NFDE(D,f), 142
Locally bounded set-valued function, 173
convex set-valued function, 165
integrably bounded set-valued function, 156
integrable set-valued function, 156
Lipschitz partition of unity, 4, 5
uniformly integrable set-valued function, 156
Lower semicontinuous function, 2
semicontinuous of set-valued function, 34
Lusin–Pliš theorem, 44
Lusin theorem, 39
Lyapunov theorem 13

M-dissipative set-valued function, 221
Mazur theorem, 8
Maximal dissipative set-valued function, 221
Mayer problem of, 149
Mean value theorem, 11
Measure of bounded variation, 13
Measurable mapping, 39
(weak measurable, Borel measurable) set-valued function, 41
Markus, L., 233
Melvin, W. R., 65
Metric space, 4
topology, 4
Michael, E., 64, 65
Michael theorem, 57
Minkowski theorem, 8
Modulus of continuity of, 15
Motyl, J., 233
Myshkis, A. D., 208

Nadler, Jr., S. B., 65
Neutral functional-differential equation, 142
functional-differential inclusion, 155
Normed linear space, 5
Norm reflexive space, 7

Olech, C., 65, 209

Papageorgiou, N. S., 65, 139
Paracompact topological space, 4
Parametrized set-valued function, 36
Parthasarathy, T., xiii
Penat, J. R., 64
Perfectly separable space, 3
Pianigiani, G., 139
Pliś, A., 65
Prikry, K., 65, 140

Quasilinear space, 25

Radom–Nikodym derivative of vector measure, 13

Radon–Nikodym derivative of set-valued measure, 56
Radon–Nikodym theorem, 13
Relative compact set, 3
compact sets in $C(I, X)$, 15
d-weak sequential compact sets, 19
l-weak sequential compact sets, 19
sequential compact set, 3
ϱ-compact sets, 19
weak sequential compact sets in AC(I, $\boldsymbol{R}^n$), 17
weak sequential compact sets in $L^1(\mu, \boldsymbol{R}^n)$, 12
Relaxation problem for, 151
theorems, 127, 174
Rzepecki, B., 65
Rybiński, L., 65
Ryll-Nardzewski, Cz., 65

Schauder fixed point theorem, 21
Schauder–Tikhonov theorem, 22
Schwartz, J. T., 22
Scorza–Dragoni–Castaing theorem, 45
Selector of set-valued function, 46
Semicontinuous extension of, 31
Separable topological space, 3
Sequential closed set, 3
compact set, 3
continuous function, 2
Set-valued function, 28
measure, 56
Šmulian theorem, 7
Solution of NFDI(D, F) through (σ, Φ), 155
Stieltjes integral of vector-valued function, 14
Strong-weak sequential continuous on, 75
upper (lower) semicontinuity of, 32
Subbase topology, 2
Support function of, 25
Suslin's operation, 40
Subtrajectory integrals of, 66
σ-selectionable set-valued function, 63

Taylor, A. E., 22
Tolstonogov, A. A., xiii
Topological space, 1
Topology, 1
Total variation of, 14
Trajectory integrals, 77
Tsoy-Wo Ma, 64

Uhl, Jr., J. J., 22
Unegaki, H., 65, 139
Uniform integrable family of set-valued functions, 68
Uniform integrable sets of, 12
Upper semicontinuous function, 2
semicontinuous set-valued function, 28, 34
Upper (lower) semicontinuous weakly mappings, 81
Urysohn theorem, 2

Valadier, M., xiii, 64
Van Vleck, F. S., 64
Variation of vector measure, 12
Vector measures, 12
Viable fixed point of, 129
Viable trajectory of, 179
Vietoris–Begle theorem, 9
Vitali convergence theorem, 11

Wagner, D. H., 65
Watkins, G. G., 208
Weak Cauchy sequence, 6
convergence in $\mathscr{A}(I, \boldsymbol{R}^n)$, 69
closed set, 6
compact set, 6
relative compact set, 6
relative sequential compact set, 6
sequential completness of, 72
sequential closed set, 6
sequential compact set, 6
-strong continuous mapping, 76
-strong upper (lower) semicontinuity of, 81
topology generated by, 6
-weak continuous mapping, 76
-weak upper (lower) semicontinuity of, 81
-* topology, 7
Weierstrass theorem, 21

Zermelo principle of choice, 46

9 780792 306757